15th International Scientific and Technical Conference "Problems of the Railway Transport Mechanics" (PRTM 2020)

AA007018

IOP Conference Series: Materials Science and Engineering
Volume 985

Dnipro, Ukraine
27 - 29 May 2020

Editors:

M. Jacyna
A. Sladkovsky
A. Zhurkovsky
O. Markova
M. Spiryagin
O. Fomin
A. Falendysh
A.O. Shimanovsky
S. Raksha
L. Neduzha
J. Kalivoda

ISBN: 978-1-7138-3299-7
ISSN: 1757-8981

Printed from e-media with permission by:

Curran Associates, Inc.
57 Morehouse Lane
Red Hook, NY 12571

Some format issues inherent in the e-media version may also appear in this print version.

This work is licensed under a Creative Commons Attribution 3.0 International Licence.
Licence details: http://creativecommons.org/licenses/by/3.0/.

No changes have been made to the content of these proceedings. There may be changes to pagination and minor adjustments for aesthetics.

Printed with permission by Curran Associates, Inc. (2021)

For permission requests, please contact the Institute of Physics at the address below.

Institute of Physics
Dirac House, Temple Back
Bristol BS1 6BE UK

Phone: 44 1 17 929 7481
Fax: 44 1 17 920 0979

techtracking@iop.org

Additional copies of this publication are available from:

Curran Associates, Inc.
57 Morehouse Lane
Red Hook, NY 12571 USA
Phone: 845-758-0400
Fax: 845-758-2633
Email: curran@proceedings.com
Web: www.proceedings.com

TABLE OF CONTENTS

STRENGTH, RELIABILITY, DIAGNOSTICS, MATERIALS SCIENCE

IOP Conf. Series: Materials Science and Engineering **985** (2020) 011001 doi:10.1088/1757-899X/985/1/011001

The materials of the conference present the results of theoretical and experimental studies of the dynamics and strength of rail transport, wear of wheels and rails, traffic safety and materials science.

Editors:

Jacyna M.
Pl. Politechniki 100-661 Warsaw, Poland
Warsaw University of Technology
maja@wt.pw.edu.pl

Sladkovsky A.
Akademicka 2A44-100 Gliwice, Poland
Silesian University of Technology
aleksander.sladkowski@polsl.pl

Zhurkovsky A.
50 Chłopickiego Józefa street, PL 04-275 Warsaw, Poland
Railway Research Institute, Warsaw
azurkowski@ikolej.pl

Markova O.
01030, Kyiv, Volodymyrska, 54 Ukraine
Institute of Transport Systems and Technology of the National Academy of Sciences of Ukraine
olgamarkova2002@gmail.com

Spiryagin M.
Norman Gardens QLD 4701, Australia
Centre for Railway Engineering
Central Queensland University
m.spiryagin@cqu.edu.au

Fomin O.
Kyiv, str. Polzunova, 6 Ukraine
State University of Infrastructure and Technology
fomin1985@ukr.net

Falendysh A.
Feuerbach Square, 7, Kharkiv, Kharkiv Region, 61000 Ukraine
Ukrainian State University of Railway Transport
ttd@kart.edu.ua

Content from this work may be used under the terms of the Creative Commons Attribution 3.0 licence. Any further distribution of this work must maintain attribution to the author(s) and the title of the work, journal citation and DOI.
Published under licence by IOP Publishing Ltd

Shimanovsky A. O.
246653, Gomel, Kirova str, 34.
Belarus Educational Institution “Belarusian State University of Transport”
tm.belsut@gmail.com

Raksha S.
2, Lazaryana street, Dnipro, Dnipropetrovsk region, 49010 Ukraine
Dnipro National University of Railway Transport named after Academician V. Lazaryan
pm@m.diit.edu.ua

Neduzha L.
2, Lazaryana street, Dnipro, Dnipropetrovsk region, 49010 Ukraine
Dnipro National University of Railway Transport named after Academician V. Lazaryan
nlorhen@i.ua

Kalivoda J.
Zikova 1903/4, 166 36 Praha 6, Czech Republic
Czech Technical University in Prague
jan.kalivoda@fs.cvut.cz

IOP Conf. Series: Materials Science and Engineering **985** (2020) 011002 doi:10.1088/1757-899X/985/1/011002

Peer review declaration

All papers published in this volume *of IOP Conference Series: Materials Science and Engineering* have been peer reviewed through processes administered by the Editors. Reviews were conducted by expert referees to the professional and scientific standards expected of a proceedings journal published by IOP Publishing.

- **Type of peer review: Double-blind**
- **Conference submission management system: the conference is being held of-line, but some participants, in connection with the COVID-19 pandemic, will take part on-line using scientific conference organization and administration software ZOOM platform**
- **Number of submissions received: 46**
- **Number of submissions sent for review: 46**
- **Number of submissions accepted: 41**
- **Acceptance Rate (Number of Submissions Accepted / Number of Submissions Received X 100): 89.1%**
- **Average number of reviews per paper: 4**
- **Total number of reviewers involved: 10**
- **Any additional info on review process: all articles have been tested for plagiarism**
- **Contact person for queries: Ursulyak Ljudmyla Viktorovna**
- **lydm.urs@gmail.com**

Content from this work may be used under the terms of the Creative Commons Attribution 3.0 licence. Any further distribution of this work must maintain attribution to the author(s) and the title of the work, journal citation and DOI.
Published under licence by IOP Publishing Ltd

IOP Conf. Series: Materials Science and Engineering **985** (2020) 012001 doi:10.1088/1757-899X/985/1/012001

Improvement of the railway track efficiency by minimizing the rail wear in curves

M Kurhan[1,4], D Kurhan[2], R Novik[3], S Baydak[1] and N Hmelevska[1]

[1]Department of Roads Design and Construction, Dnipro National University of Railway Transport named after Academician V. Lazaryan, Lazaryan St, 2, Dnipro, Ukraine
[2]Department of Track and Track Facilities, Dnipro National University of Railway Transport named after Academician V. Lazaryan, Lazaryan St, 2, Dnipro, Ukraine
[3]Department of Military Training of Specialists of the State Special Service of Transport, Dnipro National University of Railway Transport named after Academician V. Lazaryan, Lazaryan St, 2, Dnipro, Ukraine

[4] kunibor@gmail.com

Abstract. The article presents the analysis results for the parameters measurements of the rail track and the ratio between values of the vertical and side wears of rails in curves of different radii under different operating conditions. It is shown that the installed elevations of the outer rail do not meet the contemporary requirements in increasing the efficiency of the track while minimizing the rail wear. At the second stage of the research, various sections were examined by type of traction, technical equipment, plan parameters and the longitudinal profile. As an example, the results of calculations are presented in Lviv-Rava-Ruska section as a promising project that can ensure the integration of railway transport in Lviv-Warsaw direction. Analysis of the results obtained with the help of the software package – MoveRW and RWPlan programmes, allowed to determine such a combination of elevations of the outer rail in curves, so that trains of different categories realize the lowest values of cross unbalanced accelerations, which provide the minimum wear of rails in compliance with regulatory requirements for unbalanced acceleration, changing this acceleration in time, elevating speed of a wheel on the removal of the outer rail.

1. Introduction

The problem in increasing the intensity of the wheel flange wear of rolling stock and gauge face is so acute that there is a risk of loss in the efficiency of a number of sections for this reason. To date, it has not been fully resolved, as evidenced by numerous publications on the "wheel-rail" problem. This can be explained by a number of factors that affect the wear of rails to varying degrees.

Thus, factors affecting the rail track wear in curves in paper [1] are divided into uncontrolled, partially controlled and controlled. The main controlled factors, according to the authors, are track gauge, the value of elevation of the outer rail and rail lubrication. The experiment was carried out on 19 curves of Lithuanian railway lines, according to the results of which it was found that the curve radius has a decisive impact on the intensity of rails wear in curves.

In paper [2], the impact of DC electric locomotives and AC locomotives on the railway track wear in small radii curves (up to 240 m) at a constant speed of 20 km/h was studied. The simulation shows that the replacement of DC locomotives for AC locomotives with the increase in tractive effort torque

Content from this work may be used under the terms of the Creative Commons Attribution 3.0 licence. Any further distribution of this work must maintain attribution to the author(s) and the title of the work, journal citation and DOI.
Published under licence by IOP Publishing Ltd

by 31.4% will result in 68.4% quite significant increase of the maximum wear in the "wheel-rail" system. It is found that at a relatively low speed of 20 km/h, the total wear on the inner rail is always greater than on the outer rail regardless of the type of locomotive. The authors [3] deals with the parameter of detecting the wear of the wheel-rail contact based on the radiated noise. The field tests conducted by the authors showed that noise can be used as an indicator of transition from the normal wear to a critical one. This transition is accompanied by a significant increase in sound pressure and its spectrum spreading. In [4] the results of numerical research of the vehicles-railway track interaction, taking into account the existing and assumed designs of rail fastenings are presented. The authors concluded that the train motion modes and the structure of a rail base affect the speed of side wear of the railhead in small radius curves.

The authors [5] discuss the problem of reducing the wear itensity in the "wheel-rail" system through the use of lubricants that do not harm the environment. The method of reducing the friction of wheels on the rails by a decrease in temperature directly in the contact zone is considered. The idea of the method is based on the provision that at lower temperatures the wear occurs to a lesser extent. To solve the problem of reducing the wear of rolling stock wheels and rails, the authors proposed to use liquid nitrogen as a cooling agent.

The authors [6] set forth the results in research of the longitudinal and transverse displacement impact of the center of gravity for the gondola car load, in the light of the possible increase of movement speed on the figure of the wear factor of the wheels and rails surfaces.

The authors [7-9] provide data concerning the nature and amount of the rails wear. As a rule, the additional vertical wear caused by unbalanced acceleration is considered [8], and the additional side wear for the outer rail at unbalanced accelerations directed outwards of the curve is also considered. [9].

It is believed that the calculated elevation of the outer rail h at the weighted-average quadratic speed $V_{\text{weav qu}}$ provides the same average force impact on the inner and outer rails, that is:

$$\sum Q_{\text{out}\,i} = \sum Q_{\text{inn}\,i} \,. \tag{1}$$

The theoretically correct approach to determining elevation h (formula 2) by the weighted-average quadratic speed $V_{\text{weav qu}}$ (formula 3), in practical use has significant disadvantages.

$$h = \frac{12.5 \cdot k \cdot V^2_{\text{weav qu}}}{R}, \tag{2}$$

$$V_{\text{weav qu}} \cong \sqrt{\frac{\sum n_j Q_j V^2_{\text{run}\,j}}{\sum n_j Q_j}}, \tag{3}$$

where $n_j, Q_j, V_{\text{run}\,j}$ – number of trains, a mass of train, running speed of train movement j-th category on this curve.

Condition (1) does not mean fulfilling the requirement of the equal rail wear in the curved sections of the track. A simplified view of the relationship between the forces acting on the rails and the wear satisfied the engineers for a long time. At the same time, it is known that the same force acting on the outer and inner rails leads to different wear results. This is one of the reasons why in the regulatory documents of some countries the concept of weighted-average speed has been excluded in recent years.

2. Methodology

To assess the features of the rolling stock impact on the outer and inner rail lines the following methodology was implemented.

2.1 The first stage

At the first stage, the curve examination was carried out on various sections of the railways and rail wear measurements were made. It was found that the small (up to 2-3 mm) vertical wear of both rails was observed on all 25th examined curves. The side wear of the inner rail was in the range of 2-3 mm (rarely up to 5-6 mm). The side wear of the outer rail is 3-4 times higher than the wear on the inner rail. This ratio was observed in curves of different radii with the excessive elevation of the outer rail, which led to negative transverse accelerations and caused an overload of the inner rail by vertical forces.

The unbalanced accelerations always lead to the additional wear of the rails and rolling stock. A distinctive feature of the track operation in curves is the wear unevenness of the outer and inner rail lines. If the actual cant of the outer rail is greater than the calculated one, then the inevitable longitudinal sliding, caused by the rigid mounting of wheels on the axis, occurs on the outer rail, if less – on the inner one.

The main feature of the side wear (formula 4) lies in its unevenness along the length of the track:

$$h_s = h_{ave} + h_{loc}, \tag{4}$$

where h_{ave} – relatively the constant average wear along the entire length of the circular curve; depends on the passed tonnage and the curve radius; h_{loc} – the increased local wear on short sections.

It is confirmed that in places of the greatest change of the versine (20-m chord method) the maximum local wear is observed. Such zones are located, as a rule, at the beginning of the transition curves, at the conjunction of the transition and circular curves and near the middle of the circular curve, that is, where the unbalanced accelerations change (Figure 1). From the beginning of the transition curve at a distance of 40 meters, the versine on the section of about 10 meters vary from 20 to 59 mm (2.9 times). And on the same section the side wear increases from 1.7 to 8.3 mm, that is, 4.9 times (in Figure 1 the values are increased by 4 times). The next zone is located at the conjunction of the transition and circular curves. On the section of about 8 meters, the versine change from 85 to 95 mm, and the wear has increased from 9.2 to 13.3 mm (1.4 times). The same dependencies were observed on other curves.

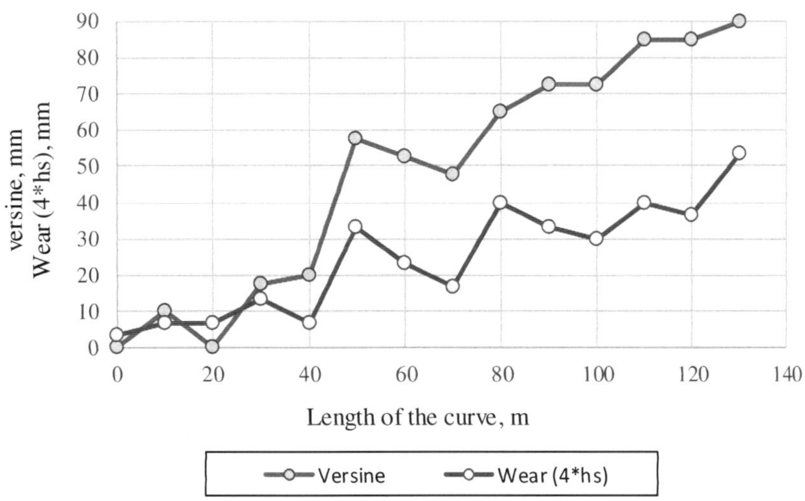

Figure 1. Change of the versine (20-m chord method) and wear along the length of the curve (curve No. 39, radius 615 m)

According to the measurements of wear and gauge parameters, the intensity of the side wear (mm/10 million tons) was determined, Figure 2.

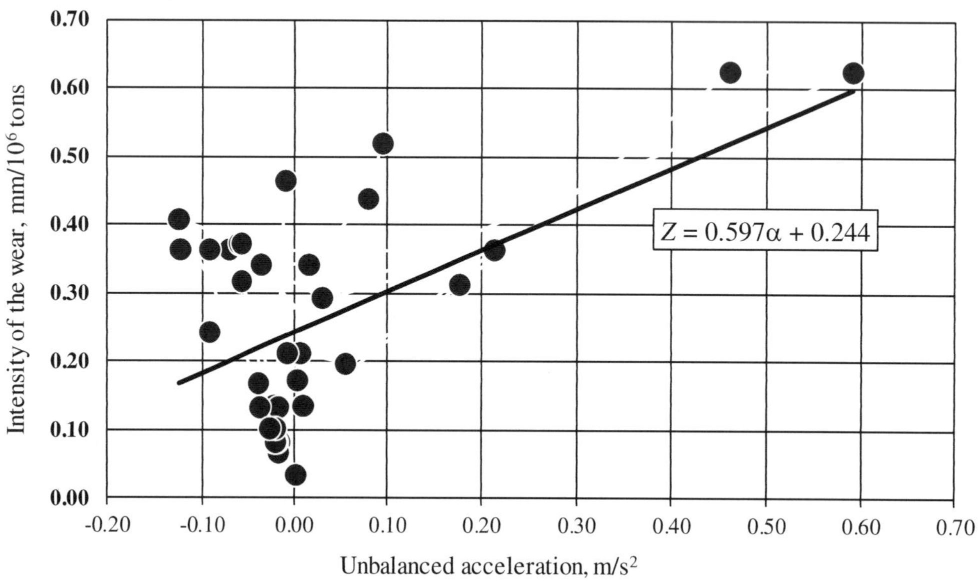

Figure 2. The intensity of the side wear, depending on unbalanced acceleration

2.2 The second stage
The second stage consists in performing calculations and analyzing the results. For this purpose, we used software package (MoveRW program is for traction calculations and RWPlan is to set the design parameters of curves, including the elevation of the outer rail by the weighted-average speed of the train movement and by the balance speed of vehicles while driving along the curve at unbalanced acceleration $\alpha \to 0$ [10]). However, the character in guiding of the vehicle in curves is such that due to the presence of sliding the wheels on rails, even when $\alpha = 0$ transverse horizontal forces appear, acting on the outer rail and causing its wear.

The research algorithm is as follows:

1. According to the results of scientific researchers at the DNURT from the previous years and similar researches of other authors [11, 12], wear indexes depending on unbalanced acceleration were accepted. Analytical expressions $Z(\alpha)$ (see as an example Figure 2) were introduced into the RWPlan program.

2. On the basis of traction calculations, authors determined speeds of movement $V_{\text{run } j}$ for trains of different mass Q_j and categories n_j that driving on the section.

3. For each segment of the way in the curve, unbalanced accelerations for all categories of trains α_j and the corresponding wear of the left Z_l and right Z_r rails were calculated.

4. As a result of calculations according to the above algorithm, the elevation of the outer rail was taken so that trains of all categories would realize the smallest values of transverse unbalanced accelerations $\alpha_j \to \min$.

5. Changing the elevation combination, we determined one that ensured the minimum wear on both rails while fulfilling all regulatory requirements (unbalanced acceleration, changing this acceleration in time, elevating speed of the wheel on the removal of the outer rail). Properly installed the elevation of the outer rail allows reducing the value of the guiding, lateral and frame forces and thereby increase

the strength, stability and reliability of the rail track.

3. Results

Various sections by type of traction, technical equipment, plan parameters and the longitudinal profile were considered for the research. As an example, the results of calculations are presented in Lviv-Rava-Ruska section as a promising project that can ensure the integration of railway transport in Lviv-Warsaw direction [13, 14]. The longitudinal profile of the Lviv–Rava–Ruska section, length 68 km, is mainly from 9 to 14‰ down-gradient. Plan of the line provides a significant impact on the maximum allowable speed. The length of curves is 31.7% of which the radius is less than 500 meters is 15.9% (Figures 3, 4).

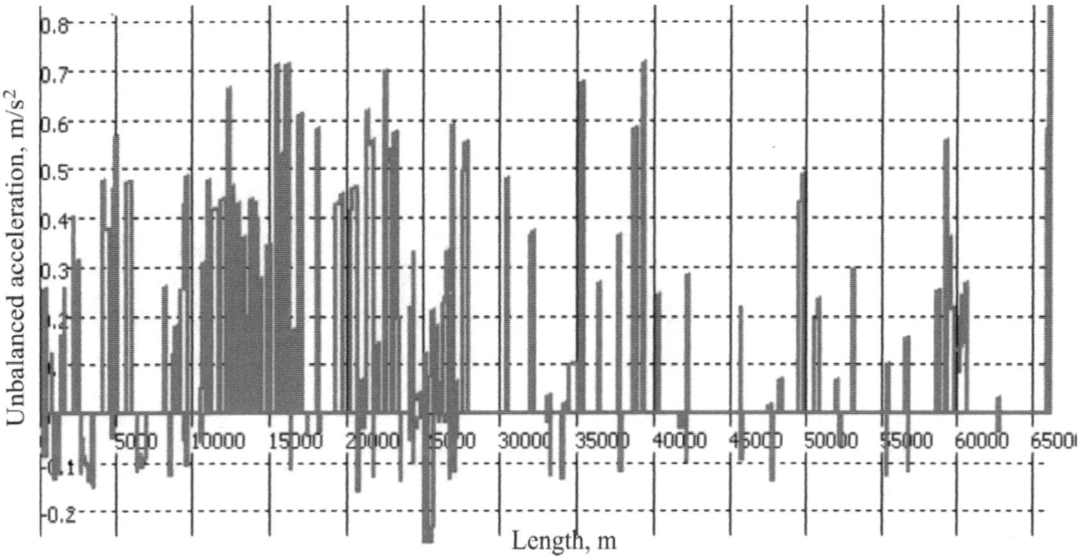

Figure 3. Unbalanced acceleration from the train movement

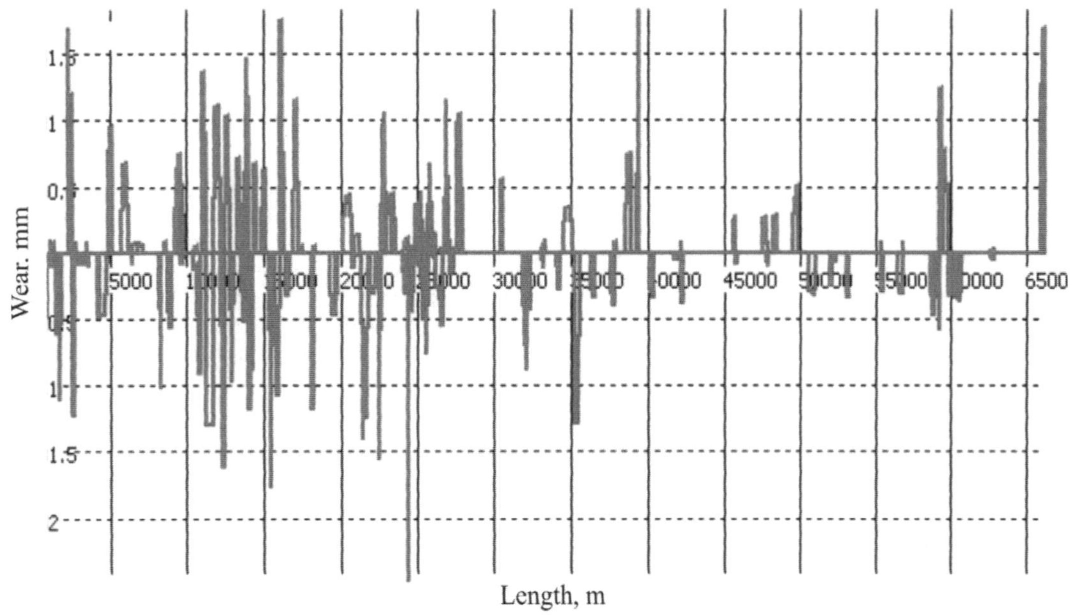

Figure 4. The wear of left and right rails from the train movement

The movement of trains in two categories is considered: passenger j=1 (locomotive M62, the mass of a train is 600 t) and freight j=1 (locomotive 2M62, the mass of a train is 2 200 t). The number of trains for the future ranged from 2 to 8 passengers and from 5 to 20 freight ones. On the basis of traction calculations, speeds of movement $V_{\text{run } j}$, unbalanced acceleration α_j for all categories of trains (Figure 3) and the corresponding wear of the left and right rails were determined (Figure 4).

Properly installed the elevation of the outer rail allows reducing the value of the guiding, lateral and frame forces and thereby increase the strength, stability and reliability of the rail track [15]. Therefore, as a result of calculations, the elevation of the outer rail was assumed that trains of all categories would realize the lowest values of transverse unbalanced accelerations.

Calculations are made for different values of the passed tonnage, and the results are presented in Figure 5.

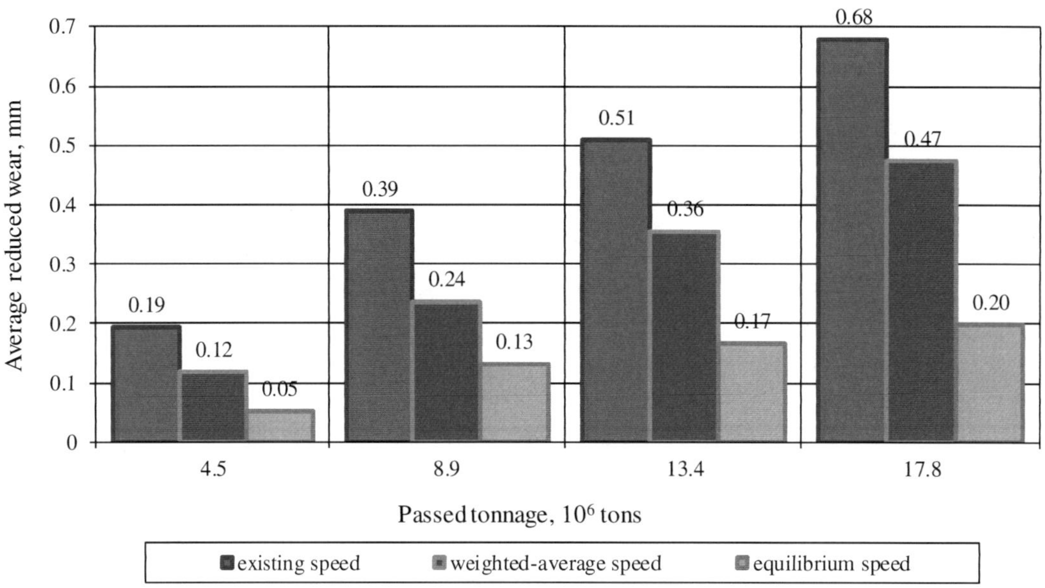

Figure 5. The average reduced wear depending on passed tonnage for existing, weighted-average and equilibrium speed

4. Conclusions

Taking into account the significant volumes of transportation direction as well as the favorable geographical position of Ukraine, it can be argued that creating an efficient modes for organizing the cargo and passenger transportation in international traffic with the EU countries with properly installed the railway track parameterswill allow attracting the additional volumes of transit traffic [13, 16, 17].

Analysis of the research results allows us to make the following conclusions:

- on the basis of analysis for the processes of increasing the rails side wear it is established that there is a fairly close correlation between the side wear, radius and unbalanced transverse acceleration along the length of the curve;
- with growth in unbalanced acceleration and the speed of train movement, the wear intensity increases, and with growth in the cant of the outer rail up to $\alpha = -0.3$ m/s^2 emergence, it decreases;
- zones of the high wear are located where there is a change of unbalanced acceleration: on the transient curves, at the conjunction with the circular curves and near the middle of the circular curves;

- one of the reasons for the uneven side wear is the violation of the smoothness of rail lines alignment. Therefore, the local side wear can be significantly reduced by systematically correcting curves. There is a relationship between the changing the curve of the versine along the length and the side wear: within the circular curve, the larger the versine, that is, the smaller the radius, the greater the wear;
- the use of the weighted-average speed to calculate the elevation of the outer rail to set the elevation of the outer rail in the curves does not allow minimizing the reduced wear from the train movement; with prevailing freight traffic, the elevation should be calculated based on the minimum of the wear, which will reduce the railway track deterioration;
- by changing the elevation, you can largely control the reliability of the track, train traffic safety and, to some extent, the costs associated with the wheel flange and rails wear.

References

[1] Povilaitienė I and Podagėlis I 2003 Research into rail side wearing on curves *Transport* **18** 3 pp 124–129
[2] Sun Y, Mandal N and Wu Q 2017 Rail wear modelling *The 25th International Symposium on Dynamics of Vehicles on Roads and Tracks* (At: Central Queensland University, Rockhampton, Australia) pp 1–6
[3] Bergseth E, Höjer M, Lyu Y, Nilsson R and Olofsson U 2019 A Wear Detection Parameter for the Wheel–Rail Contact Based on Emitted Noise *TribologyTransactions* pp 1–8
[4] Potapov D, Tuley Y, Kulik S, Manuilenko V and Safoshkina L 2019 Effects of train driving modes and the rail foundation structure on the rail side wear in small-radius curves *IOP Conf. Series: Materials Science and Engineering* **708** 012022
[5] Milovanova E and Egorova M 2020 The development of a method to reduce wear in the wheel-rail system *IOP Conf. Series: Materials Science and Engineering* **760** 012044
[6] Shvets A, Bolotov O, Saparova L and Shvets A 2019 Wear Wheels and Rails at the Uneven Loading of Gondola Cars *Railway Transport Certification Bulletin* pp 4–17
[7] Cerniauskaite L, Podagelis I, and Sakalauskas K 2001 New methods of calculation of railway exploitation expenses *Transport Engineering* **16** 2 pp 82–91
[8] Podagelis I. 1998 Factors acting on points and inner edges of rails wear *Civil engineering and environment* (Vilnius) pp 91–94
[9] The rules of railway service 2000 *Lietuvos geldinkeliai* p 213
[10] Korzhenevych I 2007 Advanced design of the realignment plan for the high speed trains in the RWplan 1.3.4 program *Science and Transport Progress* **19** pp 67–77
[11] Karpushchenko N 1996 Track gauge and rail wear *Railway and track facilities* **8** pp 6–7
[12] Povilaitene I, Sakalauskas K and Podahelis I 2004 The influence of geometrical parameters of railway track in curves side wear on outer rail *Science and Transport Progress* **3** pp 46–49
[13] Kurhan M and Kurhan D 2019 Providing the railway transit traffic Ukraine-european union *Pollack Periodica* **14** 2 pp 27–38
[14] Kurhan M and Kurhan D 2018 Problems of providing international railway transport *MATEC Web Conf.* **230** 01007
[15] Korzhenevych I 2008 Theoretical calculations are needed overall expansion between the rails in curves *Science and Transport Progress* **21** pp 285–292
[16] Pittman R, Jandová M, Król M, Nekrasenko L and Paleta T 2020 The effectiveness of EC policies to move freight from road to rail: Evidence from CEE grain markets *Research in Transportation Business & Management* 100482
[17] Kurhan M. and Kurhan D 2018 *Scientific-technical support of the railway Ukraine - European Union* (Dnipro, Ukraine) p 268

Study of the dynamic behavior of rolling stock using a computer experiment

A Kuzyshyn[1,2,4], **A Batig**[2], **S Kostritsa**[3], **J Sobolevska**[1], **S Dovhaniuk**[3] **and V Dzhus**[1]

[1]Lviv branch of Dnipro National University of Railway Transport named after Academician V. Lazaryan, 79052 Lviv, Ukraine
[2]Lviv Research Institute of Forensic Expertise, 79000 Lviv, Ukraine
[3]Dnipro National University of Railway Transport named after Academician V. Lazaryan, 49000 Dnieper, Ukraine

[4]kuzyshyn1993@gmail.com

Abstract. The authors of the article analyzed existing software systems that are used to study the dynamic behavior of rolling stock, and also emphasized the importance of using a computer experiment to implement this task. An improved spatial mathematical model of the dynamic behavior of a freight wagon when it interacts with a rail track is considered, and it is proposed to study it using an object-oriented approach using the Maple software package. To accomplish this task by writing procedures, separate modules (subprograms) were created that included differential equations of motion for the individual components of the object under study. Based on existing procedures, the main module was created, which reflected the sequence of performing certain functions (calling up the necessary information, especially the use of logic, methods for solving the problem, as well as visualizing the results). Using the created modules in the Maple software package, a computer experiment was used to study the dynamic behavior of a freight car when it interacted with a rail, and also showed the possibility of taking into account the influence of individual parameters of its technical condition on its dynamic behavior.

1. Introduction

The railway is an area of high danger, so the transportation of passengers and goods, the safety of traffic during the rolling stock operation are under special consideration.

One of the criteria of traffic safety during the design and operation of the rolling stock is the allowable level of its main dynamic indicators. The determination of such indicators consists in the study of dynamic processes in the system "vehicle-track" [1-3].

The study of dynamic processes in the mechanical systems boils down to generation and analysis of differential equations that describe these processes. Analytical methods, electronic modelling methods and approximate numerical integration are used to solve and analyse differential equations. The choice of the method for solving these equations depends on the ultimate goal of the study, the type and structure of differential equations of the system. They often use qualitative analytical methods that allow to determine natural frequencies and oscillation patterns for the vehicle.

In the study of the dynamic interaction of the system "vehicle-track", given complex system is usually simplified, and the idealized mathematical model is obtained that has basic dynamic properties of a given system. Then the parameters and characteristics of the elements of the calculation scheme are specified; differential equations are made up in such quantity that their number corresponds to the

Content from this work may be used under the terms of the Creative Commons Attribution 3.0 licence. Any further distribution of this work must maintain attribution to the author(s) and the title of the work, journal citation and DOI.
Published under licence by IOP Publishing Ltd

number of system freedom degrees. The most widely used method is D'Alembert method, which consists in the direct addition of equilibrium equations of the system, taking into account the forces of mass inertia. For the selected calculation scheme and the accepted coordinate system, differential equations of oscillations according to D'Alembert method are recommended to be made according to the geometric-physical-static rule (Popov A.A. rule).

Usually, the "vehicle-track" calculation scheme, even after simplification, has 10 to 40 degrees of freedom, and with complex connecting characteristics, the differential equation system is non-linear. Therefore, the integration of such complex systems is only possible through a computer experiment, which will reduce errors likelihood, greatly automate the process of mathematical model research and solve complex problems in a few actions.

2. Literature analysis and problem statement

At present, a considerable number of software products and complexes have been created for computer experiments related to the study of the dynamics of the rolling stock, which make it possible to simulate realistically the behaviour of the railway transport vehicles when driving along the track.

Examples of such software systems are: ADAMS / Rail, Medyna, Nucars, Simpack, Vampire, and Universal Mechanism (UM).

The ADAMS / Rail software package allows to create virtual prototypes of rail vehicles and to analyse them with the help of the same approaches that are used for real objects analysis [4]. The suspension system of a railway vehicle is divided into a primary and secondary suspension systems. These subsystems can be built by the user with the help of templates of creation or predefined templates. An illustrative example of a created bogie prototype in ADAMS / Rail is shown in Figure 1.

One of the best-performing complexes is MEDYNA software complex developed by a number of German manufacturers and scientific organizations. One of the significant advantages of this software complex is the simulation of interaction in the wheel-rail contact, as well as the successful experience application [5].

In order to describe the wheel-rail contact, MEDYNA software complex offers a choice of elements from kinematic connection to elastic multi-point contact, from a contact spot description as an ellipse to formation of a real interaction surface, from linear formulation of creep forces to nonlinear one. It is also possible not only to simulate wheel rolling on a rail, but also wheel rolling on a roller that allows to model the conditions of a bench experiment.

To simulate the dynamics of the rolling stock in the mode of traction and run, NUCARS software complex is used, which allows to apply different types of connections of individual elements of the studied object: vibration dampers, series and parallel springs, dampers (Figure 2) [6].

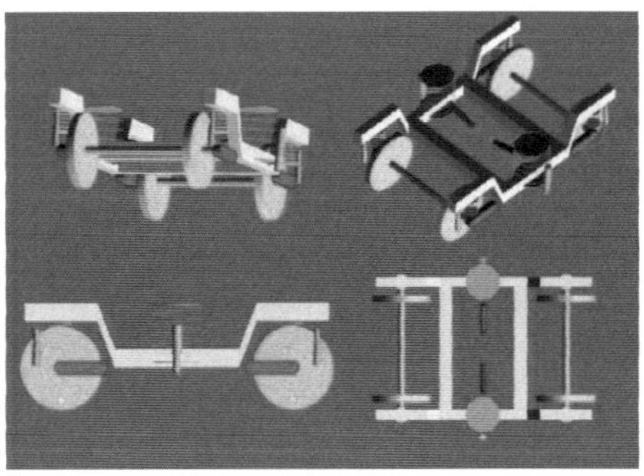

Figure 1. Model of a bogie in ADAMS/Rail.

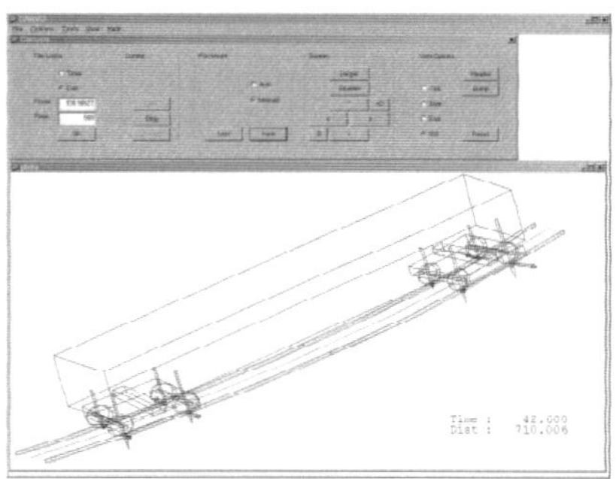

Figure 2. Modeling of the object under study in NUCARS software complex.

VAMPIRE software complex is a development of the British Railways Research Division, with the help of which a significant number of passenger vehicles have been investigated (Figure 3). However, VAMPIRE does not have the ability to simulate the processes occurring in wedge dampers, which does not allow to investigate widespread three-element bogies in the USA, Australia, South Africa and Ukraine.

The main features of SIMPACK Rail software include the possibility of track modeling – both rigid and elastic, the database availability of typical profiles of rails and wheels, with the ability to import profiles from in-situ measurements or regulatory documents (Figure 4) [7].

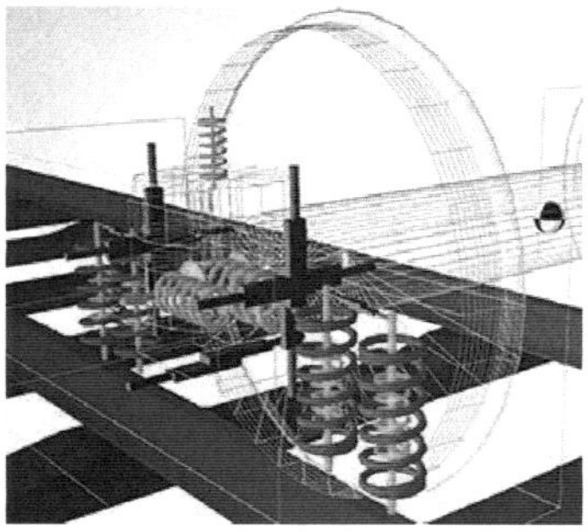

Figure 3. Suspension System Modeling in the VAMPIRE software complex.

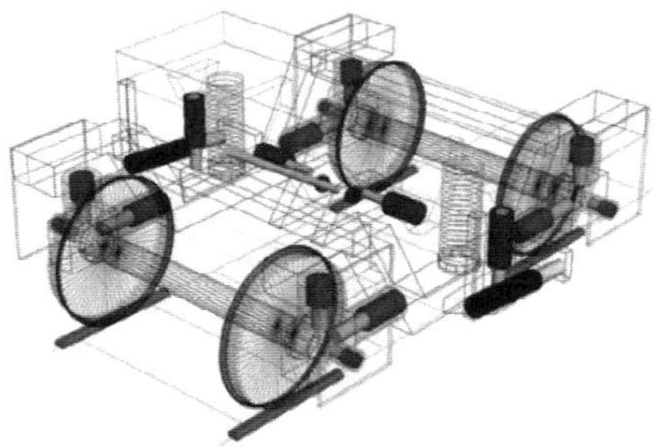

Figure 4. Model of a bogie in SIMPACK Rail software.

«Universal Mechanism» (UM) software complex is designed to simulate the dynamics and kinematics of planar and spatial mechanical systems that can be represented by a system of completely rigid or elastic bodies interconnected by kinematic and force elements. Systems of this type include: locomotives, wagons, various machines and mechanisms [8].

Whereas, in recent years, the number of applications for railway expertise implementation for the cases related to the rolling stock derailment to Lviv Research Institute of Forensic Expertise is significant and, given the current state of domestic rolling stock, with an upward trend, then the use of the above software packages to investigate derailment causes is a prerequisite. The peculiarity of such expertise is the need for its rapid execution, with the possibility of determining the appropriate causes of rolling stock derailment depending on its dynamic behaviour, which cannot be done by above software complexes in full measure.

Also, given the strategic partnership between the EU and Ukraine, which is a prerequisite for the successful integration of Ukraine into the European railway area, the issue of harmonization and formation of common approaches to the implementation of safe operation of railway vehicles is important. This peculiarity necessitates the development of a software complex with the ability to implement and compare different approaches to determine possible causes of rolling stock derailment.

Therefore, given the above, the development of a universal software system for modeling the rolling stock dynamics is an urgent task and needs further investigation.

3. The purpose and objectives of the study

The purpose of this work is to develop a software module to study the dynamic behaviour of a freight wagon in Maple computer environment and to conduct a computer experiment on the mathematical model.

Research objectives:

- to analyse necessary functions in Maple for software modules creation;
- to develop the basic software module of dynamic behaviour of a freight wagon in its interaction with the track;
- to carry out a computer experiment on a mathematical model of a freight wagon in the created main program module.

4. The basic program module of dynamic behaviour of a freight wagon

In this work, Maple software package was taken as the basis for creation of software modules of a freight wagon dynamic behaviour in its interaction with the track.

To display the algorithm for solving this problem in Maple software package, the authors of the article constructed a block diagram, which is presented in Figure 5 [9].

Blocks of differential equations of a body, bogie bolsters, lateral frames and wheel-sets are described by the procedure [10]. The procedure is called a separate module program (subroutine), which has an independent value and performs one or more operations, usually quite complex and different from the operations performed by built-in operators and functions.

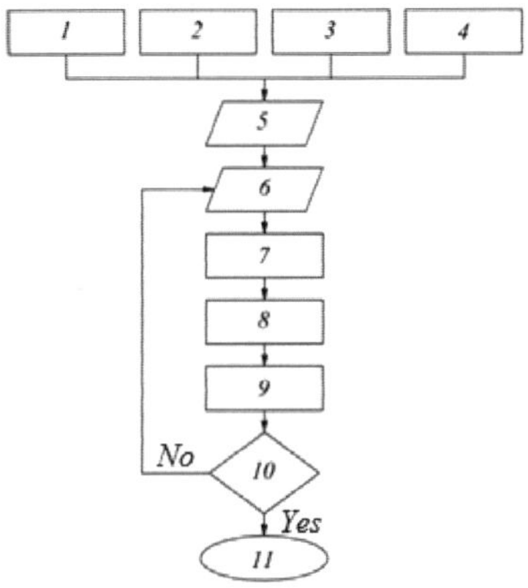

1 – block differential equations of the body of the rolling stock;
2 – block differential equations of the bolster;
3 – block differential equations of the side frames;
4 – block differential equations of wheelsets;
5 – unchanged input data;
6 – variable input data;
7 – a system of differential equations;
8 – solution of differential equations;
9 – obtaining results;
10 – verification of the fulfillment of certain conditions;
11 – output and visualization of the results

Figure 5. Flowchart of the algorithm.

The sequence of a procedure call is as follows (Figure 6).

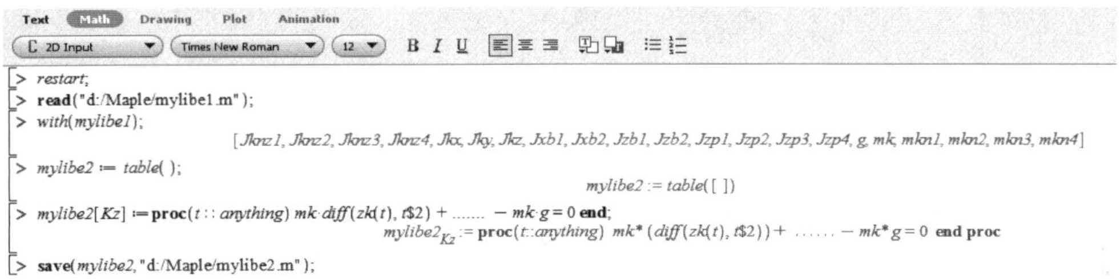

Figure 6. The sequence of a procedure call.

The input data are set by a separate block through the operator *table* (*F*), where *F* – is a table or an array (Figure 7).

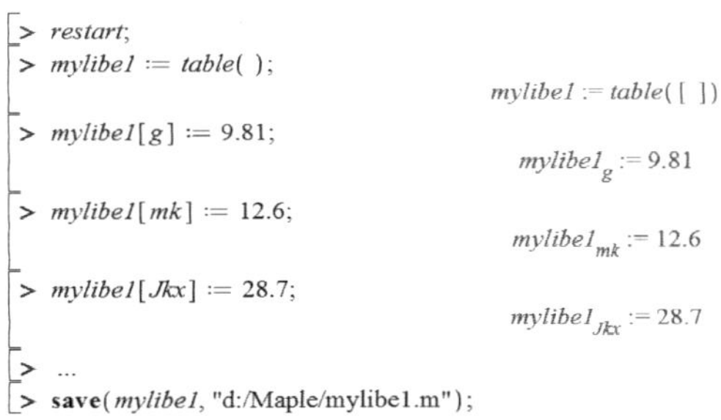

Figure 7. An array of input data.

For analytic and numerical solution of differential equations, Maple uses *dsolve* command, the format of which is:

$$dsolve(\text{eqns, vars, option}) \tag{1}$$

where eqns – is the system of equations with respect to undefined functions vars;
option – are additional conditions that allow you to specify the method for solving the problem.

The mathematical model was calculated using 4-5 order numerical Runge-Kutta method (Figure 8).

```
> dsol1 := dsolve((AB_AB(0, 0, 0, 0, 0, 0, 0, 0)), numeric, maxfun = 0);
                    dsol1 := proc(x_rkf45_dae) ... end proc
> dsol1(0)
  [t = 0., θb1(t) = 0., d/dt θb1(t) = 0., θb2(t) = 0., d/dt θb2(t) = 0., θk(t) = 0., ...... ]
```

Figure 8. Numerical solutions of differential equations.

When modeling freight wagon dynamics, the gaps closure takes place in the side bearers, axle-boxes, between the flange and the rail head, and therefore, in order to take into account the forces that arise in such cases, you can use the operator *if* whose format is (Figure 9).

```
> for cikl from 0 to 5 by 0.1 do
  dsol1 := dsolve((AB_AB(S[1], S[2], S[3], S[4], S[5], S[6], S[7], S[8])), numeric, maxfun = 0);
  print(cikl, dsol1(cikl));
  A1 := op(2, op(18, dsol1(cikl)));
  A2 := op(2, op(19, dsol1(cikl)));
  A3 := op(2, op(20, dsol1(cikl)));
  A4 := op(2, op(21, dsol1(cikl)));


  if A1 > 0 then S[1] := 0 else S[1] := 1 end if;
  if A2 > 0 then S[2] := 0 else S[2] := 1 end if;
  if A3 > 0 then S[3] := 0 else S[3] := 1 end if;
  if A4 > 0 then S[4] := 0 else S[4] := 1 end if;
  if A1 > 0 then S[5] := 0 else S[5] := 1 end if;
  if A2 > 0 then S[6] := 0 else S[6] := 1 end if;
  if A3 > 0 then S[7] := 0 else S[7] := 1 end if;
  if A4 > 0 then S[8] := 0 else S[8] := 1 end if;
```

Figure 9. Operator format *if* .

Computer graphics was used to visualize the results. The main command is *plot* which is in the standard library and does not require a preliminary call (Figure 10).

The structural appearance of the command *plot* is as follows:

$$plot\left(\{\text{func1, func2...}\}, x = \text{a...b}, y = \text{c...d, option}\right) \tag{2}$$

where func1 , func2 – analytical expressions;
$x = \text{a...b}$ – is a change interval of the variable x (segment along x - coordinate);
$y = \text{c...d}$ – is the interval along y - coordinate.

```
> p1 := plot([qwe[i], ert1[i], i = 1 ..kk − 1]) :
> with(plots) :
> display({p1})
```

Figure 10. The sequence of the command call *plot* .

5. Computer experiment of a spatial mathematical model of a freight wagon

The computer experiment is an experiment on a mathematical model of an object of study on a computer, that is, according to one parameter of the model, its other parameters are calculated and, on this basis, the properties of the object described by the mathematical model are derived. This type of experiment can only conditionally be attributed to the experiment, because it does not reflect natural phenomena, but is only a numerical implementation of a human-created mathematical model. Indeed, if the mathematical model is incorrect, its numerical solution can be strictly divergent from the physical experiment.

To accomplish this task, the authors of the article use an advanced spatial mathematical model of a freight wagon consisting of 30 second order differential equations. With the help of a computer experiment, we demonstrate the ability to study the dynamic behaviour of a freight wagon using Maple software package. For example, we study the dynamic behaviour of a body at different combinations of gaps in the side bearers and forces that result from their closure (Figure 11-14).

Based on the requirements of the regulatory documents, we study the dynamic behaviour of a freight wagon body using two different combinations of gaps in the side bearers: I – all four gaps have a nominal value of 10 mm (Figure 11), II – in the first two side bearers, the total gap is 3 mm, which is not in accordance with the requirements of Instruction TSV-0043 (Figure 12).

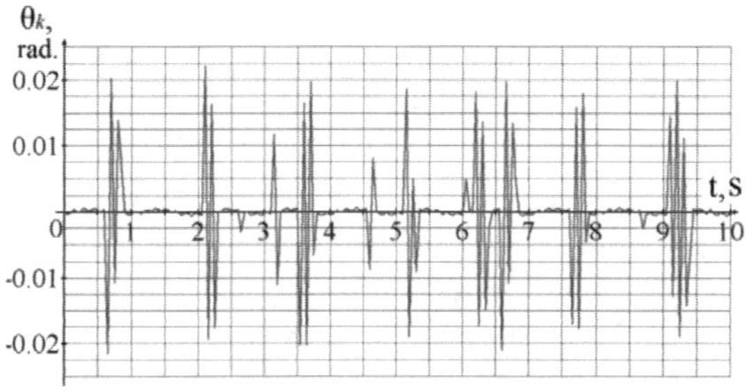

Figure 11. Lateral body oscillation at the first combination of gaps in the side bearers.

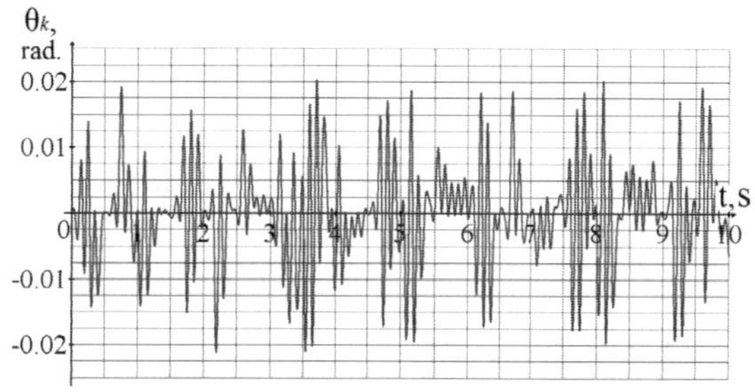

Figure 12. Lateral body oscillation at the second combination of gaps in the side bearers.

Analysing the above drawings, it is observed that the dynamic behaviour of a body is largely determined by the gaps in the side bearers, which non-normative value can lead to amplitude increase of lateral body oscillations, which is especially dangerous for curved track areas where tipping of a wagon can occur.

Having shown the possibility of studying the dynamic oscillations of a body at different combinations of gaps in the side bearers, it is appropriate to establish forces arising from these gaps closure (Figure 13).

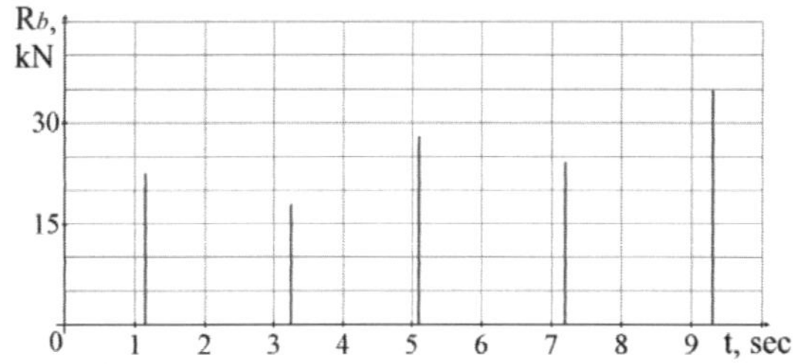

Figure 13. Forces resulting from gaps closure in the contact side bearers.

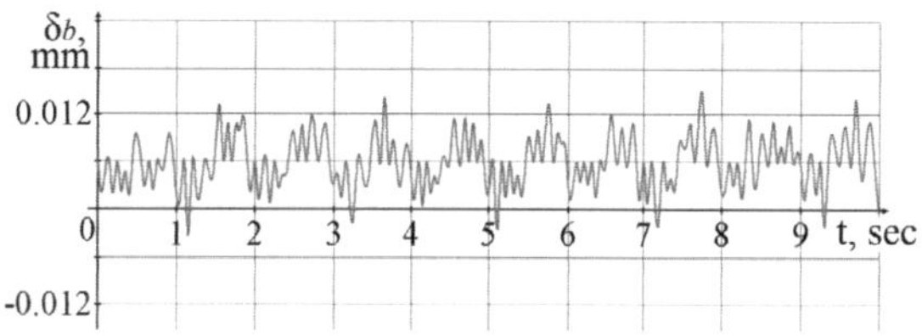

Figure 14. Dynamic change in the gap in the first contact side bearer.

After calculating the mathematical model, we obtained graphs of the dynamic change of the gaps in the contact side bearers (see Figure 14 for the first side bearer). In this case, the vertical unevenness of the track and the nominal size of the gap in the contact side bearer were adopted so that it was possible to close the gap and determine the reaction that occurs at the time of its closure. The analyses of Figure 14 shows that at certain points of time the gap in the side bearer takes on negative values, which indicates its closure, and at that moment, as it is observed in Figure 13, the reaction occurs immediately.

The authors want to point out that in this article the priority task is to test the possibility of Maple software package use in the study of dynamic processes of the rolling stock in its interaction with the track. Also, it is equally important to test the possibility of creating separate modules (subroutines) that would allow simulating any type of the rolling stock, including the rolling stock on pneumatic suspension.

To check the adequacy of the calculations performed, the forces arising in the side bearers of constant contact were analyzed. According to GOST 34387-2018 the static pressing force in the side bearers height does not exceed 27.1 kN. The work considered the movement of a wagon with small vertical irregularities of the railway track, for which it can be considered that the influence of dynamic processes is insignificant. The design forces in the side bearers were 32 kN, which exceeds the static value by 15%.

Therefore, having analysed the main functions of Maple software package and conducted the study of the dynamic behaviour of a fright wagon body, the authors of the article came to the conclusion that Maple software package could be used in the study of the rolling stock dynamics, which in the future will allow to determine the causes of the rolling stock derailment using the dynamic model of the wheel-rail interaction.

6. Conclusions

Having analysed the existing software complexes, it was emphasized the importance of their use in the study of the dynamic behaviour of the rolling stock. Due to the peculiarities of forensic railway expertise, it was decided to use Maple as an alternative.

Using «$\mathbf{proc}(t::anything)$» procedure, separate modules (subroutines) were created, with the help of which the basic module for the study of the dynamic behaviour of a freight wagon was constructed.

Having conducted a computer experiment on a mathematical model in the main module, the dynamic behaviour of a freight wagon, namely the body, was investigated at various combinations of gaps in the side bearers and the forces that result from their closure.

It has been established that the Maple software package can be used to simulate any type of rolling stock and study its dynamic behavior. Also using this software environment and a dynamic model of wheel-rail interaction, it will be possible to proceed to establish the causes of the rolling stock derailment.

References

[1] Kuzyshyn A, Batig A, Kostritsa S, Sobolevska J, Kovalchuk V, Dovhanyuk S and Voznyak O 2018 Research of safety indicators of diesel train movement with two-stage spring suspension *BulTrans 2018: 10th Intern. Sci. Conf. on Aeronautics, Automotive and Railway Engineering and Technologies* **234**-1–**234**-6

[2] Kovalchuk V, Kuzyshyn A, Kostritsa S, Sobolevska Y, Batig A and Dovganyuk S 2018 Improving a methodology of theoretical determination of the frame and directing forces in modern diesel trains *Eastern-European Journal of Enterprise Technologies* **6/7 (96)** pp 19-25

[3] Kuzyshyn A, Kostritsa S, Ursulyak L, Batig A, Sobolevska J and Voznyak O 2019 Research of the impact of geometric unevenness of the railway track on the dynamic parameters of the railway rolling stock with two-stage spring suspension *IOP Conference Series: Materials Science and Engineering* **664**-1–**664**-10

[4] Azhar I, Mohd Azman A and Khisbullah H 2013 Multibody Dynamics Models of Railway Vehicle Using Adams/Rail *Applied Mechanics and Materials* **393** pp 644–648

[5] Kortum W 1990 Analysis of railway vehicle system dynamics with the multibody program MEDYNA *DLR German Aerospace Research Establishment Institute for Flight Systems Dynamics*

[6] Wilson N, Shu X and Gonzales K 2015 *TTCI's Scientific Software Suite and NUCARS* p 34

[7] Seyed Milad Mousavi Bideleh 2016 Multiobjective Optimization and Active Control of Bogie Suspension *Department of Applied Mechanics Chalmers University of Technology Gothenburg* p 50

[8] Kovenkin D A and Karpov I G 2016 Modeling the interaction of the railway crew and the track in the Universal Mechanism software package *Laboratory workshop in the discipline Software for calculating the construction of railway tracks* p 100

[9] Kuzyshyn A, Batig A, Sobolevska J, Kostritsa S, Ursulyak L and Dovhaniuk S 2019 Determining the causes of rolling stock derailment from the track using modern research methods *2 and International Scientific and Practical Conference "Energy-Optimal Technologies, Logistic and Safety on Transport" (EOT-2019) MATEC Web of Conferences* **294**-1–**294**-8

[10] Dyakonov V P 2006 *Maple 9.5/10 in mathematics, physics and education* p 720

IOP Conf. Series: Materials Science and Engineering **985** (2020) 012003 doi:10.1088/1757-899X/985/1/012003

3D model to study transitional regimes of train motion

O Markova[1,2], H Kovtun[1] and V Maliy[1]

[1] Institute of Technical Mechanics of the NAS of Ukraine and the SSA of Ukraine, 15 Leshko-Popel street, Dnipro, 49005, Ukraine

[2]olgamarkova2002@gmail.com

Abstract. When considering issues of increasing train speeds, it is necessary to conduct a preliminary assessment of the entire train dynamic characteristics and forces acting in the elements of cars and a locomotive, as well as in inter-car connections when driving at high speeds along a path of arbitrary shape in plan and profile, and also to consider emergency situations, which may take place during the motion. This in turn requires the development of reliable mathematical models simulating the train motion. The study of dynamic characteristics of railway rolling stock is associated with the consideration of vibrations of complex mechanical systems with a large number of degrees of freedom. The vibrations of cars in the train can be described either by one-dimensional or spatial model. The aim of this work is to show that for some studied cases the modelling of train motion by 3D models more accurately reflects the processes occurring in a moving train. Some results are discussed in the paper.

1. Introduction

When designing a new and modernizing the existing freight rolling stock and choosing the conditions for cargo transportation, it is important to take into account forces acting on the car and locomotive arising during standard and emergency modes of train motion. The process of forces occurrence and distribution in a freight train depends on many reasons, the main of which are the speed and mass of the train, the plan, profile and technical conditions of the rail track. Due to the fact that there are gaps in the inter-car connections of the freight train, shock forces can occur during its motion, affecting the strength of the car structural elements. The presence of loaded and empty cars in a freight train in some cases can lead to pulling out or squeezing empty and lightly loaded cars out of the train, which in its turn leads to the car wheelsets derailment. Therefore, when considering issues of increasing freight train speeds, it is necessary to conduct a preliminary assessment of dynamic characteristics of their cars and locomotive and forces acting in inter-car connections at the motion with increased speeds along an arbitrary track shape in plan and profile, as well as to consider emergency situations that may occur in the process of motion. This, in turn, requires the development of reliable mathematical models simulating freight train motion and the corresponding software for carrying out evaluative calculations.

To study vehicles stability of motion, to determine their elements rational parameters, to investigate curvilinear motion mathematical models of separate cars or train parts spatial vibrations are used for decades already [1-6]. But to consider train motion in transitional regimes and its motion along the changes of track gradient one or two dimensional mathematical models are still used [7-9]. Similar models are often used to study trains collision with the obstacle [10-13]. Such models allow to estimate the level of forces in inter-car connections at transitional modes, and cars or locomotives

Content from this work may be used under the terms of the Creative Commons Attribution 3.0 licence. Any further distribution of this work must maintain attribution to the author(s) and the title of the work, journal citation and DOI.
Published under licence by IOP Publishing Ltd

strength under collision, but their use do not give possibility to determine the possibility of derailment at the collision of the train with the motionless or moving obstacle.

The aim of this work is to show by some studied cases that with the use of the mathematical models of train spatial vibrations one can have more accurate results describing the processes occurring in a moving train.

2. Mathematical model of train spatial vibrations

As the train vibrations model, in which a locomotive and each car are considered as fairly complete spatial systems will have a very large number of degrees of freedom the authors used the model developed earlier [6], in which to assess the quality of a freight train motion along a path of arbitrary shape in plan modelling option is considered when the calculation model of one part of cars takes into account only their longitudinal movements, and for the other part of the cars full spatial models of their vibrations are used.

Differential equations of the train motion along the track of arbitrary shape are set in the form of Lagrange's equations of the second kind:

$$\frac{d}{dt}\left(\frac{\partial T}{\partial \dot{q}_i}\right)-\frac{\partial T}{\partial q_i}+\frac{\partial \Pi}{\partial q_i}+\frac{\partial \Phi}{\partial \dot{q}_i}=Q_i^*+F_i, \qquad (1)$$

where $q_i, \dot{q}_i$ are the generalized coordinates and their velocities (i is the number of degrees of freedom); T is the kinetic energy; Π is the potential energy; Φ is the dissipative function; Q_i^* is the sum of wheel/rail interaction forces Q_i, and forces S_i acting in inter-car connections during train motion; F_i are the applied external forces.

When determining the number of degrees of freedom of the considered mechanical system, the restrictions imposed on the movements of bodies due to generally accepted assumptions and design features of the cars are taken into account. The kinetic energy of the system is obtained by adding the kinetic energies of the bodies included in the system and the track. For each of the bodies, the kinetic energy was determined by Koenig's theorem. The potential energy of the system was determined as the sum of the energy of elastic deformations and changes in energy due to the raising or lowering of the centres of masses of the bodies included in the system. The energy dissipation function in the vehicle takes into account the action of viscous friction forces as well as the effect of dry friction forces.

Generalized forces Q_i are defined as coefficients of variations of generalized coordinates in expressions of creep forces possible work. In determining the forces acting on the wheel in the horizontal lateral direction, components of the force of gravity are taken into account (in addition to the creep forces). Forces S_i simulate the inter-car connections work.

The external forces F_1, F_i acting on the locomotive and the i-th car are a tractive force F_1^* developed by the locomotive, braking forces B_1, B_i occurring when brakes are switched on, and resisting forces W_1, W_i. In total the external forces can be written as follows

$$F_1=W_1+B_1+F_1^*, \qquad (2)$$

$$F_i=W_i+B_i. \qquad (3)$$

Braking forces are applied to the wheelsets with transport delay depending on speed of braking force distribution and cars dimensions. Both braking and tractive forces are changed in time in accordance with the given characteristics. Resisting forces are divided into main ones which act all the time and additional ones acting only at the motion along separate track segments. In particular, forces occurring at the train motion along curves and grades are the additional resisting forces.

At the simulations of the train motion along the track gradients additional forces P_i acting on the locomotive $(i=1)$ and the i-th car were taken into account and were described as follows [14]:

$$P_i = \begin{cases} m_i g i_1, & if \quad x_i - L_{ci} \le L_1; \\ for \quad x_i - L_{ci} > L_1: & \\ m_i g\left(i_1 + yR_1^{-1}\right), & if \ 0 < y \le l_1 \ (y = x_i - L_{ci} - L_1); \\ m_i g i_2, & if \ 0 < y \le L_2 \ (y = x_i - L_{ci} - L_1 - l_1); \\ m_i g\left(i_2 + yR_2^{-1}\right), & if \ 0 < y \le l_2 \ (y = x_i - L_{ci} - L_1 - l_1 - L_2); \\ \cdots\cdots & \\ m_i g i_k, & if \ 0 < y \le L_k \ (y = x_i - L_{ci} - \sum_{j=1}^{k-1}[l_j + L_j]); \\ m_{ig}\left(i_k + yR_k^{-1}\right), & if \ 0 < y \le l_k \ (y = x_i - L_{ci} - \sum_{j=1}^{k-1}[l_j + L_j] - L_k), \end{cases} \quad (4)$$

where x_i is the longitudinal displacement of the i-th car; m_i is the mass of the i-th car; g is the gravitational acceleration; L_{ci} is the length of the train part before the i-th car; i_k is the gradient of the k-th track segment with the length L_k; R_k is the curve radius connecting the k-th and the $(k+1)$-th track segments; $l_k = |i_{k+1} - i_k| R_k$ is the length of the curve connecting the k-th and the $(k+1)$-th segments.

The track was taken like an inertial elastic-viscous one. It is modelled as a reduced to each wheel mass (eight reduced masses for the car and twelve masses for the locomotive) with only vertical and horizontal lateral displacements and is based in these directions on springs and dampers simulating elastic-viscous features of rails and under-rail basement. The train motion is modelled with taking into account external disturbances because of the rails geometrical imperfections. Perturbations were applied under the wheels of the vehicles with a transport delay, which depends on the distance between the wheelsets and the vehicle speed.

In cars, which motion is modelled by a spatial calculation scheme, bogies of 18-7020 type are used as running gears [15, 16]. Each car was represented by the system of solid bodies connected by rigid, elastic and dissipative elements. The calculated scheme of the separate car presents a mechanical system consisting of eleven solid bodies (a car-body, two bolsters, four side frames, four wheelsets) and eight reduced masses of the track. Characteristics of the considered bogies connections are sufficiently non-linear because of the friction elements both in the centre and axle-box suspensions and in simultaneous car-body in plan rotations in relation to the bogies. Taking into account the geometrical constrains imposed to the system each separate car has 42 degrees of freedom. To have possibility to study car motion in transitional regimes, the additions coordinate corresponding to the car absolute displacement in longitudinal direction because of train speed variation was included. So each car has 43+16=59 degrees of freedom.

A six-axle locomotive of the 2TE116 type is considered as a locomotive. The computational scheme is submitted as a branched system of nine rigid bodies (a car-body, two frames, and six wheelsets) and twelve reduced masses of the track joined by linear and non-linear connections with various rheology. Physical and geometrical interaction of wheels and rails is taken into account. With usual constrains this system has 35+24=59 degrees of freedom.

The developed model allows us to evaluate the stability and safety of motion of both separate cars and a locomotive, and the train as a whole. In total it represents a system of $(L + jK + N)$ second-order nonlinear differential equations (L=59 is the number of equations describing the motion of the locomotive, j=59 is the number of degrees of freedom for each of the K cars represented by a spatial branched system, N is the number of cars, the design scheme of which is represented by one mass: it can vary from zero to the full number of cars in the train.

Based on the developed mathematical model, an algorithm and a computer program have been designed to assess the dynamic characteristics of a moving train. The computer program allows one to simulate a wide range of different cases, covering all kinds of combinations of the number of cars in a train, their loading, technical conditions of the running gears and track, plan and profile of the track along which the train is moving. As a result of calculations, almost any set of necessary output quantities can be obtained.

First the freight train motion along grades was studied. To do it the forces acting in inter-car connections were used as the main dynamic characteristic of the train.

3. Results of studies of a freight train motion along changes of track gradient

A train consisting of a locomotive and 23 cars equipped with Sh-2-V absorbing devices [17] was considered. Here, the motion of each first and last ten cars is modelled by one rigid body, the mass of which is equal to the mass of the car as a whole. The motion of the locomotive and each of the three cars in the middle of the train was modelled using spatial models describing vibrations of these vehicles. Irregularities of the rail track in the vertical and horizontal lateral directions corresponding to a satisfactory conditions of the track are used as perturbations. Irregularities are represented as random stationary processes with a Gaussian distribution in amplitude and spectral density with monotonously decreasing character [18].

When solving the problem of assessing forces in the train moving along a variable track profile, in order to determine safe modes of motion, first of all, we consider transient processes caused by the train motion without control actions (braking, traction modes). At the same time, the influence of the rail track and train parameters, the train make-up, and the conditions characterizing the initial state of inter-car connections [6] are evaluated.

In the calculation options the train speed, the grades, the train mass, the initial state of inter-car connections (pre-compressed or pre-stretched train), the load of the train (loaded, empty) and the location of the cars of different loading in its scheme are varied. The train motion along a track section of the following shape is considered: first, there is a straight horizontal section of 400 m long, then the train moves along a downgrade of 700 m, then the motion continues along an upgrade of 700 m length and the train again goes to a horizontal section of a track.

An analysis of the calculated data showed that when the considered train moves along changes of longitudinal track gradients in couplers of those cars which vibrations are modelled by a spatial system of bodies with a large number of degrees of freedom, the largest values of longitudinal forces most often arise. As an example Figure 1 shows plots of the maximum tensile (lines 1, 2) and compressive (lines 3, 4) longitudinal forces distribution along a train moving upgrade at a speed of 60 km/h for a train consisting of all loaded cars (Figure 1, a) and a train in which all cars are empty (Figure 1, b). The spatial model represents the eleventh, twelfth and thirteenth cars. Lines 1 and 3 correspond to a pre-stretched train, and lines 2 and 4 correspond to a pre-compressed train.

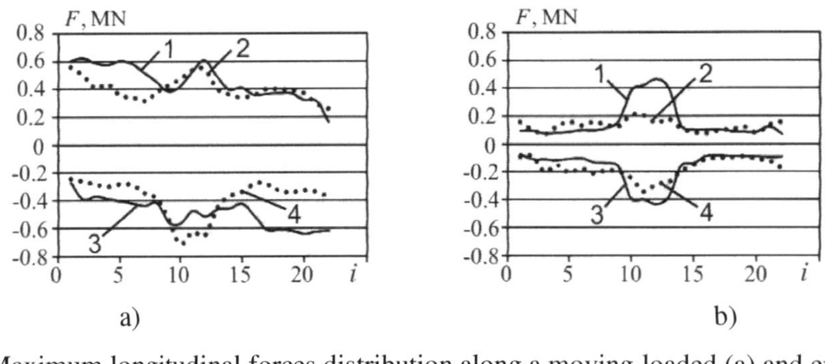

Figure 1. Maximum longitudinal forces distribution along a moving loaded (a) and empty (b) train.

To assess the influence of the characteristics of the separate car model on the magnitudes of the forces under study, the distribution of maximum forces in inter-car connections was calculated for the case when all the cars of the train are represented by bodies of the same mass and have only one degree of freedom, so they move only in longitudinal direction. The results obtained were compared with the data of the corresponding calculations, in which each of the three middle cars is represented by a spatial model of nine rigid bodies connected by bonds of various natures with a large number of degrees of freedom. As an example Figure 2 shows the distribution curves of the maximum forces in inter-car connections for the described cases during the motion of a pre-stretched train with empty cars along the downgrade (Figure 2, a) and on the upgrade (Figure 2, b). Here, lines 1 and 3 correspond to the case when the model of middle cars takes into account their spatial vibrations, and lines 2 and 4 for the model when all cars are represented by one mass model with only longitudinal motion. From the above results it follows that the calculated forces in the coupler devices of cars represented by spatial models are significantly differ from the forces obtained for a one-dimensional train model. To assess the nature of the forces change over time, Figure 3 and Figure 4 show, as an example, the time histories of the longitudinal forces changes in the connection of the 11th and 12th cars for the case when the train model is one-dimensional (Figure 3) and spatial (Figure 4). If the time histories are similar in nature for the changes of forces in time, the values of forces for different models are significantly differ from each other. Moreover, in all cases, larger values of both tensile and compressive forces arising in the inter-car connections occur when using the spatial model of train cars.

The calculations showed that when evaluating the forces arising in the inter-car connections of freight train cars during its motion along changes of track gradient, it is better to use spatial calculated schemes of train cars, which allows more accurate determination of the forces acting in automatic coupler devices.

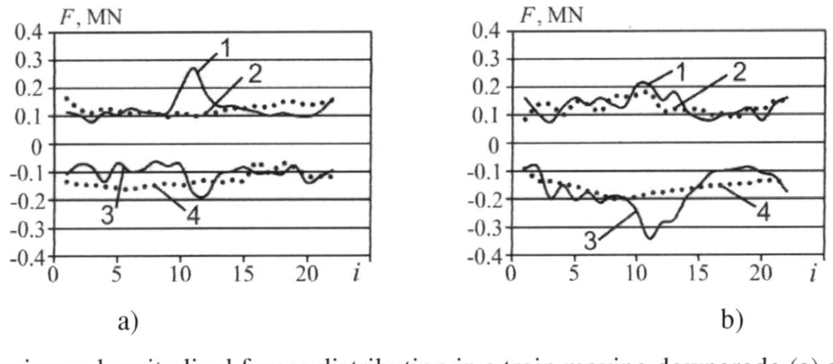

Figure 2. Maximum longitudinal forces distribution in a train moving downgrade (a) and upgrade (b).

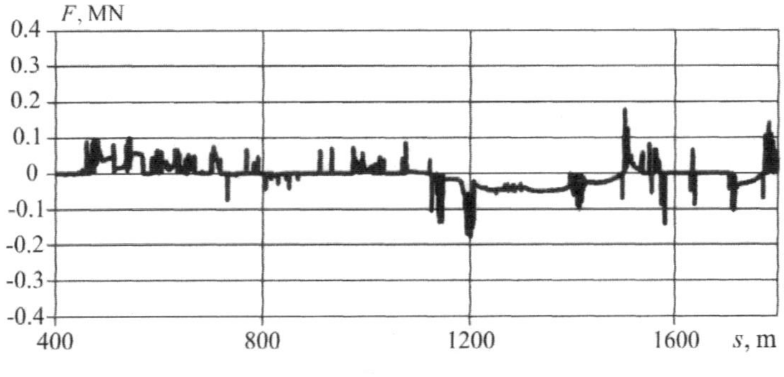

Figure 3. Time history of forces between the 11th and 12th car (one-dimensional model).

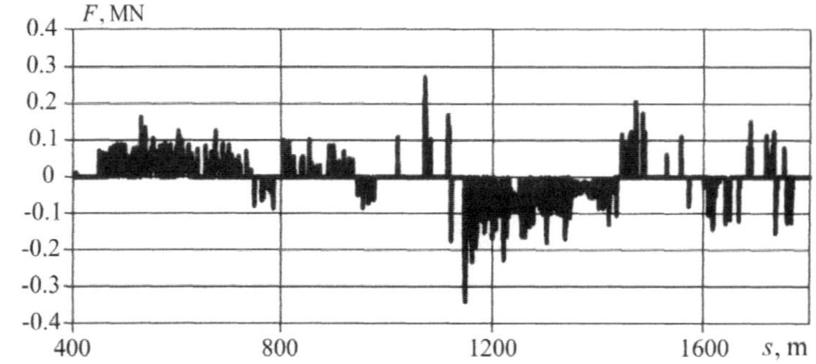

Figure 4. Time history of forces between the 11th and 12th car (three-dimensional model).

4. Train collision with an obstacle

It was also of a great interest of the authors to use the developed mathematical model for investigation of an emergency situation. As one of the most often accident for railway trains is locomotive and cars derailment, it was used to study the train collision with the obstacle and derailment possibility. An algorithm to calculate a force occurring between the train locomotive and the obstacle is described in papers [10, 11] and its shape is shown on Figure 5.

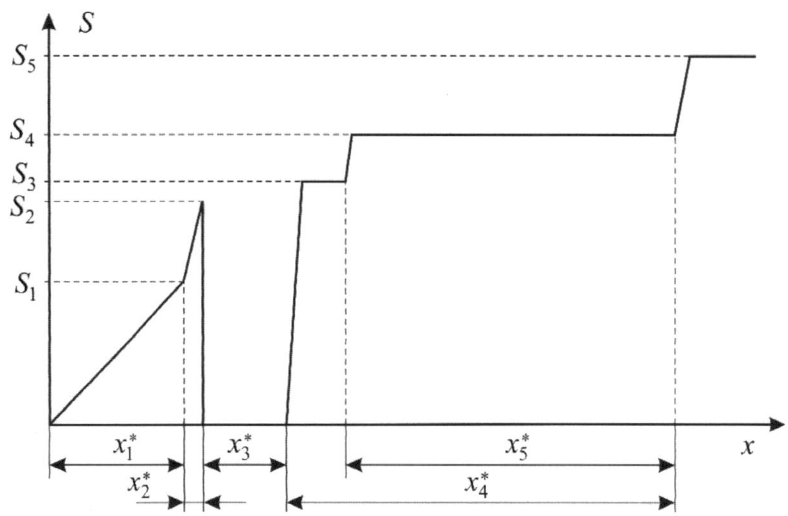

S_1 is a force of shock absorber closing; x_1^* is a shock absorber travel; S_2 is a force of locomotive automatic coupler cutting; x_2^* is the locomotive car-body structure deformation at the force increase from S_1 up to S_2; x_3^* is the distance between absorbing devices and an obstacle; S_3 is the force at which locomotive absorbing devices begin to deform; x_4^* is the locomotive absorbing devices deformation; x_5^* is the locomotive cabin sacrificial zone deformation; S_4 is the force, at which the mutual deformation of absorbing devices and locomotive cabin occurs; S_5 is the force, at which locomotive frame plastic deformation begins.

Figure 5. The force occurring between the locomotive and the obstacle.

At the initial moment of collision a standard shock absorber of the locomotive automatic coupler begins to work. Energy absorption is by the elastic deformation of the locomotive absorbing device. When it is closed, the force is acting on the locomotive frame. When the force acting on the automatic coupler reaches the given limit value, fitting elements of the automatic coupler to the locomotive frame are broken and the automatic coupler is hiding under the locomotive frame. From this moment and till locomotive passive safety devices are in contact with the obstacle the interaction force is equal to zero. After the full deformation of energy absorbing devices the sacrificial zone of the locomotive driver cabin is working. In dependence of the compression force value elastic or elastic-plastic deformation of the locomotive car-body structure occurs.

To assess dynamic behaviour of the train cars, characteristics describing car dynamic qualities were used. For the freight trains such characteristics are

– car horizontal lateral and vertical accelerations at the central plates zones;
– horizontal lateral forces related to the axle load;
– a coefficient of vertical dynamics determined like dynamic vertical forces in the axle-box suspension divided by the static axle load;
– a derailment coefficient of wheel stability.

Values of the derailment coefficient of wheel stability were determined as follows [19]:

$$k_{st} = \frac{tg\beta - \mu}{1 + \mu tg\beta} \cdot \frac{P_n(1{,}03 - 1{,}17k_{dv}^{*} + 0{,}14k_{dv}^{**}) + 0{,}515q + 0{,}705H_p}{P_n(0{,}24 + 0{,}04k_{dv}^{*} - 0{,}28k_{dv}^{**}) + 0{,}121q + 0{,}92H_p}, \quad (5)$$

where β is the angle of wheel flange inclination to the horizontal axis; μ is a friction coefficient; P_n is the static axle journal load; q is the weight of the unsuspended parts; k_{dv}^{*} and k_{dv}^{**} are coefficients of vertical dynamics for the leading and trailing wheels.

To insure train safe motion, the minimal value of the derailment coefficient of wheel stability has to be higher than 1.4 [20].

The simulation of the train collision with the obstacle has been done in accordance with the following scheme. At the initial moment the train moved with the constant speed along the straight track segment with vertical and horizontal lateral irregularities. After it has passed 100 m distance, the collision with the obstacle occurred. Vehicles of different masses were taken as the movable obstacles. They were a freight train, a similar locomotive, an empty freight car. At the initial moment the obstacle did not move. Calculations were made for the locomotive speed from 25 km/h up to 80 km/h and different levels of geometrical track irregularities corresponding to excellent, good, satisfied and permissible states of the track.

Firstly, the case was considered when collision takes place on the track segment with the excellent state (track without irregularities). In this situation, forces occurring at moving vehicle collision with an obstacle act only in longitudinal direction coinciding with the track axis. After the collision the speed of the coming locomotive reduces, and the speed of the obstacle vehicle increases. The increase of the obstacle mass leads to the greater reduction of the locomotive speed, and the increase of the locomotive speed leads to the increase of its time to full stop. It is worth to note that in the considered situation wheelset derailment does not occur.

In reality such situation does not exist. As a rule the track state can be estimated as good, satisfied or permissible [21]. Vehicle vibrations occur at the motion along such track segments because of different irregularities (gauge, cross-level, alignment, vertical profile change). It leads to occurrence of changing in time angle of the car-body rotation about its vertical axis in the horizontal plane. So not only longitudinal forces, but also lateral force component exists at the colliding vehicles interaction force. In its turn it leads to the lateral wheel/rail interaction forces occurrence. These forces values depend on the locomotive car-body yaw angle at the moment of collision.

The carried investigations have shown that all the studied dynamic characteristics of the locomotive sufficiently depend on the vehicle positioning at the moment of collision. In dependence on the locomotive car-body yaw angle, its motion speed and the obstacle vehicle mass, locomotive dynamic characteristics and the coefficient of its stability of motion along the track change dramatically. Mostly these characteristic values are in the impermissible zone, and in some cases the stability is lost and wheelset derailment occurs [22].

As one example of the large number of simulations Figure 6 shows the comparison of plots of changes in the values of derailment coefficients of stability of the locomotive wheelsets when it collides with a single freight car at a speed of 60 km/h. Here, the collision occurs on the track section with excellent conditions (Figure 6, a) and on the track section with irregularities (Figure 6, b). As can be seen from the plots, in the case when the arising collision forces act only in the longitudinal direction, the locomotive does not derail. In the case when the interacting forces have not only a longitudinal component, but also the lateral components the value of the derailment coefficient of stability becomes lower than the permissible one and stability is lost (Figure 6, b). The analysis of the results obtained has shown that the heavier is the obstacle vehicle and the higher is the collision speed, the higher is the probability of locomotive and cars derailment.

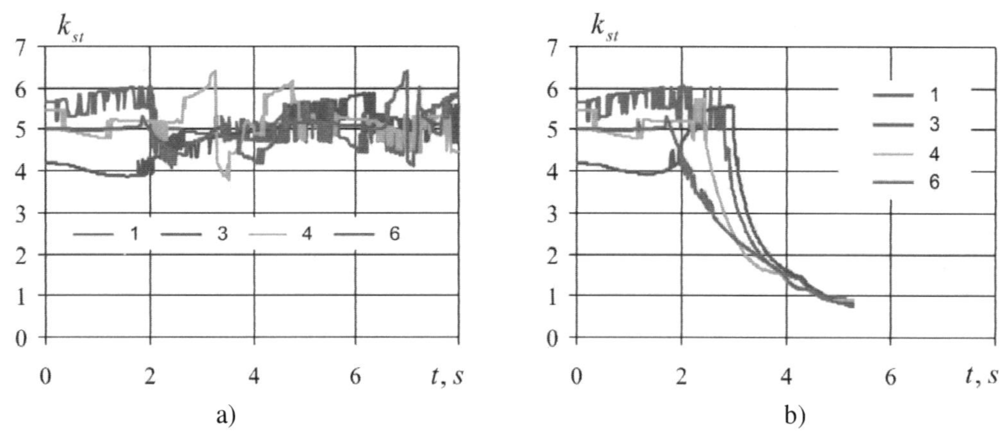

Figure 6. Dependencies of derailment coefficients of stability for locomotive wheelsets, the track section with excellent conditions (a) and with irregularities (b).

5. Conclusions

In this article the mathematical model of freight train spatial vibrations developed on the basis of the earlier presented by the authors model was used to investigate two cases of train motion. The first case was the freight train motion along track gradients. From the results obtained it followed that the calculated forces in the coupler devices of cars represented by spatial models were higher than the forces obtained for a one-dimensional train model. So, for evaluating the forces arising in the inter-car connections of freight train cars during its motion along changes of track gradient, it is better to use spatial calculated schemes of train cars, which allows more accurate determination of the forces acting in automatic coupler devices.

The second studied case was the train collision with the moving obstacle like another train, locomotive and freight car. The carried investigations have shown that the locomotive dynamics sufficiently depends on the vehicle positioning at the moment of collision. It can be changed sufficiently in dependence on the locomotive car-body yaw angle, its motion speed and the obstacle vehicle mass. In some cases the stability was lost and wheelset derailment occurred.

Thus, the obtained results for both studied cases has shown that when considering issues related to the safe motion of railway rolling stock, to have more accurate results it is better to use mathematical model describing spatial vibrations of locomotives and train cars.

As a future step, simulations of much more cases should be done to evaluate dynamics of train elements at its motion along the track of arbitrary shape in plan and profile and also in different regimes of motion.

References

[1] Demin Yu, Dlugach L, Korotenko M and Markova O 1984 *Self-excited vibrations and stability of railway vehicles* (Kyiv: Nauk. dumka) p 160
[2] Ushkalov V, Reznikov L, Ikkol V, Trubytskaya E, Red'ko S and Zalesskiy A 1989 *Mathematical modelling of rolling stock vibrations* (Kyiv: Nauk. dumka) p 240
[3] Markova O, Kovtun H, and Maliy V 2015 Spatial model of train motion in transient regimes *Russian Federation Transport* **3** pp 18–22
[4] Kuba T and Lugner P 2012 Dynamic behavior of tramways with different kinds of bogies *Vehicle System Dynamics* **50** pp 277–289
[5] Zboinski K and Golofit-Stawinska M 2018 The impact of primary suspension stiffness of 2-axle bogie of MKIII passenger car on its dynamic behavior *Transport problems* **13** pp 135–146
[6] Markova O, Kovtun H and Maliy V 2015 Modelling train motion along arbitrary shaped track in transient regimes *IMechE Part F: J. Rail and Rapid Transit* **229** 1 pp 97–105
[7] Blokhin Ye and Manashkin L 1982 *Train dynamics* (Moscow: Transport) p 222
[8] Naumenko N, Khyzha I, Sobolevska M and Bogomaz E 2010 Dynamic loading of the freight train at its motion along grades *Technical mechanics* **4** pp 86–90
[9] Sun Y, Cole C, Spiryagin M, Godber T, Hames S and Rasul M 2014 Longitudinal heavy haul train simulations and energy analysis fro typical Australian track routes *IMechE Part F: J. Rail and Rapid Transit* **228** 4 pp 355–366
[10] Naumenko N and Khyzha I 2015 Modelling cars interaction of passenger train equipped by the passive safety system at the emergency collision with an obstacle *Science and transport progress* **4** 58 pp 163–174
[11] Naumenko N, Sobolevska M and Khyzha I 2017 Dynamic loading estimation of the primary train with the passive safety system at its collision with the identical train and freight train *Technical mechanics* **3** pp 72–83
[12] Milho J, Ambrosio J and Pereira M 2003 Validated multibody model for train crash analysis *International Journal of Crashworthiness* **8** 4 pp 339–352
[13] Zhou H, Zhang C, Zhan J and Zhang J 2018 Research on the city tram collision at a level crossing *Advances in Mechanical Engineering* 2018 **10** 9 pp 1–12
[14] Lazaryan V 1985 *Dynamics of transport means* (Kyiv: Nauk. dumka) p 528
[15] Ushkalov V, Mokrii T, Malysheva I, Mashchenko I and Pasichnyk S 2007 Complex modernization of freight car running gears *Vagonnyi park* **2** pp 18–22
[16] Ushkalov V, Mokrii T, Malysheva I and Bezrukavyi N 2017 Improving the prospect freight car running gears *Technical mechanics* **4** pp 79–88
[17] Kolomiychenko V and Golovanov V 1973 *Rolling stock automatic coupler* (Moscow: Transport) p 191
[18] Ushkalov V, Lapina L and Mashchenko I 2013 Calculated irregularities to estimate freight cars dynamic features *Visn. V.Lazaryan DNUZT* **4** 46 pp 135–141
[19] Standard 24.050.37-84. Freight and passenger cars. Tests methods for strength and riding quality 1981 (Moscow: Transport) p 49
[20] Standard to calculate and design cars of 1520 mm gauge railways 1996 (Moscow: GosNIIV – VNIIHZT) p 319
[21] Rybkin V and Patlasov O 2012 Technical rules to estimate rail track state by the indices of rail measuring cars and insurance of trains safety of motion at deviation of rail track operation standards TsP-0267 (Kyiv: Polygraphservice Publisher) p 46
[22] Markova O, Kovtun H and Maliy V 2018 Determination of locomotive stability conditions from derailment at emergency collisions *Technical mechanics* **2** pp 103–112

IOP Conf. Series: Materials Science and Engineering **985** (2020) 012004 doi:10.1088/1757-899X/985/1/012004

Reduction of freight car wheel wear of 1520 mm gauge railways

V Ushkalov[1], T Mokriy[1,2], I Malysheva[1], L Lapina[1], S Pasichnik[1] and N Bezrukavyi[1]

[1]Institute of Technical Mechanics of the National Academy of Sciences of Ukraine
15 Leshko-Popel St., Dnipro 49005, Ukraine

[2]Mokriyt@gmail.com

Abstract. A brief description is given of the comprehensive modernization of standard freight car trucks through the use of devices of American companies adapted for railways with a 1520 mm gauge and wheels with ITM-73 specially developed wear-resistant profile, which allows several times to increase the resource of problematic running gears. An approximate method is proposed for solving the wheel–rail interaction problem with determining the position and size of non-elliptic contact patches, including with conformal contact. Using this method, new profiles have been developed for turning worn wheels (ITM-73-01), as well as new wheels for cars with an axial load on rails of 23.5 tf (ITM-73-02) and 25 tf (ITM-73-03). The data of experimental studies showed that the average wear rate of the wheel flanges of freight cars with complex modernized trucks equipped with wheels with ITM-73-01 profile is 3.5-5 times lower than that of a standard car with a standard wheel profile. According to forecast estimates, the use of wheels with profiles ITM-73-02, ITM-73-03 will allow to achieve even greater increase in the resource indicators of wheelsets for wear of the flanges.

In the 90s of the twentieth century on the railways of Ukraine and the CIS countries, excess wear of the wheel flanges of locomotives and freight cars began [1]. The wear rate has increased so much that on some railways the failure of the wear of the wheelsets disrupted the operation of the rolling stock. In addition, the wear rate of the side faces of the rail heads has increased significantly, especially in curved sections of a small radius track ($R \leq 650$ m) [2-4]. Since that time, this problem has been given increased attention [5-7].

Since the shape of the wheel profile directly affects not only the service life of the wheelsets and rails, but also the driving performance of the rolling stock, studies on the creation of new wheel profiles do not lose their relevance to this day [8-10].

In Ukraine, the solution of the problem of increasing the life cycle of wheelsets is carried out in several directions, including by improving the design of the truck to improve the dynamics of cars and reduce the forces of interaction of wheels and rails, as well as by changing the profiles of the contact surfaces of wheels and rails.

In freight cars of Ukraine and the CIS countries, three-element trucks of 18-100 model are widely used. These trucks have significant drawbacks: the tendency to self-excitation of yaw vibrations in straight sections of the track, poor performance when fitting into curved sections of the track, significant wear of wheels and rails. At the same time, violations of the safety conditions for train traffic, derailments are possible.

Content from this work may be used under the terms of the Creative Commons Attribution 3.0 licence. Any further distribution of this work must maintain attribution to the author(s) and the title of the work, journal citation and DOI.
Published under licence by IOP Publishing Ltd

The result of theoretical and experimental studies to improve the design of these trucks was the development at the Institute of Technical Mechanics of the National Academy of Sciences of Ukraine and the State Space Agency of Ukraine (ITM NASU and SSAU) together with A. Stucki (USA) a comprehensive modernization of freight car trucks of 18-100 model and its implementation on the railway network of Ukraine [11]. This modernization consists in replacing standard rigid side bearings with clearances by constant contact elasto-dissipative bearings; installing wedges from high-strength iron (instead of steel) and replacing friction plates with wear-resistant; laying in the center plate of the polymer layer; the use of a specially designed wear-resistant ITM-73 wheel profile instead of the standard wheel profile. As a result, along with an increase in the resource of the most problematic elements of the truck, the life cycle of the wheelsets in the wear of the flanges and the tread of the wheels was increased by more than 2-2.5 times [11, 12]. To date, 1520 mm gauge railways have comprehensively modernized trucks of more than 25 thousand freight cars.

All elements of a comprehensive modernization, including the wear-resistant ITM-73 wheel profile, were used to create a promising model 18-7020 truck [13] at the Kryukovsky Car-Building Plant, which was adopted as one of the basic models for updating the freight fleet of Ukrainian railways.

The use of constant contact side bearings in improved designs of trucks without changing the profile of standard wheels leads to increased wear of wheels and rails when curving a car, therefore the above-mentioned non-linear ITM-73 profile was developed for wheels with a flange thickness of 32 mm.

To carry out theoretical research, appropriate mathematical software has been created that allows you to take into account a sufficient number of factors affecting the performance of the interaction of wheels and rails. Design schemes and mathematical models have been developed that allow one to take into account the design features of different types of cars and modernized carriage units, as well as to investigate the spatial random vibrations of vehicles when moving along a track of an arbitrary outline. To carry out the calculations, disturbances were generated that act on the moving car from the track and consist of four components: symmetric and skew-symmetric vertical irregularities, defined respectively as half-sum and half-difference of vertical irregularities of the right and left rails, as well as horizontal irregularities of the right and left rails. The components of the disturbances are based on the processing and analysis of real irregularities in the sections of the main line of the Dnieper Railway.

The design of the wheel profile was carried out taking into account a number of requirements: the angle of inclination of the flange working face should be large enough to prevent the vehicle derailment due to the rolling of the wheel onto the rail; the wheel tread should ensure the stability of the movement of the vehicle; the shape of the fillet at the base of the wheel flange should provide favorable conditions for the vehicle to fit into curved sections of the track and at the same time not lead to excessive contact stresses on the working face of the rail head.

The experimental dependences of wheel flange wear on the mileage of cars (standard and with comprehensively modernized trucks), obtained from long-term operational tests of cars on the most difficult route in Ukraine, Krivyi Rih – Uzhgorod through the Carpathian Pass, are shown in Figure 1. As you can see, with a mileage of 280 thousand km, not a single wheelset (whs) with a wear-resistantITM-73 profile of the rim came into the regrind, while all wheelsets with a standard profile were regrinded, some several times.

In recent years, in many countries of the world, special attention has been paid to the development and implementation of optimal profiles of the wheel-rail contact pair. Two approaches are usually used in calculating the interaction parameters: the contact patch is reduced to a point and the resultant normal and tangential forces applied in it are calculated, or the shape and dimensions of the patch are determined and the acting forces distributed over it are calculated.

The point representation of the interaction of wheels and rails [14, 15] is used when the dimensions of the contact patches are relatively small, which is typical mainly for unworn surfaces of the bodies under consideration that do not have the property of conformal contact.

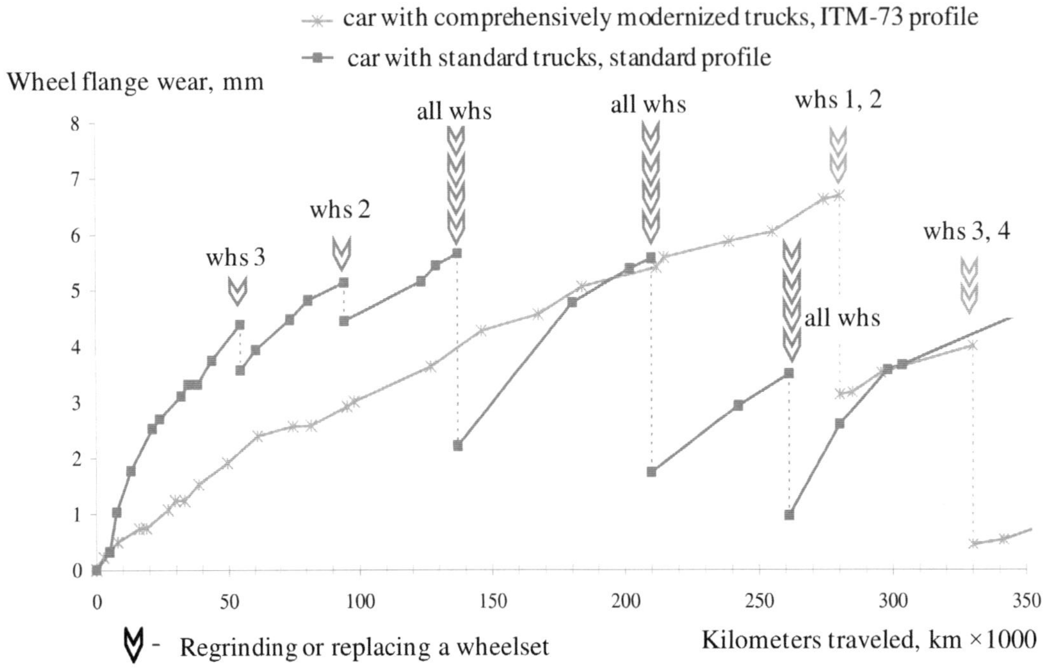

Figure 1. Experimental dependences of wear of wheel flanges on the mileage of cars.

However, conformal contact of wheels and rails is considered the most promising from the point of view of optimizing interaction processes [16]. In this case, the contact patch cannot be represented by one point and most often it has a non-elliptical shape. Therefore, it is necessary to determine its shape and size, as well as the distribution of acting forces on it. For example, the work [17] is devoted to solving the problem of non-hertz contact. It gives a mathematical model of the interaction of a wheel and a rail with a non-elliptical contact spot. The wheel and rail are represented by elastic semi-infinite bodies under pressure distributed over the contact area. The rail is adopted by a massless element, which is a cylindrical surface, and the influence of its vibrations is not taken into account. To determine the contact patch and the distribution of normal stresses at a geometrically found point of contact, the bodies are introduced one into another by a certain unknown value δ, which is found by solving a rather complex system of equations by the iteration method.

The paper [18] presents a spatial mathematical model of the interaction between the railway vehicle and the inertial elastic-dissipative track, which takes into account the size of the contact patches and the distribution of normal and tangential forces along them and proposes a less laborious way to solve the contact problem, which nevertheless gives quite acceptable results.

The contact problem is solved under the assumption of continuous motion of the wheel along the rail. The input parameters are the angles of side rolling and yaw of the wheelset, the mutual lateral displacement of the wheel and rail. The distribution of the interaction forces over the surface of the found contact patch is performed. For this, it is divided into small finite elements, in each of which the contact is assumed to be single-point and elementary forces are calculated.

Tangent elemental forces are considered creep forces. The total longitudinal and lateral components of the distributed creep force are equal to the sum of the corresponding elementary forces. The elementary vertical forces of interaction are calculated by distributing the total vertical force between the elements of the contact patch in proportion to the amount of penetration of the wheel into the rail in each element.

Using the constructed interaction model, studies were carried out to create a new wear-resistant profile of wheels with conformal contact for P65 rails, taking into account wear of their heads.

IOP Conf. Series: Materials Science and Engineering **985** (2020) 012004 doi:10.1088/1757-899X/985/1/012004

The development of the wheel profile was carried out while analyzing two functions of the target: the dynamic performance of a freight car with comprehensively modernized trucks and the wear of its wheels. From the family of constructed wheel profiles, a profile was selected, when set in the calculations, the minimum values of these functions were achieved. As a result, the ITM-73-01 repair profile was obtained for wheels with a flange thickness of 29 mm.

The operational test data of a batch of experimental gondola cars with comprehensively modernized trucks whose wheels were turned according to the ITM-73-01 profile are shown in Figure 2. As you can see, the average intensity of wear of the wheel flanges of these cars in the range up to 100 thousand kilometers is 3.5-5 times lower than that of a typical car with a standard wheel profile.

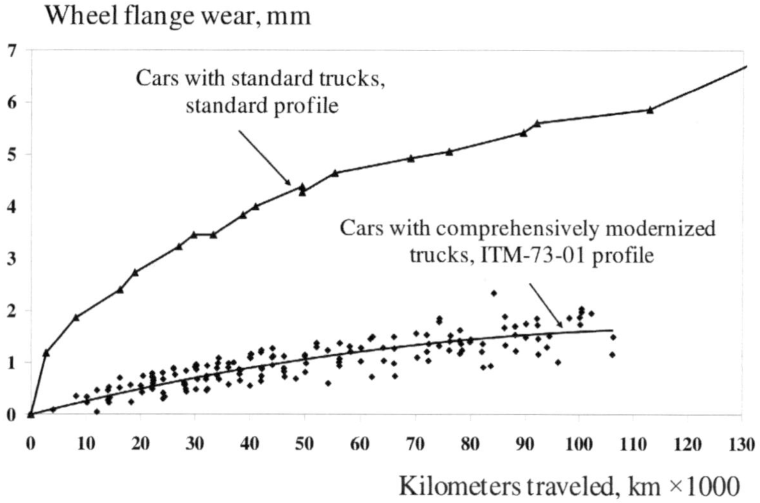

◆- average wear values of eight wheels of each of the twenty-three experimental cars

Figure2. Test data of a batch of experimental gondola cars with comprehensively modernized trucks.

With such a significant reduction in the wear of the flanges, it can be assumed that the wear of the rails along which the cars with the ITM-73-01 wheel profile will run will also noticeably decrease, which promises the railroads significant financial savings. In addition, the use of the ITM-73-01 profile allows one to increase the number of possible regrindings due to surface defects, which also significantly increases the life of the wheelset. The ITM-73-01 repair profile was implemented on the all Ukrainian railway network for regrinding worn wheels.

An ITM-73-02 profile with a flange thickness of 32 mm has been developed for turning new wheels with a similar effect on flange wear. When using this profile in the new-generation trucks of the 18-7020 model (with an axial load of 23.5 tf), according to forecast estimates, the wear intensity of the wheel flanges will decrease by more than 5 times, which will allow satisfying the requirements for wheel wear for promising freight cars. The technical council of Ukrzaliznytsya JSC decided to introduce the ITM-73-02 wheel profile in the design documentation for the new generation truck of 18-7020 model and to test cars with such wheels.

Currently, in Ukraine, as the base trucks of promising freight cars, in addition to the aforementioned trucks of 18-7020model, new three-element 18-9817 trucks with an axle load of up to 25 tf are accepted. Trucks of 18-9817 model – a joint development of the American company ASF Keystone and the Ukrainian Industrial and Investment Group InterCarGroup. Their design has a number of distinctive features [19]: constant contact side bearings with spring elastic elements (Preload Plus type), the characteristics of which can be selected depending on the type of car; elastic adapters in axle boxes; friction wedge with spaced inclined contact surfaces, etc. ITM took an active part in choosing the inertial and elastic characteristics of these trucks. As the calculation results showed, the use of theITM-73-02 profile for turning the wheels of these trucks leads to a significant increase (compared with the comprehensively modernized truck of 18-100 model and truck of 18-7020 model) the wear of the wheel flanges, which in operation can cause their undercutting. This deterioration in the conditions of contact between wheels and rails is mainly due to a change in the design of new trucks: due to the increased connection of the side frames by means of adapters, the frame of the truck becomes stiffer, which significantly increases the stability of the car motion, but at the same time makes it difficult to fit into curved sections of the track.

Studies have been carried out to improve the wheel profile for 18-9817 truck in order to improve its interaction and conditions of contact with the rail track. For this, on the basis of the ITM-73-02 profile, a family (more than 20) of wheel profiles with a flange thickness of 32 mm was built. For each profile version, the spatial contact problem of wheels and rails was solved, interaction parameters were analyzed, including the size and location of contact patches, and calculations were made of fitting a gondola car with 18-9817 trucks into a small radius circular curve (300 m) with different degrees of rail wear and its movement along straight sections of the track.

Based on the analysis of two conflicting criteria – the minimum flange wear of the wheels and the maximum stability margin of the vehicle– ITM-73-03 wheel profile was chosen [20]. As calculations showed, the use of this profile of the wheels significantly improves the process of interaction of a freight car with increased axial load and track and significantly reduces flange wear of the wheels, while ensuring high dynamic performance of the vehicle.

The dependences of the maximum values of the flange wear factor A_{max} of the wheels on the motion speed of a loaded car equipped with 18-9817 trucks with ITM-73-02 and ITM-73-03 wheel profiles are shown in Figure 3. P65 rails with different degrees of wear were considered: unworn and slightly worn (lateral wear of the outer rail 3.5 mm). For comparison, the results obtained when installing wheelsets with standard wheel profiles in these trucks are also plotted. The term "flange wear" hereinafter denotes the wear of the fillet and wheel flange.

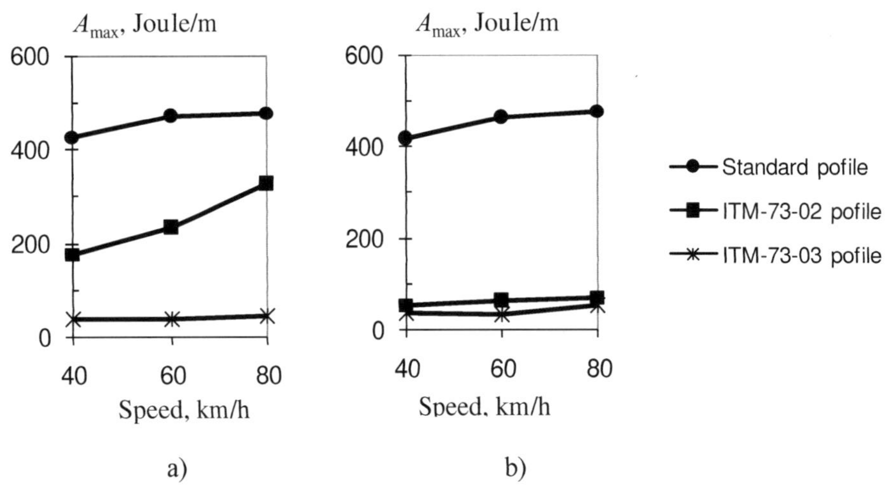

Figure3. Dependences of the maximum values of the flange wear indicator of the wheels on the speed of motion of a loaded car.

If the rails in the curve are not worn (see Figure 3, a), then the wear rate of wheels with an ITM-73-03 profile is 5-8 times less than the wheels with an ITM-73-02 profile, and 10-11 times smaller than standard profile wheels. When driving on worn rails (see Figure 3, b)), the values A_{max} of wheels with ITM series profiles are close and less than the wear of standard wheels by 7-8 times.

The Figure 4 shows examples of the distribution of vertical pressure P across the contact patches transmitted from the wheels to the slightly worn rails, and the sizes of these patches. It is seen that the contact conditions of the wheels with the ITM-73-03 profile are better than standard: the patch has extensive dimensions (conformal contact) and is located completely outside the working surface of the flange (on fillet and tread). Wheels with a standard profile with flange contact cause small patches and high pressure, i.e., such wheels wear out intensively and so-called flange cutting occurs. The level of vertical pressure on wheels with an ITM-73-03 profile is much lower.

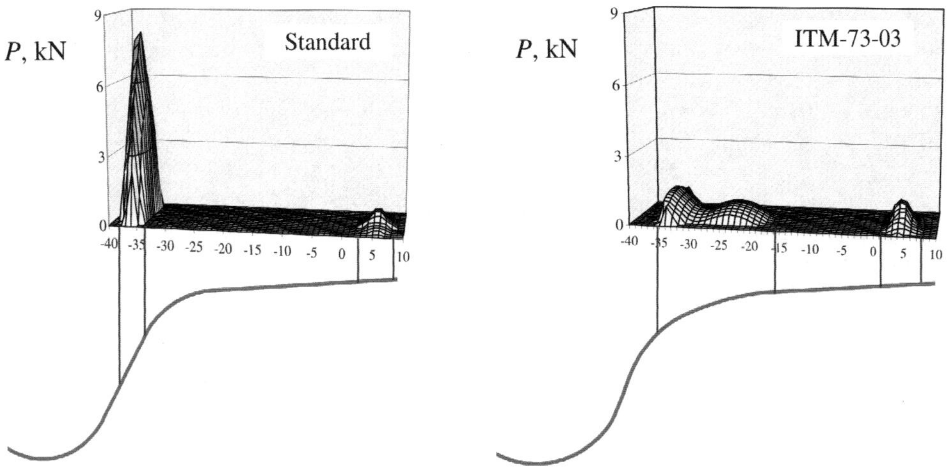

Figure4. Distribution of vertical pressure on the contact patches transmitted from the wheels to the rails, and the dimensions of these patches.

As additional studies have shown, the ITM-73-03 profile according to the criteria for wear of wheel flanges and the dynamic performance of vehicles can be used in cars equipped with promising trucks with a normal axial load on rails of 23.5 tf.

Conclusions

1. A non-linear wear-resistant ITM-73 wheel profile was developed, the use of which in freight cars with comprehensively modernized trucks of18-100 model made it possible to increase the life cycle of wheelsets by wear of wheel flanges by more than 2-2.5 times.

2. An approximate method is proposed for solving the wheel–rail interaction problem with determining the position and size of non-elliptic contact patches, including those with conformal contact, and the distribution of normal and tangential interaction forces over them.

3. Using the proposed method ITM-73-01 (for turning worn wheels) and ITM-73-02 (for turning unworn wheels)wear-resistant wheel profiles developed, taking into account the form of wear of the heads of rails P65. According to experimental data, the average wear rate of wheel flanges of freight cars with comprehensively modernized trucks equipped with wheels with ITM-73-01 profile is 3.5-5 times lower than with the same mileage of a typical car with a standard wheel profile. The use of wheels with the ITM-73-02 profile when turning new wheels in cars with comprehensively modernized trucks and new trucks of 18-7020 model (with a normal axial load on rails of 23.5 tf) will allow to increase the life cycle of wheelsets by flange wear even more.

4. A wear-resistant wheel profile ITM-73-03 was developed for promising trucks of 18-9817 model with an axial load of up to 25 tf. Using a new profile of wheels will significantly improve the process of interaction of a freight car with these trucks and significantly reduce the flange wear of the wheels, while ensuring high dynamic performance of the vehicle. This wheel profile can also be used in cars equipped with promising trucks with a normal axial load on rails of 23.5 tf.

References

[1] Minin S 1991 *J. Geleznodorognyi transport* **1** 47–50 (in Russian)
[2] Bogdanov V 1992 *J. Geleznodorognyi transport* **12** 30–34 (in Russian)
[3] Nikiforov B 1995 *J. Geleznodorognyi transport* **10** 36–40 (in Russian)
[4] Ushkalov V 1996 *Proc. of the 2nd Mini Conf. on Contact Mechanics and Wear of Rail/Wheel Systems (Budapest)* 250–58
[5] Verigo M 1997 *Vzaimodejstvie puti i podvignogo sostava v krivyh malogo radiusa I bor'ba s bokovym iznosom rel'sov I grebnej kolios (The Interaction of Tracks and Rolling Stock in Small Radius Curves and the Fight against Lateral Wear of Rails and Wheel Flanges)* (Moscow: PTKB TsP MPS) p 207 (in Russian)
[6] Nirkonen V B 2011 *J. Put' i putevoje hoziajstvo* **9** 18–21 (in Russian)
[7] Korol'kov Ye, Korshunov G and Lutsev V 1993 *J. Geleznodorognyi transport* **8** 37–38 (in Russian)
[8] Magel E and Tajaddini A 2007 *J. Geleznyje dorogi mira* **10** 62–64 (in Russian)
[9] Zakharov S, Goryacheva I, Bogdanov V, Pogorelov D and Zharov I 2008 *Problems with Wheel and Rail Profiles Selection and Optimization. Wear* **265** 1266–72
[10] Gerlici Yu and Lack T 2011 *8 Intern. Conf. on Contact Mech. and Wear of Rail/Wheel Systems (Florence, Sept. 2009)* **271** 246–58
[11] Gregg Hansen W M, Lashko A, Ushkalov V, Mokriy T, Malysheva I and Mashchenko I 2004 *Vestnik Dniepropetrovskogo natsional'nogo universiteta geleznodorognogo transporta* **5** 215–19 (in Russian)
[12] Orlova A, Lesnichij V and Haritonov B 2008 *Vestnik Dniepropetrovskogo natsional'nogo universiteta geleznodorognogo transporta* **23** 76–82 (in Russian)
[13] Biaxial truck model 18-7020 type 2 *URL http://test.kvsz.com/index.php/ru/produktsiya/gruzovoe-vagonostroenie/khodovye-chasti/telezhki/item/833-dvukhosnaya-telezhka-model-18-7020*
[14] Piotrowski J and Kalker 1988 *J. Vehicle System Dynamics* **17** 337–67
[15] Anyakwo A, Pislaru C and Andrew B 2012 *Int. Journal of Automation and Computing* **9 (3)** 237–47
[16] William J H, Ebersöhn W, Lundgren J, Tournay H and Zakharov S 2001 *Guidelines to Best Practices for Heavy Haul Railway Operations: Wheel and Rail Interface Issues* (USA: International Heavy Haul Association) p 482
[17] Kovalev R and al. 2003 *J. Mechanics Based Design of Structures and Machines* **31 (3)** 315–35
[18] Ushkalov V, Mokriy T and Malysheva I 2015 *J. Technicheskaja mechanika* **2** 79–89 (in Russian)
[19] The biaxial truck model 18-9817 with an axle load on rails of 25 tf *URL http://okb.at.ua/publ/telezhka_dvukhosnaja_modeli_18_9817_s_nagruzkoj_ot_kolesnoj_pary_na_relsy_25t/1-1-0-6*
[20] Ushkalov V, Mokriy T, Malysheva I and Bezrukavyi N 2018 *J. Technicheskaja mechanika* **1** 20–29 (in Russian)

IOP Conf. Series: Materials Science and Engineering **985** (2020) 012005 doi:10.1088/1757-899X/985/1/012005

Impact of the variable stiffness section on the conditions of track and rolling stock interaction

D Kurhan[1,3], M Kurhan[2] and M Husak[2]

[1]Department of Track and Track Facilities, Dnipro National University of Railway Transport named after Academician V. Lazaryan, Lazaryan St, 2, Dnipro, Ukraine
[2]Department of Roads Design and Construction, Dnipro National University of Railway Transport named after Academician V. Lazaryan, Lazaryan St, 2, Dnipro, Ukraine

[3] kurhan.d@gmail.com

Abstract. Railway track stiffness is one of the main parameters that determine the track and rolling stock interaction. The inelasticity of the rail base may occur for two reasons: either as a consequence of the deteriorated condition of a track or due to the structural features of a section. As a rule, areas before bridges or tunnels are treated. Today, there are several options for relevant design solutions. The main purpose of this work is to determine the characteristics of a "railway track" object with which it can be represented in the rolling stock models in the simplest way possible but adequate for sections of transient stiffness. The railway track is introduced into the rolling stock model as a sequence of elements with which the wheels of the rolling stock interact. It is assumed that a single element in its characteristics must be equivalent to the track response when passing the wheel from one inter-sleeper section to the next one, that is, to reproduce a sequence that is cyclically repeated. Such a track element will be characterized by reduced mass, stiffness and dissipation factor.

1. Introduction

Railway track stiffness is one of the main parameters that determine the track and rolling stock interaction. Moreover, it is important not only and not so much the value of stiffness but its uniformity.

The inelasticity of the rail base may occur for two reasons: either as a consequence of the deteriorated condition of a track or due to the structural features of a section. In the first case, the equal elasticity is established after the repair work, which may be accompanied by the use of reinforcing structures [1], eliminating or weakening the effect of factors that led to the deterioration of the condition. These factors include, first of all, the loss of elasticity of the ballast layer due to the destruction of its parts [2, 3], ingress of pollutants [4] or the lack of compactness [5], increase in a load of freight trains [6] and the speed of passenger traffic [7]. The second case is, as a rule, short distances, the stiffness of which is determined by the design: a zone of rail joints [8, 9], crossings, artificial structures such as tunnels, bridges [10, 11].

With the time of operation, an abrupt change in stiffness leads to the intensive development and accumulation of vertical deformations at the approach to such a place. To prevent this process, special transition sections with variable stiffness can be equipped.

Content from this work may be used under the terms of the Creative Commons Attribution 3.0 licence. Any further distribution of this work must maintain attribution to the author(s) and the title of the work, journal citation and DOI.
Published under licence by IOP Publishing Ltd

Today, there are several options for relevant design solutions. Their review and analysis are given in [12-17] and others. As a rule, areas before bridges or tunnels are treated.

There are a number of tasks that require mathematical simulation of the rolling stock passage along such a section. In general, to study the operation of rolling stock and a railway track, as a rule, not the same mathematical models are used. The main difference is to reduce the detailing of one or the other part. It is clear that not only oversimplification but also unnecessary detailing has an adverse effect on the adequacy of the results. In addition, the extension of the mathematical model requires additional output data and can significantly increase the requirements for time and characteristics of computing equipment.

If several fundamentally different approaches are used for models of a railway track, then for mathematical models of rolling stock in almost all cases the calculation of Lagrange-D'Alembert system of equations is the basic one. Two types of models are increasingly applying sequentially for the complex analysis. First, we simulate the motion of rolling stock with a simplified track representation. As a result, the dynamic forces acting from the wheels to the rails are determined. Then the track operation is simulated with a simplified representation of the rolling stock, which is reduced to a system of forces applied to the rail. As a result, the stress-strain state of the track is determined according to the load.

Thus, the main purpose of this work is to determine the characteristics of a "railway track" object with which it can be represented in the rolling stock models in the simplest way possible but adequate for sections of variable (transient) stiffness.

2. Methodology

The railway track is introduced into the rolling stock model as a sequence of elements with which the wheels of the rolling stock interact, Figure 1. It is assumed that a single element in its characteristics must be equivalent to the track response when passing the wheel from one inter-sleeper section to the next one, that is, to reproduce a sequence that is cyclically repeated. Such a track element will be characterized by reduced mass, stiffness and dissipation factor. Its vertical oscillations are described by a differential equation:

$$m\frac{d^2z}{dt^2}+\beta\frac{dz}{dt}+kz=P(t), \tag{1}$$

where m – reduced mass of the track element; z – vertical deflection; β – dissipation factor; k – stiffness; P – vertical force; t – time factor.

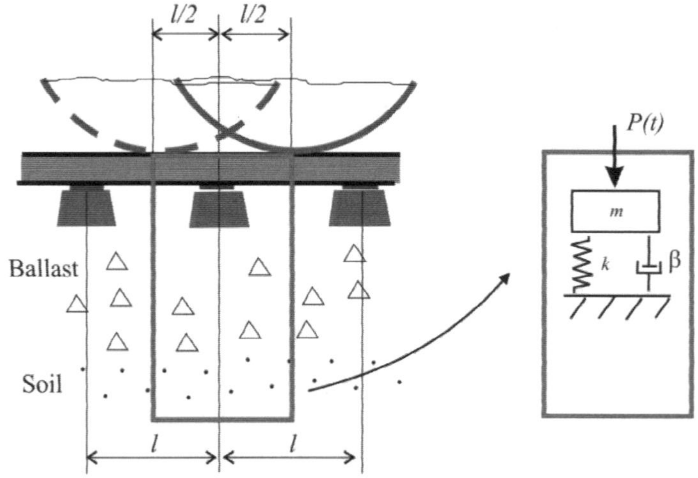

Figure 1. Design diagram of the track representation in the rolling stock model.

Equation (1) can be represented in the form:

IOP Conf. Series: Materials Science and Engineering **985** (2020) 012005 doi:10.1088/1757-899X/985/1/012005

$$z(t) = f\left(P(t), m, k, \beta\right). \tag{2}$$

Then the unknown characteristics of the element (m, k, and β) can be determined as a result of approximation by the function (2) of the results in determining the dynamic deflections of the track in the cross-section of this element from the active force. A more detailed technique for such an approximation is given in [18].

To determine the deflections, a spatial model of dynamic deformations for the railway track was applied based on the elasticity theory [19]. The basis of this model is a mathematical description of the elastic wave propagation of stresses (deformations) in the space of a railway track from the applied load. The result of the simulation is the stress-strain state of the track, depending on the time of the rolling stock passage. This approach corresponds to the physics of the process in the railway track operation as a system of objects that responds to the load by elastic deformations. The input data are the geometric dimensions of the objects (or layers) and their physical properties, such as Young's modulus, density, Poisson's ratio. Determination of the velocities for the elastic wave propagation takes into account the dynamics of the deformation process, namely, it determines the part of the rail base that is involved in the interaction at this moment in time. It provides an opportunity to determine the desired parameters (m, k, and β) as a generalized result of the system.

3. Results

In this paper, we consider the design of a section with variable stiffness, which is shown in Figure 2. The section can be divided into four zones: 1 – a ballast track; 2 – a transition zone with variable stiffness; 3 – a transition zone with constant (increased) stiffness; 4 – a superhard zone (for example, a bridge). The design of the transition section and its parameters correspond to the recommendations given in works [13, 14].

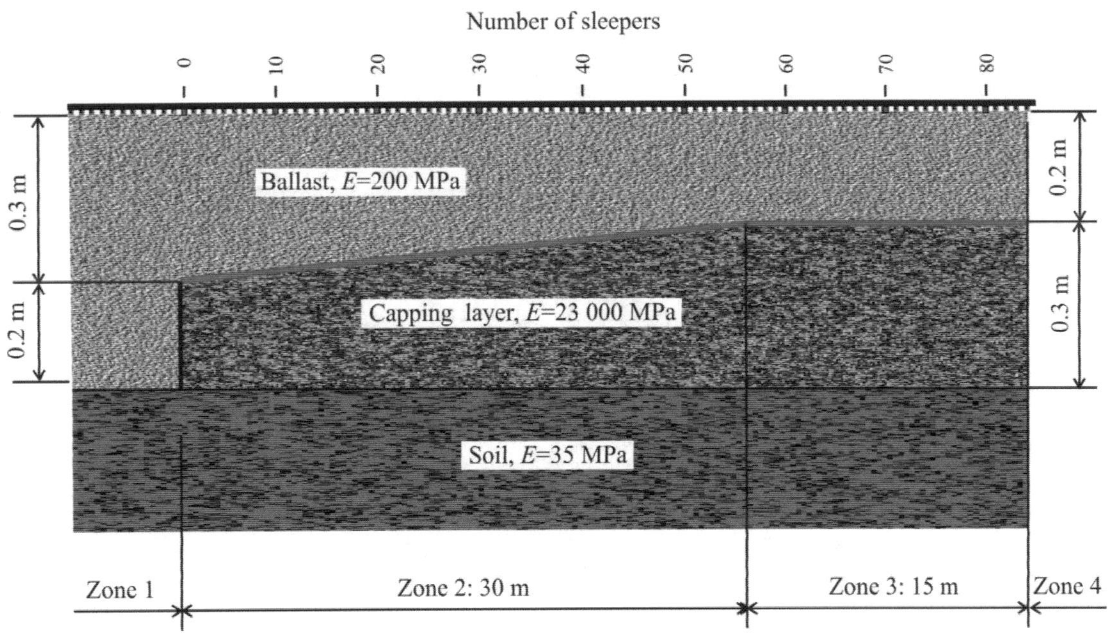

Figure 2. Design of the transition section with variable stiffness.

The results of variant simulations are shown in Figure 3 on the example of maximum stresses along the length and depth of the rail base from passing a passenger car at a speed of 160 km/h.

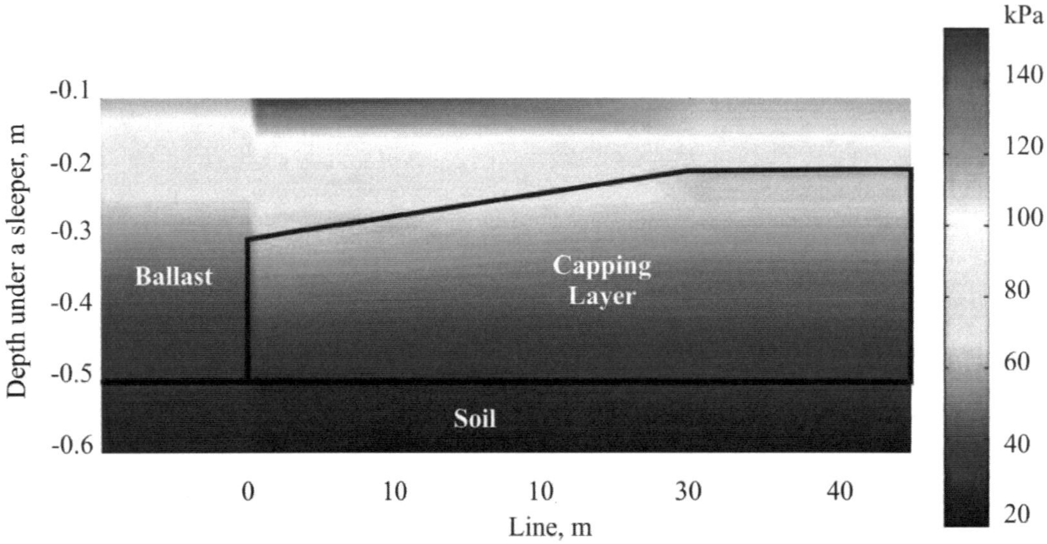

Figure 3. Maximum stresses spectrum.

Figure 4 shows examples of dynamic track deflection for several sections (it is indicated by the numbers of sleepers as in Figure 2) for passing the wheel in the interval ± 0.27 m from the axis of the corresponding sleeper, which corresponds to the sleeper density of 1 840 pieces/km. When approaching the vertical acting force (wheel), the intensity of the deformation change is greater (a lower part of the dependency on the graph) than at distant forces (an upper part of the graph).

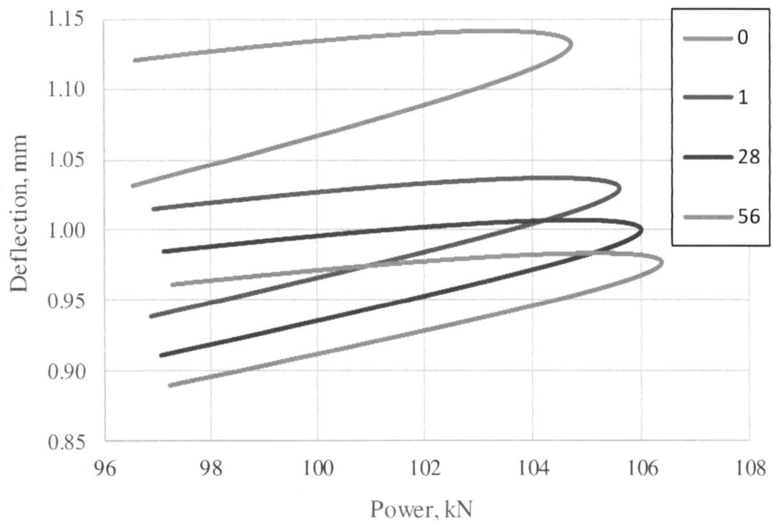

Figure 4. Dynamic deflection of the track in cross-section along the axis of the sleepers 0, 1, 28, 56.

Similar results were obtained for other cross-sections of the track. Dynamic deflections of the rail (a passenger car, speed of 160 km/h) along the length of the section are shown in Figure 5.

According to the approximation of the calculation data, the function (2) defines the stiffness characteristics of the track as an object for the rolling stock model. The change in the modulus of deformation of the railway track along the section is shown in Figure 6. The calculated values of the dynamic stiffness and the dissipation factor for the support in accordance with the sleepers numbers are shown in Figure 7. For the practical application, one can use the linear equations of the dependencies of these parameters on the number of a sleeper in accordance with Figure 2.

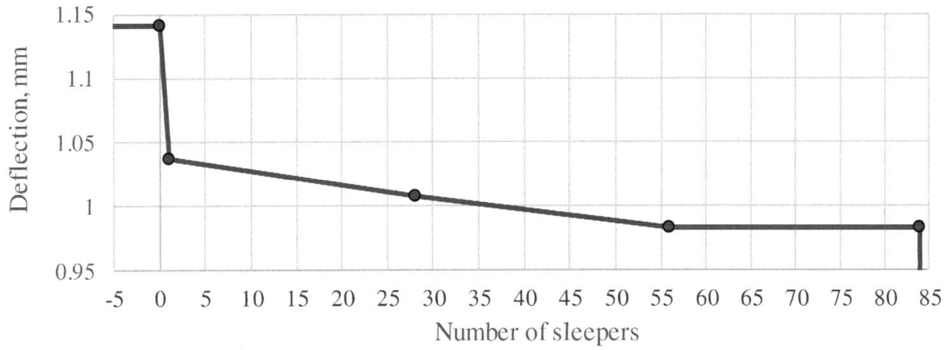

Figure 5. Deflections of the rail along the length of the section.

As conducted studies have shown, the "railway track" system did not show the presence of mass as an indicator of inertia This is, first of all, due to the significant advantage of the unsprung mass of rolling stock which presses on the track through the wheel over the mass of the track which has interacted with the wheel. Secondly, the significant manifestations of the track inertia can be expected only at high speed of movement, which will be comparable with the spreading speed of the track deflection [20].

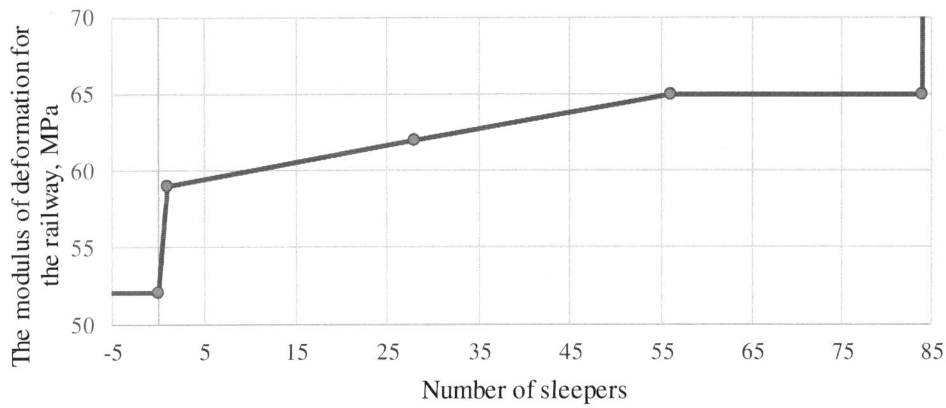

Figure 6. Change in the modulus of elasticity along the length of the section.

4. Conclusions

In this paper, we consider the transition section before the bridge as a design of a railway track with variable stiffness. Adopted characteristics of the ballast track (reinforced-concrete sleepers, crushed ballast with the modulus of deformation 200 MPa, thickness 0.5 m, soil with the modulus of deformation 35 MPa) correspond to the railway track operation as a whole with the modulus of deformation 52 MPa. The arrangement of the transition section for the considered design (see Figure 2) provides an almost linear (see Figure 6) increase of the modulus of deformation for the track up to 62 MPa – by 25%.

Moreover, in the area just before the bridge, the ballast layer is 20 cm. At the ballast thickness of 15 cm (due to a further increase in the thickness of the reinforcing layer), the total modulus of deformation of the railway track will increase to 68 MPa – by 30% compared to the primary ballast section. When using additional facilities (for example, placing a layer of reinforced ballast before the roadbed), specified indicators can be improved. In addition, one should pay attention to the characteristics of the roadbed. For example, reducing the soil deformation modulus up to 20 MPa will decrease the total modulus of deformation of a track from 52 to 32 MPa.

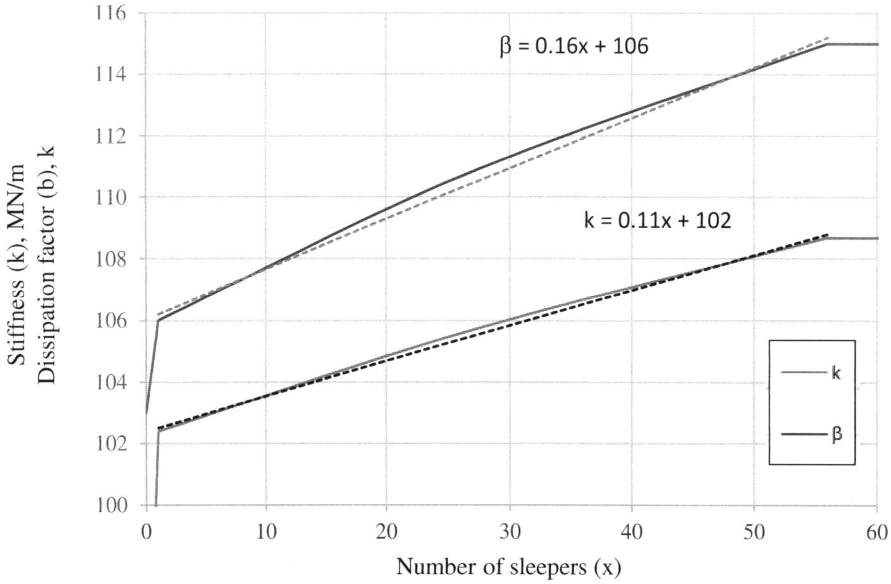

Figure 7. Changing the characteristics of point support on the section of variable stiffness.

To introduce the considered transition section of the railway track in the rolling stock model in the form of separate supports, their numerical characteristics are established. Thus, for a zone with a variable modulus of deformation (see Figure 2), the dynamic stiffness of such support along the length of the section will vary from 92.4 to 108.7 MN/m, and the dissipation factor from 103 to 115 kN s/m. Within the considered speeds of movement (up to 160 km/h) the need to take into account the reduced mass of the track has not been established. But its introduction may take a form of a conditional parameter required by the structure of the rolling stock model to divide the system into separate objects.

References

[1] Eller B and Fischer S 2019 Review of the modern ballasted railway tracks' substructure and further investigations *Science and Transport Progress* **6** 84 pp 72–85

[2] Juhász E and Fischer S 2018 Investigation of railroad ballast particle breakage *Pollack Periodica* **14** 2 pp 3–14

[3] Fischer S and Németh A 2018 Special laboratory testing method for evaluation particle breakage of railway ballast material *Science and Transport Progress* **2** 74 pp 87–102

[4] Sysyn M, Gerber U, Nabochenko O and Kovalchuk V 2019 Common crossing fault prediction with track based inertial measurements: statistical vs. mechanical approach *Pollack Periodica* **14** 2 pp 15–26

[5] Pshinko O, Patlasov O, Andrieiev V, Arbuzov M, Hubar O, Hromova O and Markul R 2018 Research of railway crashed stone use of 40–70 mm fraction *Proceedings of 22rd International Scientific Conference. Transport Means 2018* pp 170–178

[6] Patlasov O and Fedorenko Y 2019 The intensity of rail failure flow *MATEC Web of Conf.* **294** 03020

[7] Kurhan M and Kurhan D 2019 Providing the railway transit traffic Ukraine-european union *Pollack Periodica* **14** 2 pp 27–38

[8] Németh A, Major Z and Fischer S 2020 FEM Modelling possibilities of glued insulated rail joints for CWR Tracks *Acta Technica Jaurinensis* **13** 1 pp 42–84
[9] Potapov D, Panchenko S, Leibuk Y, Tuley Y and Plis P 2018 Effect of joint and isolated irregularities of the track on the wear of rails in curves *MATEC Web Conf.* **230** 01012
[10] Talavira H and Kudin A 2015 Drain base rigidity sleeper upstream of artificial structures *Bridges and tunnels: theory, research, practice* **7** pp 75–80
[11] Marochka V and Boboshko S 2018 Development of technology of arranging areas with transitional stiffness index on approaches to railway bridges *Bridges and tunnels: theory, research, practice* **13** pp 99–106
[12] Arlaud E, Costa D'Aguiar S, Balmes E and Faussurier G 2016 Numerical study of railway track dynamics: case of a transition zone *Eng Anal Boundary Elem.* **27** pp 23–38
[13] Arlaud E, Costa D'Aguiar S and Balmes E 2015 A numerical tool to assess the dynamic behaviour of different track designs *Conf.: Railway Engineering (At: Edinburgh)*
[14] Shi C, Zhao C, Zhang X and Andersson A 2020 Analysis on dynamic performance of different track transition forms using the discrete element/finite difference hybrid method *Computers and Structures* **230** 106187
[15] Alves Ribeiro C, Paixão A, Fortunato E and Calçada R 2015 Under sleeper pads in transition zones at railway underpasses: numerical modelling and experimental validation *Structure and Infrastructure Engineering* **11** 2 pp 1432–1449
[16] Paixão A, Fortunato E and Calçada R 2014 Transition zones to railway bridges: Track measurements and numerical modelling *Engineering Structures.* **80** pp 435–443
[17] Paixão A, Fortunato E, and Calçada R 2016 A contribution for integrated analysis of railway track performance at transition zones and other discontinuities *Construction and Building Materials* **111** pp 699–709
[18] Kurhan M and Kurhan D 2017 Railway track representation in mathematical model of vehicles movement *Science and Transport Progress* **6** 72 pp 40–48
[19] Kurhan D and Kurhan M 2019 Modeling the dynamic response of railway track *IOP Conf. Ser.: Mater. Sci. Eng.* **708** 012013
[20] Kurhan D 2015 Features of perception of loading elements of the railway track at high speeds of the movement *Science and Transport Progress* **2** 56 pp 130–145

IOP Conf. Series: Materials Science and Engineering **985** (2020) 012006 doi:10.1088/1757-899X/985/1/012006

The dynamics and strength of the carrying structure of a flat wagon while conducting fire from it

O Fomin[1], A Lovska[2,5], Ya Kichuk[3], N Urum[4]

[1]Department of Cars and Carriage Facilities, State University of Infrastructure and Technologies, Kyrylivska str., 9, Kyiv, Ukraine, 04071
[2]Department of wagons, Ukrainian State University of Railway Transport, Feuerbach sq., 7, Kharkiv, Ukraine, 61050
[3]Izmail State University of Humanities, Repin str., 12, Izmail, Ukraine, 68600
[4]Department of Navigation and Technical Systems Operation in Water Transport, The Danube Faculty of Marine Transport of the State University of Infrastructure and Technologies, Fanagoriyska str., 7, Izmail, Ukraine, 68600

[5]alyonalovskaya.vagons@gmail.com

Abstract. The dynamic loading of the flat wagon supporting structure during the military equipment transportation and conducting fire from it is investigated. A flat wagon model 13-401 by the construction of Dniprodzerzhirsk Car Building Plant (Ukraine) was chosen as a prototype. It was taken into account that the flat wagon was loaded with two anti-aircraft systems of muzzle energy of 89 kJ. The accelerations acting on the supporting structure of the flat wagon are determined using mathematical simulation. At the same time, it was taken into account that a simultaneous shot from two anti-aircraft systems is carried out. The solution of the mathematical model was carried out in the MathCad software. It has been found that the longitudinal accelerations of the flat wagon supporting structure were about 36 m/s^2, and the vertical ones were about 5 m/s^2. The obtained accelerations were taken into account when determining the equilibrium stability of the flat wagon, considering the conducting fire. The stability factor in this case was about 9. The calculation results of the strength of the flat wagon supporting structure are presented. The calculation was carried out in a quasi-static state. It has been established that the maximum equivalent stresses occur in the areas of frontal stops of the automatic coupling and amount to about 300 MPa. The results of the research will contribute to the creation of innovative flat wagon load-bearing structures for the military equipment transportation with the possibility of firing while moving.

1. Introduction

Nowadays Ukraine, with its significant machine-building and transit capacity and also with well-developed transport infrastructure, faces acute economic, social, engineering and national defence and security challenges. And the country must solve all of them. Thus, for rail transportation, which plays a key role in solving problems in the transport industry, the innovative resource-saving car structure is of crucial significance. And the national defence capability can be increased with special cars intended for transporting military equipment, which is of primary importance. And while designing such cars, there is a need to take into consideration refined dynamic loads from transported military equipment in combat activity (fire).

Some features of designing cars for transporting heavy loads are considered in [1]. Dynamics and capacity of a car were calculated in ProMechanica and CosmosWorks software suites. The study deals

Content from this work may be used under the terms of the Creative Commons Attribution 3.0 licence. Any further distribution of this work must maintain attribution to the author(s) and the title of the work, journal citation and DOI.
Published under licence by IOP Publishing Ltd

with designing the carrying structure for a car of various materials. But it does not touch the possibility to transport military equipment while conducting fire. The structure of a car for intermodal transportation is analyzed in [2]. The special feature of such a car is a lowered middle part and reversible sectors. The car structure allows loading/unloading automobiles by gravity. The study does not consider the possibility to transport military equipment while conducting fire.

The strength of the carrying capacity of a flat wagon with a motor semi-trailer on is modeled in [3]. The study suggests a design diagram for obtaining refined values of stresses in the carrying structure of a flat wagon. Some features of a car to transfer transport means and containers are presented in [4]. The car structure provides an easy loading and unloading of freight without any infrastructure or terminal facilities.

The study presents the dynamic characteristics obtained. The calculation is made in MSC Adams software.It should be noted that the studies do not consider the dynamic loading on a car under transporting military equipment. Research into the dynamics of a flat car under the most unfavorable operational modes is presented in [5, 6]. The study defines the maximum accelerations on the carrying structure of a flat wagon. The authors define the main strength characteristics of the carrying structure. They do not consider the dynamics under transporting military equipment.

The strength of a flat wagon under static and dynamic loading on the carrying structure is defined in [7]; the authors use experimental methods, in particular strain measurements. Dynamics characteristics of a flat wagon calculated with multibody methods are presented in [8]. Equations of motion for a flat wagon are given in the absolute coordinates with Lagrange's equations of the first kind. But strength of a flat wagon for transporting military equipment is not considered in the studies.

The spatial oscillations of a flat wagon with long cargo moving over rail joint and harmonic irregularities in horizontal and vertical planes are studied in [9]. Oscillations of the mechanic system are described by composing a twenty-differential-equation system. The authors do not define the dynamics and strength of a flat wagon loaded with military equipment.

The objective of the study presents special approaches in defining dynamics and strength of the carrying structure of a flat wagon while conducting fire. The following tasks were set to achieve the objective:

- to make a mathematical model of dynamic loads on the carrying structure of a flat wagon while conducting fire;
- to define the strength coefficient of a flat wagon while conducting fire;
- to define the strength characteristics of the carrying structure of a flat wagon while conducting fire.

2. Mathematical simulation of the dynamic loading of the flat wagon supporting structure

To define the dynamic loads on the carrying structure of a flat wagon while conducting fire the authors applied a mathematical model presented in [10]. However, the model was improved by an additional degree of freedom – oscillations in the longitudinal plane (recoils). The design diagram of a flat wagon loaded with two anti-aircraft systems is given in Figure 1.

In this case, a flat wagon is considered as a system of three solid bodies - a supporting structure and two bogies of model 18-100 with spring sets that have stiffness and a relative friction coefficient.

It is taken into account that the following links are imposed on the system:

– the movement of the body and bogies of the flat wagon along the track center are the same: $q_1 = q_4 = q_7$.

– wheelsets move without sliding: $\Psi_{rij} = x_{rij} / R,\ (r,i,j=1,\ 2)$, where R is the wheel radius;

– due to the absence of elastic elements in the axle-box suspension, the bouncing and pitching of the bogies are determined by the bouncing of the wheelsets.

The study did not consider displacements of the anti-aircraft systems while conducting fire, as they were restricted due to applying multi-screw fixation elements. Authors considered the most unfavorable loading on the carrying structure of a flat wagon, particularly, movement of the car over a joint track irregularity with simultaneous firing.

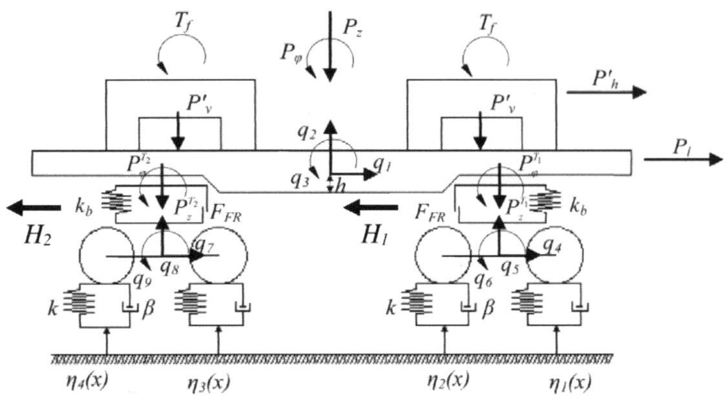

Figure 1. Design diagram of a flat wagon loaded with two anti-aircraft systems.

The equation of motion for the design model is as follows:

$$M_1' \cdot \frac{d^2}{dt^2} q_1 + M' \cdot \frac{d^2}{dt^2} q_3 = P_l + P_h', \tag{1}$$

$$M_1 \cdot \frac{d^2}{dt^2} q_2 + C_{2,2} \cdot q_2 + C_{2,5} \cdot q_5 + C_{2,8} \cdot q_8 = P_z, \tag{2}$$

$$M_2 \cdot \frac{d^2}{dt^2} q_3 + C_{3,3} \cdot q_3 + C_{3,5} \cdot q_5 + C_{3,8} \cdot q_8 = P_\varphi, \tag{3}$$

$$M_3 \cdot \frac{d^2}{dt^2} q_4 = H_1, \tag{4}$$

$$M_3 \cdot \frac{d^2}{dt^2} q_5 + C_{5,2} \cdot q_2 + C_{5,3} \cdot q_3 + C_{5,5} \cdot q_5 + B_{5,5} \cdot \frac{d}{dt} q_5 = P_z^{T_1}, \tag{5}$$

$$M_4 \cdot \frac{d^2}{dt^2} q_6 + C_{6,6} \cdot q_6 + B_{6,6} \cdot \frac{d}{dt} q_6 = P_\varphi^{T_1}, \tag{6}$$

$$M_5 \frac{d^2}{dt^2} q_7 = H_2, \tag{7}$$

$$M_5 \cdot \frac{d^2}{dt^2} q_8 + C_{8,2} \cdot q_2 + C_{8,3} \cdot q_3 + C_{8,8} \cdot q_8 + B_{8,8} \cdot \frac{d}{dt} q_8 = P_z^{T_2}, \tag{8}$$

$$M_6 \cdot \frac{d^2}{dt^2} q_9 + C_{9,9} \cdot q_9 + B_{9,9} \cdot \frac{d}{dt} q_9 = P_\varphi^{T_2}, \tag{9}$$

$$P_z = -F_{FR} \cdot \left(sign\left(\frac{d}{dt} \delta_1 \right) + sign\left(\frac{d}{dt} \delta_2 \right) \right) + P_v', \tag{10}$$

$$P_\varphi = F_{FR} \cdot l \cdot \left(sign\left(\frac{d}{dt} \delta_1 \right) + sign\left(\frac{d}{dt} \delta_2 \right) \right) + 2T_f, \tag{11}$$

$$P_z^{T_1} = F_{FR} \cdot sign\left(\frac{d}{dt} \delta_1 \right) + k\left(\eta_1 + \eta_2 \right) + \beta\left(\frac{d}{dt} \eta_1 + \frac{d}{dt} \eta_2 \right), \tag{12}$$

IOP Conf. Series: Materials Science and Engineering **985** (2020) 012006 doi:10.1088/1757-899X/985/1/012006

$$P_{\varphi}^{T_1} = -k(\eta_1 - \eta_2) - \beta \cdot a \cdot \left(\frac{d}{dt}\eta_1 - \frac{d}{dt}\eta_2 \right), \tag{13}$$

$$P_z^{T_2} = F_{FR} \cdot sign\left(\frac{d}{dt}\delta_2 \right) + k(\eta_3 + \eta_4) + \beta\left(\frac{d}{dt}\eta_3 + \frac{d}{dt}\eta_4 \right), \tag{14}$$

$$P_{\varphi}^{T_2} = -k \cdot a \cdot (\eta_3 - \eta_4) - \beta \cdot a \cdot \left(\frac{d}{dt}\eta_3 - \frac{d}{dt}\eta_4 \right), \tag{15}$$

$$M_1' = M_1 + (M_3 + M_5) + \frac{n \cdot M_w}{R^2}, \tag{16}$$

$$M' = M_1 \cdot h, \tag{17}$$

where M_1, M_2 – the mass and moment of inertia of the flat wagon supporting structure respectively; M_3, M_4 – the mass and moment of inertia of the first bogie in the direction of travel respectively; M_5, M_6 – the mass and moment of inertia of the second bogie in the direction of travel respectively; $C_{i,j}$ – elasticity characteristic of the vibration system elements defined according the values of stiffness factors of springs k_b; $B_{i,j}$ – pattern density function; a – half bogie wheelbase; k – track stiffness coefficient; β – damping factor; $\eta_i(x)$ – function that describes the track irregularity; δ_i – deformation of elastic elements of spring suspension; F_{FR} – absolute friction force in a spring complex; P_l – longitudinal loading on the automatic coupling supports; P_v', P_h' – loads on the carrying structure in vertical and horizontal planes in firing, respectively; n – the number of bogie wheelsets; M_w – the wheelset moment of inertia; H_1, H_2 – values of horizontal forces applied to body plates; T_f – moment on the carrying structure of a flat wagon in firing; h – height of the center of gravity of the carrying structure of a flat wagon (Figure 1).

The output parameters of the model were technical characteristics of the carrying structure of a flat wagon, spring suspension of bogies, perturbation action, and engineering characteristics of the anti-aircraft systems. It was considered, that a flat wagon was loaded with two anti-aircraft systems of muzzle energy of 89 kJ each. Besides, it was considered that two anti-aircraft systems were conducting fire.

The initial displacement and speeds were taken equal to zero [11-14]. The calculation was made in the plane coordinates; it considered the event of firing while the wagon was passing over a joint track irregularity. Differential equations (1)-(9) were reduced to standard Cauchy problems. Then they were integrated by the Runge-Kutta method [15, 16]. On the basis of the calculation the authors defined accelerations on a flat wagon in firing from anti-aircraft systems at various fire angles in horizontal and vertical planes. The maximum acceleration was about 36.0 m/s^2 and almost independent of the fire angle (Table 1).

Table 1. Accelerations on the carrying structure of a flat wagon at a fire angle.

Fire angle, degree	30	35	40	45	50	55	60
Acceleration, m/sec^2	35.75	35.75	35.75	35.75	35.74	35.74	35.74

The pattern of change in accelerations on the carrying structure of a flat wagon at a fire angle of 30^0 is given in Figures 2 and 3.

IOP Conf. Series: Materials Science and Engineering **985** (2020) 012006 doi:10.1088/1757-899X/985/1/012006

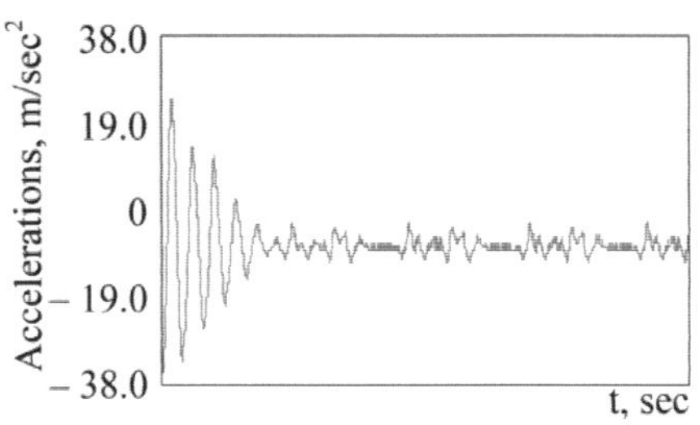

Figure 2. Longitudinal acceleration of the flat wagon supporting structure.

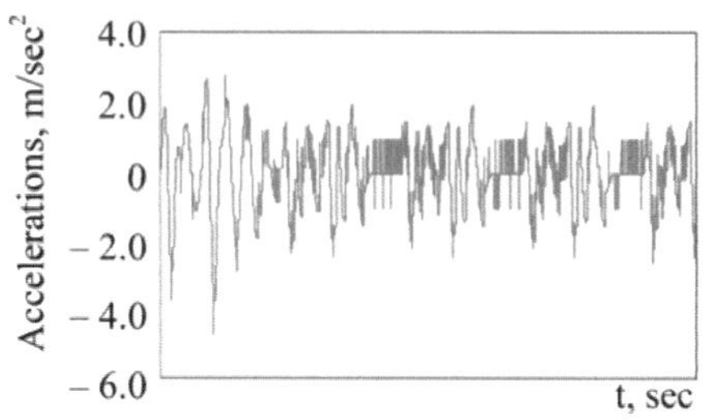

Figure 3. Vertical acceleration of the flat wagon supporting structure.

3. Determination of the stability factor of the flat wagon

The strength of a flat wagon in firing was defined with the sustainability coefficient. The design diagram is presented in Figure 4.

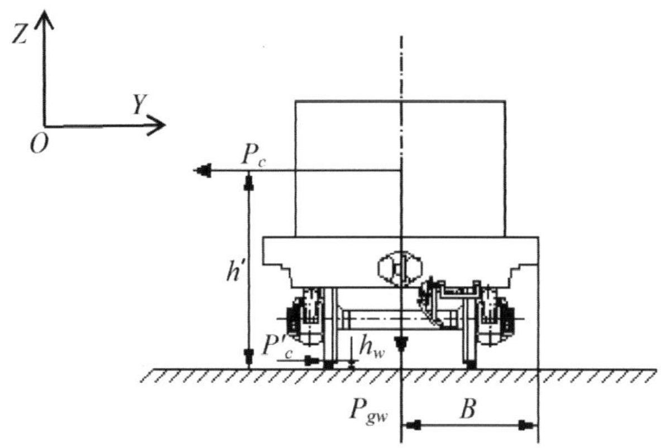

Figure 4. Design diagram of a flat wagon for studying the equilibrium stability.

The balance condition of a flat wagon in firing from anti-aircraft systems is:

$$k_s \geq \frac{T_{rest}}{2T_f}, \tag{18}$$

where T_{rest} – restoring torque, $\text{kN}\cdot\text{m}$; T_f – breakdown torque, $\text{kN}\cdot\text{m}$.

The restoring torque is defined as:

$$T_{rest} = P_{gw}\frac{B}{2} + P'_c \cdot h_w, \tag{19}$$

where P_{gw} – gross weight of a wagon; B – half-width of a wagon; $P'_c = P_c / 4$; h_w – height of the wheel flange of a bogie.

The breakdown torque is defined as:

$$T_f = P_c \cdot h', \tag{20}$$

where P_c – load on the carrying structure in firing; h' – distance from the weight center of an antiaircraft systems to the top level of a railhead.

On the basis of the calculation the authors obtained the restoring torque $T_{rest} = 466.3$ кN·m , and the breakdown torque $2T_f = 54.4$ кN·m , from where $k_s = 8.6$. Therefore, the equilibrium stability of a wagon in firing was ensured.

4. Determination of strength indicators of the flat wagon supporting structure

The accelerations on the carrying structure of a flat wagon in firing were taken into account in the strength calculation. The calculation was made by the finite element method [17-19] in CosmosWorks software suite. Spatial tetrahedrons were used as finite elements for the finite element model. The optimal number of elements was calculated with the graphical analytical method. Steel 09G2S was used as structural material for a flat wagon. The design diagram considered that the longitudinal load P_n 2.5 MN impacted frontal supports of the automatic coupling (Figure 5).

The vertical load P_v from the gross weight of anti-aircraft systems, and also loads in fixation areas impacted the carrying structure of a flat wagon. Due to angular location of fixation elements, the load P_d decomposed. The model was fixed in the areas where the model leaned on the gear parts.
The results of the calculation are given in Figure 6.

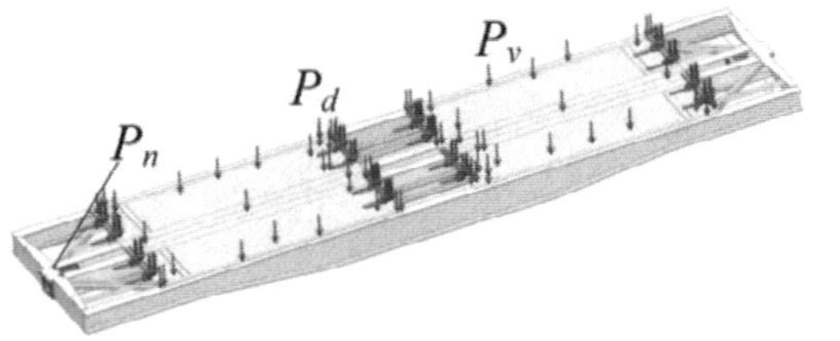

Figure 5. Design diagram of a flat wagon.

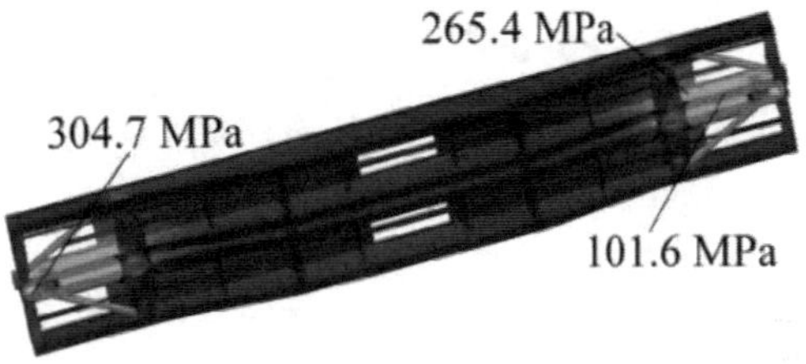

Figure 6. Stressed state of the carrying structure of a flat wagon.

On the basis of the calculation it was established that the maximum equivalent stresses emerged in the areas of frontal stops of the automatic coupling and were about $300\,\text{MPa}$. Therefore, the maximum equivalent stresses did not exceed the admissible values [20]. The maximum displacements were about $3.0\,\text{mm}$. The maximum deformations were $3.96 \cdot 10^{-3}$. Thus, it is possible to keep firing from anti-aircraft systems under rail transportation.

5. Conclusions

- The dynamic loading of the flat wagon supporting structure when conducting fire from two anti-aircraft systems placed on it has been determined. The calculation was carried out in a flat coordinate system. It was found that the longitudinal accelerations of the flat wagon supporting structure were about 36 m/s^2, and the vertical ones were about 5 m/s^2. Moreover, the acceleration rate is almost independent of the firing angle;
- The stability factor of the flat wagon while conducting fire has been determined. It has been established that the stability factor has a value of about 9. That is, the equilibrium stability of the flat wagon supporting structure while conducting fire from it is ensured;
  The main strength indices of the flat wagon supporting structure while conducting fire from it have been determined. The maximum equivalent stresses were about 300 MPa and concentrated in the area of the frontal supports of the automatic coupling, the maximum displacements were about 3.0 mm, and the maximum deformations were $3.96 \cdot 10^{-3}$. So, it is possible to conduct fire from anti-aircraft systems of given muzzle energy when transported on a flat wagon. The results of the research will contribute to the creation of flat wagon supporting structures for the transportation of military equipment and firing when moving.

Acknowledgments

This study was conducted in the framework of a scientific theme "Development of conceptual frameworks for restoring the efficient operation of obsolete freight cars" (Project registration number: 2020.02/0122).

References

[1] Divya Priya G and Swarnakumari A 2014 Modeling and analysis of twenty tonne heavy duty trolley *Intern. J. of Innovative Technology and Research* **2** **6** pp 1568–1580

[2] Krason W and Niezgoda T 2014 FE numerical tests of railway wagon for intermodal transport according to PN–EU standards *Bulletin of the Polish Academy of Sciences Technical Sciences* **62** 4 pp 843–851

[3] Bondarenko A I and Panin A Yu 2014 Teoreticheskaya i eksperimentalnaya otsenka prochnosti vagona-platformyi dlya perevozki avtomobilnyih polupritsepov *Transp. Ros. Federatsii* **3** pp 33–35

[4] Niezgoda T, Krasoń W and Stankiewicz M 2015 Simulations of motion of prototype railway wagon with rotatable loading floor carried out in MSC Adams software *J. of Kones Powertrain and Transport* **19** 4 pp 495–502

[5] Okorokov A M, Fomin O V, Lovska A O, Vernigora R V, Zhuravel I L and Fomin V V 2018 Research into a possibility to prolong the time of operation of universal semi-wagon bodies that have exhausted their standard resource *Eastern-European journal of enterprise technologies* **3/7** 93 pp 20–26

[6] Fomin O, Lovska A, Píštěk V and Kučera P 2020 Research of stability of containers in the combined trains during transportation by railroad ferry *MM Science Journal* **1** pp 3728–3733

[7] Sandu N and Zaharia N L 2014 Static and dynamic tests performed on a flat wagon *Problemy koleynictwa* **163** pp 67–77

[8] Wójcik K, Malachowski J, Baranowski P, Mazurkiewicz L, Damaziak K and Krason W 2012 Multi-body simulations of railway wagon dynamics *J. of Kones Powertrain and Transport* **19** 3 pp 499–506

[9] Anisimov P 2013 Model prostranstvennyih kolebaniy platformyi s dlinnomernyim gruzom *Mir transporta* **4** pp 6–13

[10] Domin Yu V and Chernyak G Yu 2003 Osnovi dinamiki vagoniv (Cham: Kiyiv KUETT) p 269

[11] Fomin O, Lovska A, Radkevych V, Horban A, Skliarenko I and Gurenkova O 2019 The dynamic loading analysis of containers placed on a flat wagon during shunting collisions *ARPN Journal of Engineering and Applied Sciences* **14** 21 pp 3747–3752

[12] Fomin O, Lovska A, Píštěk V and Kučera P 2019 Dynamic load computational modelling of containers placed on a flat wagon at railroad ferry transportation *Vibroengineering Procedia* **29** pp 118–123

[13] Kondratiev A V, Gaidachuk V E and Kharchenko M E 2019 Relationships between the ultimate strengths of polymer composites in static bending, compression, and tension *Mechanics of Composite Materials* **52** 2 pp 259–266

[14] Pistek V, Klimes L, Mauder T and Kucera P 2017 Optimal design of structure in rheological models: An automotive application to dampers with high viscosity silicone fluids *Journal of Vibroengineering* **19** 6 pp 4459–4470

[15] Dyakonov V 2000 MATHCAD 8/2000: spetsialnyiy spravochnik (Cham: SPb. Piter) p 592

[16] Alyamovskiy A A 2010 COSMOSWorks Osnovyi rascheta konstruktsiy na prochnost v srede SolidWorks (Cham: Moscow DMK Press) p 785

[17] Vatulia G, Falendysh A, Orel Y and Pavliuchenkov M 2017 Structural improvements in a tank wagon with modern software packages *Procedia engineering* **187** pp 301–307

[18] Kitov Y, Verevicheva M, Vatulia G, Orel Y and Deryzemlia S 2017 Design solutions for structures with optimal internal stress distribution *MATEC Web of Conferences* **133(1–3)** 03001

[19] Fomin O, Lovska A, Kulbovskyi I, Holub H, Kozarchuk I, Kharuta V 2019 Determining the dynamic loading on a semi-wagon when fixing it with a viscous coupling to a ferry deck *Eastern-European Journal of Enterprise Technologies* **8** **2/7** 98 pp 6–12

[20] GOST 33211-2014 2014 Vagonyi gruzovyie. Trebovaniya k prochnosti i dinamicheskim kachestvam p 54

Study of the interaction of the railway track and the rolling stock under conditions of accelerated movement

O Hubar[1,4], R Markul[1,5], O Tiutkin[1,6], V Andrieiev[2,7], M Arbuzov[1,8] and O Kovalchuk[3,9]

[1]Dnipro National University of Railway Transport named after Academician V. Lazaryan, Lazaryan str., 2, Dnipro, Ukraine
[2]Laboratory of Engineering and Technical Studies, Dniprovsky Scientific Research Institute of Forensic Expertise, Sichaslavskaya Naberezhna str., 17, Dnipro, Ukraine
[3]Laboratory of Railway Transport Research, Lviv Research Institute of Forensic Science, Lipinsky str., 54, Lviv, Ukraine

[4]neris@ua.fm, [5]guarangamr@gmail.com, [6]alexeytutkin@gmail.com, [7]avs_diit@ukr.net, [8]10max@ukr.net, [9]orestakovalchuk@ukr.net

Abstract. Comprehensive studies of the interaction of the railway track and the investigated rolling stock under accelerated conditions have been carried out. The basis of the research was to obtain and substantiate the process of stress state, force interaction of the railway track and the rolling stock under conditions of accelerated movement. Introduction of the new rolling stock into continuous operation is a serious task that must be solved comprehensively. The implementation of this direction significantly affects the reliability of the elements of the railway track. Within the framework of the research, the approach was used that takes into account theoretical and experimental parts of the research using the methodology of prof. O. P. Yershkov. As a result of the studies, average stress values were established: in rails, they variate within the range 44.37...70.99 MPa; in the sleeper, they reach 0.78 MPa; in the ballast layer, they have 0.13 MPa; in the body of the subgrade, they reach 0.04 MPa, which is significantly less than the maxima allowable values. While speed change from 140 to 160 km/h, the values of the studied parameters decrease: vertical force (9.39 %), lateral forces (19.79 %); stresses in the edge of the rail flange (15.86 %); stresses in the edge of the rail head (16.67 %); stresses in the neck of the rail (8.74 %). The obtained results made it possible to form trends and recommendations on the feasibility of further increasing speeds development. At the same time, this made it possible to substantiate in more detail the operation of the railway track in the case of accelerated movement.

1. Introduction

The introduction of a new rolling stock into continuous operation is a serious task that must be solved comprehensively. The possibility of further implementation of this direction substantially depends on the reliable operation of the elements of the railway track. The important component of solving this problem is the study of the interaction of the track and the rolling stock under operating conditions. Today, there is a lot of discussion as for the correctness of applying different techniques especially when studying the interaction of the track and the rolling stock under operating conditions. Each method is fundamental in its own way, and is applied depending on the specific task.

Schlumpf method [1-3] is widely used in the scientific research on the interaction of the railway track and the rolling stock. The essence of the method is to measure the dynamic stresses in the neck of a rail,

Content from this work may be used under the terms of the Creative Commons Attribution 3.0 licence. Any further distribution of this work must maintain attribution to the author(s) and the title of the work, journal citation and DOI.
Published under licence by IOP Publishing Ltd

measured in an individual cross section of the rail with strain gauge sensors [4-6]. It is established that this method, with insufficiently accurate sticking of sensors, has a significant dependence of measurement readings on the point of application of a vertical force relative to the rail axis *«e»* (eccentricity), inferior in accuracy to digital processing of experimental data.

The further development of the Schlumpf method [1–3], which increases the information content of experimental data, is the tensometric scheme of "piecewise continuous" registration. This method was applied by scientists and was tested to measure the vertical forces transmitted from the wheel to the rail [7]. An important advantage of the method is the possibility of using one recording channel for the experimental assessment of vertical force interaction. On the one hand, this provides savings in the use of channels while achieving the necessary reliable amount of experimental data. But the issues related to the registration of the lateral component of the force interaction of the railway track and the experimental rolling stock remained unresolved.

For 70 years, the method of experimental estimation of lateral forces by dynamic stresses in the edges of the flange and the outer section of the rail head has been widely used (method of prof. O. P. Yershkov) [8]. Normal stresses arising in rails from bending and torsion reach the greatest value at the edges of the rail flange. In [9], the adequacy of lateral forces estimation by the algebraic combination of measured edge stresses at the bottom of the rail flange was also noted.

To take into account spatial elasticity characteristics of rail threads, in work [10], results were proposed on the adjustment of the method of prof. O. P. Yershkov [8]. The problem is reduced to the theoretical determination of the modulus of elasticity and stiffness of a rail thread during its torsion. But the use of this approach is justified only in the presence of practical (experimental) data. Moreover, with each introduction of a new rolling stock, the depressions of the flange and the rail head will be different. As part of the research, the approach was used that takes into account the theoretical and experimental parts of the study using Yershkov technique.

2. The purpose and objectives of the study

The aim of the work is to study the interaction of the railway track and the rolling stock under conditions of accelerated movement.

To achieve this goal, it is necessary to solve the following tasks:

- in order to obtain the results of the stress state of the elements of the upper and lower track structure, comprehensive studies of the interaction of the railway track and the rolling stock under conditions of accelerated movement should be carried out;
- based on the results of the research, to discuss and justify the results of the study in order to form an assessment of the state of the railway track at accelerated train traffic;
- to substantiate trends in the further development of increasing speeds.

3. Characterization of the experimental site, the methodology for testing and data processing

The experimental site is a direct line; rail construction is rails P65; reinforced concrete sleepers, 1840 pcs/km; crushed stone ballast under the sleeper 30...35 cm thick, intermediate fasteners of КБ and КПП-5 types, maximum speed during the experiment 176 km/h.

The interpretation of the experimentally obtained data was carried out according to the indicators of each sensor at a given speed of movement of the rolling stock. The gradations of speed movement of the experimental rolling stock were taken after 10...40 km/h, and their choice was carried out with respect to possible sharp impulse changes in dynamic indicators. In this work, the gradation in speed movement changes was taken every 20 km/h. In our case, this can be explained by the fact that, in the speed range from 10 to 20 km/h, the correlation dependence between the dynamic indicators was mainly observed.

The number of runs along the experimental section of the track was established by the technical task for testing, depending on the number of instruments-sensors installed along the way and the required number of measurements resulting in reliable statistical data. In accordance with the approved programme and methodology for testing the new rolling stock in terms of impact on the track, the obtained statistical data do not exceed the probability level $p = 0.994$, which is adopted in calculating the track strength. Taking this into account and with the aim of obtaining reliable statistical data, the number of arrivals in each direction for one speed of movement was taken from 10 to 15 (but not less than 10). The experience of research and experiments [11-16] show that during the test the track and rolling stock parameters do not change significantly. It can be assumed with sufficient accuracy that the results of previous experiments do not affect subsequent ones. In this case, the observation results turn out to be independent random variables. As a rule, the work analyzed the results of the stress state and force interaction of the elements of the railway track from the axes of the rolling stock, which cause the greatest influence. For rolling stocks, this is, as a rule, the first ones along the axis in the guide trolley.

The scheme of the experimental site and the rail with the measuring devices installation is shown in Figures 1-2.

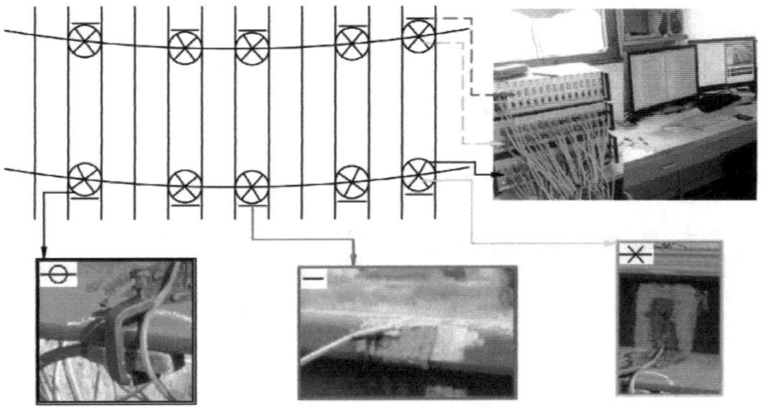

Figure 1. The scheme of measuring devices installation in the experimental site.

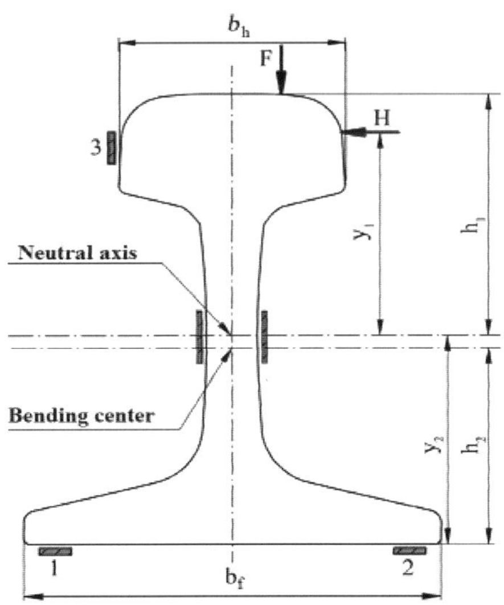

Figure 2. The calculating scheme of sensors distribution on a rail.

Figure 2 shows: 1, 2, 3 – strain gauge numbers; b_h, b_f, y_1, y_2, h_1, h_2 – geometric characteristics of the rail; H – lateral force; F – vertical force.

To register the ohmic resistance of the strain gauge when it is powered by current, Wheatstone bridge circuits are used [17]. The circuit of the electrical installation for measuring edge stresses in the rail head is shown in Figure 3. The magnitudes of the stresses that occur in the edge of the rail head at different speeds of the experimental rolling stock are shown in Figure 4.

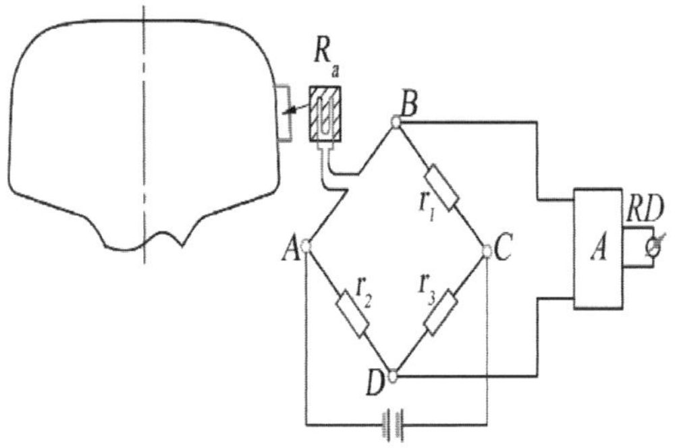

Figure 3. Half-bridge circuit of the electrical installation: R_a - active (working) strain gauge; r_1, r_2, r_3 resistance; A – amplifier; RD – recording device.

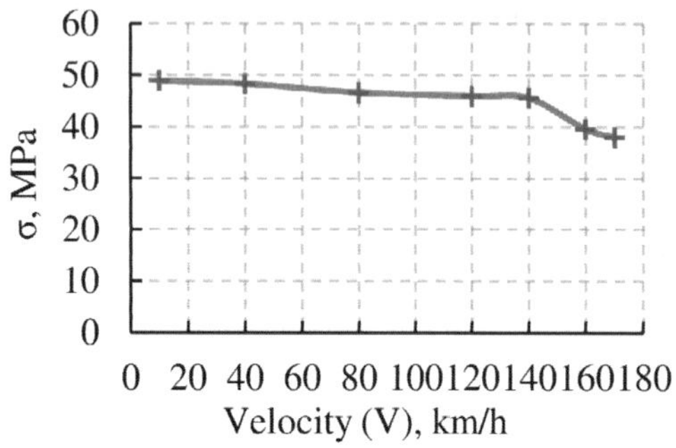

Figure 4. The dependence of stresses in the edge of the rail head on the speed of the experimental rolling stock.

The calculation of the vertical force «F» was determined by the preliminary calibration of sections where the stresses in a rail neck were recorded. Section calibration was performed when moving an experimental rolling stock at a speed of 5 km/h to 10 km/h (at a known static load of an experimental rolling stock). Stresses were recorded using strain gauges (R_a, R_c), which were stuck vertically in the neutral axis zone on both sides of the rail neck (Figure 5). The magnitudes of stresses that occur in the neck of the rail at different speeds of the experimental rolling stock are shown in Figure 6.

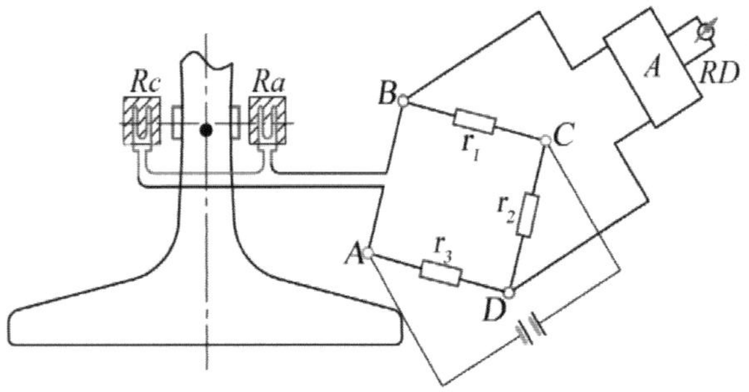

Figure 5. The bridge circuit of the electrical installation: R_a - active (working) and R_c - compensation strain gauges; r_1, r_2 r_3 – resistance; A – amplifier; RD – recording device.

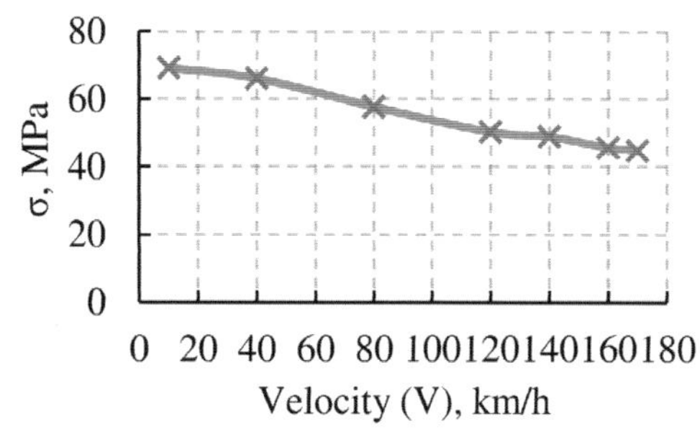

Figure 6. Dependence of stresses in the neck of the rail on the speed of the experimental rolling stock.

Since the scattering of the range of these vertical forces is insignificant, the measurement results were reduced to one sample. The sampling results are presented in Figure 7.

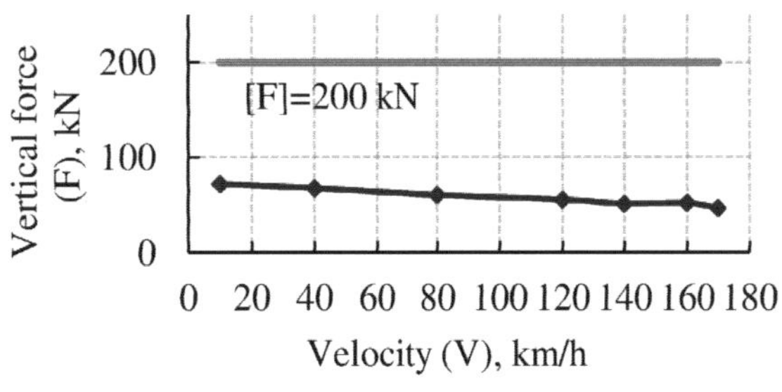

Figure 7. Dependence of vertical forces in rails on speed of movement.

To determine the horizontal force, we used the values of stresses obtained experimentally in the edges of the rail flange, as shown in Figure 8. Stresses arising along the outer edge of the rail flange from the action of lateral forces are shown in Figure 9.

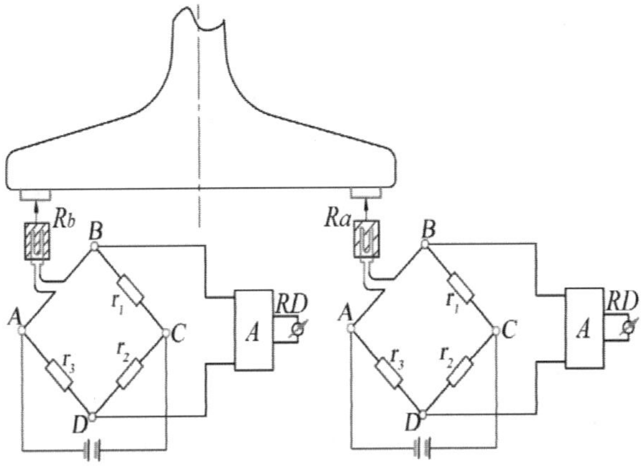

Figure 8. Half-bridge circuits of the electrical installation: R_a, R_b – active (working) strain gauge; r_1, r_2 and r_3 – resistances; A – amplifier; RD – recording device.

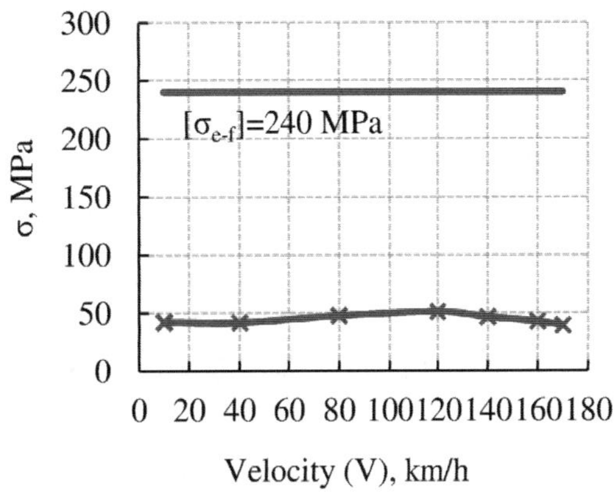

Figure 9. The dependence of stresses in the edge of the rail flange on the speed of the experimental rolling stock.

Based on the results (Figure 9), we can conclude that the stress in the edge of the rail flange does not exceed the allowable and recommended values. The values of lateral forces were calculated according to the method of prof. O. P. Yershkov [8] using experimentally obtained values of edge stresses in the flange and in the outer face of the rail head. The final expression for determining the lateral forces by this technique is:

$$H = B \cdot \sigma^{o}_{e-f} \cdot \tag{1}$$

where H – lateral force (kN); coefficient for determining forces in (cm²); σ^{o}_{e-f} – stresses at the edge of the rails flange (MPa); o – outer; e - edge; f – flange.

The results of the horizontal forces are shown in Figure 10.

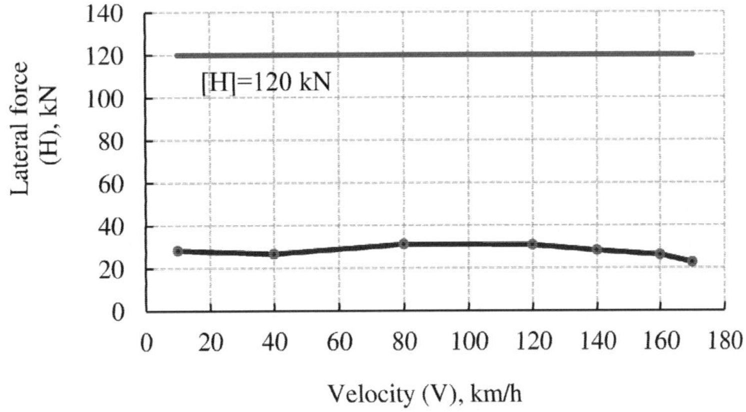

Figure 10. Dependence of horizontal forces in the rails on the speed of movement.

Ensuring the bearing capacity and operation of the railway track structure is characterized by not exceeding the design stresses of their maxima permissible values. The conditions must be met:

$$\begin{cases} \sigma_{sleep}^{calc} \leq \left[\sigma_{sleep}\right], \\ \sigma_{ball}^{calc} \leq \left[\sigma_{ball}\right], \\ \sigma_{subgr}^{calc} \leq \left[\sigma_{subg}\right]. \end{cases} \quad (2)$$

where σ_{sleep}^{calc} - calculated compressive stresses in the sleepers under the pad, MPa; $[\sigma_{sleep}]$ – allowable compressive stresses in the sleepers under the pad, MPa; σ_{ball}^{calc} - calculated compressive stress in the ballast under the sleeper, MPa; $[\sigma_{ball}]$ – allowable compressive stress in the ballast under the sleeper, MPa; σ_{subgr}^{calc} - calculated stresses of soil compression in the main site of the subgrade, kPa; $[\sigma_{subgr}]$. permissible soil compression stresses in the main site of the subgrade, kPa.

The calculation results are shown in figures 11–13.

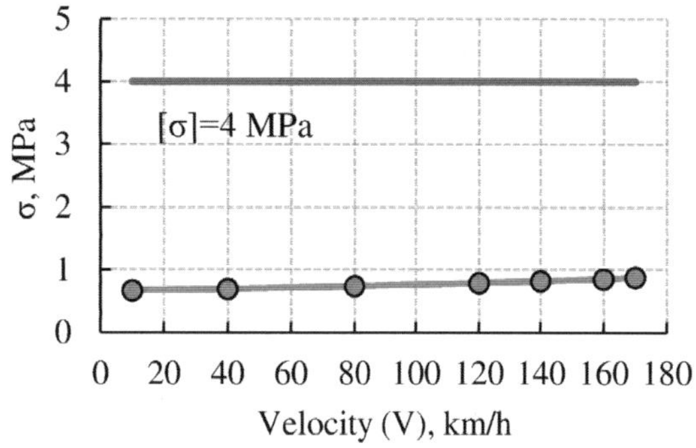

Figure 11. Dependence of stresses in the sleeper on the speed of movement.

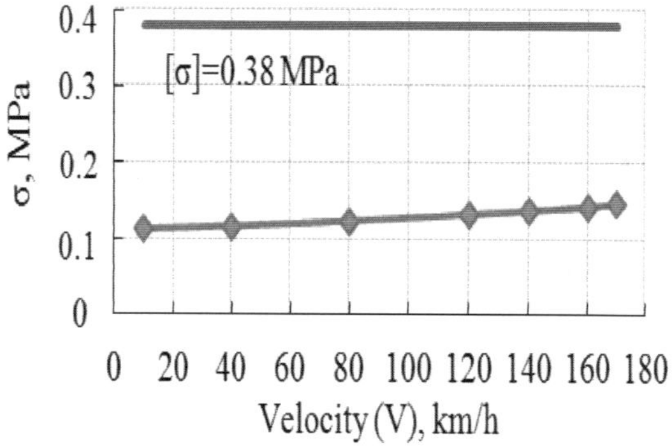

Figure 12. Dependence of stresses in the ballast on the speed of movement.

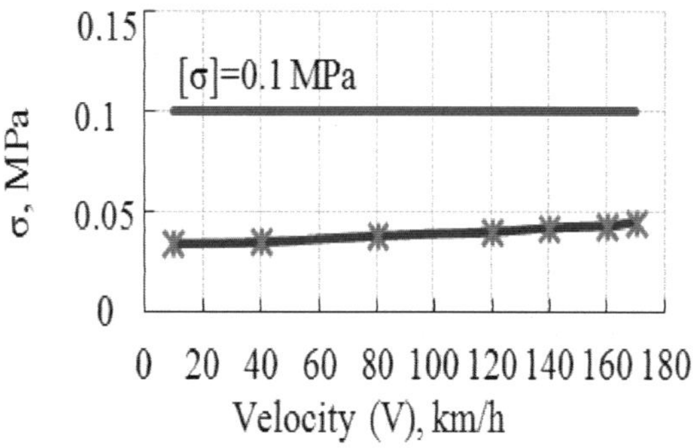

Figure 13. Dependence of stresses in the subgrade on the speed of movement.

Based on the results that are shown in Figures 11–13, the conclusion can be made that the calculated stresses in the elements of the railway track do not exceed the allowable and recommended values. This makes it possible to suggest the possibility of a further increase in the speed of movement on the railways of Ukraine. Trends on the possibility of increasing speeds, and ensuring rail transit traffic between Ukraine and the European Union have also been substantiated in [18].

4. Discussion of the results of the study with the aim of forming an assessment of the state of the railway track under condition of accelerated movement of trains

Comprehensive studies of the interaction of the track and the rolling stock under operation have been carried out. The basis of the research is an assessment of the power component, stresses in the elements of the upper and lower structure of the track from the effect of the rolling stock of a new generation. The average stress values in the elements of the upper and lower structure of the track do not exceed the maxima allowable values, which is shown graphically in Figure 14.

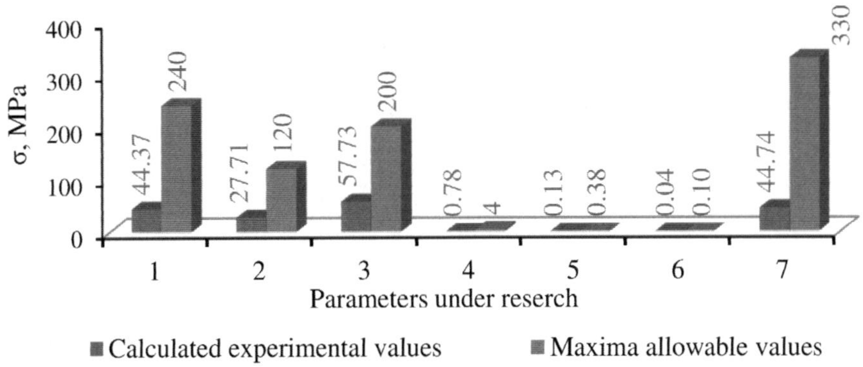

Figure 14. Comparable values of the obtained results.

Figure 14 shows: 1 – stresses in the edge of the rail flange, 2 – horizontal force, 3 – vertical force, 4 – stresses in the sleeper, 5 – stresses in the ballast, 6 – stresses in the subgrade, 7 – stresses in the edge of the rail head.

At an increase in speed hidden (dynamic) uneven paths may appear that leads to additional dynamic forces at oscillations of galloping and rolling motion of a car. In such cases, unloading of car truck wheels is observed that cause an alternating change of dynamic additives on rails. Since the sensors are placed discretely on the way, it is not possible to take into account the influence of various dynamic factors fully. In addition, a decrease of vertical force, when a passenger rolling stock is moving, can become the reason for not coinciding of an elastic line of rail deflection under the influence of dynamic forces at high speed indicators with the elastic line under a static load. One of the first who made a research in this area was prof. G. M. Shakhunyants [19].

With the aim of determining the lateral force, the method of prof. O. P. Yershkov [8] was used applying experimentally obtained values of edge stresses in the foot and head of the rail. But there is also a second widely used method of experimental determination of the lateral force (determination of the force effect on the rail by measuring the elastic deformation of the rail web) – the Schlumpf method [1-3]. In accordance with the comparative tests of JSC VNIIZHT, to which references were made in the work, it was established that the discrepancy in values of the lateral force with the use of O. P. Yershkov and Shlumpf methods did not exceed 4–9 %. This is mainly due to the accuracy of the choice of cross-sections for mounting strain gauges and their number on the rail neck (Schlumpf's method [1-3]). In order to save material resources, it was decided to use the method of prof. O. P. Yershkov [8] during the experiment.

5. Conclusions

The obtained maxima values of vertical and lateral forces from the interaction of the rolling stock are 115.5 kN and 53.38 kN, respectively, which is significantly less than the permissible value (200 kN, 120 kN).

Average stress values: in rails vary in the range 44.37…70.99 MPa; in the sleepers – 0.78 MPa; in the ballast layer – 0.13 MPa; in the body of the subgrade – 0.04 MPa, which is significantly less than the maximum allowable value.

Based on the experimental studies, it was found out that with the existing design of the railway track, it is advisable to develop trends for further increase in speed. With an increase in the speed of movement from 140 to 176 km/h, the values of the studied parameters decrease: vertical force (9.39 %), lateral force (19.79 %); stresses in the edge of the rail flange (15.86 %); stresses in the edge of the rail head (16.67 %); stresses in the neck of the rail (8.74 %). This indicates a sufficiently large margin of safety of the elements of the railway track. Knowing the sections of the railway track where the increase in speed will be implemented, it is necessary to carry out control and elimination of deviations from norms timely.

References

[1] Ward C, Weston P, Stewart E, Li H, Goodall R, Roberts C, Mei T, Charles G and Dixon R 2011 Condition monitoring opportunities using vehicle-based sensors *Proc. Inst. of Mech. Eng. Part F: Journal. of Rail and Rapid Transit* **225** 2 pp 202-218

[2] Mei T and Li H 2008 Measurement of vehicle ground speed using bogie-based inertial sensors *Proc. of the Inst. of Mech. Eng. Part F: Journal of Rail and Rapid Transit* **222** 2 pp 107-116

[3] Lunin A A 2017 Force impact of freight wagons with axial loads of 23.5 and 25 tnf on the track structure *Transport of the Russian Federation* **6** 73 pp 73-75

[4] Gapanovich V A 2016 Questions of the interaction of rolling stock and infrastructure under heavy traffic *Railway transport* **10** pp 10-15

[5] Romen Y S, Suslov O A, Balyaeva A A 2017 Determining the force of interaction in a wheel - rail system based on measuring stresses in rails neck *Vestnik of the Railway Research Institute* **76** 6 pp 354-361

[6] Brzhezovskiy A M 2017 Methods of experimental evaluation of lateral forces (review) *Vestnik of the Railway Research Institute* **76** 1 pp 10-18

[7] Zhou L, Brunskill H, Pletz M, Daves W, Scheriau S and Lewis R 2019 Real-Time Measurement of Dynamic Wheel-Rail Contacts Using Ultrasonic Reflectometry *Journal of Tribology* **141** pp 061401-1-061401-9

[8] Gerlici J, Nozhenko O, Cherniak G, Gorbunov M, Domin R, and Lack T 2018 The development of diagnostics methodological principles of the railway rolling stock on the basis of the analysis of dynamic vibration processes of the rail *MATEC Web of Conf.* **157** 2 03007

[9] Romen Yu S and Tikhov M S 2007 The volume of information for establishing the permissible speeds of movement based on the results of complex tests *Rolling stock of the XXI century: ideas, requirements, projects. Coll. of scientific papers. St. Petersburg, (PGUPS)* **4** pp 87-94

[10] Danilenko E I 2016 Calculation of characteristics of stiffness and elasticity of rail threads when torsion under combined action of vertical and horizontal forces *Science and Transport Progress* **5** 65 pp 79-91

[11] Pshinko A, Petrenko V, Tiutkin A, Andreev V, Gubar A, Ihnatenko D and Markui R 2019 Comparative analysis of calculation results of supporting structure of soil-cement piles *Proc. of 23rd Int. Sci. Conf. Transport Means 2019* pt. II (Kaunas, Lithuania: Kaunas Univ. of Technology) pp 820-828

[12] Lunys O, Neduzha L and Tatarinova V 2019 Stability research of the main-line locomotive movement *Proc. of the 23rd Int. Sci. Conf. Transport Means 2019* pt. III (Palanga, Lithuania: Kaunas Univ. of Technology) pp 1341–1345

[13] Pshinko O, Patlasov O, Andrieiev V, Arbuzov M, Hubar O, Hromova O and Markul R 2018 Research of railway crashed stone use of 40-70 mm fraction *Proc. of the 22rd Int. Sci. Conf. Transport Means 2018* pt. I (Trakai, Lithuania: Kaunas Univ. of Technology) pp 170-178

[14] Bondarenko I 2016 Development of algorithm for calculating dynamic processes of railroad track deformability work *Eastern-European Journal of Enterprise Technologies* **84** 6/7 pp 28-36

[15] Kovalchuk V, Markul R, Bal O, Milynych A, Pentsak A, Parneta B and Gajda O 2017 The study of strength of corrugated metal structures of railroad tracks *Eastern-European Journal of Enterprise Technologies* **2** 7(86) pp 18-25

[16] Bondarenko I, Lunys O, Neduzha L and Keršys R 2019 Dynamic track irregularities modeling when studying rolling stock dynamics *Proc. of the 23rd Int. Sci. Conf. Transport Means 2019* pt. II (Palanga, Lithuania: Kaunas Univ. of Technology) pp 1014-1019

[17] Dolidze D E 1975 *Test of constructions and constructions* (Moscow: Higher School) p 252

[18] Kurhan M and Kurhan D 2019 Providing the railway transit traffic Ukraine-European Union *Pollack Periodica* **14** 2 pp 27-38

[19] Shakhunyants G M 1959 *Calculation of track superstructures* (Moscow: Transzheldorizdat) p 264

Ways to improve the mathematical model of a freight car for the execution of forensic railway-transport expertises

A Batig[1,4], A Kuzyshyn[1,2], J Sobolevska[2], A Milyanych[2] and P Hrytsyshyn[1,2,3]

[1]Lviv Research Institute of Forensic Expertise, 79000 Lviv, Ukraine
[2]Lviv branch of Dniprovsk National University of Railway Transport named after Academician V. Lazaryan, 79052 Lviv, Ukraine
[3]Western center of the Ukrainian branch of the World laboratory, 79021 Lviv, Ukraine

[4] batigasha1992@gmail.com

Abstract. According to statistics, receipts of forensic railway-transport expertises related to the derailment of freight wagons annually grows. At the same time, an analysis of the available mathematical models of freight wagons for the study of such cases showed that at present they are not sufficiently detailed to the extent that it is necessary to analyze the causes of rolling stock derailment. Therefore, taking into account the main reasons that occurrence of such railway accidents over the last five years on the railways of Ukraine, the article selects the main directions for improving the mathematical model of a freight wagon. These directions, in turn, should cover the whole plural of factors (explicit and hidden) and identify the most significant ones regarding the circumstances of the derailment rolling stock off the track, established on the basis of a computer experiment. An algorithm and a mathematical model are proposed taken into account the influence on the dynamics of the wagon of process of closing gaps in the axle-boxes and slides, and the equations of motion of the wheelset are supplemented by a guiding force, whose value is one of the main indicators of the stability of the rolling stock.

1. Introduction

Traffic safety is a priority on the railways. To ensure it, the regulatory framework is being improved, measures are being taken to prevent the occurrence of railway accidents and the factors that pose a threat to the traffic safety and tend to grow as well as other necessary measures are being explored.

However, as statistics show, the number of the railway accidents is steadily increasing every year. Of the total number, the majority of cases are related to the rolling stock derailment. The reasons for this kind of railway accidents are, first of all are the deviations in the technical state of the track and the running gear of the rolling stock of their maintenance norms [1].

Many years of experience in conducting expertises shows that there are more and more cases when it is impossible to unambiguously determine the reasons for the rolling stock derailment. Such reasons are not obvious and require an in-depth analysis of the complex combination of many factors of the force interaction of the rolling stock and the rail track, which form the most unfavourable (worst) combination of circumstances occurring under particular state of the rolling stock and the railway track [2, 3].

Hence, there is the need to substantiate proposals for scientific approaches improvement in the

Content from this work may be used under the terms of the Creative Commons Attribution 3.0 licence. Any further distribution of this work must maintain attribution to the author(s) and the title of the work, journal citation and DOI.
Published under licence by IOP Publishing Ltd

study of the interaction of the rolling stock and the track, the calculation of the indices that evaluate the dynamic properties of its running gear and traffic safety indicators. This, in turn, will not only facilitate the investigation of the causes of the rolling stock derailment, but in most cases, it will allow to set the limit values of its technical state, at which there is a threat of an accident.

2. Purpose and objectives of the research

The purpose of the work is to highlight the main problems that arise when investigating the causes of the rolling stock derailment, and to develop ways to overcome them with the help of an advanced mathematical model of a freight wagon.

Research objectives:

- to analyse the main causes of freight wagon derailment on the railways of Ukraine over the last five years;
- for these reasons, to establish the main ways of improving the mathematical model of a freight wagon;
- to improve the mathematical model of a freight car for the possibility of its use in forensic railway-transport expertise.

3. Statement of the problem and review of scientific sources, which are devoted to its solution

During the period 2015 to 2020, the railways of Ukraine experienced about 165 cases of freight wagons derailment in train formations, of which 20% of cases was obvious running gear malfunction, 31% of other cases was malfunctions of the track, 15% was the rolling stock collisions [4].

The percentage of malfunctions of the running gear technical state that caused their derailment in train formations are shown in Figure 1

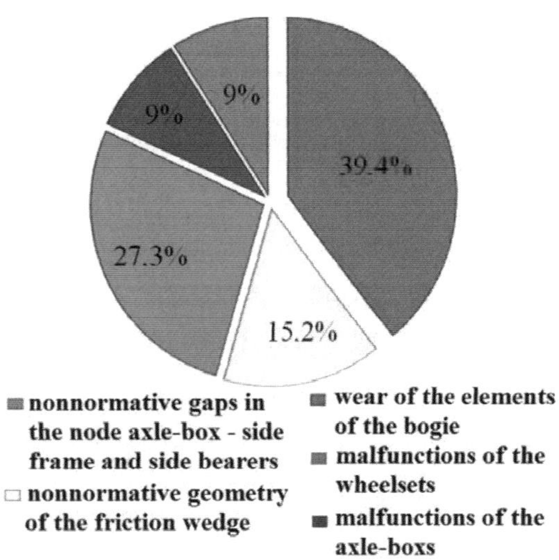

Figure 1. Diagram of the distribution of freight wagons running gear malfunctions that caused derailment for the last 5 years.

Other reasons for the loss of stability of the rolling stock are those that were not definitely established during railway investigations, and as a percentage, it accounted for about 7% of the total cases of the rolling stock derailment.

Obviously, the study of such cases is impossible without a detailed mathematical model of a freight wagon, which could give a reliable quantitative assessment of the parameters that affect the safety traffic, namely, friction wedges, spring suspension, gaps in the boxes and side bearers, etc.

Therefore, we analyzed the existing mathematical models of freight wagons and complex developments in the field of rolling stock dynamics, which served as a basis for further development of methods and means of this problem research.

The technique of constructing mathematical models of wagons spatial vibrations for the study of reasons of their stability loss during movement was substantially enriched due to works of M.L. Korotenko [5] and V.F. Ushkalov [6]. The researches performed by them allowed to solve the problem of optimization of the parameters of wagons running gears under condition of stability of their undisturbed movement.

In work [7], with the use of the developed mathematical model of a freight wagon, the influence on the stability of its movement a considerable difference in the base of bogies and wheel diameters was investigated. However, in the course of the research, the authors of this paper did not take into account the influence of horizontal irregularities of the railway track and a significant number of parameters of the bogies of freight wagons provoking their derailment.

In work [8], a detailed mathematical model of a freight wagon was developed, which allows to estimate the influence of the magnitude of angles of friction wedges, gaps in the boxes and side bearers of the wagon affecting its derailment. The disadvantage of this work is taking not into account the guide force, whose value is one of the main indicators of the stability of his motion.

The strategic method of analysis of operational safety of the rolling stock preventing its derailment was proposed in works [9, 10]. The mathematical model of a freight wagon offered in these works allowed to establish the influence of vertical irregularities of the railway track on the stability of wagon's movement. The disadvantage of the developed mathematical model of a wagon in works [9, 10] is the neglect of deviations influence on the parameters of the technical state of wagons' running gear from normative values in case of their derailment.

In work [2], many hidden causes of rolling stock derailment in the Russian Federation are outlined and a simplified model of a freight wagon is proposed to investigate such cases. The main disadvantage of this mathematical model of a wagon is that it gives the possibility to take into account only the irregularities of the track as a factor that impacts wagon's derailment.

With regard to the implementation of investigations of real cases of railway accidents, the work of Sokol E.M. should be noted [11], in which analytical expressions were derived to determine sufficient conditions for the rolling stock derailment on the basis of the developed mathematical model of motion a wheel-set of a freight wagon. However, this model allows to determine the influence of only a part of required parameters of a freight wagon and the railway track on wagon's stability. In addition, the equations of motion of a wheel-set of a wagon drawn up by Sokol E.M. do not take into account the action of creep forces that can have a significant effect on the stability of wagon's motion.

As can be seen, at the present time, many scientific works, both domestic and foreign, are devoted to the problem of the rolling stock derailment. However, each of the above works has its advantages and disadvantages when studying the dynamics of the rolling stock, and the mathematical models of freight wagons offered in them are not sufficiently detailed to investigate the reasons for their derailment.

4. Improved mathematical model of a freight wagon

Due to the fact that at present the experts of the railway research laboratory of Lviv Scientific Research Institute of Forensic Expertise undertake examination of cases of rolling stock derailment, which occur in the absence of obvious deviations of the rolling stock and the track from the requirements of the regulatory documents, there is a necessity to select such a mathematical model of a freight wagon that would take into account a larger number of parameters of its running gear than those currently available.

Therefore, the mathematical model of a freight wagon, which was developed in work [8], was selected as a basis, and in order to choose ways to improve such a model, the assumptions made during its creation were analyzed.

Of all the assumptions adopted in [8] when creating a mathematical model of a freight wagon, it should be noted from the fact that the rigidity of the rail track in the horizontal plane was not taken into account in the equations of motion of the wheelset. So, the authors of the mentioned work did not take into account the influence of the force, arising as a result of pressing the wheel flange to the rail head due to the exhaustion of the gap between them, which can lead to unreliable of the quantitative assessment influence of the parameters of the railway track and running gears of wagons on their derailment

With this in mind, the mathematical model of a freight wagon was improved by the authors of this article by taking into account the force, arising as a result of pressing the wheel flange to the rail head due to the exhaustion of the gap between them, in the equations of wheel-set motion. In addition, to create a computer model on its basis, the algorithm for determining the forces that arise during gaps closure in the side bearers and axle-boxes of a freight wagon was developed.

Since this mathematical model is quite cumbersome, we will give some of its equations (1-4) in which improvements and additions were made.

$$m_b H\ddot{y}_b + \left(m_b H^2 + J_{bx}\right)\ddot{\theta}_b + \sum_{i=1}^{2} R_{ti} e_{fti} + R_{ck1}B_1 - R_{ck2}B_2 + R_{ck3}B_3 - R_{ck4}B_4 = 0, \quad (1)$$

$$J_{xbl1}\ddot{\theta}_{bl1} + N_{z1}a_1 - N_{z2}a_2 + N_{z3}a_3 - N_{z4}a_4 + \sum_{j=1}^{5} P'_{fzj}b'_{fj} - \sum_{j=1}^{5} P''_{fzj}b''_{fj} - R_{ck1}B_1 + R_{ck2}B_2 - \sum_{j=1}^{7} P'_{fyj}h_{б1} - \sum_{j=1}^{7} P''_{fyj}h_{bl1} = 0, \quad (2)$$

$$J_b\ddot{\psi}_b + \sum_{i=1}^{7} P'_{fyi}\left(L_i \pm e'_{fi}\right) + \sum_{i=1}^{7} P''_{fyi}\left(L_i \pm e''_{fi}\right) - \sum_{i=1}^{7} P'_{syi}\left(L_i \pm e'_{si}\right) - \sum_{i=1}^{7} P''_{fyi}\left(L_i \pm e''_{si}\right) + \sum_{i=1}^{7} T_{fyi}\left(L_i \pm d_i\right) -$$

$$-\sum_{i=1}^{7} T_{syi}\left(L_i \mp d_i\right) - X_9 \cdot T_{ck1}B_1 + X_{10} \cdot T_{ck2}B_2 - X_{11} \cdot T_{ck3}B_3 + X_{12} \cdot T_{ck4}B_4 - M_{fr1} + M_{fr2} = 0, \quad (3)$$

$$m_{wh1}\ddot{y}_{wh1} + F_{y1} + F_{y2} - X_{13} \cdot T_{py1} - X_{14} \cdot T_{py2} + R_1 n_1 - R_2 n_2 + N_{aby11} - N_{aby12} - N_{aby21} + N_{aby22} + N_{g1} - N_{g2} = 0, \quad (4)$$

where m_b, m_{wh1} – mass of the body and mass of first wheelset, *t*; H – height of center of weight of the body, *m*; y_b, y_{wh1} – transverse displacement of the body and wheelset, *m*; J_{bx} – body moment of inertia about an axis *x t·m*²; R_{t1-2} – vertical reactions on body center plate, *kN*; e_{fi}, e_{si} – transverse displacements relative to the center of application of the vertical reactions on body center plate, *m*; R_{ck1-4} – vertical reactions on the side bearers, *kN*; B_{1-4} – the distance from the center pin to the middle of the side bearer, *m*; $P'_{fzj}, P''_{fzj}, P'_{fyj}, P''_{fyj}$ – vertical and transverse reactions of the springs of the first and second spring sets of the first bogie, *kN*; P'_{syj}, P''_{syj} – transverse reactions of the springs of the first and second spring sets of the second bogie, *kN*; T_{pyi} – transverse friction forces between the side frame and the axle-box, *kN*; N_{abyij} – lateral forces, arising when closing transverse and longitudinal gaps in axle boxes, *kN*; N_{g1-2} – force, arising as a result of pressing the wheel flange to the rail head due to the exhaustion of the gap between them, *kN*; R_{1-2} – reaction in wheel-rail contact, *kN*; n_{1-2} – taper of the wheel rim at the points of contact with the rail; F_{y1-2} – the transverse forces of the creep, *kN*; θ_b, θ_{bl} – generalized coordinates of the lateral oscillations of the body and the bolsters, *rad*; N_{z1-4} – projection on the axis of the normal forces acting between the bolster and friction wedges, *kN*; a_{1-4} – distance from the longitudinal symmetry of the bogie to the wedge spring, *m*; h_{bl1-4} – height of the center of gravity of the bolster above the upper supporting surface of the springs of the spring sets of the first and second bogie, *m*; T_{fyi}, T_{syi} – the friction force between the bolster and the friction wedge of the first and second bogie, *kN*; M_{fr1-2} – moments of friction forces in the bolsters, *kN ·m*; T_{ck1-4} – friction forces the side bearers, *kN*; X_{9-14} – auxiliary coefficients.

In the equations, the upper signs in the brackets at horizontal reactions $P'_{fyj}, P''_{fyj}, P'_{syj}, P''_{syj}$ ′ correspond to the spring No 1, 2, 6 of the spring set of a bogie, low signs – for No 3, 4, 7, and for the forces of transverse friction T_{fyi}, T_{syi}, which arising between the friction wedge and the bogie bolster – the upper signs for friction wedges of a bogie No 1, 2, 6, 7, low signs – for wedges No 3, 4, 5 and 6.

As we see equation (4) was supplemented by the force, arising as a result of pressing the wheel flange to the rail head due to the exhaustion of the gap between them N_g, which acting on the right and left wheels of a wagon wheel-set, and her value proposed to be determined by the following formula:

$$N_g = C_y\left(z_a^{\varphi} + z_t\right), \tag{5}$$

where C_y – is the horizontal rigidity of the track, *kN/m*; z_a^{φ}, z_t – angular and transverse movement of the rail when being exposed to the wheel-set of a wagon, *m* [11].

To establish the influence of the force, arising as a result of pressing the wheel flange to the rail head due to the exhaustion of the gap between them, on the transverse movement of the wheel-set within the gap of the rail track, we will conduct study with and without respect of its value in Maple software environment.

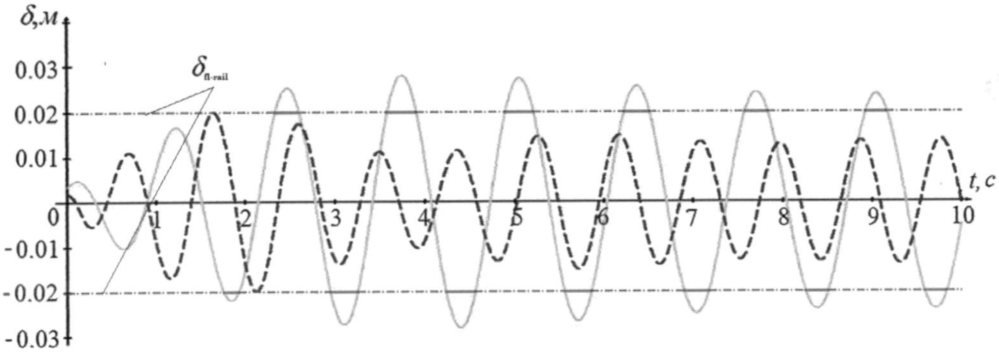

Figure 2. Transverse movement of the wheel-set with and without the guiding force.

As can be seen from Figure 2, the lack of regard of the force, arising as a result of pressing the wheel flange to the rail head due to the exhaustion of the gap between them, affects the value of the transverse movement of the wheel-set and may be greater than the value of the boundary clearance between the wheel flange and the rail, that is, leads to false results. When accounting the value of this force, the value of the wheel-set movement does not go beyond the limit gap.

Now let's consider the algorithm development for establishing the forces that arise during gaps closure in the side bearers and axle-boxes of a wagon. Since there are four gaps in the side bearers wagon and thirty-two gaps in the axle-boxes, we present formulas for determining several of them, and expressions for the remaining gaps are given in detail in work [8].

The gaps in the side bearers of a wagon are determined by the following formula:

$$\begin{cases} \delta_{ck1} = \delta_{ckn1} - \theta_b B_1 + \theta_{bl1} B_1, \\ \ldots \\ \delta_{ck4} = \delta_{ckn4} - \theta_b B_4 + \theta_{bl2} B_4, \end{cases} \tag{6}$$

where $\delta_{ckn1}, \delta_{ckn4}$ – nominal value of the gaps in the side bearers, *mm*.

Transverse clearances in the axle-box of a wagon are determined by the formula:

$$\delta_{aby1} = \delta_{abn1} - y_{f1} + y_{wh1} - \psi_{f1} l_1, \tag{7}$$

where δ_{abn1} – nominal value of the operational transverse clearances in the axle-boxes, *mm*; y_{f1}, y_{wh1} –

lateral movement of the side frame of a bogie and the first wheel-set, *mm*; ψ_{f1} – generalized coordinates of the frame of the first bogie, *mm*; l_1 – half of the base of the first bogie, *mm*.

As can be seen from expressions (6, 7), the gaps closure in the side bearers and axle-boxes is possible when their value becomes equal to or less than zero, that is $\ddot{\delta}_{abyi}, \ddot{\delta}_{ckj} \le 0$.

To determine interaction forces in the side bearers when their gaps are closed, in work [8] a shock model was proposed, which assumes that the second derivative at the time of the strike, when inequality is fulfilled $\ddot{\delta}_{abyi}, \ddot{\delta}_{ckj} \le 0$, becoming zero, that is $\ddot{\delta}_{abyi} = 0, \ddot{\delta}_{ckj} = 0$.

Having differentiated expression (6) twice and having derived from equations (1, 2) $\ddot{\theta}_b$ and $\ddot{\theta}_{bl1}$, we obtain the value of reactions in the side bearers of a wagon when the gaps are being closed.

As an example, here are the analytical expressions for the forces that occur when the first and fourth gaps are being closed:

$$\begin{cases} R_{ck1} = X_2 \cdot R_{ck2}\left(\dfrac{B_2}{B_1 J_{xbl1}} + \dfrac{B_2}{B_1(m_b H^2 + J_{bx})}\right) - \dfrac{X_3 \cdot R_{ck3} B_3}{B_1(m_b H^2 + J_{bx})} + \dfrac{X_4 \cdot R_{ck4} B_4}{B_1(m_b H^2 + J_{bx})} + (A_1 - B_1)\cdot X_5, \\ \dots \\ R_{ck4} = X_3 \cdot R_{ck3}\left(\dfrac{B_3}{B_4 J_{xbl1}} + \dfrac{B_3}{B_4(m_b H^2 + J_{bx})}\right) + \dfrac{X_1 \cdot R_{ck1} B_1}{B_4(m_b H^2 + J_{bx})} - \dfrac{X_2 \cdot R_{ck2} B_2}{B_4(m_b H^2 + J_{bx})} - (A_4 - B_4)\cdot X_8, \end{cases} \quad (8)$$

where $$A_1 = \frac{\left(m_b H^2 + J_{bx}\right)\left(N_{z1}a_1 - N_{z2}a_2 + N_{z3}a_3 - N_{z4}a_4 + \sum_{j=1}^{5} P'_{fzj} b'_{fj} - \sum_{j=1}^{5} P''_{fzj} b''_{fj} - \sum_{j=1}^{7} P'_{fyj} h_{61} - \sum_{j=1}^{7} P''_{fyj} h_{bl1}\right)}{B_1 J_{xbl1}}, \quad (9)$$

$$B_1 = \frac{\left(m_b H \ddot{y}_b + R_{t1} e_{f1} + R_{t2} e_{f2}\right)}{B_1(m_b H^2 + J_{bx})}, \quad (10)$$

X_{1-8} – auxiliary coefficients; A_4, B_4 – coefficients are derived similarly to coefficients A_1, B_1.

Similarly, it is possible to determine the values of forces (reactions) that arise when the gaps are being closed in the axle-boxes of a wagon.

As noted above, the reactions in the side bearers and axle-boxes occur only when they are being closed, so in order to perform computer simulation of a freight wagon dynamics, the algorithm of their calculation should be fulfilled in the following steps (for example gaps in the side bearers):

- the value of the gaps is determined by the formulas (6) and if one of four gaps is closed, then the coefficient of X at its constituents A and B becomes equal to one, and the coefficients of X for the other reactions in the side bearers are zero;
- expression (8) defines the reaction in the closed side bearer;
- in case of closing of two gaps, the coefficients A and B at their constituents and their coefficients of X become equal to one, the other coefficients are equal to zero and the system of equations eventuates, from which the corresponding reactions in the side bearers are determined.

Since more than two gaps cannot be closed at the same time, then two of four equations (8) can be in the real situation.

It should be noted that when one or two gaps are being closed under the condition of relative movement of a body and bolsters, the friction forces T_{ckj} additionally arise and in equation (3) the value of coefficient X_i i becomes equal to one at these forces.

Therefore, the solution of the system of equations (8) will allow to evaluate how wagon movement dynamics is being affected by not only the forces that arise when the gaps are being closed in its side bearers, but also by the friction forces on their surface, which is especially relevant for curved sections of the track where their increased wear occurs.

Having done the research with the help of an advanced mathematical model of a freight wagon in Maple software environment, we give the graphs of the lateral oscillations of its body and bolster obtained during closed gaps in the side bearers.

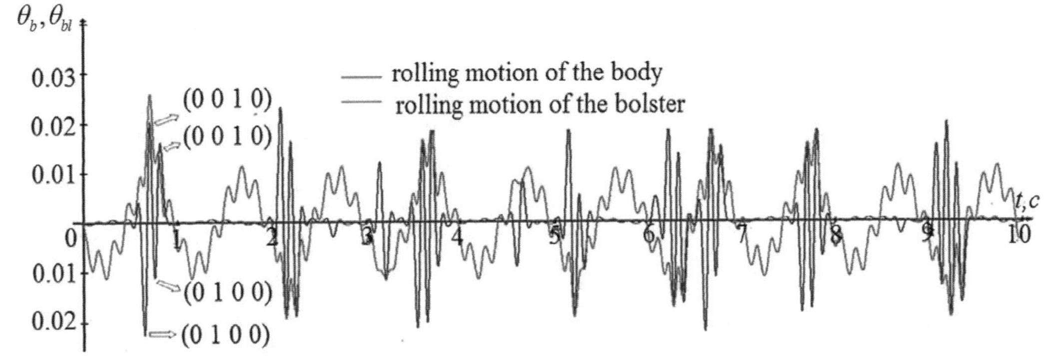

Figure 3. Lateral oscillations of a body and a bogie bolster when gaps are being closed in the slides.

As can be seen from Figure 3, the reactions occurring in the side bearers during their closure make a significant impact on the amplitude of a body and a bogie bolster movement. In this case, the moments of gaps closure can be traced from the matrix that reflects such a process ($\delta_{ck1}, \delta_{ck2}, \delta_{ck3}, \delta_{ck4}$) (Figure 3).

The similar algorithm and model can be used to determine the reactions that occur during gaps closure in the axle-boxes. However, unlike the gaps in the side bearers, when the gaps are being closed in the axle-boxes, the friction force between the axle-box and the side frame T_{pyi} becomes zero, since in this case they will move inseparably. Besides, in the system of differential equations (formula 4), the coefficients near the expressions, which determine the friction forces between the axle-box and the side frame T_{py1}, T_{py2}, which gap is depleted, must be equal zero, that is $X_{13}, X_{14} = 0$ (the case of two transverse gaps closure between the axle-box and the side frame).

Regarding the possibilities of the improved mathematical model of a freight wagon, it can be used for establishing causes of increased wear of the side bearers, side frames, axle-boxes, and in the study of the process of wheels of wheelsets rolling on the rail head with improper regulation of gaps under operational conditions.

Due to the fact that the main task in solving the problem of rolling stock safety is to find effective methods and tools that allow, with high probability, to describe the dynamic processes connected with rolling stock movement along the railway track. The proposed ways to improve mathematical models of freight wagons provide not only qualitative but also quantitative information about such processes and are adequate to real analogues of wagons movement in the real sections of the track.

5. Conclusions

1. The analysis of the reasons of rolling stock derailment on the railways of Ukraine over the past five years has revealed that in about 20% of cases they are obvious and in 7% are not obvious and implicit. The research of such cases in the course of an official investigation by the railway and during the performance of forensic examinations require the use of an advanced mathematical model of a freight wagon, which would allow quantify the influence of the parameters and malfunctions of

running gears of wagons on their derailment.

2. It was found that among the malfunctions of wagons' running gears, which caused their derailment, 39.4% were nonnormative wear of structural elements, and 29.3% – were nonnormative gaps in the axle-boxes and side bearers.

3. The main ways of improving the mathematical model of a freight wagon were proposed, namely, accounting when studying its dynamic behaviour of gaps in the axle-boxes and side bearers and the force, arising as a result of pressing the wheel flange to the rail head due to the exhaustion of the gap between them.

4. Performed research calculations with the use of the advanced mathematical model of a freight wagon showed that the lack of consideration of forces that arise during gaps closure in the slides, axle-boxes and guiding forces acting on the wheel-sets can lead to erroneous results when performing research in the course of forensic railway-transport expertises.

References

[1] Klimenko I V 2015 Development of theoretical foundations and methods for assessing and improving the safety of rolling stock of railways: diss. dr. tech. sciences: 05.22.07 pp 17–18
[2] Gracheva L O, Khamoev A D, Kosarev L N and Martynyuk A V 1995 *Reasons for derailment of rolling stock in trains and measures for the prevention of accidents and wrecks* p 93
[3] Kuzyshyn A Y, Batih A V, Kostritsa S A, Sobolevska J G, Kovalchuk V V, Dovhanyuk S S and Voznyak O M 2018 Research of safety indicators of diesel train movement with two-stage spring suspension *BulTrans: 10th Intern. Sci. Conf. on Aeronautics, Automotive and Railway Engineering and Technologies* 234-1–234-6
[4] Analysis of the railway traffic safety inthestructure of Ukrzaliznytsia for 2016-2019
[5] Korotenko M L 1974 Investigation of the stability of the movement of rail crews and the determination of their rational parameters: diss. dr. tech. sciences: 05.22.07 p 234
[6] Ushkalov V F, Reznikov L M and Ikkol V S 1989 *Mathematical modeling of rail vehicle vibrations* p 240
[7] Permyakov A A 2004 Carriage safety in curved sections of the track under various technical conditions of the car-track system: diss. dr. tech. sciences: 05.22.07 p 175
[8] Chan F T 2010 Dynamics of a freight wagon with nonlinear connections of a body with bogies: diss. dr. tech. sciences: 05.22.07 p 287
[9] *Dynamic modeling of Freight wagons* 2011 p 84
[10] Dos SantosG F M and R S Barbosa 2016 Safety analysis of a railway car under the periodic excitation from the track *Mechanical engineering* pp 1–12
[11] Sokol E N 2004 *Derailment and collisions of rolling stock (Forensic Expertise Elements of the theory and practice)* p 386

IOP Conf. Series: Materials Science and Engineering **985** (2020) 012009 doi:10.1088/1757-899X/985/1/012009

Mathematical model of a motor-grader movement in the process of performing working operations

V Shevchenko[1,3], **O Chaplyhina**[1,3,4], **I Pimonov**[1,3], **O Reznikov**[1,3] **and S Ponikarovska**[2,3]

[1]Department of construction and road machines
[2]Department of foreign languages
[3]Kharkiv National Automobile and Highway University, st. Yaroslava Mudrogo, 25, Kharkiv, 61002, Ukraine

[4]olexandrachaplygina@gmail.com

Abstract. The design features of a motor-grader help it perform a wide range of working operations, which puts it on a par with the main machines in the construction industry. Unlike other earth-moving machines, the motor-grader can perform work operations not only in a cyclic mode, but also in a continuous action mode. Considering these features, not only productivity should be addressed as the main indicator of the efficiency of this machine, but also the indicators of the quality of the working operations performed, in particular, the indicators of road-holding ability in the process of performing continuous working operations. The work has substantiated and developed a mathematical model of the motor-grader movement during working operations. This model makes it possible to predict the expected trajectory of the motor-grader on the basis of deterministic dependencies, which enables to assess the indicators of road-holding ability. The difference between the proposed model and the existing ones is the consideration of a two-stage mode of machine movement: plane motion at the first stage and rotational around the stop point at the second. When describing the external forces acting on the motor-grader, the differences in the formation of resistance forces on the blade during various working operations are taken into account.

1. Introduction

It is typical for a motor-grader to perform working operations with the main blade shifted or turned relative to the longitudinal axis. As a rule, this applies to operations for planning and profiling working surfaces, as well as the operations related to cutting out troughs or cuvettes. In such situations, external working resistances act on the machine, applied asymmetrically with respect to the longitudinal axis of the machine. In addition to them, in situations where the angle of the main blade capture differs from 90°, in addition to longitudinal resistance forces transverse forces also act from the side of the developed medium. The specified features of the formation of the force effect in the process of moving the motor-grader during a working operation can cause its deviation from the planned trajectory of movement. The consequence of this is the need for additional machine passes in the working face, which ultimately leads to a drop in the machine performance. In order to understand the processes of the motor-grader deviation from the planned trajectory of movement, it is necessary to substantiate, develop and study a mathematical model of the motor-grader movement in the process of performing working operations.

Content from this work may be used under the terms of the Creative Commons Attribution 3.0 licence. Any further distribution of this work must maintain attribution to the author(s) and the title of the work, journal citation and DOI.
Published under licence by IOP Publishing Ltd

2. Analysis of publications

The issues of modeling the movement of machines are most fully considered in the works of specialists in such sectors of the national economy as the automobile industry and tractor building, as well as in agricultural machine building. The main issues related to the dynamics of movement and road-holding ability of vehicles are considered in [1-3].

In these works, the dynamic processes are considered that are associated with turning the vehicles, performing maneuvers, movement on a support surface with a transverse slope. Destabilizing forces capable of violating the parameters of road-holding ability are inertial (centrifugal, etc.) forces, as well as the components of gravity when moving on transverse slopes.

The loss of road-holding ability of agricultural machinery is primarily associated with the action of additional lateral resistance forces on the working bodies (plows, cultivators), and is also caused by the design features of most wheeled tractors with articulated frames.

The development and analysis of mathematical models of the motion process for agricultural machines and articulated transport are considered in [4-6].

In the field of earthmoving engineering, the issues related to ensuring the road-holding ability of the machine have not been fully investigated, which is explained by the design variety of both the machines themselves and their working bodies. To date, the works of V.D. Musiyko, A.B. Koval and others are known in which the issues of road-holding ability of a trench excavator with a swinging working body are studied [7]. With regard to motor-graders, the issues of road-holding ability were considered in the works of L.V. Nazarov, A.V. Voronovich and others [8-9]. At the same time, a detailed analysis of the motor-grader road-holding ability in the process of performing various working operations has not been fully performed.

3. The purpose and objectives

The purpose of the article is to substantiate and develop a mathematical model of the motor-grader movement in the process of performing working operations.

To achieve this goal the following tasks must be solved:

- analysis and substantiation of the structure of the mathematical model based on exploratory experimental research;
- drawing up a generalized mathematical model of the motor-grader movement based on the generally accepted concepts of analytical and theoretical mechanics;
- substantiation and presentation in mathematical form of all components included in the mathematical model of the motor-grader movement in the process of performing working operations.

4. The main part

4.1. Substantiating the structure of the mathematical model based on the analysis of the search experiment results

To assess the features of generating the trajectory of the motor-grader movement when performing working operations in the conditions of the KhNAHU test site, an exploratory experimental study was prepared and carried out on the Dzk-251 motor-grader. During the experiments, the process of cutting the soil of category II by the angle of the main blade with the chips of constant cross-section was implemented. The initial speed of the motor-grader and the angle of the blade rotation in the plan, the depth of cut, etc. were chosen as the variable parameters [10]. The digging process was carried out from the sump hole.

The studies have shown that the reasons for the deviation of the real trajectory of the motor-grader motion from the planned one are: asymmetric application of an external load, an increase in the values of external resistances as the motor-grader moves in the result of accumulation of the soil prism in front of the blade, the effect of one-time dynamic loads due to an intensive growth of resistance at the initial stage of digging.

The experiments have shown that the trajectory of a motor-grader movement can be formed in various ways (Figure 1-3).

Figure 1. Single turn at the beginning of the cutting process.

Figure 2. Patch-rotary motion in the process of soil cutting.

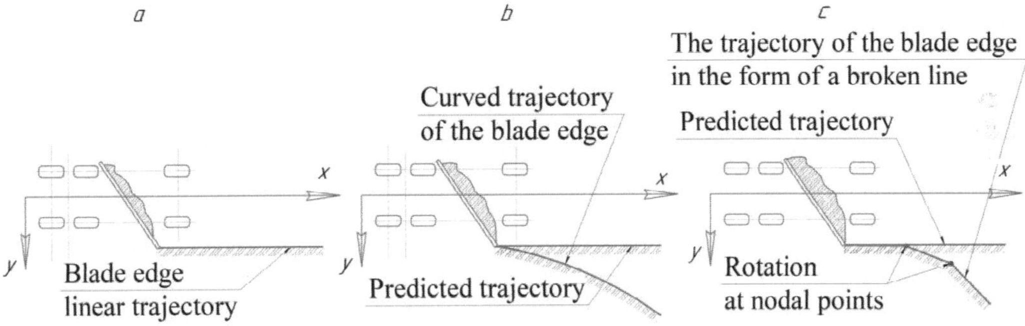

Figure 3. Types of trajectories of the motor-grader movement when performing working operations.

- the linear trajectory of the motor-grader movement (Figure 3, a) is formed in cases when the working resistances on the blade are relatively small. In this situation, the deviation of the real trajectory of the motor-grader motion from the planned one is not registered;
- in situations where the coefficient of the propellers' adhesion with the supporting surface is insignificant, the movement of the motor-grader is recorded when performing working operations along a curved trajectory (Figure 3, b);
- most often in the process of cutting the soil, a patch-linear trajectory of the motor-grader is recorded (Figure 3, c); at the same time, at the initial stage the machine moves straight-line, then, as external resistances grow, it slows down, turns around the blade stop point and then continues the straight-line movement.

The features of formation of the motor-grader movement trajectories in the process of performing working operations make it possible to determine the structure of a generalized mathematical model that describes this process:

- the mathematical model should include two systems of differential equations:
  - mathematical model of the plane motion of the machine in the support plane under the action of external forces;
  - mathematical model of the machine turning relative to the blade stop point under the influence of external forces;
- a generalized mathematical model should include the conditions for the transition from one type of movement to another and back again;
- the model should take into account the redistribution of the values of the support reactions on the driving and driven wheels, as well as the change in the coordinate of application and the magnitude of external resistance forces as a result of the accumulation of the soil prism in front of the blade.

4.2. Development of a mathematical model of the motor-grader movement

The dynamic design diagrams of a motor-grader for the process of performing working operations, corresponding to various types of movement, are shown in Figure 4,5.

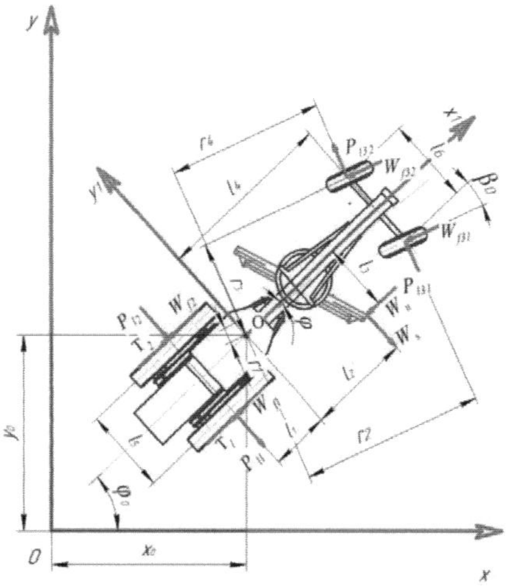

Figure 4. Design diagram of a motor-grader for the stage of plane motion.

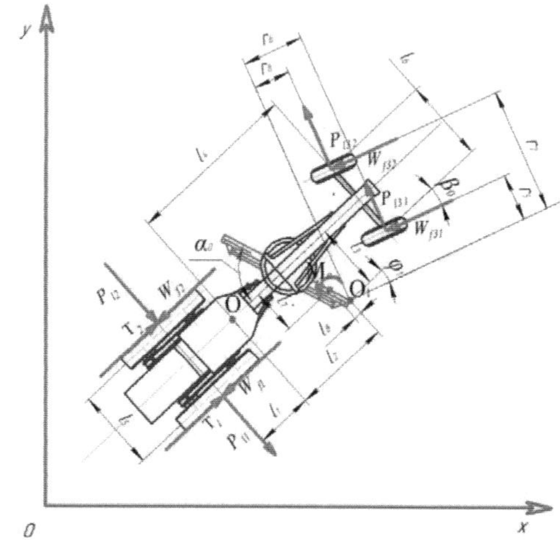

Figure 5. Design diagram of a motor-grader for the stage of turning around the blade stop point.

Since the object of research is the trajectory of a motor-grader movement, the problem of synthesizing a mathematical model is reduced to analyzing the motion of a free body relative to a fixed coordinate system *x0y*. According to Shal's theorem, any motion of a body can be represented as a combination of translational motion of an arbitrary point of the body, called a pole, and rotational motion of the body around the same point [11] To simplify calculations, it is recommended to choose the center of the body mass as a pole (*p.O*, Figure 4). In this case, based on the well-known laws of dynamics, the motion of the body is described by a system of differential equations [11]:

$$\begin{cases} m\ddot{x} = (T_1 + T_2)\cdot\cos\varphi - (W_{f1} + W_{f2})\cdot\cos\varphi + (P_{l1} + P_{l2})\cdot\sin\varphi - W_h\cdot\cos\varphi + \\ +W_l\cdot\sin\varphi - (W_{f31} + W_{f32})\cdot\cos(\beta_0 + \varphi) - (P_{l31} + P_{l32})\cdot\sin(\beta_0 + \varphi); \\ m\ddot{y} = (T_1 + T_2)\cdot\sin\varphi - (W_{f1} + W_{f2})\cdot\sin\varphi - (P_{l1} + P_{l2})\cdot\cos\varphi - W_{hz}\cdot\sin\varphi - \\ -W_l\cdot\cos\varphi - (W_{f31} + W_{f32})\cdot\sin(\beta_0 + \varphi) + (P_{l31} + P_{l32})\cdot\cos(\beta_0 + \varphi); \\ I\ddot{\varphi} = (T_2 - T_1 + W_{f1} - W_{f2})\cdot\dfrac{l_5}{2} - (P_{l1} + P_{l2})\cdot l_1 + W_h\cdot l_3 + W_l\cdot l_2 - W_{f31}\cdot r_1 - \\ -P_{l31}\cdot r_2 - W_{f32}\cdot r_3 - P_{l32}\cdot r_4. \end{cases} , \quad (1)$$

where x, y, φ are the generalized coordinates for the conditions of the motor-grader plane motion; m is the mass of the grader; I is the moment of inertia of the motor-grader relative to the center of mass; T_1, T_2 is the tractive effort on the right and left leading sides; N_1, N_2 are the support reactions on the corresponding sides of the balance carriage; N_{31}, N_{32} are the support reactions on the front driven wheels; W_{f1} is the resistance to rolling of the right leading side; W_{f1} is the resistance to rolling of the left leading side; W_{f2} is the resistance to rolling of the front driving wheels (equal to each other since the front axle is hinged to the center beam and has the ability to freely rotate in the vertical plane at an angle of ±20°); P_{l1} is the lateral force on the right leading side; P_{l2} is the lateral force on the left leading side; P_{l31}, P_{l32} is the lateral force on the right and left wheels of the front axle; W_h is the horizontal component of resistance to digging; W_l is the lateral component of digging resistance.

When performing working operations with a motor-grader, it is possible to lock the blade by increasing working resistance forces (cutting resistance of the developed medium and resistance to the movement of the prism). In this case, the experiments recorded the rotation of the motor-grader relative to the blade stop point. It is noteworthy that a change in the nature of the machine's motion causes a change in both the dynamic scheme and the equations of motion. The equation of motion of the motor-grader during the turn around the blade stop point has the following form [12]:

$$I_{O1}\ddot{\varphi}_1 = \left(T_1 - W_{f1}\right)\cdot(l_3 - \frac{l_5}{2}) + \left(T_2 - W_{f2}\right)\cdot(l_3 + \frac{l_5}{2}) - (P_{l1} + P_{l2})\cdot(l_1 + l_2) - \\ -W_{f31}\cdot r_5 - P_{l31}\cdot r_6 - W_{f32}\cdot r_7 - P_{l32}\cdot r_8 - M_{res}, \tag{2}$$

where M_{res} is the moment of resistance forces from the side of the developed soil.

Formally, the condition for the transition from the first stage of movement to the second is the equality of the sum of the driving forces projections along the longitudinal axis of the motor-grader and the sum of the projections of the resistance forces along the same axis:

$$\sum P_{dr} \le \sum P_{res}\,. \tag{3}$$

Based on the dynamic scheme (Figure. 4), this condition takes the form

$$T_1 + T_2 + m\cdot\sqrt{\ddot{x}^2 + \ddot{y}^2}\cdot\cos\left(arctg\frac{\ddot{y}}{\ddot{x}}\right) \le W_{f1} + W_{f2} + W_h + (W_{f31} + W_{f32})\cdot\cos\beta_0\,. \tag{4}$$

It is noteworthy that if during the transition to the second stage of movement the blade remains blocked (the motor-grader cannot rotate relative to the stop point), the experiments recorded the development of uncontrolled oscillations of the motor-grader in the longitudinal plane, which have an increasing amplitude. Such loading modes are dangerous and can lead to the destruction of the metal structure of both the working equipment and the main frame of the machine.

4.3. Substantiation and determination of analytical dependences of forces that form the process of a motor-grader moving

Support reactions on the driving and driven wheels determine the magnitude of most of the forces acting in the considered dynamic system. Since the operating speeds of the motor-grader when performing working operations are not high (from 0.7 m/sec to 1.5 m/sec), the inertial forces can be neglected in determining the support reactions due to their smallness. In this case, based on the analysis of the static equilibrium of the machine, taking into account the vertical force $W_v = \pm k_2 W_h$ acting on the blade from the soil side, and also considering the possible separation of individual wheels from the support surface, we obtain:

$$\begin{cases} N_{31} = N_{32} = \dfrac{m\cdot g - (\pm k_2\cdot W_h)\cdot(l_1 - l_2)}{2(l_1 + l_4)} & \text{at} \quad N_{31} + N_{32} \ge 0; \\ N_{31} = N_{32} = 0 & \text{at} \quad N_{31} + N_{32} < 0. \end{cases} \tag{5}$$

$$\begin{cases} N_2 = \dfrac{\left[mg - (N_{31} + N_{32})\right]\cdot\frac{l_5}{2} + (\pm k_2\cdot W_h)\cdot(l_3 - \frac{l_5}{2})}{2(l_1 + l_4)} & \text{at} \quad N_2 \ge 0; \\ N_2 = 0 & \text{at} \quad N_2 < 0. \end{cases} \tag{6}$$

$$\begin{cases} N_1 = mg - (N_{31} + N_{32}) - N_2 - (\pm k_2\cdot W_h) & \text{at} \quad N_1 \ge 0; \\ N_1 = 0 & \text{at} \quad N_1 < 0. \end{cases} \tag{7}$$

In the presented dependences g is the acceleration of gravity; k_2 is the empirical coefficient, adjustable in the range from 0.1 to 0.3. The sign in front of the coefficient determines the direction of the vertical reaction vector from the soil side.

Almost all modern motor-graders are equipped with overpowered engines. The experiments show that even during a sharp increase in the level of drag forces on the blade, the number of engine crank shaft rotations decreases by less than 5%. At the same time, the engine has enough power to put the cars in the mode of 100% slipping of the driving wheels. Considering this fact, in order to determine the values of the traction forces on the driving wheels, the known dependences proposed by Professor N.A. Ulyanov, can be converted to the form [10, 11]:

$$\begin{cases} T_1 = N_1 \cdot \varphi_{ad} \cdot (1 - a\dot{x}_1 - b\dot{x}_1^5), \\ T_2 = N_1 \cdot \varphi_{ad} \cdot (1 - a\dot{x}_1 - b\dot{x}_1^5), \\ \dot{x}_1 = \sqrt{\dot{x}^2 + \dot{y}^2} \cdot \cos\left[arctg\dfrac{\dot{y}}{\dot{x}} - \varphi \right], \end{cases} \qquad (8)$$

where φ_{ad} is the coefficient of the wheel adhesion with the supporting surface; a, b are the empirical coefficients.

To determine the rolling resistance forces, the resistance coefficient f is taken into account, which depends on the type of the supporting surface:

$$W_{f1} = f \cdot N_1; \quad W_{f2} = f \cdot N_2; \quad W_{f31} = f \cdot N_{31}; \quad W_{f32} = f \cdot N_{32}. \qquad (9)$$

To describe the forces of external resistance acting on the blade on the part of the developed medium, it is advisable to use the well-known theories of cutting and digging soils. In particular, based on the principle of superposition, the horizontal W_h and lateral W_l components of the main resistance vector can be represented as the sums:

$$W_h = W_c + W_{pr} + W_{up} + W_{al}, \qquad (10)$$

$$W_l = W_c^l + W_{pr}^l + W_{up}^l + W_{al}^l, \qquad (11)$$

where W_c, W_c^l are the longitudinal and transverse components of the soil cutting resistance; W_{pr}, W_{pr}^l are the longitudinal and transverse components of resistance to the movement of the soil prism in front of the blade; W_{up}, W_{up}^l are the longitudinal and transverse components of resistance to the movement of the cut off soil chips up the blade; W_{al}, W_{al}^l are the longitudinal and transverse components of resistance to the movement of the soil prism along the blade.

Analytical dependencies for determining all these forces are given in Table 1.

The following letters are used in the formulas: k is the specific coefficient of soil cutting resistance; F is the cross-sectional area of the cut off chips; φ_α is the coefficient depending on the value of the cutting angle; x_A, y_A are the displacements along the corresponding axes of the point of application of the external resistances main vector; δ_s is the soil density in natural bedding; k_{loos} is the coefficient of soil loosening; V_{pr} is the volume of the soil prism in front of the blade; μ_1 is the coefficient of internal friction of the soil; μ_2 is the coefficient of external soil friction.

Table 1. Analytical dependencies for determining the components of the main vector of resistance on the part of the developed medium.

Plane motion of the motor-grader. Scheme Figure 3		Rotation of the motor-grader around the blade stop point Diagram Figure 4
Horizontal force components	Lateral force components	
$W_c = k\cdot F\cdot\varphi_\alpha\cdot\sin(\alpha_0 + arctg\frac{y_A}{x_A})$	$W_c^l = k\cdot F\cdot\varphi_\alpha\cdot \cdot\cos\left(\alpha_0 + arctg\frac{y_A}{x_A}\right)$	$W_c = k\cdot F\cdot\varphi_\alpha$
$W_{pr} = \frac{\delta_s}{k_{loos}}\cdot g\cdot V_{pr}\cdot\mu_1\cdot \cdot\sin(\alpha_0 + arctg\frac{y_A}{x_A})$	$W_{pr}^l = \frac{\delta_s}{k_{loos}}\cdot g\cdot V_{pr}\cdot\mu_1\cdot \cdot\cos\left(\alpha_0 + arctg\frac{y_A}{x_A}\right)$	$W_{pr} = \frac{\delta_s}{k_{loos}}\cdot g\cdot V_{pr}\cdot\mu_1$
$W_{up} = \frac{\delta_s}{k_{loos}}\cdot g\cdot V_{pr}\cdot\mu_2\cdot\cos^2\delta\cdot \cdot\sin(\alpha_0 + arctg\frac{y_A}{x_A})$	$W_{up}^l = \frac{\delta_s}{k_{loos}}\cdot g\cdot V_{pr}\cdot\mu_2\cdot\cos^2\delta\cdot \cdot\cos\left(\alpha_0 + arctg\frac{y_A}{x_A}\right)$	$W_{up} = \frac{\delta_s}{k_{loos}}\cdot g\cdot V_{pr}\cdot\mu_2\cdot\cos^2\delta$
$W_{al} = \frac{\delta_s}{k_{loos}}\cdot g\cdot V_{pr}\cdot\mu_1\cdot\mu_2\cdot \cdot\cos(\alpha_0 + arctg\frac{y_A}{x_A})$	$W_{al}^l = \frac{\delta_s}{k_{loos}}\cdot g\cdot V_{pr}\cdot\mu_1\cdot\mu_2\cdot \cdot\sin\left(\alpha_0 + arctg\frac{y_A}{x_A}\right)$	-

The analysis of the methods to perform working operations by a motor-grader showed that the values of these forces significantly depend on the cross-sectional area of the cut off soil chips. The values of this parameter for different ways of developing the medium are given in Table 2.

The volume of the soil prism that is formed in front of the blade is variable at the initial stage of digging, after which, upon reaching its limiting value V_{ult}, it remains constant due to the soil loss into the side rolls. To determine this parameter, an analytical dependence of the form is offered:

$$V_{pr} = \begin{cases} F\int_0^t \sqrt{(\dot{x}_A)^2 + (\dot{y}_A)^2}\,dt & \text{at } V_{pr} \le V_{\text{ult}} \\ V_{\text{ult}} & \text{at } V_{pr} > V_{\text{ult}} \end{cases}. \quad (12)$$

Lateral forces $P_{l1}, P_{l2}, P_{l3}, P_{l4}$ determine the ability to keep the motor-grader on the planned trajectory of movement during work operations. The nonlinearity of the lateral force characteristics on the wheel is explained by the fact that with forces lower than the lateral slip forces, the wheel experiences elastic deformation, otherwise it is displaced along the supporting surface. The general dependence for the lateral holding force on the i – th wheel is as follows:

$$P_l = \begin{cases} C_{el}\cdot\Delta S_i + \lambda_{el}\cdot\dot{S}_i\cdot sign(\dot{S}_i) & \text{at } C_{el}\cdot\Delta S_i + \lambda_{el}\cdot\dot{S}_i\cdot sign(\dot{S}_i) \ll N_i\cdot\varphi_{ad.l} \\ N_i\cdot\varphi_{ad.l} & \text{at } C_{el}\cdot\Delta S_i + \lambda_{el}\cdot\dot{S}_i\cdot sign(\dot{S}_i) \ge N_i\cdot\varphi_{ad.l} \end{cases}. \quad (13)$$

Table 2. Analytical dependencies for determining the components of the main vector of resistance on the part of the developed medium.

Soil digging pattern	Chip cross-sectional area, F	Moment of resistance on the part of the developed medium
Digging with the full width of the blade 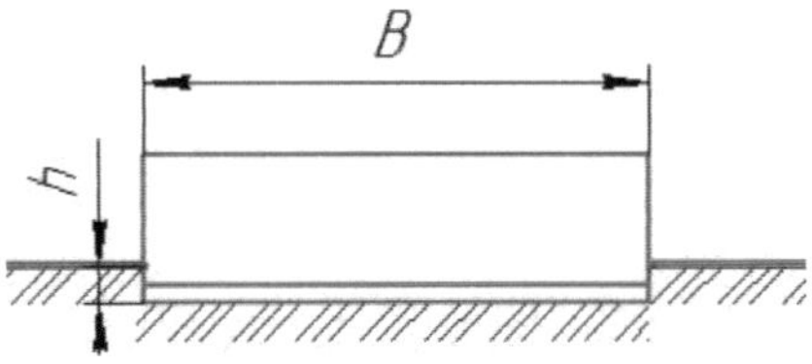	$F = B \cdot h$	$M_{res} = \frac{B}{4} \cdot (W_c + W_{pr} + W_{up})$, $W_c = \frac{1}{2} \cdot B \cdot h \cdot k \cdot \varphi_\alpha$, $V_{pr} = \begin{cases} \frac{1}{2} V_{pr1} + \frac{B^2 \cdot h \cdot \varphi_1}{8} & \text{at } V_{pr} \le \frac{1}{2} V_{\text{ult}} \\ \frac{1}{2} V_{\text{ult}} & \text{at } V_{pr} > \frac{1}{2} V_{\text{ult}} \end{cases}$.
Linear digging pattern 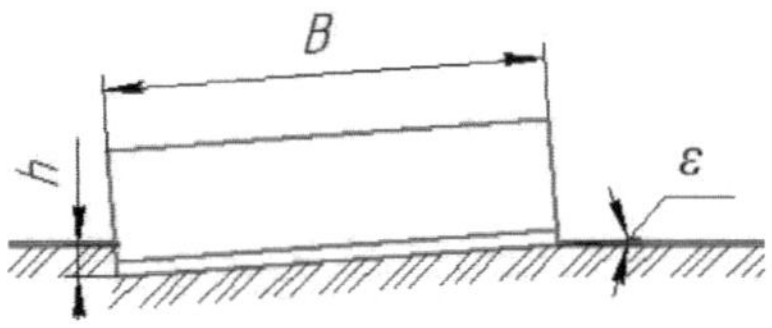	$F = \frac{B \cdot h}{2 \cos \varepsilon}$	$M_{res} = \left(\frac{W_c + W_{up}}{3} + \frac{W_{pr}}{2} \cdot \frac{V_{pr}}{V_{\text{ult}2}} \right) \cdot \left(\frac{B}{2} \cdot \pm b + \frac{l_3}{\sin \alpha_0} \right)$, $W_c = \frac{1}{2} \left(\frac{B}{2} \pm b + \frac{l_3}{\sin \alpha} \right) \cdot k \cdot \varphi_\alpha \cdot tg\varepsilon$, $V_{pr} = \begin{cases} V_{pr1} \cdot k_1 + \frac{1}{6} \cdot \left(\frac{B}{2} \pm b + \frac{l_3}{\sin \alpha} \right)^3 \cdot \varphi_1 & \text{at } V_{pr} \le V_{\text{ult}2} \\ V_{\text{ult}2} & \text{at } V_{pr} > V_{\text{ult}2} \end{cases}$
Digging with blade angle 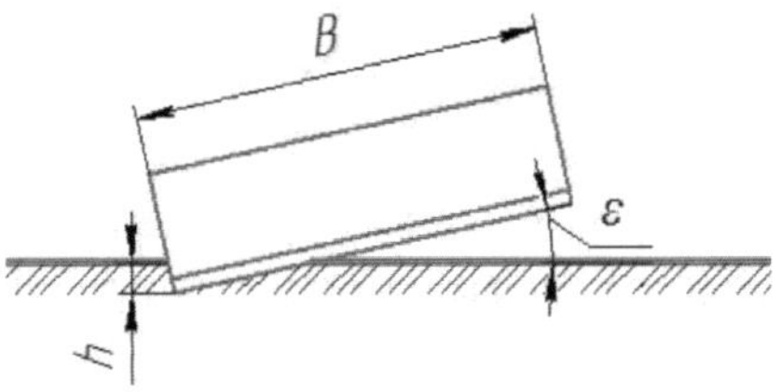	$F = \frac{h^2}{\sin 2\varepsilon}$	$M_{res} = \left(W_c + W_{pr} + W_{up} \right) \cdot \frac{h}{3} \cdot \left(\frac{1}{tg\varepsilon} - \frac{2}{3tg2\varepsilon} \right)$, $W_c = k \cdot F \cdot \varphi_\alpha = k \cdot \varphi_\alpha \left[\frac{h^2}{2} \cdot tg\varepsilon \left(\frac{1}{tg\varepsilon} - \frac{2}{3tg2\varepsilon} \right)^2 \right]$, $V_{pr} = \begin{cases} V_{pr1} \cdot k_1 + \frac{h^3 \cdot tg\varepsilon}{6} \cdot \left(\frac{1}{tg\varepsilon} - \frac{2}{3tg2\varepsilon} \right)^3 \cdot \varphi_1 & \text{at } V_{pr} \le V_{\text{ult}2} \\ V_{\text{ult}2} & \text{at } V_{pr} > V_{\text{ult}2} \end{cases}$

In the given dependence: C_{el} is the coefficient of the wheel elasticity in the lateral direction; λ_{el} is the coefficient of the wheel damping in the lateral direction; $\varphi_{ad.l}$ is the coefficient of adhesion of the wheel with the support surface in the lateral direction; ΔS_i is the increment of displacement of the $i-$th wheel in the lateral direction; $\dot{S}_i$ is the speed of displacement of the $i-$th wheel in the lateral direction; N_i is the support reaction on the $i-$th wheel.

Changing the mode of the motor-grader motion from plane to rotational relative to the blade stop point leads to a change in the structure of resistance forces on the part of the developed medium. The studies carried out enabled to record the analytical dependences to determine the value of the main moment of external resistances acting on the blade (Tables 1, 2).

Thus, the mathematical model of the motor-grader movement on a plane support surface in the process of performing working operations is fully specified.

5. Conclusion

The analysis of the conducted experimental studies showed that in the process of performing working operations, at least three options of the motor-grader movement are possible: along a straight-line trajectory without deviations from the planned one; along a curved trajectory simultaneously with lateral sliding and forward movement; along a patch-linear trajectory, when the machine moves straight ahead, then stops, independently turns in place and continues moving straight-line again.

All three variants of motion can be adequately described using a dynamic diagram, which includes two groups of second-degree differential equations of motion. The first group of equations describes the plane motion of the machine along a horizontal support plane. The second equation describes the rotational motion of the machine on the support plane around the blade stop point. The transition from one variant of motion to another is carried out as a result of the analysis of an additional condition in the form of an inequality that determines the balance between the driving forces and the external resistance forces.

The values of the force factors included in the equations of movement depend on the type of working operation being performed and are described using analytical dependencies that are different from each other. The developed mathematical model of the motor-grader movement takes into account all the specified features.

References

[1] Podrigalo M, Turenko A, Bogomolov V, Klets D, Sergiyenko O, Karpenko V, Gritsuk I V, Turenko O, Korobko A, Bulgakov N and Boboshko O 2018 Increase of Stability for Motor Cars in Service Braking *SAE Technical Paper* **2018-01-1880**

[2] Klets D, Podrigalo M, Kholodov M, Klimenko V, Rudzinskyi V and Kholodov A 2019 Analysis of the Tractor-Trailer Dynamics during Braking *SAE Technical Paper* **2019-01-2144**

[3] Podrigalo M.A. and Karpenko, V.A. 2001 Nonuniformity of vertical reactions on the automobile wheels and its stability in braking *Avtomobil'naya Promyshlennost* **2** pp 1921

[4] Podrigalo M., Getsovich E, Artiomov N and Holodov M 2019 Study of the tractor's driving wheels slipping during moving on a deformed surface *Bulletin of Sumy National Agrarian University. The series: Mechanization and Automation of Production Processes* **3** 37 pp 13–18

[5] Chieh C. and M. Tomizuka 1995 Dynamic modeling of articulated vehicles for automated highway systems *Proceedings of 1995 American Control Conference* (Seattle, Publisher: IEEE) pp 653–657.

[6] Melnyk V, Antoshchenkov R and Antoshchenkov V 2020 Determination of Mobile Machine Wheel Dynamics *Control and Signal Processing Applications for Mobile and Aerial Robotic Systems* pp 125

[7] Koval A 2014 Definition of the conditions assuring the course stability of the multipurpose earth-moving machine *Dissertation* (Dnipropetrovsk: Publisher MP Lesya) **21**

[8] Nazarov L V, Istomin V P, Ovsyannikov S I, Razaryonov L V, Gingera A I and Lesthenko A V 2007 Theoretical investigations into the dynamics of compact pneumatic frontal loaders under self-induced vibrations *Bulletin of Kharkov National Automobile and Highway University* **38** pp 13–17

[9] Voronovich A 2007 Improvement of motor-graders with mass 15-16 t. by means of completeness with power modules of heightened reliability *Dissertation* (Kharkiv: Publisher KhNAHU) **22**

[10] Chaplygina O, Shevchenko V and Beztsennaya Zh 2015 Methods to determine measures providing a motor-grader road-holding ability *Machines, Technologies, Materials* **12/2015**, pp 78–83

[11] Markeev A P 1990 Theoretical mechanics (Moscow: Publisher Science) **416**

[12] Torok J 2000 Analytical mechanics with an introduction to dynamical systems (New York: Publisher John Wiley & Sons) **373**

IOP Conf. Series: Materials Science and Engineering **985** (2020) 012010 doi:10.1088/1757-899X/985/1/012010

Research of dynamic indicators and influence of different types of rolling stock on railway track

A O Shvets[1,3,4], O M Bolotov[2,3], A K Percevoj[2,3], V V Ghlukhov[2,3], O O Bolotov[2,3] and L S Saparova[2,3]

[1]Department «Theoretical and Structural Mechanics»
[2]Department «Branch Research Laboratory of The Dynamics and Strength of Rolling Stock»
[3]Dnipro National University of Railway Transport named after Academician V. Lazaryan, Lazaryan St., 2, Dnipro, Ukraine, 49010

[4]angela_Shvets@ua.fm

Abstract. The technical level of rolling stock of railway transport directly affects the economic indicators of the transport industry and the economy of the country as a whole, which leads to the need for improvement of the control, quantitative assessment of the dynamic loading of rolling stock to ensure safe and reliable communication at the railways. Therefore, in the process of designing and operating the rolling stock the quantitative assessment of dynamic loadings is a relevant scientific and technical problem. The purpose of this work is to study the influence of different types of rolling stock, taking into account the speed of movement on their main dynamic indicators and indicators of interaction of track and rolling stock. The basis of the research is the method of mathematical and computer simulation of the dynamic loading of a freight car using the model of spatial variations of the coupling of five cars and a software complex developed in the branch research laboratory of dynamics and strength of rolling stock (BRL DSRS). Theoretical studies have been carried out under condition of movement of several types of freight cars: a gondola car of the model 12-532, a hopper car for coal transportation of the model 12-4034 and a flat car with a long cargo of the model 13-401 and typical bogies 18-100 with the speeds from 50 to 90 km/h in the curves with radii 350 and 600 m, with superelevation 130 and 120 mm, correspondingly.

1. Introduction

Among the main directions of activity of the railway industry are the following: the development of high-speed trains; increased traffic safety; development of a new rolling stock and modernization of the existing fleet. The safety of freight trains, the values of permissible speeds and load capacity, the costs for maintenance of rolling stock and track facilities, increase of the runs between repairs essentially depend on the design of the freight rolling stock of the railways [1-3].

Increasing the speed of railway carriages will allow strengthening the integration processes between countries, satisfying the demand of the population in transportation, reducing the electricity and resources consumption, using the production capacity of the leading industries. However, this leads to the need for improvement of the control, quantitative assessment of the dynamic loading of rolling stock to ensure safe and reliable communication at the railways. Therefore, in the design process of rolling stock quantitative assessment of dynamic loadings is a relevant scientific and technical problem. Operation of the morally obsolete rolling stock with low dynamic properties is one

Content from this work may be used under the terms of the Creative Commons Attribution 3.0 licence. Any further distribution of this work must maintain attribution to the author(s) and the title of the work, journal citation and DOI.
Published under licence by IOP Publishing Ltd

of the reasons for insufficient safety level of train traffic and high operating costs due to increased repairs, as well as increased energy consumption for traction of trains [4-6].

The purpose of this work is to study the influence of various types of rolling stock, taking into account the loading mode and the movement speed on their main dynamic indicators and indicators of interaction of the track and rolling stock.

2. Methodology

During operation of trains, primarily the ones with increased length and weight, particular attention is paid to the assessment of dynamic performance of carriages, among which the most important is the indicator characterizing the traffic safety of the carriage – the derailment stability coefficient of the wheel. In order to do this, the mathematical models of spatial oscillations of the car (or group of cars) moving in a train are used. The models of the rolling stock, as a rule, represent the variations of the system of differential equations in the second-order partial derivatives compiled according to the Lagrange–d'Alembert principle, which correspond to the set task. Using such models, the problem can be found analytically. Mathematical modeling makes it possible to determine the dynamic indicators of cars during their movement along the straight and curved track sections with real irregularities in the vertical and horizontal planes, taking into account the actual rolling surface of the wheel and the rail head profile [7-9].

The oscillations of a single car and its interaction with the railway track are considered on sufficiently full calculation schemes. Figure 1 presents the calculation scheme of a freight cat and shows positive directions for all displacements and rotation angles [10-12].

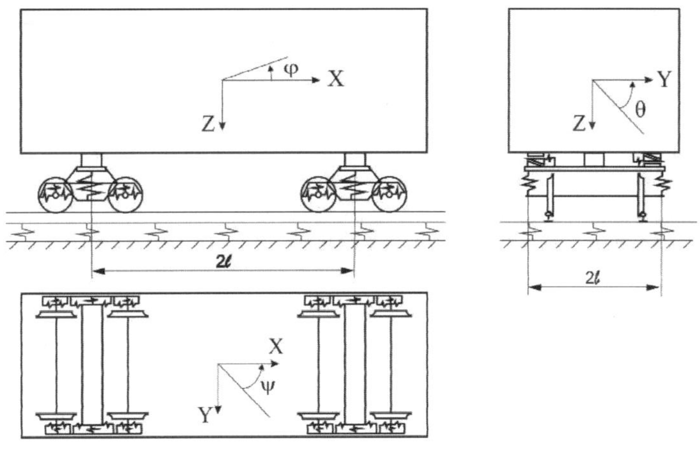

Figure 1. Calculation scheme of 4-axle gondola car.

The work [12] proposes a mathematical model describing the spatial oscillations of the car coupling in a train (Figure 2), one rail carriage of which is considered according to the fullest calculation scheme (called "zero"), and the calculation schemes of neighboring cars, depending on the task setting, are simplified as further is the distance from the "zero" carriage on both sides. A mechanical system with 58 degrees of freedom is taken as a calculation scheme of the "zero" carriage. The cars, adjacent to the "zero" one, are represented by a system with 12 degrees of freedom. In the calculation schemes describing the oscillations of these cars, the main features of freight car bogies are preserved – the side frames lozenging.

During the study of spatial oscillations of cars adjacent to the "zero" one, considered by the simplified calculation scheme, the following assumptions are introduced. It is assumed that the cars have single-stage spring suspension. Each of them consists of eleven solid bodies: a body, two bolsters, four side frames of a bogie and four-wheel sets. Unlike "zero" car, the track under adjacent cars is considered to be absolutely rigid in the vertical direction and elastic in the horizontal transverse direction. This assumption does not lead to increase in the number of degrees of freedom, since the speed of rails displacement in the expressions for transverse forces can be neglected. Outer cars of coupling, these cars are the systems with six degrees of freedom.

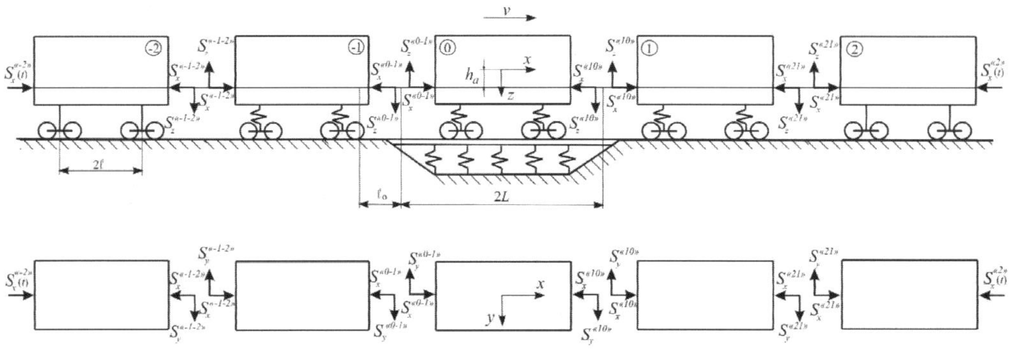

Figure 2. The forces arising from the action of longitudinal forces in the auto-coupling of cars.

The complement of the mathematical models of spatial oscillations by the initial data with the specified inertial characteristics of the car and cargo elements makes it possible to approximate the results of calculations to the real state of objects and thereby increase the objectivity of mathematical and computer simulation. It was for solving this problem that the "Software complex for determining the moments of inertia of car bodies" was created [11, 13].

Several types of freight cars are considered: gondola car of the model 12-532, hopper car for coal transportation of the model 12-4034 and flat car with a long cargo of the model 13-401 on typical bogies 18-100 under different loading conditions. In this case, the weight of the cars, the inertia moments of the body I_x, I_y, I_z and the height of the mass center of the body h above the level of rail heads (LRH) vary. Preparatory calculations, which are given in Table 1, are carried out with the help of "Software complex for determining the moments of inertia of car bodies".

Table 1. Inertia and geometric characteristics of cars at different loading modes.

Parameters	Designations	Name of cars		
		Gondola car of the model 12-532	Hopper for coal of the model 12-4034	Flat car of the model 13-401
Car base	$2L$, m	8.66	7.84	9.72
Car weight	M_o, t	76.5 (13.3)*	88.34 (18.72)	63.6 (13.6)
Inertia moments	I_x, t·m^2	75 (20)	160 (33.6)	22 (13.5)
	I_y, t·m^2	1050 (300)	1190 (300)	1194 (240)
	I_z, t·m^2	1100 (300)	1245 (307)	1223 (260)
Mass center height above the LRH	h, m	1.843 (1.6)	2.7 (1.77)	2.1 (1.1)

* – in brackets the values of cars` parameters in the empty mode are specified

3. Calculation results

Calculations can be made with sufficient accuracy for practice, limited to consideration of the movement of a group of five cars (Figure 2). The basis of the research is the method of mathematical simulation using the model of spatial oscillations of the coupling of five cars and a software complex developed by the BRL DSRS of the Dnipro National University of Railway Transport named after Academician V. Lazaryan. Theoretical studies have been carried out under the condition of movement of several types of freight cars: gondola car of the model 12-532, hopper car for coal transportation of the model 12-4034 and flat car with a long cargo of the model 13-401 on typical bogies 18-100 with speeds in the range from 50 to 90 km/h along the curves with radii 350 and 600 m, with superelevation 130 and 120 mm, correspondingly. The rails – P65, wooden sleepers, broken stone ballast [13, 14].

In this study, the influence of several types of freight rolling stock on the railway track at different loading modes was considered. The graphs of changes of indicators under analysis when moving in the curved track sections of $R = 600$ m and 350 m are presented in Figure 3-6.

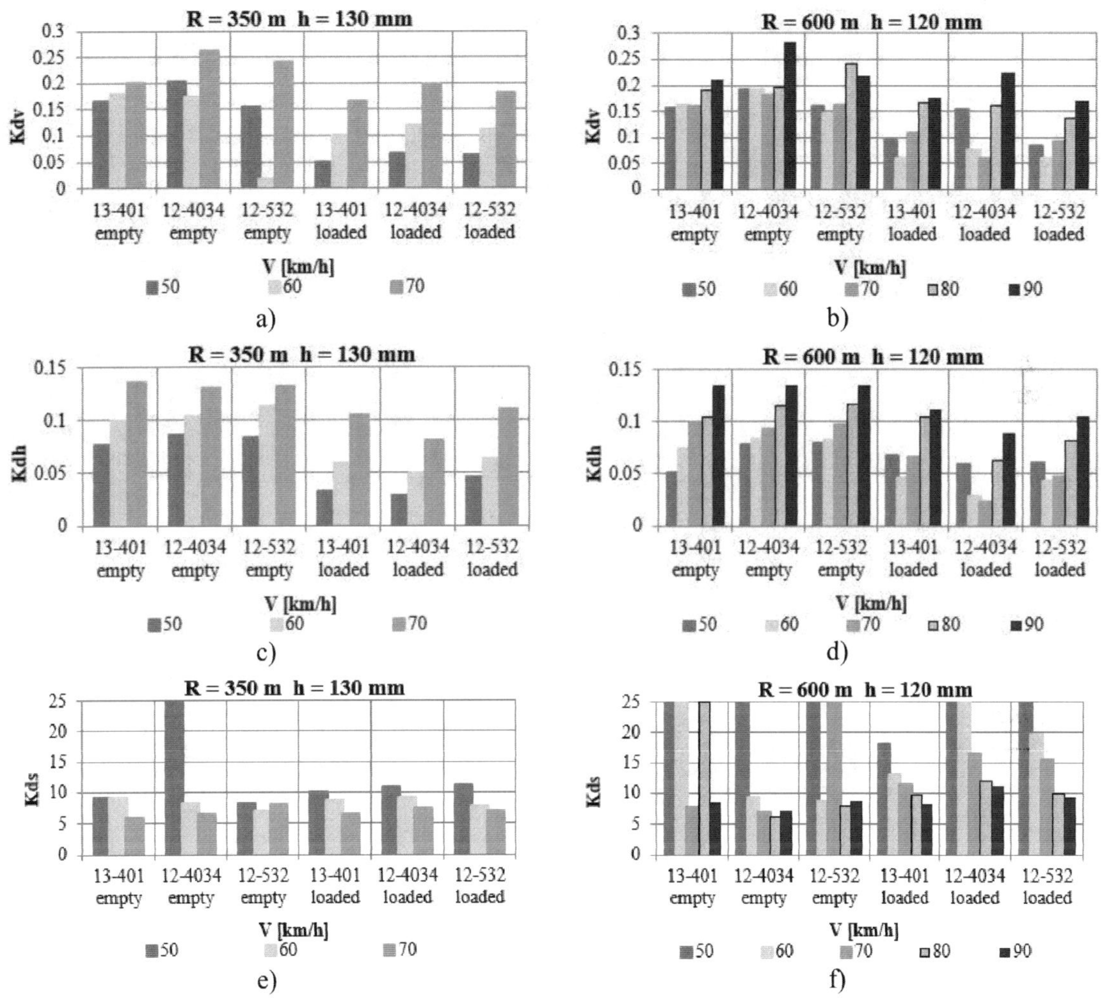

Figure 3. Dependence graphs on the loading mode when moving in the corresponding curve (a, b) coefficients of vertical dynamics; (c, d) coefficients of horizontal dynamics; (e, f) derailment stability coefficient.

As one can see from Figure 3 (a, b), in general, the coefficients of vertical dynamics increase. Thus, in the whole range of investigated speeds, the indicators K_{dv} in both loading modes do not exceed the permissible norm both in the curve of $R = 600$ m and in the curve of $R = 350$ m. The level of assessment in both curves corresponds to the "excellent" one. Only a gondola car in the curve of $R = 350$ m at the speed of 60 km/h has a much better vertical dynamic. Figure 3 (e, f), presents the coefficients of horizontal dynamics K_{dh} when moving in curves with a radius $R = 350$ m and 600 m, correspondingly. One can see that with increasing speed and loading mode, the coefficients of horizontal dynamics K_{dh} increase and remain in the curve of $R = 600$ m and in the curve $R = 350$ m at the excellent level of assessment [14]. The loaded hopper car has the best coefficients of horizontal dynamics in both curves.

The coefficients of the derailment stability in the curve of $R = 350$ m (Figure 3, d) have little dependence on the loading mode, in contrast to the curve of $R = 600$ m (Figure 3, f), but in both cases do not exceed the minimum permissible value determined by the norms [1]. It follows from the obtained results, that the loaded hopper car in case of increase in the movement speed has higher indicators of coefficients K_{ds}, which can be explained by the increased car weight. In the curve of $R = 350$ m, the coefficient of derailment stability for hopper car at the speed of 50 km/h is the greatest. Flat car has the smallest indicators K_{ds}, which is explained by the large car base, which worsens curve negotiation.

Figure 4 presents the coefficients of vertical K_{vdt} and horizontal dynamics K_{hdt} of the track for the forces of interaction of wheels and rails, as well as the displacement stability coefficient of track panel when moving in the corresponding curved track sections.

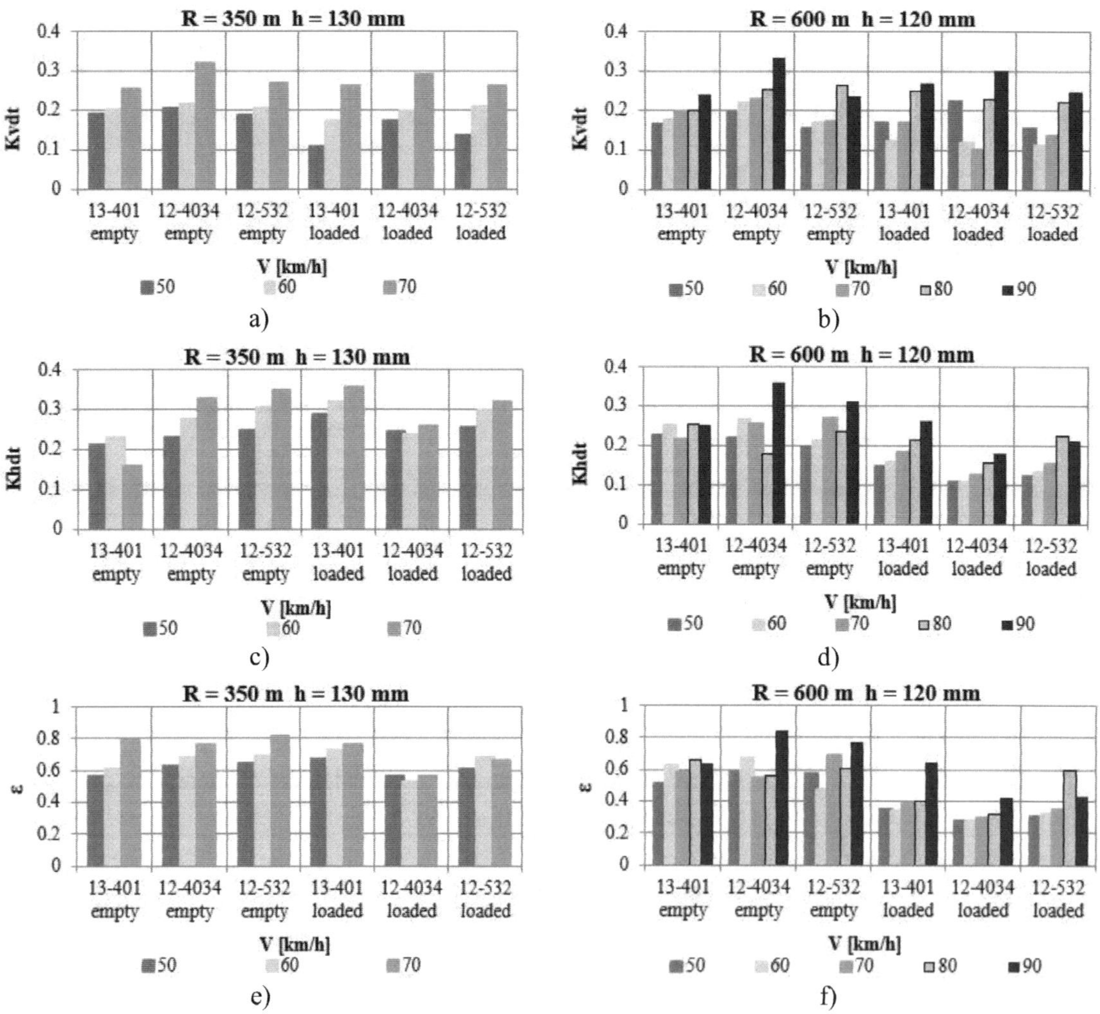

Figure 4. Graphs of dependence on the loading mode when moving in the corresponding curve: (a, b) coefficient of vertical dynamics of the track for the forces of interaction of wheels and rails; (c, d) coefficient of horizontal dynamics of the track for the forces of interaction of wheels and rails; (e, f) displacement stability coefficient of the track panel.

IOP Conf. Series: Materials Science and Engineering **985** (2020) 012010 doi:10.1088/1757-899X/985/1/012010

Permissible value of the vertical dynamics coefficient of the track K_{vdt} is calculated in accordance with the permissible dynamic loading per unit length on the railway track from the group of axles of the bogie 168 kN/m and for these types of rolling stock on the bogies of the model 18-100 is $[K_{vdt}]=0.45$ [1]. The coefficient of vertical dynamics of the track K_{vdt} in Figure 4 (a, b) does not exceed the permissible value in curves with radius $R=350$ m and 600 m.

The coefficient of horizontal dynamics of the track K_{hdt} in Figure 4 (d, e), which is considered a safety criterion for the displacement of track panel, does not exceed the permissible value $[K_{hdt}]=0.4$ in the curves with a radius $R=350$ m and 600 m.

The value of the stability coefficient of the track panel ε (Figure 4, c, d) in the track with broken stone ballast is much less than the permissible value. One should take $[\varepsilon]$ for the track with broken stone ballast at normal speeds of movement $[\varepsilon]=0.85$ [14].

Figure 5 shows the influence of the speed on the indicators of interaction with the rolling stock in the curves with $R=350$ m and $R=600$ m, correspondingly – the side force acting from the track on the wheel, the edge stress at the rail base, the wear factor of the side edge of the wheel tread.

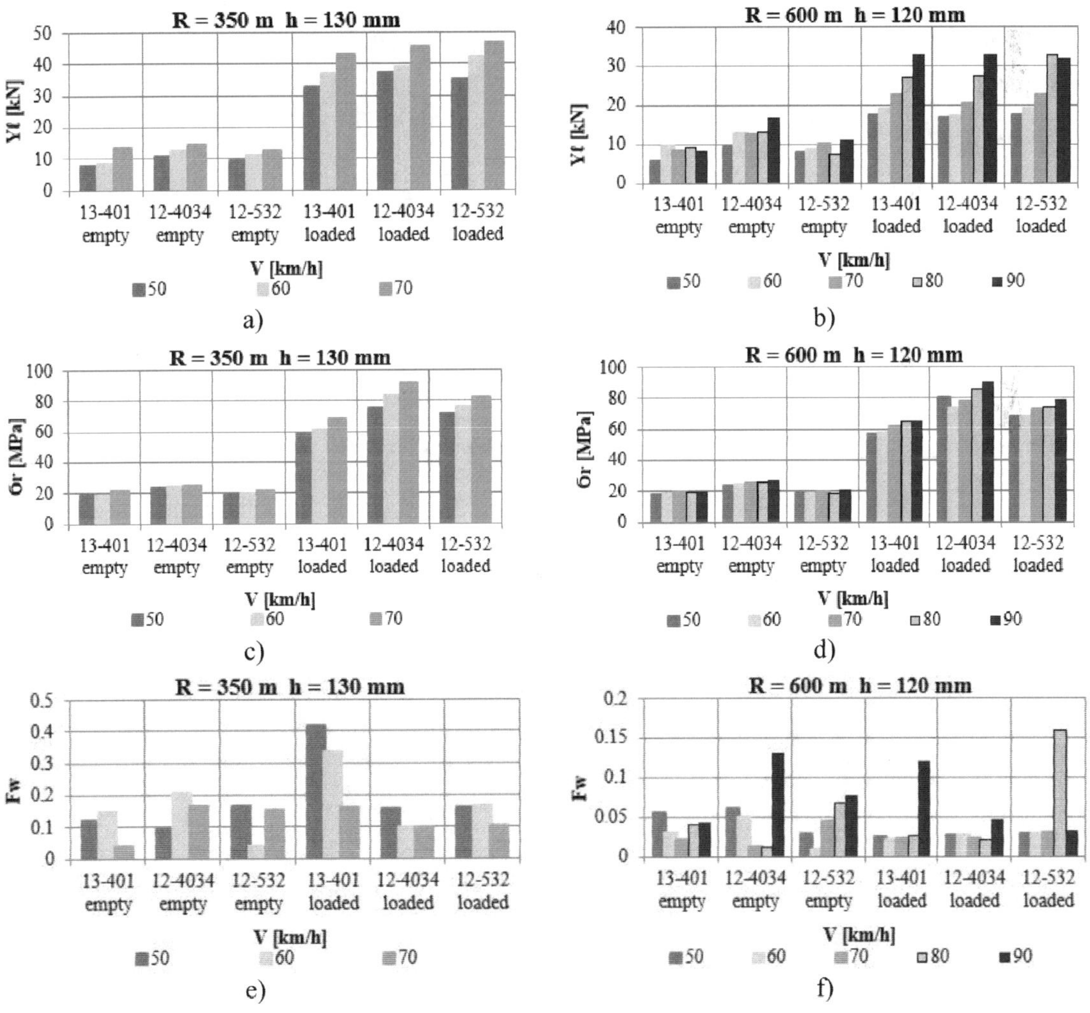

Figure 5. Graphs of dependence on the loading mode when moving in the corresponding curve: (a, b) side force acting from the track side on the wheel; (c, d) edge stress at the rail base; (e, f) the wear factor of the side edge of the wheel tread.

The side forces acting from the track on the wheel (horizontal forces) Y_L (Figure 5, a, b) increase and, in comparison with the permissible values of 100 kN, have no excess. The values Y_L for the gondola in both curves are on average higher than the corresponding values for the flat car and hopper.

The dynamic influence of rolling stock on the track increases with increasing the speeds of train movement, and, as a result, the stresses at the rail edges increase (Figure 5, c, d). The maximum stresses occurring at the edges of the rail bases are used as a criterion for establishing the permissible speeds and should not exceed 240 MPa for the track with rails P65 [14].

According to the results of calculations, the edge stresses increase with increasing the movement speed and do not exceed the permissible values for both loading modes in curves with a radius of $R = 350$ m and 600 m. The largest edge stresses arise when a hopper car moves in both empty and loaded state.

In the curve of $R = 350$ m at the speed of movement 50-60 km/h, the wear factor of the side edge of the wheel tread of the loaded flat car increases significantly. At the speed of 80 km/h, the wear factor of the side edge of the wheel tread F_w significantly increases for a gondola in the curve of $R = 600$ m. In turn, there is a significant F_w increase at the speed of 90 km/h in the curve $R = 600$ m for an empty hopper and a loaded flat car.

The wear factor of the side edge of the wheel tread F_w is determined as a characteristic equal to the product of the guiding force Y_d at the angle of hunting (climbing) ψ_{ws} of the wheel on the rail. Figure 6 shows the influence of speed on the indicators of rolling stock interaction with the track in the curves of $R = 350$ m and $R = 600$ m, correspondingly, guiding force acting on the wheel from the track side on the wheel and the wheel set hunting.

Guiding forces acting on the wheel Y_d from the track (Figure 6, a, b), with increasing the movement speed, significantly increase in the curve of $R = 350$ m and, on average, have identical values for all types of rolling stock. The values Y_d in the curve of $R = 600$ m at the speeds of 80-90 km/h vary significantly for all types of studied rolling stock.

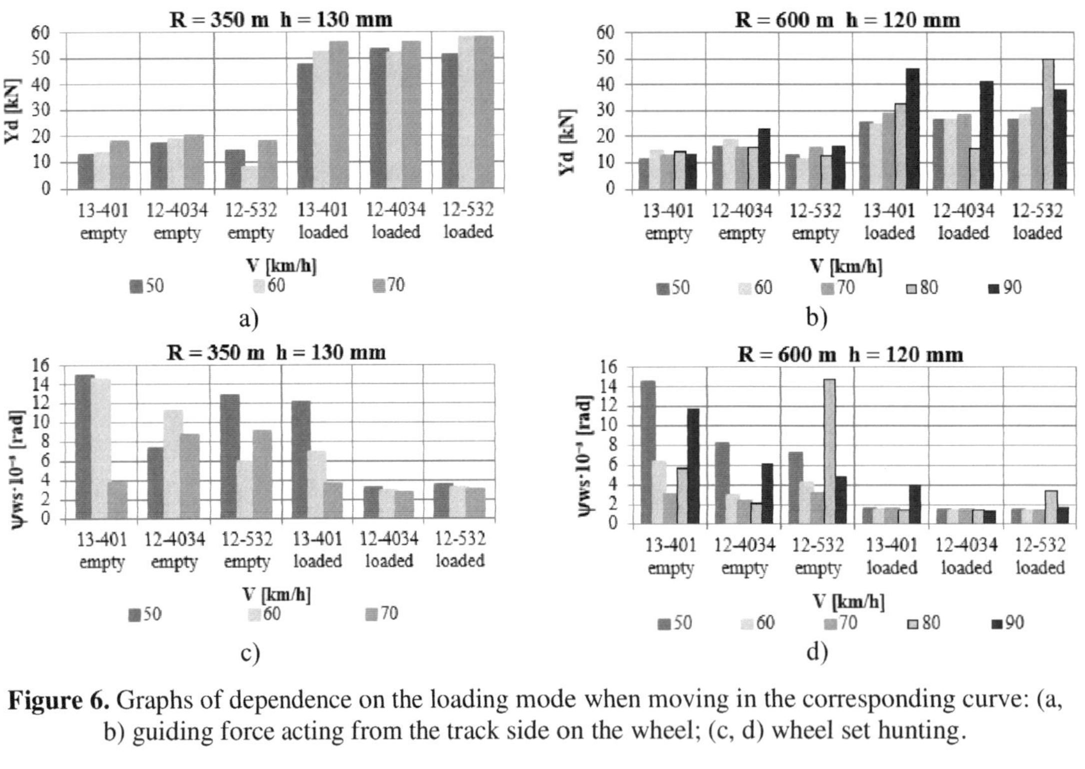

Figure 6. Graphs of dependence on the loading mode when moving in the corresponding curve: (a, b) guiding force acting from the track side on the wheel; (c, d) wheel set hunting.

The results of calculations show (Figure 6, c, d) that in the curves of $R = 350$ m and $R = 600$ m in the empty mode, at the speeds up to 90 km/h, a flat car has the largest angle of wheel set hunting ψ_{ws}. At the speeds of 80 km/h in the curves of the average radius $R = 600$ m, the values ψ_{ws} increase for the gondola and significantly differ from the other range of the investigated speeds. For gondola-cars on the bogies of the model 18-100, the speed of 80 km/h is critical and associated with the loss of movement stability, when the dynamic transverse oscillations of the hunting of the car parts stop damping, becoming steady (self-oscillations) [15].

4. Conclusions

In papers presents an analysis of theoretical studies of dynamic qualities of various types of rolling stock; the calculations were performed with the use of a package of applied programs. Based on the study, the following conclusions can be made:

- in all range of speeds the indicators K_{dv} and K_{dh} in both loading modes increase and do not exceed the permissible norm both in the curve of $R = 600$ m and in the curve of $R = 350$ m. The gondola in the curve of $R = 350$ m at the speed of 60 km/h has considerably better vertical dynamics, and the loaded hopper car has better coefficients of the horizontal dynamics in both curves;
- the derailment stability coefficients of the wheels in the curve have little dependence on the loading mode, in contrast to the curve of $R = 600$ m, but in both cases they do not exceed the minimum permissible value determined by the norms. The loaded hopper car, in case of the speed increase, has higher indicators of the coefficients K_{ds} which can be explained by the increased car weight. In the curve of $R = 350$ m, the derailment stability coefficient of the wheels for hopper at the speed of 50 km/h is the greatest. Flat car has the smallest indicators K_{ds} which is explained by the large car base that worsens curve negotiation;
- the coefficients of the vertical K_{vdt} and horizontal dynamics K_{hdt} of the track do not exceed the permissible value in the curves with radius of $R = 350$ m and 600 m. The loaded flat car has the smallest values of the track panel`s stability coefficient ε in the track with broken stone ballast;
- side forces Y_L increase linearly and have no excess, and the values Y_L for gondola cars in both curves on average are larger than the corresponding values for the flat car and hopper;
- edge stresses increase with increasing speed and do not exceed the permissible values for both modes of loading in the curves with radius of $R = 350$ m and 600 m. The largest edge stresses arise when hopper car moves in both empty and loaded condition;
- the wear factor of the wheel tread`s side edge of the loaded flat car significantly increases in the curve of $R = 350$ m at the speed of 50-60 km/h. The wear factor $F_W$ of the wheel tread`s side edge significantly increases for gondola car in the curve of $R = 600$ m at the speed of 80 km/h. In turn, there is a significant F_W increase for the empty hopper and the loaded flat car at the speed of 90 km/h in the curve of $R = 600$;
- guiding forces acting on the wheel from the track side Y_d significantly increase with the speed increasing in the curve of $R = 350$ m and have on average equal values for all types of rolling stock. The values Y_d in the curve of $R = 600$ m at the speeds of 80-90 km/h vary considerably for all types of the studied rolling stock;
- flat car has the largest angles of wheel set hunting ψ_{ws} in the curves of $R = 350$ m and $R = 600$ m in the empty mode at the speeds up to 90 km/h. At the speeds of 80 km/h in the curves of the average radius $R = 600$ m, the values ψ_{ws} increase for the gondola car and significantly differ from the other range of the studied speeds.

The paper presents an analysis of theoretical studies of dynamic qualities of various types of rolling stock; the calculations were performed with the use of a package of applied programs.

The obtained results have a practical orientation. During the theoretical studies and after the simulation, taking into account the oscillation processes of freight car, the dependences of the main dynamic indicators and interaction indicators of rolling stock and track, taking into account the speed and loading mode, were obtained. Application of the obtained results will increase the traffic safety of freight cars and will improve the technical and economic performance of railway transport.

References

[1] Pshinko O, Ursulyak L, Kostrytsia S, Fedorov Ye and Shvets A 2019 The influence of the «train-track» system parameters on the maximum longitudinal forces' level *Transport Problems* **4** 14 pp 161–172
[2] Kurhan M B and Kurhan D M Providing the Railway Transit Traffic Ukraine–European Union *Pollack Periodica* **14** 2 pp 27–38
[3] McKinnon A C 2016 Freight Transport Deceleration: Its Possible Contribution to the Decarbonisation of Logistics *Transport Reviews* **36** 4 pp 418–436
[4] Kostritsa S A, Sobolevska Y H, Kuzyshyn A Y and Batih A V 2018 Mathematical model of DPKR-2 dyzel train car *Science and Transport Progress* **1** 73 pp 56–65
[5] Shvets A O 2020 Stability of freight cars under the action of compressive longitudinal forces *Science and Transport Progress* **1** 85 pp 119–137
[6] Lunys O, Neduzha L and Tatarinova V 2019 Stability research of the main-line locomotive movement *Proc. of the 23rd Int. Sci. Conf. Transport Means 2019* pt III (Palanga, Lithuania: Kaunas Univ. of Techn.) pp 1341–1345
[7] Kurhan D 2016 Determination of Load for Quasi-static Calculations of Railway Track Stress-strain State *Acta Technica Jaurinensis* **9** 1 pp 83–96
[8] Ursulyak L V and Shvets A A 2017 Improvement of Mathematical Models for Estimation of Train Dynamics *Science and Transport Progress* **6** 72 pp 67–79
[9] Klimenko I, Černiauskaite L, Neduzha L and Ochkasov O 2018. Mathematical Simulation of Spatial Oscillations of the "Underframe-Track" System Interaction *Proc. of 12th Int. Conf. ITELMS'2018* (Panevėžysm, Lithuania: Kaunas Univ. of Techn.) pp 105–114
[10] Myamlin S, Neduzha L and Urbutis Ž 2016 Research of Innovations of Diesel Locomotives and Bogies *Proc. Eng.* **134** (Vilnius, Lithuania: Vilnius Gediminas Techn. Univ.) pp 469–474
[11] Shvets A O and Bolotov O O 2019 Influence of loads from the axis of a gondola car on its dynamic indicators and railroad tracks *Science and Transport Progress* **1** 79 pp 151–166
[12] Danovich V D and Malysheva A A 1998 Mathematical model of spatial oscillations of the coupling of five cars moving along a rectilinear section of the track *Transport. Stress loading and durability of a rolling stock* (Dnepropetrovsk: Dnipro National University of Railway Transport named after Academician V. Lazaryan) pp 62–69
[13] Shatunov O V and Shvets A O 2019 Study of dynamic indicators of flat wagon with load centre shift *Science and Transport Progress* **2** 80 pp 127–143
[14] Danilenko E I 2010 *Railway track* [*Zaliznychna kolija*] (Kiev: Inpres) p 456
[15] Lazaryan V A, Dlugach L A and Korotenko M L 1972 *Ustoychivost dvizheniya relsovykh ekipazhey* (Kiev: Naukova dumka) p 197

Problems of the use of renewable energy sources in the structure of railway power supply

Oleksandr Ostapchuk[1,5], Mykola Kuznietsov[2], Valeriy Kuznetsov[3] and Vitaliy Kuznetsov[4]

[1]Associate Professor at the Department of Renewable Energy National Technical University of Ukraine «Igor Sikorsky Kyiv Polytechnic Institute» 37, ave. Peremohy, Kyiv, Ukraine, 03056
[2]Head of Department Department № 1 Integrated Power Systems Institute of Renewable Energy of the National Academy the Sciences of Ukraine, 20-a, st. Gnata Hotkevycha, Kyiv, Ukraine, 02094
[3]Professor at Electric Power Department, Railway Research Institute, 50, Chlopickiego str., Warsaw, Poland, 04-275
[4]Associate Professor at the Department of the electrical engineering and electromechanic, National metallurgical academy of Ukraine, 4, Gagarina ave., Dnipro, Ukraine

[5]O.Ostapchuk@kpi.ua

Abstract. The paper presents the main indicators of the renewable energy sources in the railway transport structure of European countries. A new structure of railway power supply systems, energy storage systems and control algorithms under the conditions of stochastic mode of renewable energy sources operation are defined as the main problematic issues. Further directions for sustainable development of such systems and factors that influence the modes of operation of traction substations and electric networks were identified. Average power and a random component are two criteria that were used to build local systems. The paper presents the real schedules of daily load on the traction substation terminals and solar insolation with determined power distribution changes. The power deviations indicators depending on the time interval and the type of generation were determined. A normal law of distribution of a random component was determined as a result of all calculations. It is possible to determine the shortage or excess of electric power in the local system in a certain area and to plan the direction of its development in the near future by analyzing the nature of the determined average power change.

1. Introduction

According to the latest statistical data on the level of power generation in Europe – the best dynamic is in renewable energy sources (RES) [1]. Their market share was 34.6% in the total energy production by 2019. Providing further stable increase of their share can be achieved by expansion of the field of use. One of the promising directions of use of distributed generation sources (DGS) is in the railway transport [2]. The main problem of further development of such systems is to evaluate the possibility of using 100% renewable energy for different transport systems [3]. The development of a new hierarchy structure of traction power supply based on "smart grid" technologies [4], Electric Storage System [5] and control algorithms in terms of the stochastic nature of RES pertain to problem issues [6]. A number

Content from this work may be used under the terms of the Creative Commons Attribution 3.0 licence. Any further distribution of this work must maintain attribution to the author(s) and the title of the work, journal citation and DOI.
Published under licence by IOP Publishing Ltd

of issues characterizing the impact of DGS by connecting electricity to the overall structure of railways been considered, namely, the degree of influence of solar generation in operation traction substations and stability of the traction power systems, system's stability reserve (level of robustness) [7] and the work of relay protection [8, 9]. However, the amount of research on development strategies for certain railway structures remains insufficient, leading to a complete lack of regulatory documents. In order to improve this situation, it is necessary to analyze the differences related to the characteristics of certain traction grids and renewable energy systems and to develop strategic directions for its further development. When analyzing the structure of railways traction grids, their geographical location must be considered, with further justification of possible integration with the systems of distributed generation, introduction of a comprehensive assessment of the impact of DGSs on the modes of operation of traction substations and electric networks. Further development and improvement of load simulation methods and DRG characteristics, modes and decision-making are needed. They allow to take into account the multicriteria nature of these tasks and the inherent uncertainty of information.

2. The basic material

The key feature of Ukraine's renewable energy system is using of solar and wind power (approximately 90% of total renewable power generation) [10]. The peculiarity of these systems is the stochastic nature, which depends on the time of day or the season and weather conditions. This circumstance can lead to a negative impact on the grid operation, including the stability of dynamic processes in power systems and supervisory control. Since the work of strong wind and solar power stations in the national grid has legal regulations, the simplest way would be to integrate these into the overall power network. Railways power supply thus would not depend directly on the renewables as their stochastic nature would be compensated by the total power capacity. However, the significant part of electricity consumption for the needs of transport, railways dispersion and large neighboring areas, it would be appropriate to consider independent renewables usage. In this case, wind or solar power stations, together with the consumer (railway transport structural element) can be considered as the local power grid. It can be defined as a combination of power-limited generators, power converters and load interconnected according to the network topology with the same electromagnetic processes. The following condition should be carried out:

$$\sum P_{load}(t) = \sum P_{gen}(t). \tag{1}$$

In this case we should consider variable generation and consumption of energy as a single process. However, the proportion of energy consumed from the global network decreases, while the balancing demand for local network increases. Minimization of the uncontrolled spread of energy balance as the difference between generated and consumed energy is the natural demand in the construction of the electrical network based on RES. Since the stationer component of generation and consumption can be predicted and planned by the work schedule, there is a need to evaluate the random component of power fluctuations. Actual data about generation and consumption in form of time series can be the source for this evaluation.

Actual data about wind speed, solar insolation and energy consumption is the information for the study and for real mathematical modeling. Work schedule for alternating current traction substation (Poltava South) in winter period and weather data related to the central region of Ukraine for the same season were used in this example. The assumption that the short-term (less than one hour) fluctuations in generation and consumption are independent random variables is proven by authors. Analysis of statistical data shows that operation of WES and SES have some average power and a random component, which are a function of time. The average power can be considered as well-predictable for the short terms and controlled setting. The expression for determination of power balance will be [11]:

$$p_{ij} = (a_{ij} - a_i) - [(w_{ij} - w_i) + (s_{sj} - s_i)], \tag{2}$$

where a_x – energy consumption; w_x and s_x – WES and SES power respectively; i – time index (time interval number); j – day number, p_{ij} – deviations from the load schedule.

One-index RES indicators are averaged for a certain hour of the day (called daily variation). a_i corresponds to the planned schedule of consumption, while $p_i = a_i - w_i - s_i$ corresponds to the planned so-called "pure" load. The length of time interval selection (Δt) is determined by the characteristics of energy sources – for wind farms10 minutes interval is usually taken. For SES – 1 minute. Short-term load changes can have a seconds-scale dimension. Available weather data make it possible to assess the capacity of a typical wind and solar power plants in conventional units, or the fate of the nominal units (maximum achievable) capacity. The same way the load can be described. Examples of typical modes of power change is shown on Figure 1, 2.

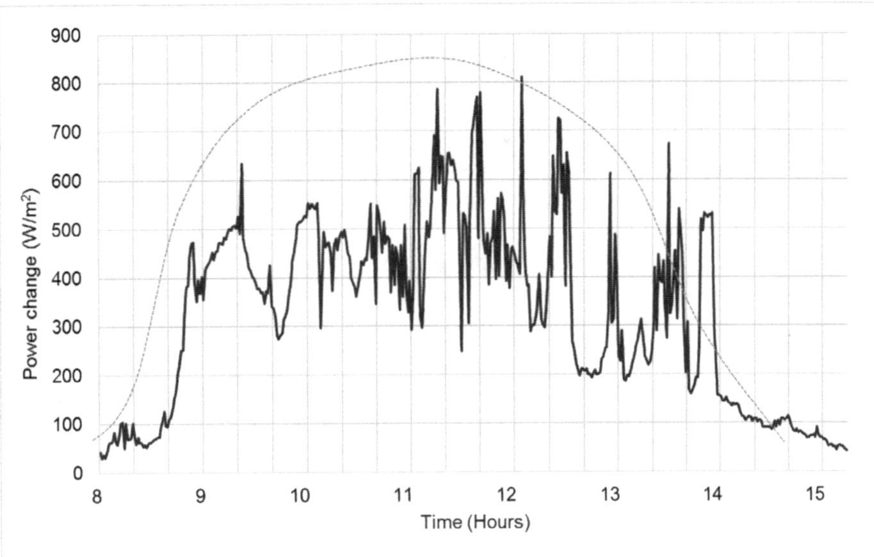

Figure 1. The diurnal variation of solar radiation (the maximum attainable and actual).

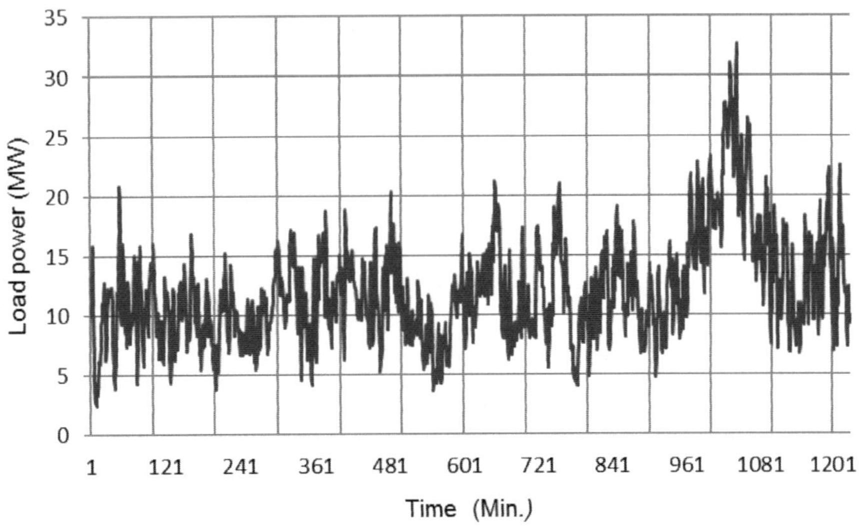

Figure 2. The load power (time interval – 1 min).

Selection of an elementary time interval determined by the properties of an object, such as rate changes of specific parameter. When comparing process of different variability, they must be given for the same time period. The change of interval length causes a change of variance values of the time series. When averaging consolidation scope is used rule that the population variance is the sum of interval mean and variance interval variance: $\sigma^2 = \sigma_0^2 + M\{\sigma_k^2\}$, where σ_0^2 – variance of average values at each interval, σ_k^2 - variance on the k-th interval M - symbol of mathematical expectation. It should be borne in mind that the number of intervals must provide a valid statistical error [12] and as a general population can be considered a long process in which its basic properties are sufficiently manifested.

Since RES's critical factor is the balancing process variability, differentiated (in finite differences) series, elements of which are the difference of successive values of power: $dp_{ij} = p_{(i+1)j} - p_{ij}$ is considered. Such a series can be built for a total power balance as well as for the generation and consumption: da_{ij}, dw_{ij}, ds_{ij}. Distribution of random variables in rows corresponds to the received hypothesis of normality. Since the received series of successive changes (fluctuations) power may vary considerably, it is necessary to remove cases of extreme emissions, for example to evaluate the range of possible values with confidence probability of 0.99 and 0.95, respectively rejecting one and five percent of the sample. Analysis of the actual power fluctuations in conventional units while changing the scatter range for different network parts, increasing time interval Δt and giving in the form of nominal power (extreme, confidence intervals, standard deviations σ_d) is shown in Figure 3. From the comparison of variability is clear that with the time's interval the fluctuation increases as well. Nonlinear dispersion growth is limited from the above and changes significantly depending on the size of the confidence interval and significantly reduces after some extreme values rejected. Proper load changes for the same time intervals have greater relative values than RES power change (Figure 3, the horizontal scale is nonlinear).

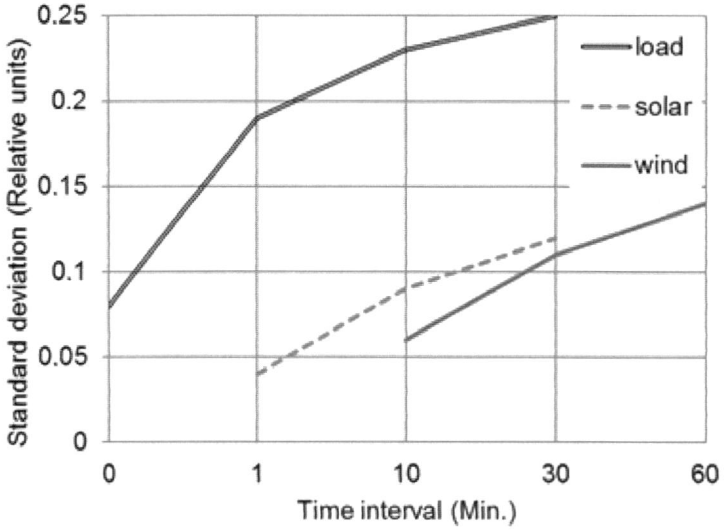

Figure 3. The standard deviation of changes in power depending on the time interval.

Current variances of local power grid components have different relative values; the total variance of power balance must be defined as the sum of covariation components including their weights. Difference series variance that characterizes the rate of power balance change defines similarly:

$$D\{dp_{ij}\} = D\{da_{ij}\} + wD\{dw_{ij}\} + sD\{ds_{ij}\}, \tag{3}$$

where w, s – weights that determine the share of wind and solar power plants respectively. The mutual independent power fluctuations are taken into account.

For example, if the nominal capacity of WES and SES are equal and are equal to the maximum load power, the normalized standard deviation of fluctuations in the power range for averaging interval of 10 minutes will increase from 0.23 (without RES) to 0.25. Similarly, in the averaging interval of 30 minutes deviations increase from 0.25 to 0.30. This is a potential increase in the standard deviation of the current power balance changes. If the standard deviation is known, confidence intervals for current power fluctuations with required accuracy can be defined. The distribution of these fluctuations as random variables described by normal (Gaussian) law (Figure 4).

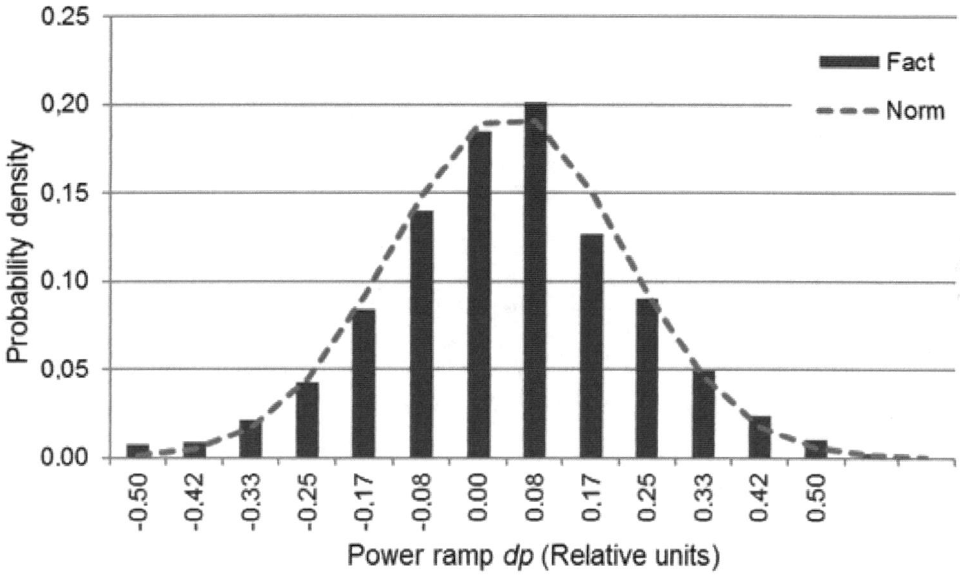

Figure 4. Current power balance changes histogram.

It should be noted that in this example the weights coefficients reflect the nominal RES capacity while effective (or averaged) power indicators are significantly lower. Thus, for the wind power stations utilization coefficient of rated power is 0.3-0.4 depending on location and season and for solar power stations 0.14-0.18. This fact should be taken into account when calculating the energy contribution of RES. The economic components thus are determined by standard means [13]. For stability and reliability ensuring current power fluctuations are important.

Another indicator is representing energy and power curves and further operations on them. All generation and consumption graphics must be set as functions for later use. Curve extremes are determined graphically (Figure 5).

After that the curve is divided into a number of extreme negative areas. The equation of the curve for each section is determined by formula:

$$\begin{vmatrix} x^2 & xy & y^2 & x & y & 1 \\ x_1^2 & x_1y_1 & y_1^2 & x_1 & y_1 & 1 \\ x_2^2 & x_2y_2 & y_2^2 & x_2 & y_2 & 1 \\ x_3^2 & x_3y_3 & y_3^2 & x_3 & y_3 & 1 \\ x_4^2 & x_4y_4 & y_4^2 & x_4 & y_4 & 1 \\ x_5^2 & x_5y_5 & y_5^2 & x_5 & y_5 & 1 \end{vmatrix} = 0, \quad (4)$$

where $(x_1; y_1)$, $(x_2; y_2)$, $(x_3; y_3)$, $(x_4; y_4)$, $(x_5; y_5)$ – points that are in this part of the curve. In this case, two of them are extreme points of the investigated area.

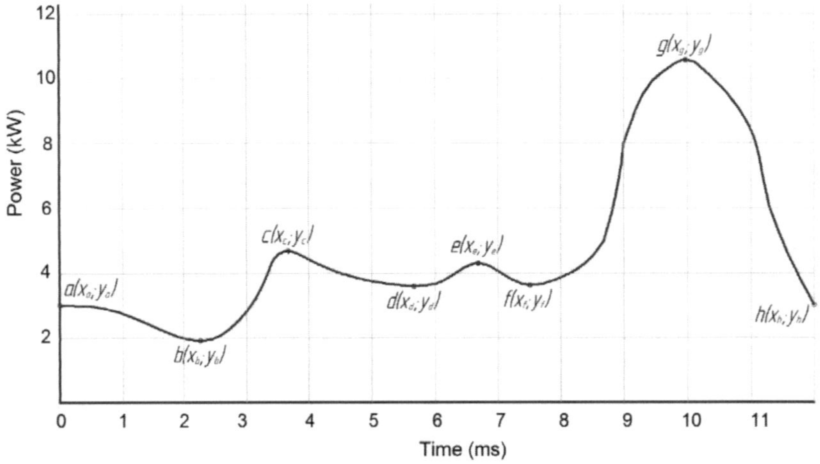

Figure 5. Schedule and its extremes example.

Equation (3) is valid only when the three given points are collinear, and if none of the four of them are collinear. Each function that has been defined by the formula (4), will correspond to a certain section of the main graph (Figure 6).

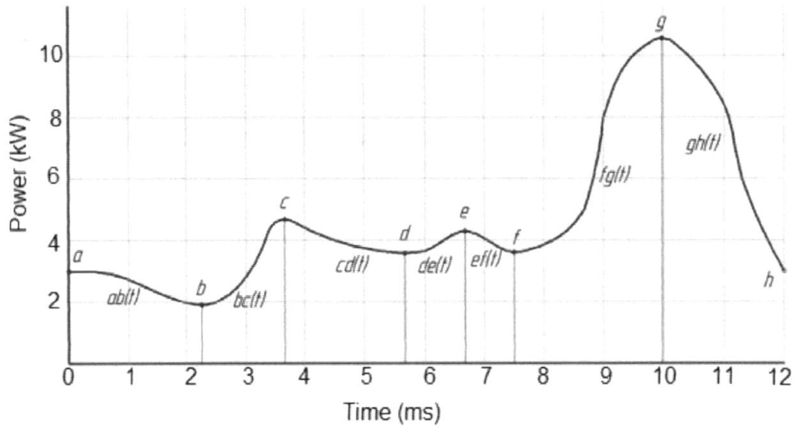

Figure 6. Example of schedule, which consists of several functions.

According to Figure 6, each function has its own definition range.

The next step is to find the sum of all load schedules:

$$(f+g+\ldots+i)(t)=f(t)+g(t)+\ldots+i(t). \tag{5}$$

The total energy schedule for generators is defined similarly.

After determining functions describing the two major graphics (power generation and power consumption) model calculates their difference and makes characteristic graph. An example of the difference between two energy curves is shown in Figure 7.

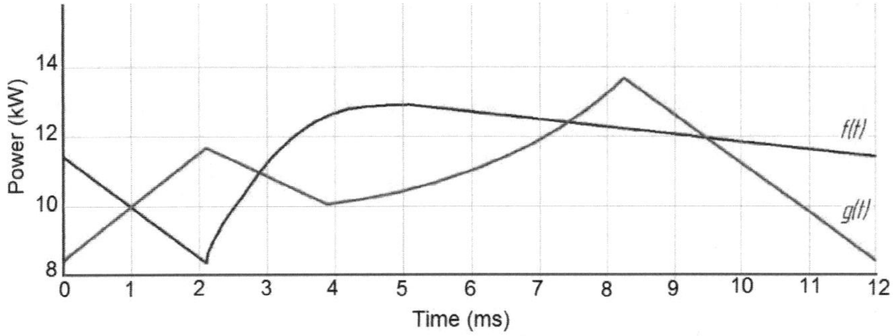

Figure 7. Major energy load f(t) and generators g(t) graphics.

To simplify the example, each graph in Figure 8 described by one function. In real case, the number of features describing the energy curve can be much larger, depending on the number of objects (load and generators) and the curvature of the energy graphs of each object.

According to Figure 7, the difference of two functions mathematically could be written as:

$$(f-g)(t)=f(t)-g(t). \tag{6}$$

The resulting graph *(f-g) (t)* is shown in Figure 8.

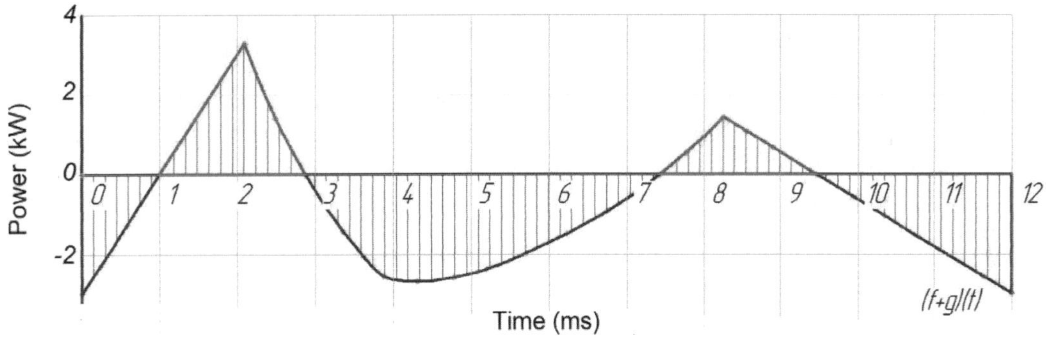

Figure 8. The resulting energy graph.

This curve (f-g) (t) is proposed as the end result of the model. Based on this schedule, the expert can make a conclusion about the shortage or excess of electricity at a certain time, and make a constructive decision to enable a new energy source or shut off the load at a certain time. Expert can also change load indicators of load (if projected). After correcting the outputs, the model calculates everything from the beginning and produces a corresponding result that is comparable to the previous one. The example shows the calculations of the electricity supply system during the year. The model can perform calculations for smaller periods of time, but due to the large amount of information uncertainty, the accuracy of such calculations will decrease proportionally to the time interval reducing.

3. Conclusions

Therefore, when choosing the type and capacity of RES (namely wind and solar) in the local grid, it is necessary to take into account their real energy characteristics in these climate conditions, which allow them to assess their impact on the stability of electricity supply. Another factor is the estimation of electricity needs and the consumption changes over time. Combined analysis of these data allows us to evaluate the real possibilities of using RES while ensuring the proper quality of electricity supply. These findings relate to arbitrary local energy systems, but in the case of railway transport should be based on load features that are different from other load.

Acknowledgments: This paper is elaborated in the framework of the project co-financed by the Polish National Agency for Academic Exchange.

References

[1] Bórawski, P., Bełdycka-Bórawska, A., Szymańska, E. J., Jankowski, K. J., Dubis, B., & Dunn, J. W. 2019 Development Of Renewable Energy Sources Market And Biofuels In The European Union *Journal of Cleaner Production.* **228** pp 467-484
[2] Xinjun Zhou 2019 A Study on Potential for Using New Energy and Renewable Energy Sources in Railways *International Journal of Energy and Power Engineering.* **8** 4 pp 45-51
[3] García-Olivares A, Solé J and Osychenko O 2019 Transportation in a 100% renewable energy system *Energy Conversion and Management* **185** p 891
[4] Rehmani M, Reisslein M, Rachedi A, Erol-Kantarci M and Radenkovic M 2018 Integrating Renewable Energy Resources Into the Smart Grid: Recent Developments in Information and Communication Technologies *IEEE Transactions on Industrial Informatics* **14** 7 pp 2814–2825
[5] Aguado J, Sanchez Racero A and de la Torre S 2018 Optimal Operation of Electric Railways With Renewable Energy and Electric Storage Systems *IEEE Transactions on Smart Grid* **9** 2 pp 993–1001
[6] Sengor I, Kilickiran H, Akdemir H, Kekezoglu B, Erdinc O and Catalao J 2018 Energy Management of a Smart Railway Station Considering Regenerative Braking and Stochastic Behaviour of ESS and PV Generation *IEEE Transactions on Sustainable Energy* **9** 3 pp 1041–1050
[7] Sychenko V, Kuznetsov V, Kosariev Y, Hubskyi P, Belozyorov V, Zaytsev V and Pulin M 2018 Development of an approach to ensure stability of the traction direct current system *Eastern-European Journal of Enterprise Technologies* **5** 295 pp 47–56
[8] Sichenko V, Kosariev Y, Danilov O, Gubsky P, Zubenko V and Pulin M 2018 Features of relay protection function in distributed systems of type electrical supply *Bulletin of the National Technical University «KhPI» Series: New solutions in modern technologies* **9** 1285 pp 70–75
[9] Pivniak G, Shkrabets F and Ostapchuk O 2014 Operation of relay protection devices at substations in distributed generation systems *XV Anniversary International Scientific Conference "Renewable Energy of XXI Century"* pp 104–106
[10] Renevable energy development. The official portal of the Ministry of energy and environment protection of Ukraine *Menr.gov.ua* 2020 [Online] Available: https://menr.gov.ua/timeline/VDe.html
[11] Kuznietsov M, Vyshnevska Y, Brazhnyk I and Melnyk O 2019 Modeling of the Generation-Consumption Imbalance in the Heterogeneous Energy Systems with Renewable Energy Sources *IEEE 6th International Conference on Energy Smart Systems (ESS)* pp 196–200
[12] Miller B and Pankov A 2002 *Theory of Random Processes in Examples and Problems* (Moscow: Fizmatlit) p 320
[13] Bahramara S, Moghaddam M and Haghifam M 2016 Optimal planning of hybrid renewable energy systems using HOMER: A review *Renewable and Sustainable Energy Reviews* **62** pp 609–620

IOP Conf. Series: Materials Science and Engineering **985** (2020) 012012 doi:10.1088/1757-899X/985/1/012012

Study of railway traffic safety based on the railway track condition monitoring system

S Kliuiev[1,3], Ie Medvediev[1] and N Khalipova[2]

[1]Volodymyr Dahl East Ukrainian National University, Department of logistics management and traffic safety in transport, Tsentralnyi ave., 59-a, 93406, Severodonetsk, Ukraine

[2]University of Customs and Finance, Department of Transport Systems and Technologies, Volodimir Vernadskiy Ave., 2/4, 49000, Dnipro, Ukraine

[3]sergistreet@gmail.com

Abstract. The solution to the problem of monitoring the railway track, as well as potentially dangerous objects and phenomena in the adjacent territories, is determined by the need to implement increased requirements for traffic safety (primarily to the geometric parameters of the devices of a long railway track under construction and functioning load-bearing highways), prevention of technological disasters and large-scale negative effects of natural phenomena on the railways. The continuous process of integrating methods and means of monitoring and preventing emergencies into a single information management system is shown. The possibility of determining a dynamic model of the adverse effect of the train on the track and the most unfavorable for the track train and traction parameters, which must be avoided when forming trains in this section, is established. Three key factors are formulated, the use of which will reduce the negative impact on the track from trains. The concept of information technology improvement of the monitoring system for potentially dangerous sections of the railway on the basis of new innovative technologies has been developed.

1. Introduction

Ensuring the continuity of the transport process and the prevention of emergencies is an urgent task for rail transport. Emergencies pose serious anthropogenic risks for railway workers and the population living in the surrounding areas. The innovative technical solutions used for the monitoring of long stretches of track make it possible to achieve the maximum integrated effect for railway safety.

Thus, the urgency of solving the problem of monitoring the railway track, as well as potentially dangerous objects and phenomena in the adjacent territories, is determined by the need to meet the increased requirements for traffic safety (first of all to the geometrical parameters of longitudinal railway track devices on under construction and functioning freight lines), prevention of man-made catastrophes and large-scale negative effects of natural phenomena on railways.

An effective means of implementing new forms of track monitoring is the remote sensing space technologies, which allows simultaneous assessment of the geometric parameters of the path and adjacent territories over extended sections [1]. With the widespread introduction of heavy and high-speed traffic, there is a need for an operational comprehensive assessment of long sections of the railway [2]. This and a number of other factors required the search for fundamentally new principles for monitoring and diagnosing the railway track, which led to the use of satellite-based geolocation systems in railway transport [3].

Content from this work may be used under the terms of the Creative Commons Attribution 3.0 licence. Any further distribution of this work must maintain attribution to the author(s) and the title of the work, journal citation and DOI.
Published under licence by IOP Publishing Ltd

2. Formulation of the problem

Simultaneously with technological development, there is a gradual integration of national railway systems into a single world transport system [4], which requires a unified scientific and technological approach to monitoring and maintenance of the railway track and, first of all, unification of measurement methods, forms of presentation of information and regulatory requirements to its completeness and accuracy [5, 6].

Foreign rolling stock, primarily high-speed and speed trains from Skoda, Hyindai, and General Electric, is gaining ground on Ukrainian railways. Their effective and safe operation requires the generalization and integration of foreign and Ukrainian experience in maintaining the railway track, as well as the regulatory requirements for its technical condition and, first of all, for the geometric parameters of the rail track.

The purpose of the article. Study of railway traffic safety based on the railway track condition monitoring system by improving the structural scheme of a system for monitoring extended deformations of a railway track.

3. Analysis of literary data

Currently, the world practice of monitoring the geometric parameters of the path [7] distinguishes two main types of control - expert assessment (technical inspection) and automated control. In the case of expert assessment, the person himself acts as a diagnostic system, and the effectiveness of control depends on the experience and analytical qualities of the inspector. An experienced engineer determines the geometry of the path according to physical sensations during an examination trip on a train. In the case of automated control, most of the functions for collecting and processing diagnostic information are transferred to electronic computer technology [8, 9, 10].

The fundamental difference between global control means is the ability to assess the condition of the railway on a macro-territorial scale, in a single coordinate-time measurement system [11, 12].

The existing system of information and technological support for monitoring the railway track includes 2 levels (Figure 1): a complex of local infrastructural technical solutions – reference networks, travel patterns, geodetic equipment, etc. and the second level – the so-called continuous monitoring means (of which the track vehicle is the most striking representative and performs periodic measurements of the geometric parameters of the rail gauge).

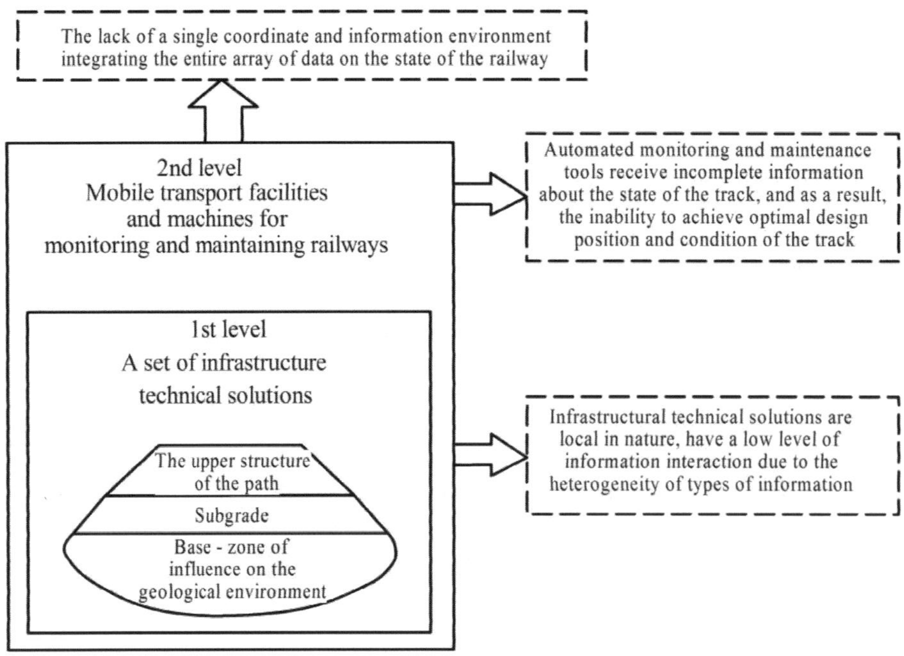

Figure 1. The main technological problems of monitoring.

Moreover, all these tools give a discrete and heterogeneous picture [13], which does not allow, in principle, to build an adequate spatial model of the state of the railway track.

To solve the presented technological problems, it is advisable to use global control means – space, aviation, and ground-based remote sensing means with high-precision geolocation equipment [14].

Rail transport has entered the active phase of the application of new technical means, and, first of all, of new rolling stock (long-haul freight and high-speed passenger trains) which cause the track to be subjected to a new form of impact - volumetric deformation (displacement of embankments and other structures, appearance of multi-radius curves, etc.) [15].

The use of satellite navigation systems for locomotives, according to experts of the locomotive economy "Ukrzaliznytsya", is the most effective tool for identification of vehicles and creation of a digital model of the railway track of Ukraine. On mainline locomotives, satellite radio navigation systems (SRNS) have been widely used [16, 17].

The causes of crashes and crashes are, first of all, deviations in the condition of the track and undercarriage of rolling stock from the norms of their contents [18, 19, 20].

4. Research results

In the case of mixed traffic on the Ukrainian railways on most sections of the track, it is not possible to single out one type of train that can be oriented in calculations. In the case of following trains with different lengths, masses, speeds, traction distribution and other parameters, it is advisable to build a different calculation algorithm (Figure 2). First of all, it is necessary to identify sections of the path subject to deformation. Based on the result, it becomes possible to determine a dynamic model of the adverse effect of the train on the track and the most unfavorable for the train parameters of the train and traction, which must be avoided when forming trains in this section.

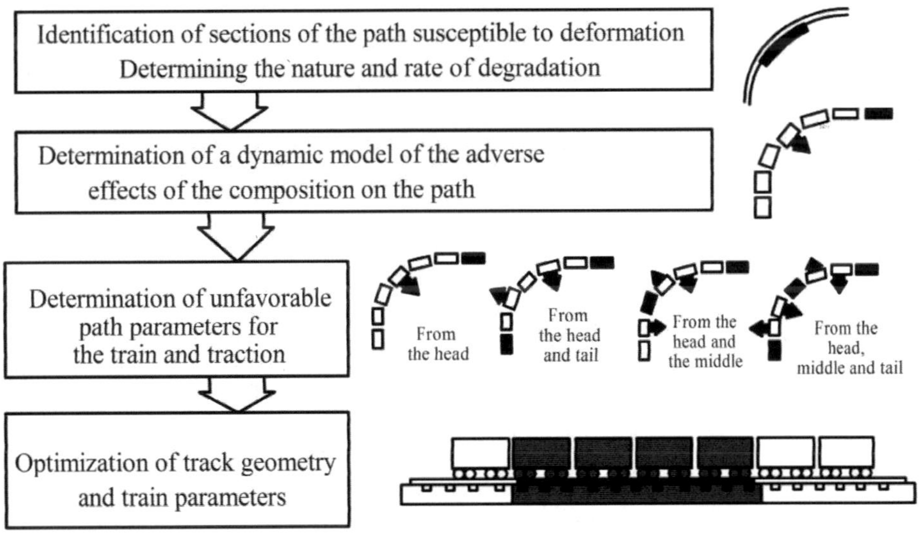

Figure 2. Building the optimal train model and track for mixed traffic.

Integrating the results of the studies, a modernized block diagram of a system for monitoring the longitudinal deformations of the railway track is proposed, providing for 4 main levels (Figure 3).

The results of specific studies by European organizations in the area of creating a single coordinate space for railways implementing the "Europe without Borders" principle confirm that the use of geolocation satellite navigation technologies provides solutions for monitoring and maintenance of the railway track. One example is the widespread use of spatial data in the German railway infrastructure, where route information is constantly updated as a result of maintenance, repair and upgrading.

The first step in implementing a promising track monitoring system is to form a single, high-precision rail coordinate space (Figure 3 – Level 1) based on satellite navigation systems and their differential additions.

The second stage is the introduction of a unified technology for the technical diagnostics of the railway track (Figure 3 – level 2-3) with the use of high-precision satellite positioning and inertial system. This will optimize the timing of the road works while improving their quality, reduce the number of restrictions on the movement of trains, increase their speed, and will also save money on the maintenance of the railway track.

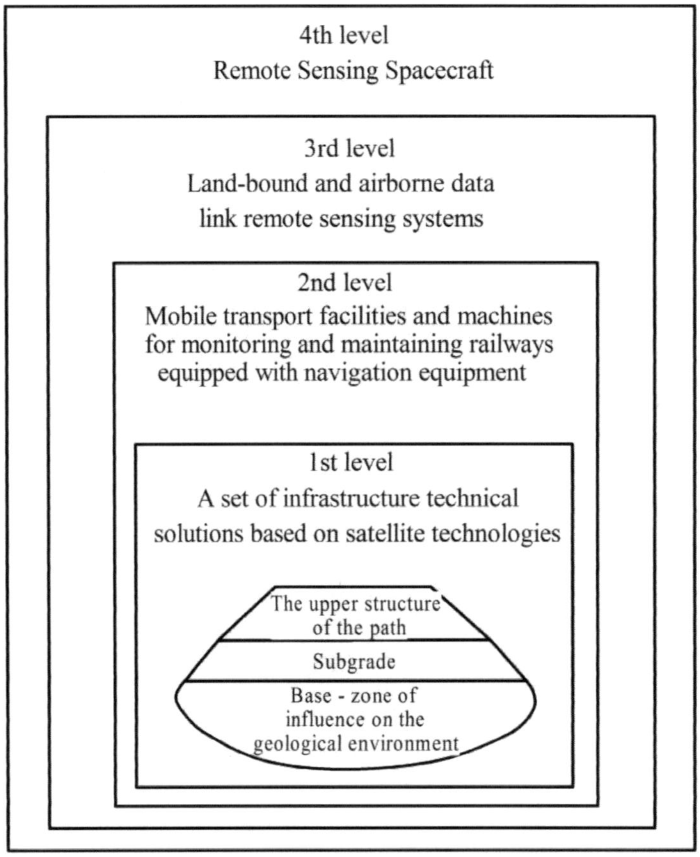

Figure 3. Upgraded monitoring system.

The third stage is the introduction of railway monitoring technology, artificial structures and negative natural-technogenic processes affecting traffic safety on the basis of modern information-measuring systems and technologies of remote sensing of the Earth (Figure 3 – Level 4). It will allow to reduce difficult costly engineering and geodetic works at technical diagnostics of artificial constructions and monitoring of negative natural-technogenic processes on railways, to prevent possible accidents and catastrophes on railway transport.

5. Discussion

Analyzing the relationship of the track and the dynamic load on it (Figure 4), three key factors can be formulated, the use of which will reduce the negative impact on the track from trains and the negative natural and anthropogenic phenomena in the territories adjacent to the track [2]:

- optimization modeling of the geometric parameters of the path over extended sections [22];
- monitoring of the geometric parameters of the track and infrastructure over extended sections;

- monitoring the influence of natural and technogenic factors that worsen the condition of the track and infrastructure (tracking changes in the geometry of potentially dangerous objects in the adjacent territories to the track).

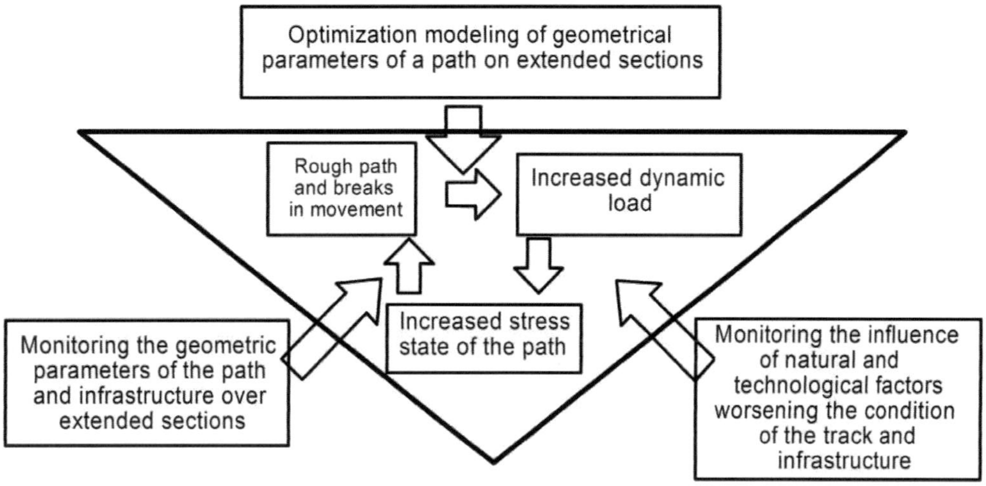

Figure 4. New tools to counteract the negative impact on the track of trains and natural and man-made phenomena.

6. Conclusion

The concept of information technology improvement of the monitoring system for potentially dangerous sections of the railway on the basis of new innovative technologies has been developed.

For track facilities, the following new technological opportunities arise to evaluate the geometric parameters of the track in combination with infrastructure objects:

- assessment of the spatial position and geometric parameters of the railway track in combination with artificial structures on a macro-territorial scale (displacement of embankments, approach / removal, etc.);
- adaptive control of monitoring and routine maintenance of the track (control of the monitoring frequency during monitoring, less time to eliminate defects by direct transmission of information from the monitoring equipment to the track machines, speed control of the track machines during repairs, etc.);
- identification of the epicenters of potentially dangerous phenomena over large areas adjacent to the railway line (the formation of water bodies, the growth rate of ravines in the direction of the track, flooding and waterlogging of soils, etc.).

References

[1] Lyovin B 2017 Integrated monitoring of transport infrastructure *Science and Technology of Railways* **1** pp 14–21
[2] Farid M, Ariff M, K. Chong A, Majid Z and Setan H 2013 Geometric and radiometric characteristics of a prototype surveillance system *Measurement* **46** 1 pp 610–620
[3] Bondavalli A, Ceccarelli A, Gogaj F, Seminatore A and Vadursi M 2013 Experimental assessment of low-cost GPS-based localization in railway worksite-like scenarios *Measurement* **46** 1 pp 456–466
[4] Kampczyk A 2015 Technical specifications for interoperability and Polish regulations in the design of the geometry of railway line *Bauingenieur* **90** pp 229–234
[5] Maciuk K and Lewińska P 2019 High-Rate Monitoring of Satellite Clocks Using Two Methods of Averaging Time *Remote Sensing* **11** 23 2754
[6] Chen D, Tang T, Cao F and Cai B 2012 An integrated error-detecting method based on expert

knowledge for GPS data points measured in Qinghai–Tibet Railway *Expert Systems with Applications* **39** 2 pp 2220–2226
[7] Morais P, Morais J, Santos C, Paixao A, Fortunato E, Asseiceiro F, Alvarenga P and Gomes L 2019 Continuous Monitoring and Evaluation of Railway Tracks: Proof of Concept P*rocedia Structural Integrity* **17** pp 419–426
[8] Bock Y and Melgar D 2016 Physical applications of GPS geodesy: a review *Reports on Progress in Physics* **79** pp 1–119
[9] Xiong J and Han F 2020 Positioning performance analysis on combined GPS/BDS precise point positioning *Geodesy and Geodynamics* **11** 1 pp 78–83
[10] Chrobak T, Lupa M, Szombara S and Dejniak D 2019 The use of cartographic control points in the harmonization and revision of MRDBs *Geocarto International* **34** pp 260–275
[11] Han S, Kwon J and Jekeli C 2001 Accurate absolute GPS positioning through satellite clock error estimation *Journal Geodesy* **75** 1 pp 33–43
[12] Zeng J, Wei L and Wu P 2016 Safety evaluation for railway vehicles using an improved indirect measurement method of wheel-rail forces *Journal of Modern Transportation* **24**(2) pp 114–123
[13] Isaev A, Bartalev S, Lupyan E and Lukina N 2014 Earth observations from satellites as a unique instrument to monitor Russia's forests *Herald of the Russian Academy of Sciences* **6** pp 413–419
[14] Kliuiev S and Brahin N 2018 Choosing a method for determining the direction of rounding a section of a track passing a locomotive *Bulletin of VDEUNU* **1** 242 pp 60–64
[15] Spiryagin V, Kliuiev S and Zubar Ye 2013 The choice of a method for determining the coordinates of a locomotive's location with controlled movement of a pair of wheels in a rail *Bulletin of VDEUNU* **5** 194(2) pp 144–146
[16] Ling L, Dhanasekar M, Thambiratnam D and Sun Y 2016 Lateral impact derailment mechanisms, simulation and analysis *International Journal of Impact Engineering* **94** pp 36–49
[17] Kliuiev S 2016 Improving railway safety *Bulletin of VDEUNU* **1** 225 104–107
[18] Kapitsa M, Mikhailov E, Kliuiev S, Semenov S and Kovtanets M 2019 Study of rail vehicles movement characteristics improvement in curves using fuzzy logic mechatronic systems *MATEC Web of Conferences* **294** 03019
[19] Sapronova S, Tkachenko V, Fomin O, Hatchenko V and Maliuk S. 2017 Research on the safety factor against derailment of railway vehicles *Eastern-European Journal of Enterprise Technologies* **6** 7(90) pp 19–25
[20] Fomin O, Gerlici J, Lovskaya A, Kravchenko K, Prokopenko P, Fomina A and Hauser V 2018 Research of the strength of the bearing structure of the flat wagon body from round pipes during transportation on the railway ferry *MATEC Web of Conferences* **235** 00003
[21] Fuxun Ma, Ruijie Xi and Nan Xu 2016 Analysis of railway subgrade frost heave deformation based on GPS, *Geodesy and Geodynamics* **7** 2 pp 143–147
[22] Kardas-Cinal E 2015 Selected problems in railway vehicle dynamics related to running safety *Archives of Transport* **31** 3 pp 37–45

Accumulation of solar energy to heat greenhouses

M Savytskyi[1], **V Danishevskyy**[2] **and M Bordun**[1,3]

[1] Department of Reinforced Concrete and Masonry Structures, State Higher Education Establishment «Prydniprovska State Academy of Civil Engineering and Architecture», 24-A, Chernishevskogo str., Dnipro, Ukraine
[2] Department of Structural Mechanics and Strength of Materials, State Higher Education Establishment «Prydniprovska State Academy of Civil Engineering and Architecture», 24-A, Chernishevskogo str., Dnipro, Ukraine

[3]klmari@ukr.net

Abstract. The primary task of a greenhouse is to create optimal microclimate, which doesn't depend on external climatic conditions, to grow plants. We can observe significant air temperature and humidity difference in a traditional greenhouse during 24 hours. The stable microclimate in greenhouses is provided by different engineering systems such as heating, ventilation, lighting, etc. These systems require a lot of energy. Saving energy and reducing the mode of temperature amplitude inside the operating greenhouse during exploitation period can be achieved by using solar energy in the greenhouse heating system, namely, by using heat accumulators. However, the test method of optimal heat accumulator to equalize the temperature amplitude inside the greenhouse wasn't described in any research. The purpose of this research is to determine the optimal heat accumulator for equalizing temperature amplitude inside the greenhouse under atmospheric effects in spring. The experimental research of the heat-humid regime was carried out for the greenhouse without any heat accumulation. Relationship between the volume and the surface area of the accumulator and its efficiency was identified.

1. Introduction

The primary purpose of a greenhouse is to create optimal microclimate, which doesn't depend on external climatic conditions to grow plants. Greenhouses are usually used to grow plants in areas where geographical factors or seasonal climatic variations don't allow growing plants on the field. However, the structure of the greenhouse itself doesn't provide optimal microclimate for growing plants.

In a traditional greenhouse, significant air temperature and humidity difference is observed over 24 hours. The difference between day and night temperatures can achieve 40°C and more.

A stable microclimate in greenhouses can be provided by different engineering systems such as heating, ventilation, lighting, etc. These systems require a lot of energy, especially when the greenhouse is used around the year.

The large areas of transparent surface in greenhouses also contribute to heat losses. During any warm season the air temperature in greenhouse increases quickly to the level significantly higher than required. Sometimes natural ventilation isn't enough to escape excess heat. During the cold season heat losses in a greenhouse are very high through transparent surfaces. The microclimate stability in greenhouses is constantly supported by the heating systems during this period, and as a result, everything leads to

Content from this work may be used under the terms of the Creative Commons Attribution 3.0 licence. Any further distribution of this work must maintain attribution to the author(s) and the title of the work, journal citation and DOI.
Published under licence by IOP Publishing Ltd

IOP Conf. Series: Materials Science and Engineering **985** (2020) 012013 doi:10.1088/1757-899X/985/1/012013

significant energy losses. Thus, greenhouse agriculture production is the most energy-intensive and expensive kind of agricultural industry.

Therefore, more and more attention is focused on the search of design solutions to create sustainable and energy efficient greenhouses. Such greenhouses can be highly productive with minimal energy consumption but have a high productivity.

One of the ways to increase energy efficiency and reduce energy costs is to accumulate and to use the excess solar energy in greenhouses. The heat accumulator can be used to level the temperature fluctuation in greenhouse.

2. Statement of the problem

Heat accumulators can be classified according to their properties of heat-accumulation materials. There are accumulators of capacitive type, which use the heat capacity of the accumulating material. The material of this accumulator is heated (cooled) without any change to its aggregate state [1 - 3]. There are also accumulators with phase transition of substance. Such accumulators use melting heat (solidification) of a substance [1, 4].

The accumulators that based on a phase transition are more efficient than accumulators of capacitive type because they have a much higher the enthalpy value of the phase transitionary than the heat content. However, these accumulators also are expensive. Additional research on the economic feasibility of their use in the greenhouse have to carry out for each case [4, 5].

The accumulators of capacitive type are mostly employed in greenhouses. Concrete, stone, water, brick, other materials and their combination can be used as heat accumulation materials [3, 6]. As heat accumulators in a greenhouse element of its structures (floor, walls, soil and other structural elements) can be used. The separate objects (for example, tanks with water, stones and other subject, which can accumulate heat) can be also used [2, 6, 7].

The heat accumulation capacity of materials depends on their thermal storage properties and temperature difference. The materials with higher density have got generally the highest heat storage properties, but at the same time these materials have got a higher thermal conductivity. The properties of some materials and substances to create heat accumulators are presented in the Table 1 [8].

Table 1. The heat accumulation capacity and thermal conductivity of some materials and substances.

Material / substance	Density, ρ_0, kg/m^3	Specific heat, C_0, kJ / kg · K	Thermal conductivity, λ, W / (m · K)
Water	1000	4.2	0.55
Ground	1600	0.8	1.02
Natural stone (granite, rubble)	2800	0.88	3.49
Concrete	2400	0.84	1.86
Brick	1800	0.88	0.81
Gravel	800	0.84	0.23
Sand	1600	0.84	0.58
Timber	500	2.3	0.18

Water is characterized by the highest heat capacity among all heat accumulation materials. Different water tanks can be used for heat accumulation. The key distinctive feature of the water heat accumulator is that a big accumulator tank saves the heat longer, but the deeper water layers are heated slowly. The small tanks quickly respond to temperature change that makes possible to avoid overheat inside a greenhouse. However, they give the accumulated heat back very quickly. Water-filled passive solar hoses were used to heat Canary greenhouses as an additional heat source [3].

One of the cheapest and accessible ways to accumulate thermal energy is to use the soil. However, the soil has a lower heat storage capacity than water. Therefore, systems of ventilators and underground air tubes are used to increase the efficiency of the soil accumulator [4, 9].

Natural and artificial stones are of high heat storage capacity and such accumulators, along with additional equipment (ventilators and air tubes) can be used as seasonal heat accumulators.

The existing studies don't provide any methodology to choose the optimal heat accumulator in order to reduce the internal temperature fluctuations in a greenhouse.

3 Materials and methods

The purpose of this research is to determine the optimal heat accumulator to level the temperature fluctuation in a greenhouse with atmospheric effects during an early spring.

The experimental investigations of temperature and humidity were carried out in the greenhouse without any heat accumulator. The investigations were conducted in climatic conditions of Dnipro.

This experimental greenhouse is located in Dnipro (Ukraine) and it has a transparent gable roof and the following dimensions – $7.9\ x\ 7.3\ m$. The height of the greenhouse is $3.1\ m$. The greenhouse frame is made of metal pipe profile $40\ x\ 40\ mm$. Transparent material of the envelope is $8.0\ mm$ thick polycarbonate. The general view of the experimental greenhouse is presented in Figure 1.

Figure 1. General view of the experimental greenhouse.

The continuous measurements of the temperature and humidity were carried out at a height of $0.3\ m$ and $2.2\ m$ above the ground inside and outside the greenhouse. The measurements were taken in the period from 23.03.2019 to 06.04.2019. The period for the study was chosen and based on the weather forecast. According to it, significant changes in the weather were expected during the measurement period – from cloudy weather with precipitation to a significant increase in temperature to $+20°C$.

The temperature and relative humidity measurements were performed by the digital sensors Tinytag Ultra 2 TGU-4500 (Figure 2). These sensors control temperature from $-25°C$ to $+85°C$ and relative humidity from 0 to 95%. Tinytag Ultra 2 sensors is usually used for internal monitoring. However, the sensor can be used for outside monitoring and is equipped with a special security screen – Stevenson screen. Software and a USB cable (CAB-0007-USB) should be used for Tinytag Explorer monitor. The full technical specification of the sensors employed is available at https://www.geminidataloggers.com/data-loggers/tinytag-ultra-2/tgu-4500.

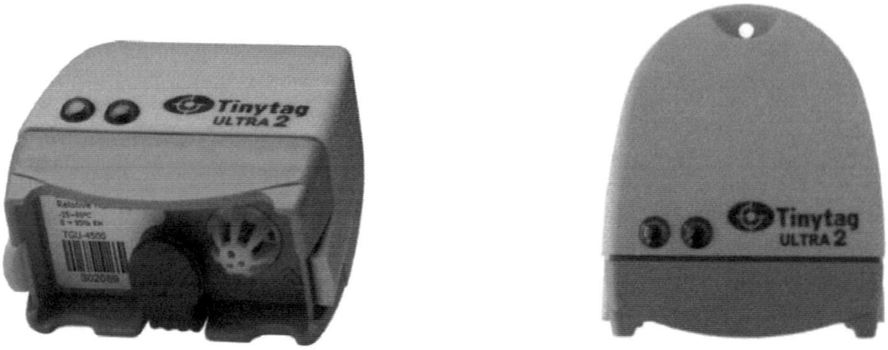

Figure 2. Tinytag Ultra 2 data logger.

The measurements were made during the established period around the clock at 1-hour interval. Changes in weather conditions (cloudiness, precipitation, wind speed and direction) were also recorded throughout the study to determine their effect on the microclimate in the greenhouse. The experimental measurements are presented in Figure 3.

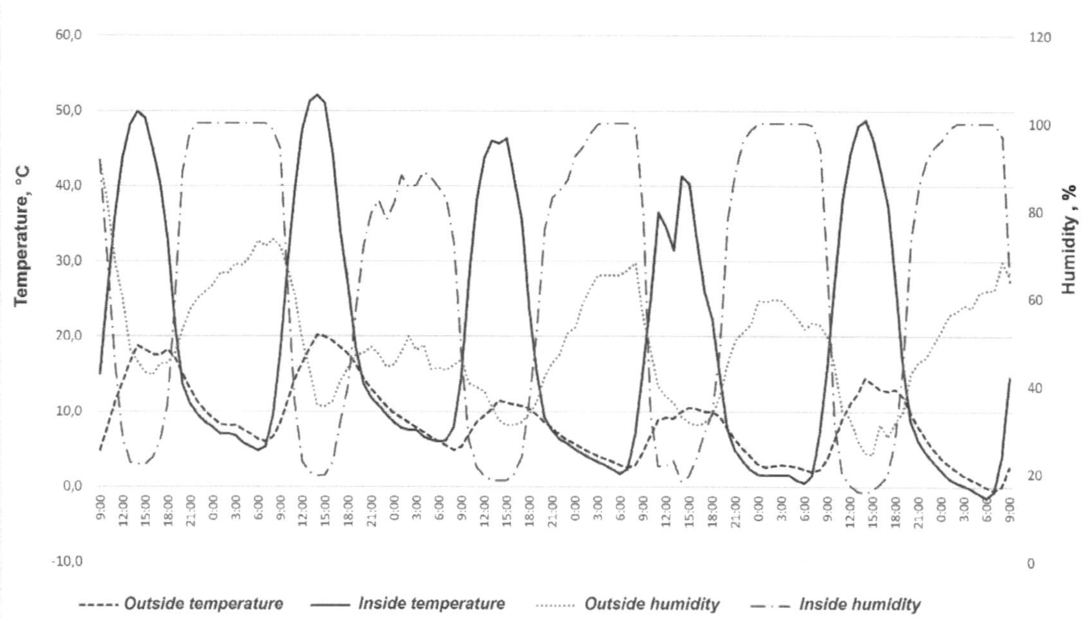

Figure 3. The experimental measurements of the temperature and humidity inside and outside the greenhouse during the period from 31.03.2019 to 05.04.2019 (without a heat accumulator).

The temperature curve shows that the temperature fluctuations inside the greenhouse are from $-1°C$ to $52°C$ in the experimental period. The indoor air temperature in the greenhouse was reduced to a level of the outside air temperature or a few degrees below at night. The analysis of similar studies shows that the same problems often occur in ordinary greenhouses. This fact can be explained by technological space between the frame and the glazing material of the greenhouse enclosure structures. Therefore, the cold outside air gets into a greenhouse very easily and stays inside, especially when it is windy [10]. In

IOP Conf. Series: Materials Science and Engineering **985** (2020) 012013 doi:10.1088/1757-899X/985/1/012013

the daytime, the inside air temperature of a greenhouse rises to $50°C$, that is far to exceed the comfortable temperature for a plant. The excess heat is usually removed from the greenhouse by ventilation, but in general, it is simply lost.

This indicates that the indoor temperature in the greenhouse does not meet the established temperature to grow plants, which is from $15...18°C$ (at night) to $26...30°C$ (in the afternoon), even in spring [11].

A thermal physical model was proposed to investigate the efficiency of the heat accumulator use in the greenhouse. The model takes into account the heat input from solar radiation Q_3, the heat losses through to enclosing structures Q_2, and the heat that was accumulated in a heat accumulator Q_1 (Figure 4).

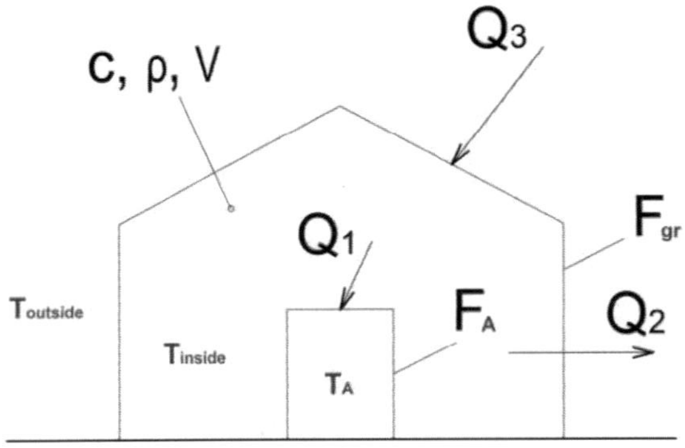

Figure 4. Thermal model of the greenhouse with a heat accumulator.

The heat balance equation for the greenhouse with the accumulator reads:

$$\Delta Q_{gr} = Q_3 - Q_1 - Q_2, \tag{1}$$

where ΔQ_{gr} is the heat increment into the greenhouse, W.

The heat accumulated by the heat accumulator is:

$$Q_1 = \alpha_A \left(T_{inside} - T_A\right) \cdot F_A, \tag{2}$$

where α_A is the heat transfer coefficient on the accumulator surface, $W/\left(m^2 \cdot K\right)$; T_{inside} is the indoor temperature in the greenhouse, $°C$; T_A is the accumulator temperature, $°C$; F_A is the surface area of the accumulator, m^2.

The heat losses through the enclosing structures are:

$$Q_2 = \frac{1}{R}\left(T_{inside} - T_{outside}\right) \cdot F_{gr}, \tag{3}$$

where R is the heat transfer resistance of the greenhouse enclosing structures, $m^2 \cdot K/W$; $T_{outside}$ is the outdoor air temperature, $°C$; F_{gr} is the area of the greenhouse envelope constructions, m^2.

The heat input from solar radiation depends on the time and surface orientation:

$$Q_3 = Q_3(\tau). \tag{4}$$

For this model of the greenhouse, the accumulator temperature T_A, the indoor T_{inside} and the outdoor $T_{outside}$ temperatures are considered as time-dependent functions:

$$T_A = T_A(\tau), \tag{5}$$

$$T_{inside} = T_{inside}(\tau), \tag{6}$$

$$T_{outside} = T_{outside}(\tau). \tag{7}$$

The dependence (7) of the outdoor air temperature on time is numerically determined according to the experimental data. Hence, there are two unknown functions: the indoor air temperature T_{inside} and the accumulator temperature T_A.

The heat balance equation for the indoor air can be written as follows:

$$c\rho V_{gr} T_{inside} = \Delta Q_{gr} d\tau, \tag{8}$$

and the heat balance equation of the accumulator is:

$$c_A m_A dT_A = \alpha_A F_A (T_{inside} - T_A) d\tau, \tag{9}$$

where c is the heat capacity of the air, $J/(kg \cdot K)$; ρ is the air density, kg/m^3; V_{gr} is the internal volume of the greenhouse, m^3; c_A is the heat of the heat accumulator, $J/(kg \cdot K)$; m_A is the mass of the heat accumulator, kg.

Substituting expressions (2)–(8) into equation (1) and rewriting equation (9), we obtain the system of two coupled differential equations, which allows us evaluating the unknown functions $T_{inside}(\tau)$, $T_A(\tau)$:

$$\frac{dT_{inside}}{d\tau} = \frac{1}{c\rho V_{gr}} \left[Q_3(\tau) - \alpha_A F_A \left[T_{inside}(\tau) - T_A(\tau) \right] - \frac{F_{gr}}{R} \left[T_{inside}(\tau) - T_{outside}(\tau) \right] \right], \tag{10}$$

$$\frac{dT_A}{d\tau} = \frac{\alpha_A F_A}{c_A m_A} \left[T_{inside}(\tau) - T_A(\tau) \right]. \tag{11}$$

Equations (10), (11) describe the dependencies between the parameters of the accumulator (i.e., the volume and the surface area) and the fluctuations of the indoor temperature. These dependencies cannot be expressed explicitly in the analytical form, because the functions $T_{inside}(\tau)$ and $Q_3(\tau)$ are determined by the experimental data and, therefore, are purely non-linear. Meanwhile, these dependencies are revealed by the developed numerical solution. System (10), (11) were integrated numerically by the Euler method. The calculations were performed for the 5 days period with the 1 minute integration step, which provides a good convergence of the procedure. The software implementation was developed in the open-source computer algebra system Maxima.

4. Results

The calculation of the internal temperature change in the greenhouse without taking into account the heat accumulator was performed to verify the greenhouse mathematical model. The comparison of experimental and calculated data are presented in Figure 5.

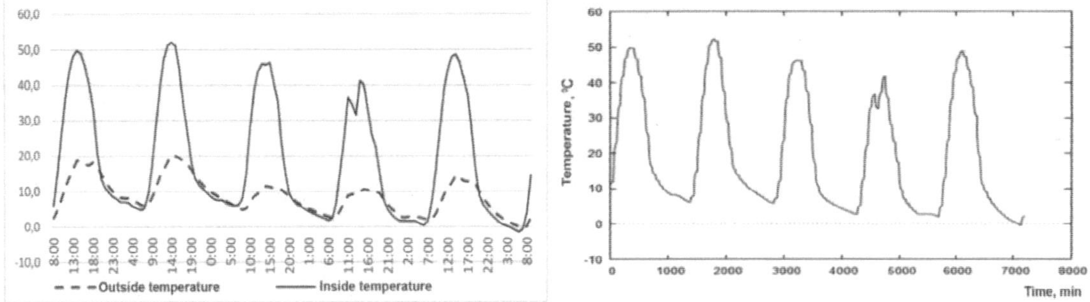

Figure 5. Experimental verification of the greenhouse mathematical model: left - graph of internal temperature change according to the experimental data; right - graph of internal temperature according to the calculation results.

Having analyzed the graphs, it is possible to conclude that the calculation results are in a good agreement with the experimental data. It confirms the accuracy of the greenhouse mathematical model which was proposed.

The calculations of the temperature change of the heat accumulator and the internal temperature in the greenhouse were worked out to determine the optimal parameters of the heat accumulator for the experimental greenhouse by the mathematical model which was proposed. The water tanks were used as the heat accumulators. Initial heat accumulator temperature for calculation was taken as $T_A = 10°C$.

The calculation results of the heat accumulator temperature changes and the internal temperature in the greenhouse during five days were shown in Figure 6. The calculations were carried out for the heat accumulator with the volume of $V_A = 3\ m^3$ and the accumulator surface area of $F_A = 32.1\ m^2$, which is like fifteen 200-liter water tanks.

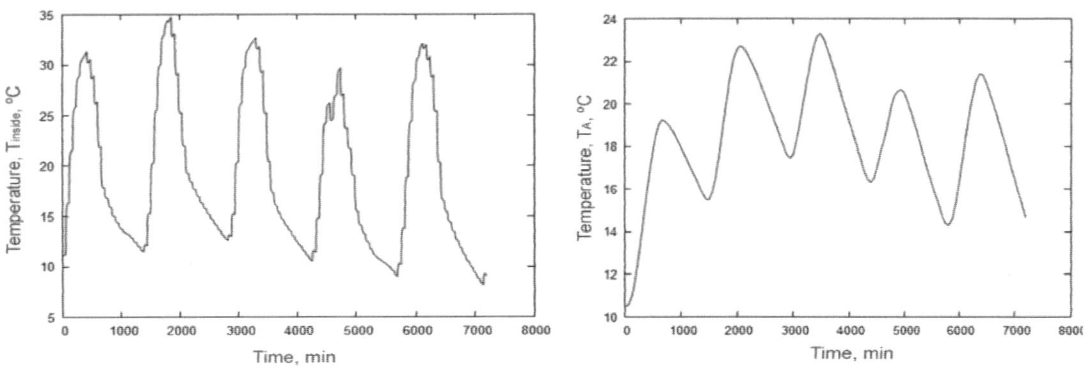

Figure. 6. Temperature change of the heat accumulator and the internal temperature in the greenhouse during five days for the heat accumulator with volume of 3 m^3 and the accumulator surface area of 32.1 m^2: left – internal temperature in the greenhouse; right – temperature of the accumulator.

The graphs show that the interval of temperature change is decreased. The night temperature was reduced to $8°C$, and the day temperature was increased to almost $35°C$, which isn't suitable for plants. When we increased the volume of the heat accumulator to $V_A = 4\ m^3$ and the accumulator surface area to $F_A = 42.8\ m^2$ (it corresponds to twenty 200-liter water tanks), the interval of temperature change didn't almost change (Figure 7). The temperature dropped to $10°C$ during the night and rose to $32°C$ in the daytime. This temperature range is not also friendly and appropriate for plants.

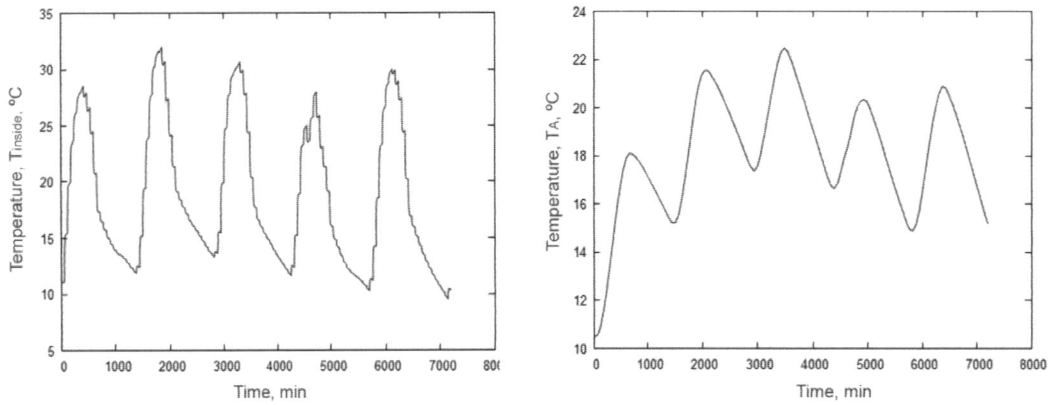

Figure 7. Temperature change of the heat accumulator and the internal temperature in the greenhouse during five days for the heat accumulator with volume of 4 m^3 and the accumulator surface area of 42.8 m^2: left – internal temperature in greenhouse; right – temperature of the accumulator.

Moreover, it can be concluded that in case of a higher accumulator volume, it is not possible to get necessary results anyway.

The following calculations were conducted for a flat-shaped heat accumulator. The volume of the accumulator adopted was $V_A = 4\ m^3$, the accumulator surface area was $F_A = 200\ m^2$. The calculation results were shown in Figure 8.

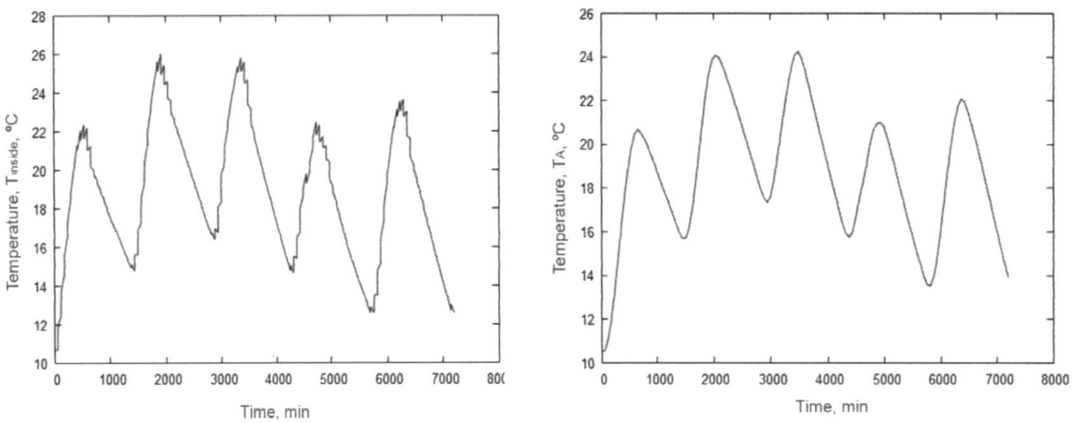

Figure 8. Temperature change of the heat accumulator and the internal temperature in the greenhouse during five days for the heat accumulator with the volume of 4 m^3 and the accumulator surface area of 200 m^2: left – internal temperature in greenhouse; right – temperature of the accumulator.

The graph shows that the internal temperature range in the greenhouse was from $14°C$ to $26°C$, which is optimal for the growth of plants.

5. Conclusions

The optimal daily heat accumulator to support the necessary temperature in the greenhouse was determined by the proposed mathematical model for an early spring.

Dependences between the accumulator volume and the accumulator surface and its efficiency were identified. The efficiency of the heat accumulator increases when we extend the accumulator surface area with the same volume accumulator.

The proposed mathematical model of the greenhouse allows determining the optimal volume and dimensions of the daily heat accumulator for any greenhouse. Let us also note that the particular design and the shape of the accumulator may depend on a number of technological factors, e.g., on the internal planning of the greenhouse, types of plants, a space available, etc. The problem of the design of the accumulator may be a topic for a further research.

Acknowledgements

The authors are grateful to the anonymous reviewers, whose valuable comments and suggestions helped to improve the paper.

References

[1] Kharchenko N V 1991 *Individual Solar Installations* (Moscow: Enerhoatomizdat) p 208

[2] Erat B and Vulston D 1994 *Greenhouse in your house (Handbook* Vol 2*)* ed N V Obolenskiy (Moscow: Stroyizdat) p 191

[3] Bazgaou A, Fatnassi H, Bouharroud R, Elame F, Ezzaeri K, Gourdo L, Wifaya A, Demrati H, Tiskatine R, Bekkaoui A, Aharoune A and Bouirden L 2020 *Solar Energy Performance assessment of combining rock-bed thermal energy storage and water filled passive solar sleeves for heating Canarian greenhouse* **198** pp 8–24

[4] Lukyanov A, Ostapenko V and Aleksandrov V 2010 *Bulletin of DonNACEA Thermal energy accumulators based on phase transition* **6** 86 pp 64–68

[5] Budlianskyi S, Redko A and Chaika Y 2014 *Scientific Bulletin of Civil Engineering Analysis of heat accumulation systems in thermal installations of economic sectors* **1** pp 144–146

[6] Gorjian S, Tavakkoli Hashjin T and Ghobadian B 2011 10th Proc. Int. Conf. on Sustainable Energy Technologies (Istanbul) *Solar Powered Greenhouses* pp 4–7

[7] Ivanko A, Kalinichenko A and Shmat N 1996 *The Solar Vegetarium* (Kuyv: MChP "Anfans") p 112

[8] DSTU B V.2.6-189:2013 2014 *The Method to choose the Thermal Insulation Material for Building Insulation (National Standard of Ukraine)* (Kuyv: Minregion of Ukraine) p 51

[9] Santamourists M, Mihalakakou G, Balarasj C, Lewis J, Vallindras M and Argiriou A 1996 *Energy Energy conservation in greenhouses with buried pipes* **21** pp 353–360

[10] Whitcher C, Taber M, Basehore S and Kiely W 2013 *Retrofitting the Future: Preserving Native Crops with Sustainable Greenhouse Technology (Final report for the Interactive Qualifying Project submitted to the faculty of Worcester Polytechnic Institute)* (Worcester, MA: the faculty of Worcester Polytechnic Institute) p 80

[11] DBN V.2.2-2-95 1995 *Buildings and Structures. Greenhouses and Hotbed (State Construction Rules of Ukraine)* (Kuyv: Ukrarkhbudinform) p 29

IOP Conf. Series: Materials Science and Engineering **985** (2020) 012014 doi:10.1088/1757-899X/985/1/012014

To the problem of train running safety

O M Pshinko[1,3], L V Ursulyak[2,3,5], K I Zhelieznov[4] and A O Shvets[2,3]

[1]Department «Architectural Design, Land Management and Building Materials»
[2]Department «Theoretical and Structural Mechanics»
[3]Dnipro National University of Railway Transport named after Academician V. Lazaryan, Lazaryan St., 2, Dnipro, Ukraine, 49010
[4]Software Engineer AMC Bridge LLC, 303 Wyman Street, Suite 300 Waltham, MA 02451, USA

[5]lydm.urs@gmail.com

Abstract. The paper is presenting the procedure used for establishing the cause of the wagon derailment. To do this, the computer simulations and the computational software, developed in the Dnipro National University of Railway Transport (DIIT) were used. The level of longitudinal forces and the wagons dynamic performance have been evaluated using the mathematical models of longitudinal oscillations of a train and the spatial vibrations of wagons, in particular of tank wagons. As a result of modeling we obtained oscillograms of longitudinal forces in each inter-wagon connection, the dependence of the largest longitudinal forces on travel time and distance traveled, the distribution of the maximum longitudinal forces along the train length, the speed dependence on travel time and track coordinates. We also obtained the dynamic performance of wagons: the vertical dynamics coefficients of the axle-box and central suspension, the horizontal dynamics coefficients and the derailment stability coefficient. The influence of the movable load in the tank wagons and the characteristics of rail irregularities on the stability coefficient against wheel climbing onto the rail is also considered. The presented methodology was used to determine the cause of the tank wagon derailment in a non-homogenous freight train consisting of 50 wagon tanks on an existing track section of the Lithuanian railways. When simulating the train movement, it was assumed that the train was equipped with elastic-friction absorbing devices and air distributors, turned on to the average operation mode. As a result of numerical experiments, an assumption was made about the cause of the train derailment.

1. Introduction

Despite the large number of solved problems related to analysis of the dynamic loading of wagons, which affects the traffic safety, situations leading to derailment still take place at the railways of the world [1-4]. As a result of such accidents, not only rolling stock but also infrastructure can be damaged. It is important to determine the causes of such accidents in order to prevent their repeating. As one knows, driving non-homogenous tanker trains is connected usually with some difficulties [5, 6]. In a traditional train the wagons' vibrations take place under the action of longitudinal forces, originated from the driver's control actions as well as because of the forces caused by track profile and its curvilinearity in plan. In a train with tank wagons, those vibrations take place additionally under the action of the fluid, moving in a tank wagon relatively to the shell.

Content from this work may be used under the terms of the Creative Commons Attribution 3.0 licence. Any further distribution of this work must maintain attribution to the author(s) and the title of the work, journal citation and DOI.
Published under licence by IOP Publishing Ltd

2. Methodology

This article provides a methodology for determining the causes of derailment using computer simulation of train movement.

Determining a cause of wagon derailment is based on the materials provided by the official investigation:

- the consignor list;
- modes of train driving;
- the data on the tank shells calibration;
- the certificate of the train's speedometer tape transcript;
- the certificate on the wagons' brakes;
- the data on the longitudinal track profile and on its curvilinearity in a horizontal plane;
- freight wagons' technical passports;
- the scheme of the derailment place.

In general, the wagon derailment is possible for the following causes:

- technical condition of wagons;
- improper driver's actions;
- deviations in the track maintenance.

If, according to the technical passports of wagons and the certificate on brakes, all wagon parameters comply with regulatory requirements, then the first derailment cause can be excluded. To evaluate the driver's actions that can lead to wagon derailment, especially as a result of braking, it is first advisable to implement provided parameter chart (train driving modes that preceded wagon derailment) using computer simulation and compare the obtained trajectory using the train's speedometer tape provided for transcript. During simulation, the train is considered as a one-dimensional array of bodies (carriages) interconnected by non-linear deformable elements, taking into account the presence of gaps in the inter-wagon connections. If the train includes tank wagons, then it is taken into account the fact that in case of tank ullage, part of the fluid moves in the longitudinal and transverse (horizontal) directions. In this case, in the design scheme, the fluid is represented as one moving mass and one stationary mass rigidly connected to the body. It is assumed that the moving mass is located above the stationary mass and is "articulated" with it and the tank by a "spring." Fluid oscillations caused by perturbations are described by differential equations having a certain specificity [8]. Braking force, resisting force from track profile and plan, the force of the main resistance to the translational movement of the train, the traction force or the electric braking force of the locomotive and the longitudinal forces arising in the inter-wagon connections were taken as disturbances acting on each carriage [5, 6]. As a result of simulation, oscillograms of longitudinal forces in each inter-wagon connection, the dependences of their maximum values on the travel time and the distance traveled, the distribution of the maximum values of longitudinal forces along the train length, the movement speed dependence on the travel time and the track coordinates are obtained. If the obtained movement speed curve satisfactorily coincides with the results of transcript of speedometer tape, then we can analyze the level of longitudinal compressive forces. As it is known [9], for lift-off of a 4-axle fully loaded open wagon with 18-100 bogies, the dangerous level of quasistatic longitudinal compressive forces (acting during 1.5-2 sec) is 1000 kN, and for empty wagon – 500 kN. If the level of longitudinal compressive forces obtained as a result of simulation exceeded the indicated values, the wagon lift-off is possible. In order to make sure that such a control mode has led to the emergence of dangerous longitudinal forces, another control option should be implemented (change the track coordinate in which braking started or perform braking with less discharge of the brake line). If the level of longitudinal forces does not exceed the permissible value, it can be argued that the control mode implemented by the driver has led to the wagon derailment.

Next, you need to evaluate the dynamic indicators of the wagons, characterizing the movement stability. To do this, we simulate the spatial vibrations of the wagon, arising not only from external forces, but also from rail irregularities.

The controlled simulation parameter is the derailment stability coefficient.

If the derailment stability coefficient does not exceed the permissible value, then it can be argued that the driver`s actions could not lead to the wagon derailment.

Next, you need to evaluate the derailment stability coefficient taking into account the rail irregularities. If the stability coefficient has exceeded the permissible value, then the cause of derailment is a deviation in the track maintenance.

To simulate the spatial vibrations of the wagon, a spatial model of the carriage is used. The wagon is divided into separate objects with the corresponding connections between them.

In this work, for the first time, when simulating the spatial vibrations of tank wagons, a movable load was also taken into account. Therefore, the wagon (tank) was divided into the following objects:

- movable load;
- body + fixed load;
- two truck bolsters
- four side frames;
- four wheel sets.

These bodies are regarded as solid and non-deformable. In addition, the inertial and viscoelastic properties of the track in vertical and horizontal transverse directions were taken into account. Therefore, the above list of objects also includes the reduced mass of the rails.

All objects have at least one inertial parameter – mass (for example, movable load and rail). In addition, objects can also have other inertial parameters – mass moments of inertia relative to their main central axes (if the angular displacements of objects are taken into account). Position of any object at any time moment is determined by six coordinates. The movement equations of the object under consideration in Cartesian coordinates have the following form:

$$\begin{cases} m\cdot\ddot{x} = F_x + \sum_{k=1}^{n} R_{xk} \\ m\cdot\ddot{y} = F_y + \sum_{k=1}^{n} R_{yk} \\ m\cdot\ddot{z} = F_z + \sum_{k=1}^{n} R_{zk} + W \\ J_x\cdot\ddot{\theta} = M_x + \sum_{k=1}^{n} MR_{xk} \\ J_y\cdot\ddot{\varphi} = M_y + \sum_{k=1}^{n} MR_{yk} \\ J_z\cdot\ddot{\psi} = M_z + \sum_{k=1}^{n} MR_{zk} \end{cases} \quad (1) \qquad (1)$$

where F_x, F_y, F_z – projections of external forces on the x, y, z axes, respectively, R_{xk}, R_{yk}, R_{zk} – projections of connection reactions on the x, y, z axes, respectively, M_x, M_x, M_x – moments of external forces F relative to the x, y, z axes, respectively, MR_{xk}, MR_{yk}, MR_{zk} – reaction moments R_k relative to the x, y, z axes, respectively, $\ddot{x}$, $\ddot{y}$, $\ddot{z}$ – objects acceleration along the x, y

, z axes, $\ddot{\theta}$, $\ddot{\varphi}$, $\ddot{\psi}$ – angular accelerations of objects when rotating them around the x, y, z axes; J_x, J_y, J_z – inertia moments relative to the main central axes x, y, z respectively, m – object mass, W – object weight. As external forces F_x, F_y, F_z, acting on the tank shell, the components of the longitudinal forces in the inter-wagon connections are taken. Vertical components are caused by the difference in height between the longitudinal axes of automatic couplers of neighboring wagons; the horizontal ones are caused by the relative rotation of automatic couplers of two neighboring wagons. External forces acting on wheel sets are caused by rail irregularities. In equations (1), n is the number of connections applied to the object. As can be seen from the movement equations of body (1), the right parts include the connection reactions R_{xk}, R_{yk}, R_{zk}, which depend on the type of connection (linear elastic-viscous, bilinear elastic-viscous, "dry friction"). Between wheels and rails, friction and interaction forces arise due to the elastic movements of the rails. The friction forces (creep) and their components along the axes x and y are determined by Carter's theory. When creating a carriage model, the type of connection and its numerical values are indicated for each pair of objects [10].

In our computations the coefficient of stability against derailment (sometimes called coefficient of resistance to derailment or stability factor) has been computed by the following formula: $C_S = \dfrac{tg\alpha - \mu}{1+\mu tg\alpha}\dfrac{S_p}{S_y}$, where: α – is the flange inclination angle, μ – friction coefficient, S_p – vertical interaction force between the leading wheel and rail, S_y – lateral interaction force between the leading wheel and rail [4, 10-12].

Based on mathematical models of the longitudinal train dynamics, fluid oscillations in a tank, and spatial vibrations of wagon [5–8], a computer program was developed that was used for calculations.

The above methodology was used to determine the derailment cause of the first tank wagon, in the direction of travel, in non-homogenous train consisting of 50 tank wagons with one 2M62m diesel locomotive, which is located in the head of the train. The train was formed according to the following scheme: 19 loaded tank wagons with tank ullage of 0.59 m and a mass of 86 tons, 24 empty tank wagons and 7 loaded tank wagons with tank ullage of 0.49 m and a mass of 88 tons.

Figure 1 presents the site diagram, in the lower part of which the distances to the control points relative to the point of computation beginning are given. The same distances are shown in the graphs of the computation results. In the above figure, L_1 and L_2 are the lengths of the transition curve at the entrance to the corresponding curve, R_1 и R_2 are the radii of the circular curves, respectively. During computations, an element 800 m long with a slope of 0.008 was added before the first element of the track profile, on which the train was located at the time of brake test.

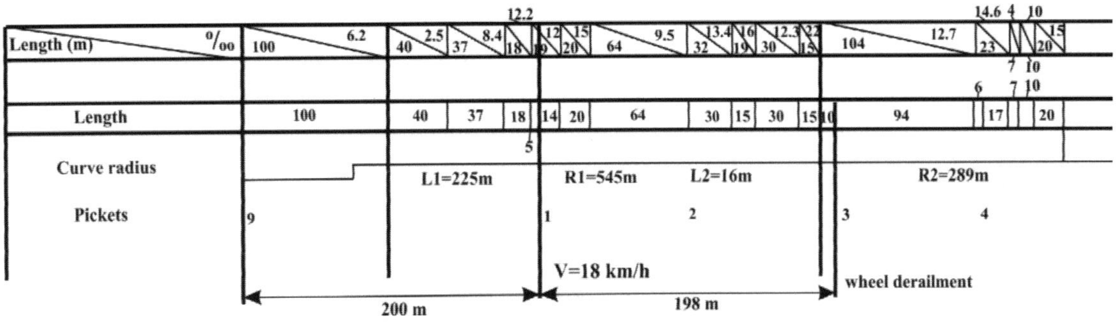

Figure 1. The track section scheme.

As appears from the official investigation materials, immediately before the tank wagon derailment the driver run brake test (on 38 km/h speed) on a descending grade and in a curve. Then, after brake release, he allowed the train to speed-up till 24 km/h and performed the emergency braking till the train's stop. The longitudinal compressive forces arising in the train during such a maneuver can cause lift-off of wagons. To determine the cause of the derailment, the computer modelling was performed. Its results are presented below.

3. Study case

3.1. Determination of longitudinal forces in a train

The proposed methodology was used to determine the cause of the tank wagon derailment in a non-homogenous freight train consisting of 50 wagon tanks on an existing track section of the Lithuanian railways. The motion trajectory (dependences of movement speed on running time and track coordinate) was evaluated first (see Figures 2 and 3). It was performed at the set modes of train driving, starting from the brake test at 38 km/h, the following brake release, speeding-up from 18 km/h at coasting retardation to 24 km/h due to the longitudinal profile configuration and then emergency braking. In these Figures the track coordinates are shown for the locomotive.

As a result of the computations, the satisfactory coincidence between the train motion trajectory and the data presented in the speedometer tape transcript has been obtained. In particular, according to this source of information, speeding up of the train from 18 km/h to 24 km/h after the brake release, took 200 m of the track, and the modelling gave the value of 185 m. Braking distance after applying the emergency brakes at 24 km/h was 80 m according to the investigation materials that fully coincides with the simulation results (Figures 2 and 3). Therefore, we can proceed to the analysis of the longitudinal forces obtained as a result of the computation.

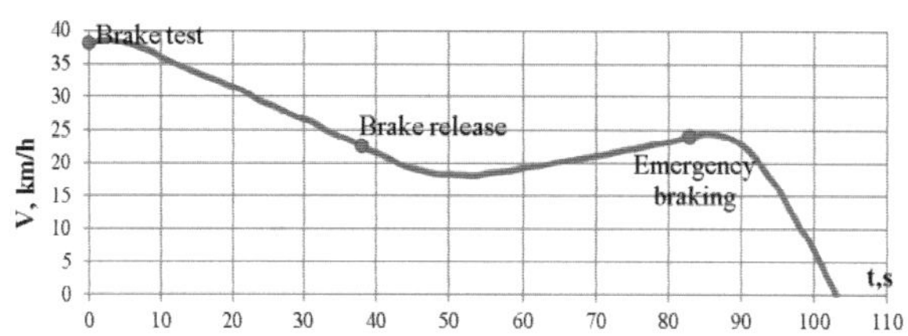

Figure 2. Dependence of movement speed on travel time.

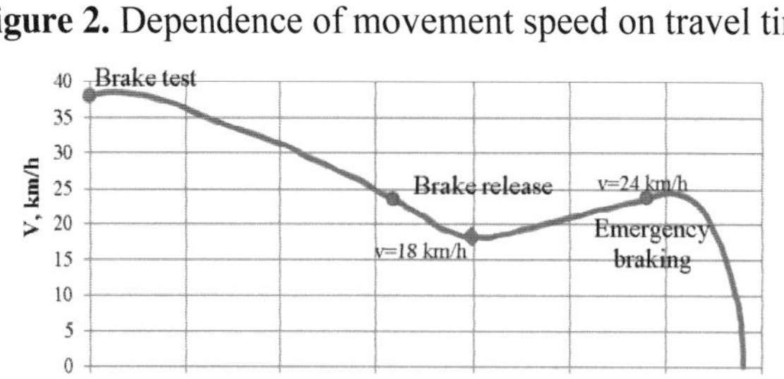

Figure 3. Dependence of speed on the coordinate of the path.

During computations, oscillograms of the longitudinal forces in the front and rear automatic couplers of the derailed tank-wagon (Figure 4) and the distribution of the maximal longitudinal forces maxS along the train length (Figure 5) were obtained.

The level of quasistatic longitudinal compressive forces acting on the 1st tank wagon after brake release was approximately 500 kN (Figure 4), and the maximum shock compressive forces did not exceed 600 kN. This result does not give reason to make assumptions about wagon lift-off due to braking.

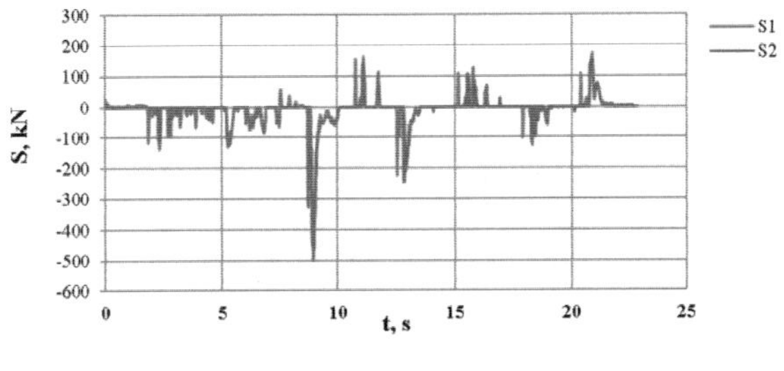

Figure 4. Longitudinal forces acting on the derailed tank wagon.

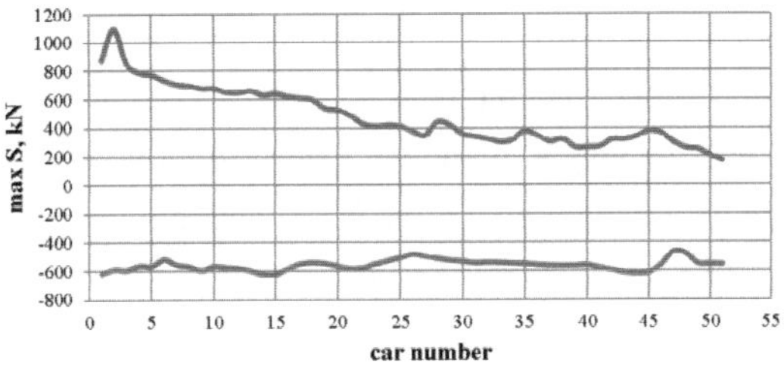

Figure 5. Distribution of the forces' maxima along the trainset length.

3.2. Determination of the dynamic indicators of the wagon tank without taking into account the irregularities of the railway track

Now we evaluate the derailment stability coefficient. As a result of modeling the spatial oscillations of the first (derailed) tank wagon after the locomotive, the following was obtained:

- vertical dynamics coefficients of the axle-box $Kdv1$ and central $Kdv2$ suspension stages,
- derailment stability coefficient,
- horizontal dynamics coefficients Kdh.

As an example, the values of the coefficients of vertical and lateral dynamics as well as the derailment stability coefficient are shown in Figures 6 and 7. The track coordinates are given for the first tank wagon, also on the track section preceding the derailment site.

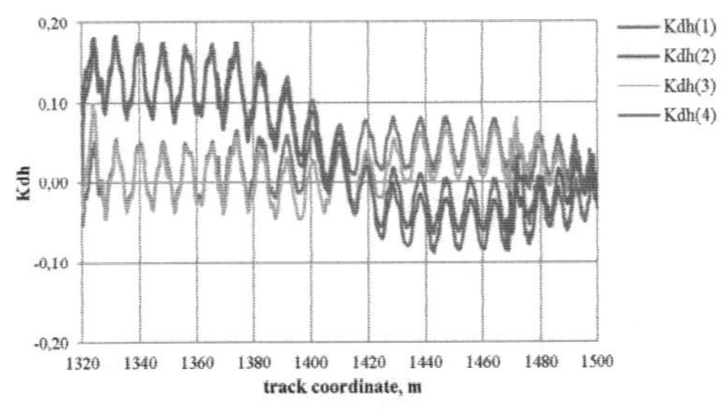

Figure 6. Coefficients horizontal dynamics 4 wheelsets.

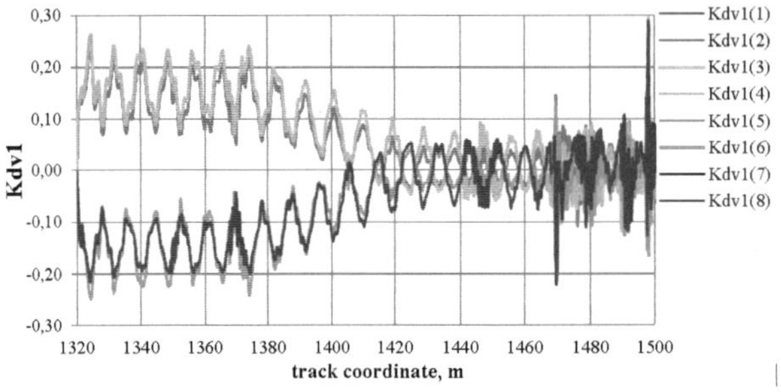

Figure 7. Coefficients of the vertical dynamics of the axle box.

In Figure 6, the notation $Kdh(i)$ correspond i -th to the wheel pair number. In Figure 7, the graphs $Kdv1(1) - Kdv1(4)$ – correspond to the change in the coefficients of the vertical dynamics in the left axle boxes of all wheel sets, $Kdv1(5) - Kdv1(8)$ – in the right axle boxes of all wheel sets. The nature of the obtained coefficients of the horizontal dynamics of the wheel sets is explained by the curve running in the track plan. As can be seen from the above Figures, the values of the coefficients of horizontal and vertical dynamics do not exceed the permissible values of 0.38 and 0.6, respectively [13, 14].

The performed modelling gave the value of 1.8 for the coefficient of stability for the first tank wagon. That is not less than the permissible value of 1.4 [10, 11].

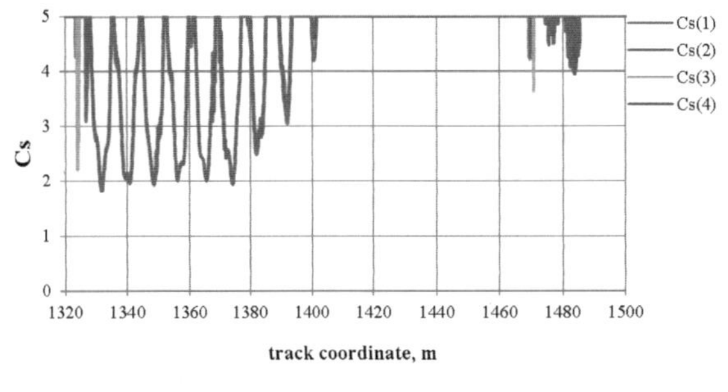

Figure 8. Derailment stability coefficient.

Thus, the obtained results allow to say that the train driving mode and the drivers' actions could not lead to the derailment.

3.3. Determination of the dynamic indicators of the wagon tank taking into account the irregularities of the railway track

Let us now evaluate the influence of the track condition on the derailment stability coefficient. Based on the given information the displacement of the left and the right rails in relation to their axes in the plan and profile were determined. These data were used as the rail irregularities for determination of the tank wagon motion stability (Figure 9, 10).

Below in Figure 9, deviations in the vertical plane of each rail from its longitudinal axis are shown. Positive values of Z correspond to downward deviation from the longitudinal axis of the track. Figure 10 presents deviations in the horizontal plane of each rail from its longitudinal axis. In this Figure, positive Y values correspond to the rail deviation to the right of its longitudinal axis in the direction of travel.

The results of calculations for determining the tank wagon running safety taking into account the above mentioned rail irregularities in the vertical and horizontal planes, are shown in Figure 11.

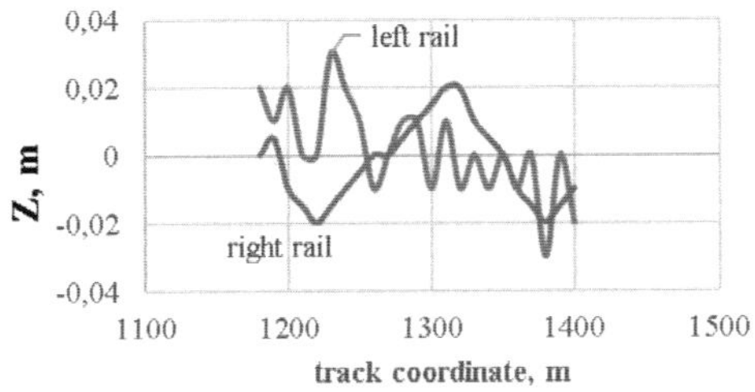

Figure 9. Vertical track irregularities.

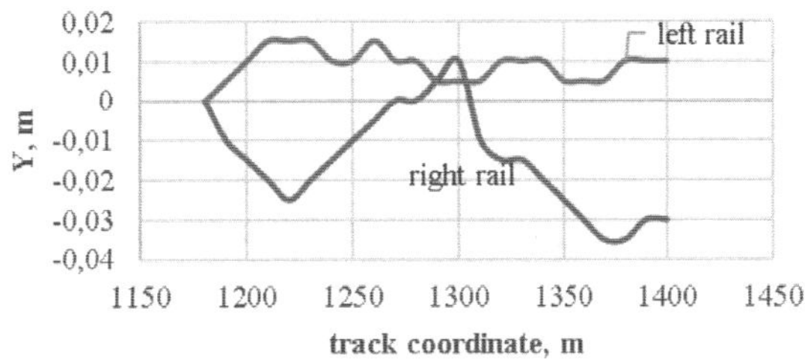

Figure 10. Lateral track irregularities.

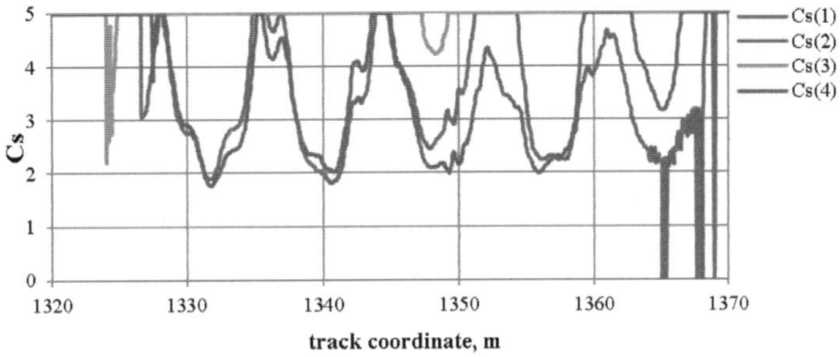

Figure 11. Derailment stability coefficient.

The coefficients of vertical dynamics in the axle-box Kdv-1 and in the central suspension Kdv-2 correspondingly, as well as the coefficient of horizontal dynamics Kdh are shown in Figures 12-14.

Taking into account the rail irregularities significantly changed the form of the dependences of the dynamics coefficients on the track coordinate (Figure 12-14), although their maximum values are within the normal range. It is seen in Figure 11, that the stability coefficient has fallen to zero. Thus, one can claim with confidence, that the studied derailment case was caused by non-compliance of the track with the Norms of Track Maintenance as a result of badly performed repair works.

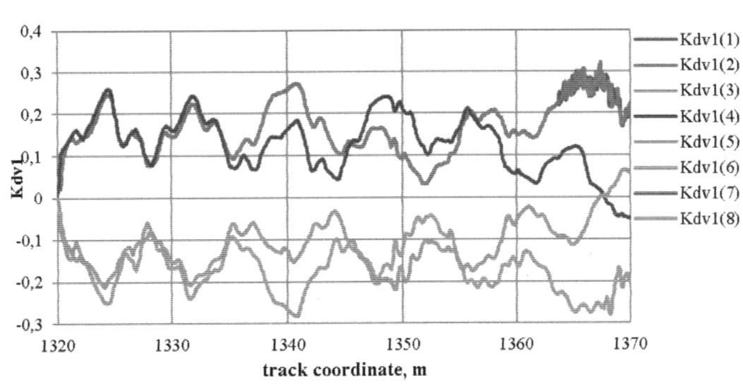

Figure 12. Coefficients of vertical dynamics in the axle-box unit.

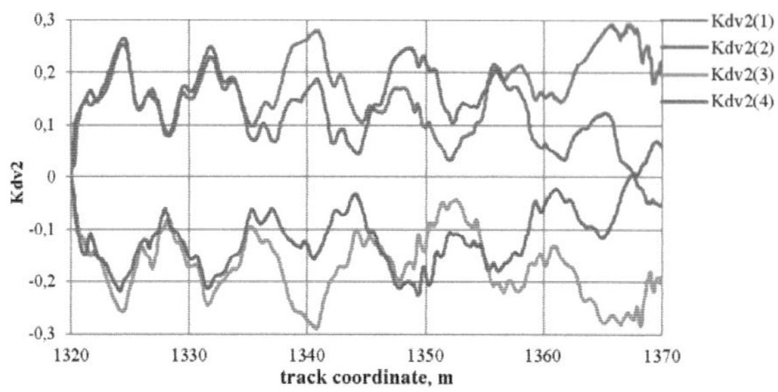

Figure 13. Coefficients of vertical dynamics in the central suspension.

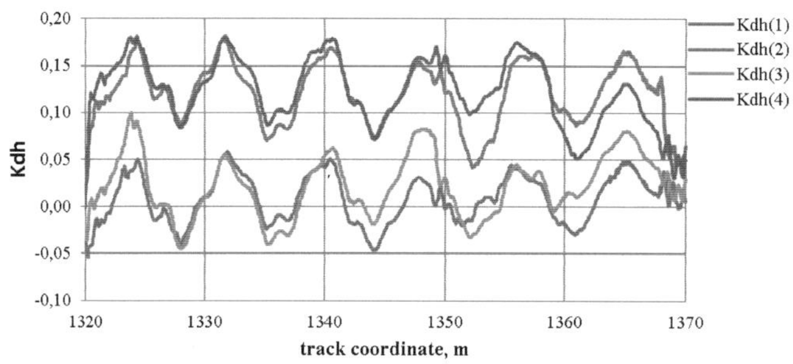

Figure 14. Coefficients of lateral dynamics.

To evaluate innovations in the spatial model of the tank wagon (load movability) for the stability of wagon movement, a similar simulation of the movement of a freight non-homogenous train was carried out, in which load movability was not taken into account. The results of the calculations for determining the stability coefficient of the leading tank wagon taking into account given rail irregularities in the vertical and horizontal planes are shown in Figure 15.

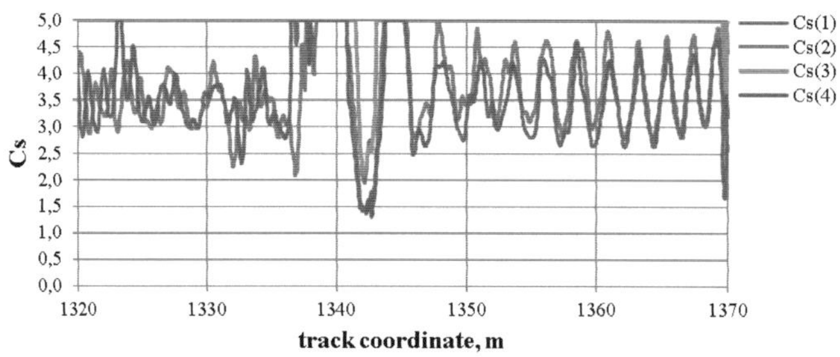

Figure 15. Derailment stability coefficients for open wagon freight train.

As one can see from the plot, the derailment stability coefficients in this case are greater than 1.4. From this, it is possible to say that not only deviations in the track maintenance, but also fluid vibration in the tank wagons were the cause of derailment.

4. Conclusions

When analysing the results of modelling, one can come to the following conclusions:

- the given methodology can be used to establish the causes of wagon derailments, after providing all materials of the official investigation, using computer simulation;
- for determining the derailment causes for the first time its spatial model, which takes into account the load movability was used;
- as a result of simulation of train movement and the spatial oscillations of the tank wagon, the oscillograms of the longitudinal forces, the distribution of their maximum values along the train length, the dependence of speed on the travel time and the track coordinate, as well as the wagon dynamic indicators and the derailment stability coefficient should be obtained and analyzed;
- the above methodology was used to determine the derailment cause in a real non-homogenous freight train formed from 4-axle tank wagons;
- based on the analysis of the results, it was concluded that the cause of the tank wagon derailment was not only the deviation from the track maintenance norms, but also the fluid fluctuations in the tank wagons.

References

[1] Romen Yu S, Klebanov I M and Soldusova E A 2013 Simulation of railcar body vibrations *Vestnik samarskogo gosudarstvennogo tekhnicheskogo universiteta. seriya «tekhnicheskie nauki»* **3** pp 141–147
[2] Vasilev A S, Boldyrev A P, Keglin B G and Gurov A M 2014 Research of freight car's longitudinal loading equipped new frictional absorbing devices *Bulletin of Bryansk State Technical University* **1** 41 pp 12–17
[3] Powell J P and Palacín R 2015 Passenger Stability Within Moving Railway Vehicles: Limits on Maximum Longitudinal Acceleration *Urban Rail Transit* **1** pp 95–103
[4] Korotenko M L, Klimenko I V and Panasenko V Ya 2013 Safety from wheel derailment and improvement of rail carriage designs (Dnepropetrovsk: Makovetsky Yu V) p 221
[5] Blokhin Y P, Zheleznov K I and Ursuliak LV 2007 Computing complex for solving problems of safety and stability of the movement of rolling stock of railways *Bulletin of the Dnipropetrovsk National University of Railway Transport* **18** pp106–114
[6] Blochinas E, Dailydka S, Lingaitis L and Ursuliak L 2015 *Nonstationary and quasi-static modes of railway train movement* (Vilnius: Technika) p 167
[7] Ursuliak L V, Vaiciunas G, Romanyuk Ya N, Petrenko V and Stepchenkova E S 2016 About waveform of braking cylinder filling in freight cars *Science and Transport Progress* **34** pp 165–173
[8] Blokhin E P, Garkavi N Ya, Zheleznov K I, Ursuliak L V and Romanyuk Ya N 2010 Software and computer complex for studying the longitudinal loading of bulk trains *Science and Transport Progress* **2** 62 pp 6–16
[9] Vershynskiy S V 1970 Dynamics, durability and the stability to heavy wagons and high-speed trains *Collection of works VNIIZhT* (Moscow: Publisher Transport) **425** p 208
[10] Myamlin S V 2002 *Modeling the dynamics of railway vehicles* (Dnepropetrovsk: New ideology) p 240
[11] Iwnicki S 2006 *Hand book of Railway Vehicle Dynamics* (CRC Press: Taylor & Francis Group) p 527
[12] Garg V K and Dukkipati R V 1991 *Rolling stock dynamics* (Moscow: Publisher Transport) p 359
[13] Vahony vantazhni. Zahalni vymohy do rozrakhunkiv ta proektuvannia novykhi modernizovanykh vahoniv kolii 1520 mm (nesamokhidnykh) 2017 DSTU 7598:2014 p 158
[14] Normy dlya rascheta i proektirovaniya vagonov zh. d. MPS kolei 1520 mm (nesamohodnyh) 1996 (Moscow: GosNIIV-VNIIZHT) p 319

IOP Conf. Series: Materials Science and Engineering **985** (2020) 012015 doi:10.1088/1757-899X/985/1/012015

Contemporary principles for solving the problem in noise reduction from railway rolling stock

Yu Zelenko[1,2], D Zelenko[1] and L Neduzha[1,3]

[1]Dnipro National University of Railway Transport named after Academician V. Lazaryan, Lazaryan str., 2, 49010, Dnipro, Ukraine

[2]j.v.zelenko@gmail.com, [3]nlorhen@i.ua

Abstract. The statistical and monitoring data confirm the significant impact of the operational mode on the most important noise and vibration parameters, which is relevant for ensuring the efficiency and quality of the operational process with regard to ergonomic and environmental requirements. The article analyzes the acoustic parameters of rolling stock of the railway, emphasizes the peculiarities of noise emission from various elements of rolling stock and a track. Existing approaches towards solving the problem in the organization of railway noise control are outlined. On the basis of existing modern methods for measurement and analysis of acoustic parameters, the authors propose conceptual approaches toward monitoring the acoustic parameters, predicting noise characteristics and constructing the noise maps of railways. The algorithm of the developed acoustic model for the operational monitoring of noise load and the possibility to compile a case of noise maps for Ukrzaliznytsia is presented. The proposed algorithm allows at given basic design parameters and speed of rolling stock, known sound-absorbing and dissipative characteristics of all elements to carry out a rational selection of noise-protective measures in accordance with the sanitary noise regulations for a specific receptive facility.

1. Introduction

Along with higher safety, railway transport has obvious advantages over other modes of transport in terms of environmental safety, it is characterized by high energy efficiency and lower hazardous substances emissions [1-7].

Simultaneously, one of the current priority directions for the development of transport systems is to ensure the efficiency and quality of the operational process, taking into account ergonomic and environmental requirements [7-18]. Herewith, statistics and monitoring data confirm that the operating mode significantly affects the most important noise parameters arising from the rolling stock movement in curves [7, 12-15, 17, 18], and the basic operational requirements are determined by the infrastructure, in particular, the presence curves and their minimum radii.

2. Evaluation the degree of the railway noise impact

To meet the requirements of sanitary and environmental safety [7, 19, 20, 22-25], it is necessary to calculate the expected noise levels in territories adjacent to the railway lines at the stages of railway design, development and commissioning of rolling stock.

If necessary, one should develop noise reduction measures, for example, through specific technologies for noise reduction in the source or installation of the noise screens along the railway

Content from this work may be used under the terms of the Creative Commons Attribution 3.0 licence. Any further distribution of this work must maintain attribution to the author(s) and the title of the work, journal citation and DOI.
Published under licence by IOP Publishing Ltd

tracks. In this regard, the reliability of methods for calculating noise from railway transport and the correct determination of noise characteristics are important.

As the noise generated by the railway transport has a pronounced non-permanent nature of changes over time, its evaluation for compliance with the permissible levels is carried out simultaneously according to the equivalent and maximum sound levels, which are determined at a distance of 25 m from the axis of the close track at a height of 1.5 m from the ground level.

Unfortunately, there are still no specific technical regulations that require monitoring and control of noise from railway transport at the nationwide level. Table 1 compares data upon noise levels from domestic rolling stock with European standards.

Table 1. Comparative analysis of noise levels of rolling stock compared to European standards.

Train class	Maximum noise level at a distance of 25 m, dB(A)		Exceeding norms, dB
	Ukraine	European standards	
Electric train	87	83	4
High-speed train type ICE	88	85	3
Freight train	89	84	5
Passenger train	93	90	3
Track equipment	95	85	10

As indicated in the data analysis, the recorded noise levels at one of the sections of the Prydniprovska railway line exceed the established European standards by 3 – 10 dBA.

3. Analysis of the acoustic parameters of rolling stock

The main sources of noise on railway transport are classified as follows:

- train locomotive noise (traction noise);
- noise generated by rolling vibrations;
- noise from the impact of car wheels in contact with the rails;
- aerodynamic noise at the train movement, which depends on the speed and a set of cars.

Such factors as traffic intensity, speed and length of trains, etc. [7, 19, 21] affect the noise level from the traffic of railway transport.

To evaluate the noise levels in the surrounding area, calculation techniques based on current regulatory documents are applied. They take into account all of the above-described factors.

In the practice of estimating the transport noise, it is accepted as a model train to consider a linear source of finite length (ℓ) over an acoustically rigid surface, consisting of uniformly distributed incoherent non-directional, inphase- oscillating point sources. The mechanism of noise generation and emission demonstrates the distribution of energy between the main emitting surfaces (Figure 1) and has a clearly expressed frequency character.

The physical mechanism of rolling noise from rolling stock is characterized by the following parameters:

- the propagation of sound waves along the rail causes noise emission in the range up to 1000 Hz;
- the vibration of the wheel arises from 1600 Hz, in this, the main contribution to the noise level of the train occurs at the frequency of 2000 – 4000 Hz and can reach 8000 Hz;
- the frequency range of the noise emitted by the elements of rolling stock is 100 – 500 Hz;
- the noise emission from sleepers occurs in the frequency range up to 400 Hz;
- the surface irregularities of the wheel and rails (roughness) generate vibration upon their contact. The vibration is transmitted through the pad between a rail and a sleeper.

In general, there is a characteristic noise dependence on speed: at speeds up to 50 km/h internal sources: elements of rolling stock (electric motors, compressors) prevail; at a speed of 50 – 300 km/h noise from the "wheel-rail" interaction prevails; more than 300 km/h one can observe that typical aerodynamic processes are manifested [7, 20].

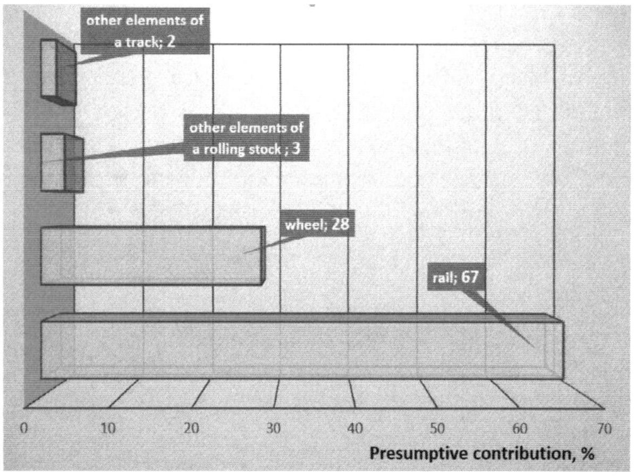

Figure 1. Noise emission characteristics for the high-speed ICE train (at ν =185 km/h).

Today, many analytical and numerical methods are used to solve the task of rolling noise analysis. The most common methods of calculation are based on a refined Remington-Thompson model. In this model, in addition to vibration caused by the interaction of wheel and rail irregularities, the processes of sound emission during their force interaction are taken into account: vibrations from elastic wheel and rail deformation. This allows obtaining detailed vibroacoustic parameters for the recipient accounting point in order to develop specific measures concerning the sound damping in the source [7].

Thus, the analysis of the experimental data (Figure 2) confirms that at small distances the indices of noise emission from the wheel (W) and the rail (R) is approximately the same. At the same time, with increasing distance, they increasingly differ. This is because the wheel emits a spherical sound wave and the rails emit cylindrical one. At distances greater than 50 m this difference can reach 7 – 10 dB.

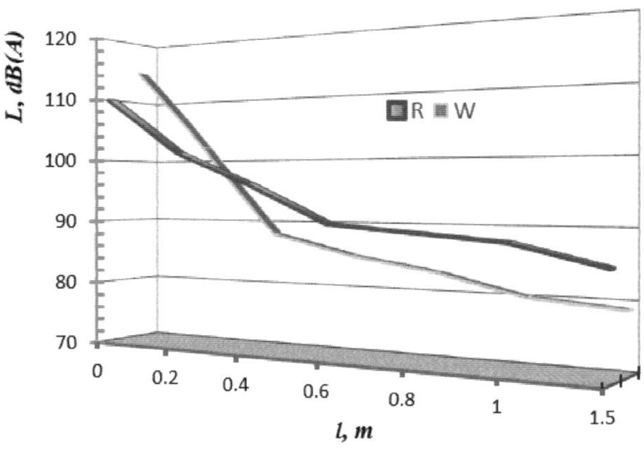

Figure 2. Sound pressure distribution L, dB(A) depending on the distance to the noise source: R is rail noise, W is wheel noise.

Thus, at these distances, the rails are the main source of rolling noise and noise reduction measures should be aimed at eliminating irregularities on the surface of a rail, increasing the impedance of the rail, reducing its noise-emitting ability.

4. Development of measures for monitoring of acoustic parameters on railway transport

The concept of noise monitoring has been introduced as an overall system for monitoring environmental noise, estimating and predicting changes in environmental noise due to technogenic activity [22-25]. This definition allows to provide the noise monitoring using special calculation methods. The inclusion of calculation methods in the noise monitoring system makes it possible to monitor large linear-type and magistral facilities at visible time intervals. This would be practically impossible if monitoring was limited only by long-term noise measurements in a representative number of points necessary for compiling noise maps. At the same time, it is necessary to emphasize the necessity of measurements when checking the calculation results for compiling operational noise maps, developing action plans to reduce noise, as well as evaluating the effectiveness of their implementation.

For many complex projects, such as linear mainline facilities, nodes, bridges, and stations, a noise map is the only possible tool toward reliable monitoring and solving the noise question. Noise maps enable:

- to select recipient points for acoustic monitoring and perform acoustic monitoring;
- to estimate background noise levels at any point of the mainline when designing a new housing estate;
- to justify the evaluation of individual sites for building;
- to evaluate the list of necessary noise-protection measures when designing objects;
- to simulate changes in acoustic impact when changing the characteristics of transport nodes and traffic.

The stringency of domestic environmental requirements for rolling stock, their approximation to European standards also confirms the need to use more extensively the results of numerous foreign studies in estimating the processes of formation, propagation of transport noise, its prediction.

The most comprehensive understanding of the acoustic situation near the railways lines is represented by noise maps, which are developed on the basis of the current regulations and calculation techniques for noise level evaluation. In Europe, the legal basis for compiling the noise maps is represented by European Parliament Directive 2002/49/EC concerning the environmental noise. According to this Directive, the development of noise maps for cities, motorways and railways, as well as the development of action plans for heavy railways passing near residential areas, are envisaged [7, 22-25].

Estimation of existing and comparison of planned noise situations is mainly performed by means of special acoustic programs for the analysis of ambient noise. Software packages can effectively reduce the load and allow performing a more in-depth analysis. Here are some basic requirements for successful selection of the appropriate program:

- functions in processing efficiency of computational models;
- open software interface and modular system;
- configuration capability for types of objects and their attributes;
- support for customizable interface features programmable by the client.

To compile noise maps, the authors apply new approaches and methods that have been developed in the light of the worldwide experience of noise evaluation and monitoring [7]. According to which, the authors created a database filled with information about the geometry of the railways line and the source of traffic noise (it makes it possible to construct an acoustic model and make calculations to determine the noise levels at the recipient point). The construction of the acoustic model is presented in the form of a block-diagram, which is shown in Figure 3. The presence of this algorithm allows with the known basic design parameters and the speed of rolling stock, sound-absorption and dissipative characteristics of all elements by calculation, to carry out their rational selection accordingly to compliance with sanitary noise standards for the receptive facility.

In fact, the name "noise map" means not one single map, but a massif of preassigned digitized cartographic objects, which are applied with colored zones of all noise indices ($L_{full\,day}$, L_{day} and L_{night}), as well as all existing sources of noise that fall into the impact zone on the recipient calculated point (main roads, streets, railway and tram tracks, airport, industrial facilities, etc.).

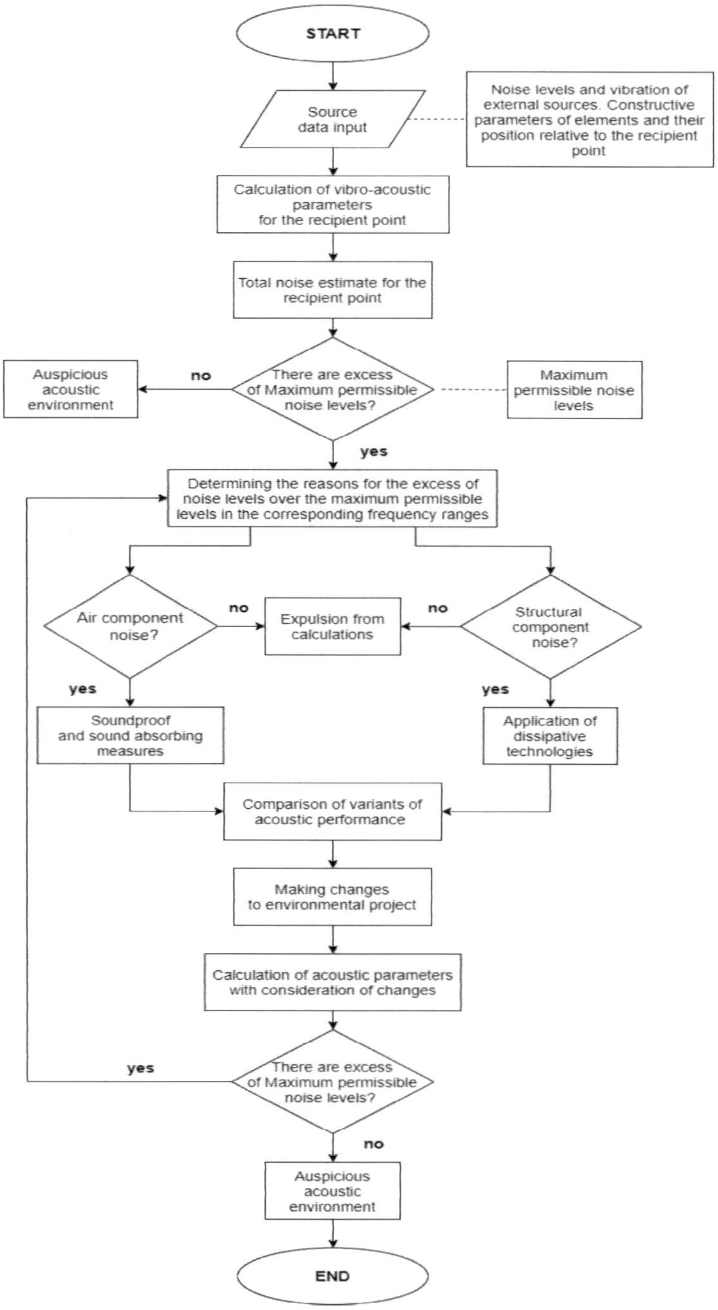

Figure 3. The algorithm of the proposed acoustic model.

Figure 4 shows visualization in the distribution of the indicator value ($L_{full\,day}$) according to the proposed by authors acoustic model for the section of the railway line in the area of the large rail junction at the Regional Branch "Prydniprovska zaliznytsia" – Nyzhnodniprovsk-Vuzol. Analysis of the representative data confirms that at a designated railway facility in the area of Nyzhnodniprovsk-Vuzol sections with above-permitted indices of noise load ≥ 85 dB were identified (A), which marked with red color.

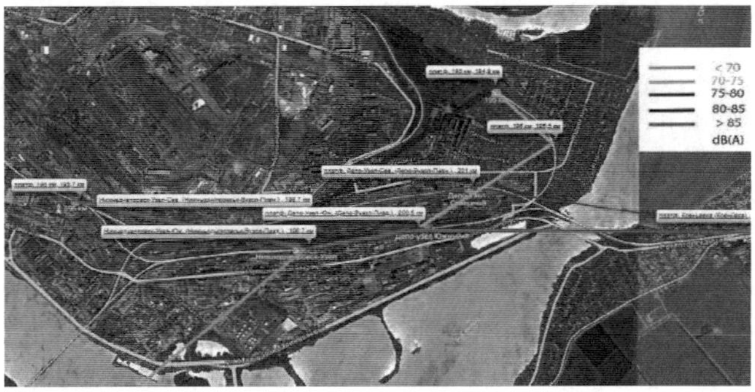

Figure 4. Visualization of the indicator value distribution for the section of the railway line in the area of Nyzhnodniprovsk-Vuzol, L_{day}, dB(A).

Based on the analysis of the representative data obtained, it is proved that determining the dominant type of noise generated is of great importance for improving the effectiveness of measures against the skirr in curves. In addition, the typical characteristics of the track (curve radii, slopes, requirements for adhesive coefficient), rolling stock and operating process must also be clearly defined. It is on the basis of these data that we can determine which measures are most appropriate.

The application of visualization models allows to identify linear and precise objects, in which an excess of permissible norms in noise loads is recorded, to conduct monitoring for specified objects in dynamics, to determine the causes of acoustic loads and to develop priority measures to handle the acoustic situation at the facility.

5. Conclusions

European Directive 2002/49/EC on the evaluation and oversight of noise in the environment emphasizes that EU Member States need to develop a noise reduction plan (the so-called Action Plans), and in the current European integration context, this is already true for Ukraine. In this connection, the authors analyzed static and research data, confirming the availability of a wide range of noise reduction systems and technologies that can be included in the database to select and test specific measures for a specific recipient point.

Since measures to prevent or reduce noise can be applied at different stages in the process of occurrence of the described types of noise, the search and development of contemporary mathematical models that can take into consideration the type of noise, features and phasing of its formation is an important monitoring process. Such models will make it possible to rationalize the choice of the activity or justify applying the range of noise-protective methods, which is a timely and most urgent task of acoustic ecology on transport.

The presence of the algorithm proposed by the authors will allow, at the given basic design parameters and speed of rolling stock, known sound-absorbing and dissipative characteristics of all elements to carry out a rational selection of noise-protective measures in accordance with the sanitary regulations of noise for a specific receptive facility.

6. References

[1] Zelenko Y, Malovanyy M and Tarasova L 2019 Optimization of heat-and-power plants water purification *Chemistry and Chemical Technology* **13** 2 pp 218–223

[2] Bannikov D O 2011 Analysis of the causes of accidents of steel capacitive structures for bulk materials *Metallurgical and Mining Industry* **3** 5 pp 243–249

[3] Kunah O M, Zelenko Y V, Fedushko M P, Babchenko A V, Sirovatko V O and Zhukov O V 2019 The temporal dynamics of readily available soil moisture for plants in the technosols of the Nikopol Manganese Ore *Basin Biosystems Diversity* **27** 2 pp 156–162

[4] Zelenko Y and Bezovska M 2019 Development of an environmentally friendly scheme for the recovery of used engine oils *New stages of development of modern science in Ukraine and EU countries: monograph* 3rd ed (Riga: Baltija Publishing) pp 143–164

[5] Plakhotnyk V M, Yaryshkina L A, Boychenko A M 2005 Environmental aspects of transport accidents with dangerous freights *Czasopismo Techniczne. Mechanika* **10** pp 33–38

[6] Zelenko Yu, Lunys O, Neduzha L and Steišūnas S 2019 The assessment of negative impact of oil products on railroad track and rolling stock constructions *Proc. of the 23rd Intern. Sci. Conf. Transport Means 2019* pt III (Palanga, Lithuania: Kaunas Univ. of Techn.) pp 1300–1306

[7] Myamlin S V 2014 *Parametric environment in railway transport. Principles, assessment, monitoring, security: monograph* (Dnipropetrovsk: Lithographer Publ) p 203

[8] Burdzik R and Słowiński P 2020 Images of Vibrations of a Passing Railway Vehicle *Proc. of the: Transportation Science and Technology. Lecture Notes in Intelligent Transportation and Infrastructure* (Vilnius: Springer, Cham) pp 47–56

[9] Kardas-Cinal E and Gągorowski A 2019 Application of Statistical Analysis to Investigate the Relation Between Road Roughness and Vehicle Vibrations *Proc. of the 23rd Int. Sci. Conf. Transport Means 2019* pt I (Palanga, Lithuania: Kaunas Univ. of Technology) pp 285–289

[10] Klimenko I, Černiauskaite L, Neduzha L and Ochkasov O 2018. Mathematical Simulation of Spatial Oscillations of the "Underframe-Track" System Interaction *Proc. of 12th Int. Conf. ITELMS'2018* (Panevėžysm, Lithuania: Kaunas Univ. of Techn.) pp 105–114

[11] Myamlin S, Neduzha L and Urbutis Ž 2016 Research of Innovations of Diesel Locomotives and Bogies *Proc. Eng.* **134** (Vilnius, Lithuania: Vilnius Gediminas Techn. Univ.) pp 469–474

[12] Klimenko I, Kalivoda J and Neduzha L 2018 Parameter optimization of the locomotive running gear *Proc. of 22nd Int. Sci. Conf. Transport Means 2018* pt III (Trakai, Lithuania: Kaunas Univ. of Techn.) pp 1095–1098

[13] Lunys O, Neduzha L and Tatarinova V 2019 Stability research of the main-line locomotive movement *Proc. of the 23rd Int. Sci. Conf. Transport Means 2019* pt III (Palanga, Lithuania: Kaunas Univ. of Techn.) pp 1341–1345

[14] Kostritsa S A, Sobolevska Y H, Kuzyshyn A Y and Batih A V 2018 Mathematical model of DPKR-2 dyzel train car *Science and Transport Progress* **1** 73 pp 56–65

[15] Raksha S V, Anofriev P G, Bohomaz V M and Kuropiatnyk O S 2019 Mathematical and S-models of cargo oscillations during movement of bridge crane *Naukovyi Visnyk Natsionalnoho Hirnychoho Universytetu* **2** pp 108–115

[16] Shashenko O, Shapoval V, Khalymendyk O, Andrieiev V, Arbuzov M, Hubar O and Markul R 2019 Features of the nonlinear calculation of the stress-strain state of the "rock massif-excavation support" system taking into account destruction *Proc. of the 23rd Int. Sci. Conf. Transport Means 2019* pt III (Palanga, Lithuania: Kaunas Univ. of Techn.) pp 1356–1363

[17] Bondarenko I, Lunys O, Neduzha L and Keršys R 2019 Dynamic track irregularities modeling when studying rolling stock dynamics *Proc. of the 23rd Intern. Sci. Conf. Transport Means 2019* pt II (Palanga, Lithuania: Kaunas Univ. of Techn.) pp 1014–1019

[18] Ignatenko D, Tiutkin O, Petrenko V and Alkhdour A 2019 Application of centrifugal modeling for the study of landscape structure stability *International Journal of Civil Engineering and Technology (IJCIET)* **10** 01 pp 2179–2187

[19] Gerlici1 J, Gorbunov M, Kravchenko K and Lack T 2018 Noise and temperature reduction in the contact of tribological elements during braking *MATEC Web of Conf.* **157** 4: 02010

[20] Matsiuk V, Myronenko V, Horoshko V, Prokhorchenko A, Hrushevska T, Shcherbyna R, Matsiuk N, Khokhlacheva J, Biziuk I and Tymchenko N 2019 Improvement of efficiency in the organization of transfer trains at developed railway nodes by implementing a "flexible model" *Eastern-European Journal of Enterprise Technologies* **2** 3-98 pp 32–39

[21] Bodnar B, Kapitsa M, Bobyr D and Kyslyi D 2019 Defining the Limits of Application and the Values of Integration Variables for the Equations of Train Movement *Naukovyi Visnyk Natsionalnoho Hirnychoho Universytetu* **6** pp 59–65

[22] ISO 2631-1. 1985, 1997 Mechanical vibration and shock. evaluation of human exposure to whole-body vibration. Part 1: International Organization for Standardization

[23] ISO 8608:1995 Mechanical vibration – road surface profiles – reporting of measured data

[24] SSR 3.3.6.037-99 "Sanitary regulations for industrial noise, ultrasound and infrasound"

[25] Directive 2002/49/EC of the European Parliament and of the Council of 25 June 2002 Relating to the Assessment and Management of Environmental Noise

IOP Conf. Series: Materials Science and Engineering **985** (2020) 012016 doi:10.1088/1757-899X/985/1/012016

Development of passive protection devices for a power head of a high-speed multiple unit train at its collisions

M Sobolevska[1,2], D Horobets[1] and S Syrota

[1]The Institute of Technical Mechanics of the National Academy of Sciences of Ukraine and the State Space Agency of Ukraine (ITM NASU and SSAU) 15, Lyashko-Popelya, Dnepr, 49005, Ukraine

[2]sobmb@i.ua

Abstract. Ukrainian passenger multiple
power heads with passive safety systems
damage and to minimize consequences at collisions
energy absorption devices (EADs) for power head
EAD 2, EAD 3 designs have been developed
locomotive EADs creation and EAD prototype
elements which are located sequentially.
hexagonal honeycombs inside. Element
triangular cells. EAD 2 design has three
developed on the basis of element 1. It has been
head frontal part at the coupling level, two E
EAD 3 in the inter-car connection at the coupling level
plastic deformation at an impact has been
. Finite element simulation of EAD
carried out for determination of design parameters.

1. Introduction

Increasing the speed and improving safety of passenger railway transportation are actual problems in Ukraine. Their solution directly influences on the technical and economic performance indicators of railway transport and determines its competitiveness in the world transport service market. The passenger multiple-unit train upgrading supposes on creation of new power heads which are equipped with an effective active protection to prevent collisions and with passive safety systems to operate automatically at accident collisions of trains with obstacles and to create conditions for passengers and train staff safety, to reduce rolling stock damage and to minimize emergency consequences.

This article is devoted to development energy absorption devices which are included in a passive safety system of a high-speed multiple-unit train power head at collisions on the 1520 mm gauge railways.

The analysis of the normative requirements which act in EU countries, Ukraine, the Customs Union countries to regulate high-speed train vehicles strength in operation and their passive protection at collisions has been carried out. It has been shown that such standards in EU countries are EN 12663 [1] and EN 15227 [2] respectively, in the Customs Union countries – Technical Regulation of the TC "On the safety of railway rolling stock" [3], GOST R 53076-2008 (structural requirements of railway vehicle bodies) [4] and GOST 32410-2013 (requirements for crash- systems) [5]. In 2016, Ukraine adopted standards DSTU EN 12663: 2015 (EN 12663-1: 2010, IDT) [6] and DSTU EN 15227: 2015 (EN 15227: 2008 + A1: 2010), IDT) [7]. Now these standards are in action in Ukraine [8] and [7].

Content from this work may be used under the terms of the Creative Commons Attribution 3.0 licence. Any further distribution of this work must maintain attribution to the author(s) and the title of the work, journal citation and DOI.
Published under licence by IOP Publishing Ltd

In accordance with DSTU EN 12663, the driver's cab design should be capable to stand compressive longitudinal loads without plastic deformation: 1.5 MN at the coupler level (or buffer level), 0.4 MN at 150 mm above the top of the structural floor at head stock, 0.3 MN at the level of the window-sill and 0.3 MN at the level of the cant rail.

DSTU EN 15227 regulates the mandatory availability of passive safety systems in the multiple-unit train vehicle designs. DSTU EN 15227 defines collision scenarios, reference trains, and conformity criteria for railway vehicles with PSS. For experimental and computational verification of PSS standard DSTU EN 15227 defines four collision scenarios, which are characterized by a set of collision conditions (mass values, initial impact speeds and other parameters of collision objects). Scenario 1 is a frontal collision of two identical reference trains at a speed of 36 km/h. Scenario 2 is a frontal collision of a reference train at a speed of 36 km/h with an 80 t freight wagon. Scenario 3 is a frontal collision of a reference train at a speed of 110 km/h with a 15 t large road vehicle as a deformable obstacle on a level crossing. Scenario 4 is a collision of a train with a low obstacle (car, animal, rubbish) on a level crossing.

A reference train for power head design development consists of a power head and four vehicles behind it. It is necessary to demonstrate the fulfillment of tasks of ensuring the reference train passive safety in scenarios of collisions according to DSTU EN 15227 criteria. The mandatory DSTU EN 15227 requirement is to save intact space for survival of passengers and the train crew (safety zone) throughout the deformation of energy absorption devices which are included in the PSS. Local plastic deformation and local buckling are acceptable if it is demonstrated that they are sufficiently limited, so as not to reduce the survival spaces. The mean longitudinal deceleration in the survival spaces shall be limited to 5 g (g is the acceleration of gravity) for Scenario 1 and Scenario 2 and 7,5 g for Scenario 3.

At present, an actual and important task for Ukraine is to create a domestic high-speed multiple-unit train, especially its power head with its own develop PSS according to DSTU EN 12663 and DSTU EN 15227 requirements. It should be noted that the use of DSTU EN 15227 is impossible without rectification of reference trains and obstacles characteristics and without taking into account such features as combined draw-and-buffer gear devices, overall dimensions and mass requirements for power heads and coaches operating on 1520 mm gauge railways.

2. Analysis of existing energy absorption devices for power head passive protection

Today, we can see an active development of power head passive safety systems in Europe, China, South Korea, the Russian Federation, Kazakhstan and other countries. Patent-bibliographic researches of existing energy absorption devices for power head of high-speed passenger multiple-unit train at emergency collisions have been carried out [9-15]. Power head EADs absorb the main part of the accident collision energy. The front end of European power head is equipped with a lightweight push-back automatic coupler containing energy absorbing elements, an obstacle deflector, a multi-level system of energy absorbing devices which is usually placed outside the driver's cab and anti-climb units. The driver's cab frame is a safety zone for the driver's survival and evacuation at a collision. As a rule, lower-level energy absorbing devices are placed on the coupling level. EAD design is a box or a tubular type design with 1 MN energy capability. At accident collisions, impact energy is absorbed as a result of plastic deformation, flaring, or cutting of EAD design elements. An upper-level EAD design is placed in driver's cab window-sill part above the coupling level. It is a box design with various strain initiators and filling. The rear end of a European power head is equipped with couplers and anti-climb units usually including energy-absorbing components.

An upper-level EAD is not used in Russian power head designs unlike European ones with PSS. It is connected with different requirements of EN 15227 and GOST 32410-2013 in the scenarios of a train collision with an obstacle on a level crossing. According to EN 15227, 15 t deformable obstacle with a height of 3.2 m is used. According to GOST 32410-2013 10 t low non-deformable obstacle with a height of 2.2 m is considered.

There are no real projects on creation of the domestic power head with in-house design PSS in Ukraine now. The first Ukrainian electric train ("Tarpan") with PSS was developed in 2013 [15]. Only its power head had a PSS. It was developed by the Polish company IC engineering in accordance with the specification and orientation to the EN 15227 requirements. The power head front end was equipped with two steel box EADs at the coupling level, two steel box EADs at the driver's cab floor level and five steel box EADs in the window-sill part. The train power head had a stainless steel support design equipped with the SA-3 automatic coupler, which could move back into the under coach space at a collision. Other end of the power head was equipped with clearance-free couplers [20] with absorbing devices R-2P. Power head PSS effectivity has not been confirmed with crash tests.

3. Proposals on passive protection of a domestic power head

The proposals on passive protection of a domestic power head are developed. It is advisable:

- to integrate in the power head design lightweight push-back automatic coupler containing energy absorbing elements which do not create obstacles for the passive safety system work at collisions, an obstacle deflector, EADs with anti-climb units in a power head front end and in its rear end;
- to develop designs of power head body and underframe which have to be stronger in the longitudinal direction. These designs should have power elements to limit the safety zone for the survival and evacuation of passengers. These designs should allow to operate EADs adequately without loss of their total bearing capacity in the collision scenarios;
- to take such allowed values of longitudinal dynamic loads at collisions according to DSTU EN 15227 scenarios as 4,2 MN for a power head underframe, 3,0 MN for a coach power head underframe and values up to 0.7 MN for driver's cab front wall;
- to place two lower-level EADs with about 2 MJ total energy absorbing capacity in the power head front end at the coupling level for protection at the identical trains collision (Scenario 1) or the train collision with freight wagon (Scenario 2). The recommended length of such EAD should be from 1.0 m to 1.5 m;
- to place two upper-level EADs with about 0.25 MJ total energy absorbing capacity in the driver's cab window-sill part primarily for power head passive protection at the train collision with a large road vehicle on a level crossing (Scenario 3). The recommended length of the upper-level EAD should be from 0.8 m to 1.0 m;
- to place two lower-level EADs with about 0.3 MJ total energy absorbing capacity in the power head rear end part at the coupling level;
- to use box designs in the form of a truncated pyramid with cellular packages inside as lower/upper level EAD designs;
- to develop a EAD design using successful experience of EP20 locomotive EAD development and EAD prototype crash test results [17-18];
- to take into account that a driver's cab frame shall be a power design to limit the safety zone for the driver's survival and evacuation at accident collisions. The driver's cab frame should have a length of at least 750 mm, withstand operating loads in accordance with DSTU EN 12663 and maximum impact loads at collisions according to the scenarios of standard DSTU EN 15227.

4. Mathematical simulation

A task of EAD design plastic deformation at collision with 80 t impactor at 36 km/h speed is considered. The scheme of interaction between the fixed EAD design (rear end EAD wall is fixed rigidly) and the moving impactor is shown in Figure 1.

The scientific methodology and mathematical models have been developed to analyze EAD designs plastic deformation at impact. Finite-element simulations have been carried out taking into account geometrical and physical nonlinearities, dynamic hardening of the steel depending on the impact speed, nonlinear contact interaction between elements of the considered mechanical system of colliding bodies. The Krieg and Kay incremental model of plasticity [19] has been used to describe the nonlinear elastic-plastic material properties under impact taking into account the kinematic hardening. This model has been based on the bilinear approximation of the true stress-strain diagram.

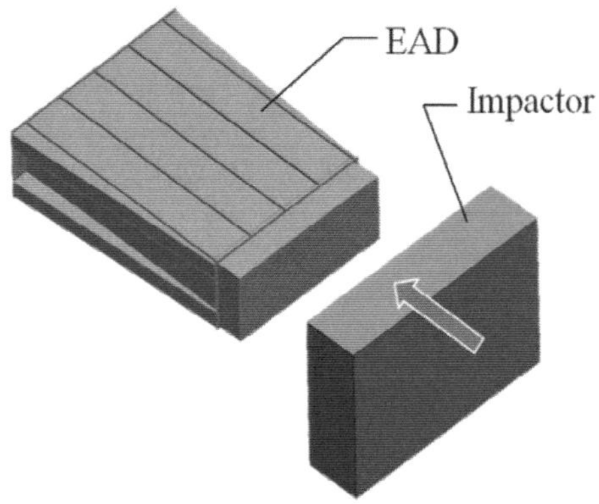

Figure 1. The scheme of interaction between the EAD and the impactor.

The breaking point of such a two-part piecewise linear curve corresponds to the dynamic yield stress, which depends on the strain rate. Modeling of nonlinear material characteristics has been based on the use of a true stretching diagram. The Simonds-Cooper dependence has been used to calculate the true dynamic yield point [20]. The obtained system of differential equations has been solved by the successive loading method [21]. As a result of the solution nodal displacements, velocities, accelerations, deformations, stresses and nodal forces at different moments of time at a given initial collision speed have been determined. The deformation diagram (dependence of the contact force F on the longitudinal displacement u_b of the impactor mass center) has been plotted using the obtained values u_b and the corresponding values of the contact force (integral over the contact area from the distributed contact stresses) between the investigated design and the impactor at the current moments. The low-pass filter with a cutoff frequency of 180 Hz [22] has been used for plotting this diagram according to DSTU EN 15227.

The finite-element mathematical models for the nonlinear dynamic analysis of the EAD design plastic deformation at impact have been developed using special plate elements with four or three nodes, each of them has three linear and three angular displacements, as well as three linear velocities and accelerations relative to the node coordinate system. The impactor (rigid vertical wall) has been simulated by solid elements with three linear displacements, velocities and accelerations at each node. This approach has previously been used and has shown good results when comparing simulation results with field test data [17-18].

5. Energy absorbing devices for a power head

Lower-level EAD 1 and upper-level EAD 2 for placing in the power head front end part and lower-level EAD 3 for placing in its rear end part were developed.

EAD 1 geometric model is shown in Figure 2 (all dimensions are given in millimeters). EAD 1 design includes two elements located sequentially. Element 1 is a parallelepiped form box with a single-layer package of variable-height hexagonal honeycombs inside. Element 2 is a truncated pyramid of honeycombs with triangular cells. The thickness of the box walls is 2 mm, the thickness of the front end plate is 4 mm. The length of the hexagonal cell side is 30 mm. The thickness of plates which are forming hexagonal cells is 0.6 mm. The thickness of the truncated pyramid end plates is 8 mm. The thickness of the plates which are forming triangular cells is 2.2 mm. The thickness of the outer longitudinal ribs is 6.6 mm and their height is 2.2 mm. EAD 1 deformation diagram is shown in Figure 3.

EAD 1 design energy absorbing capacity is 0.95 MJ, its stroke is 700 mm. 1.9 MJ impact energy can be absorbed as a result of two EAD 1 deformations. The maximum contact force which is transmitted to the power head underframe is 3.0 MN not exceeding the allowed value of 4.2 MN at considered collision.

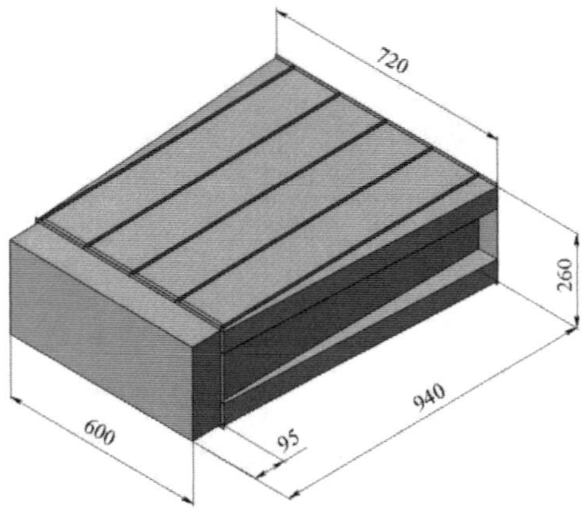

Figure 2. EAD 1 geometric model.

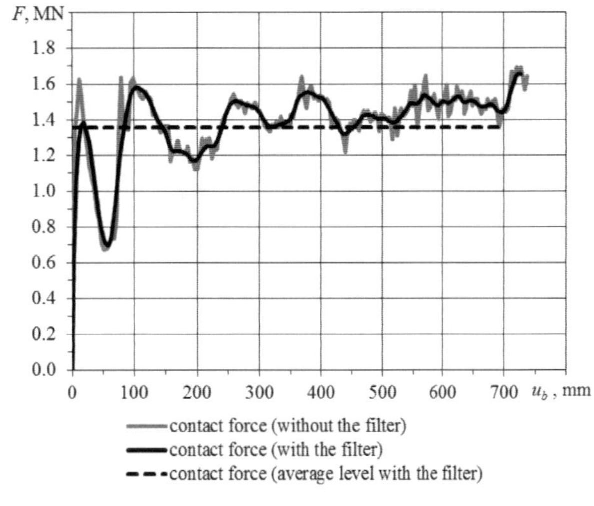

Figure 3. EAD 1 deformation diagram.

EAD 2 geometric model is shown in Figure 4, where all dimensions are given in millimeters. EAD 2 design has three steps in element 2 form. The possibility of placing two EAD 2 designs in a power head nose window-sill part has been taken into account. The thickness of plates which are forming the triangular cells is 2.2 mm. EAD 2 deformation diagram is shown in Figure 5.

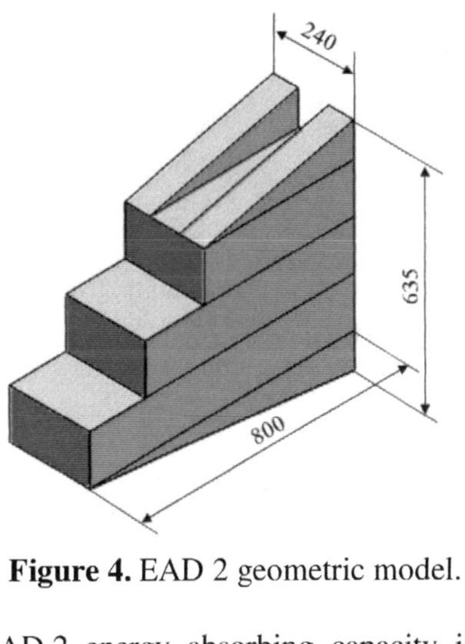

Figure 4. EAD 2 geometric model.

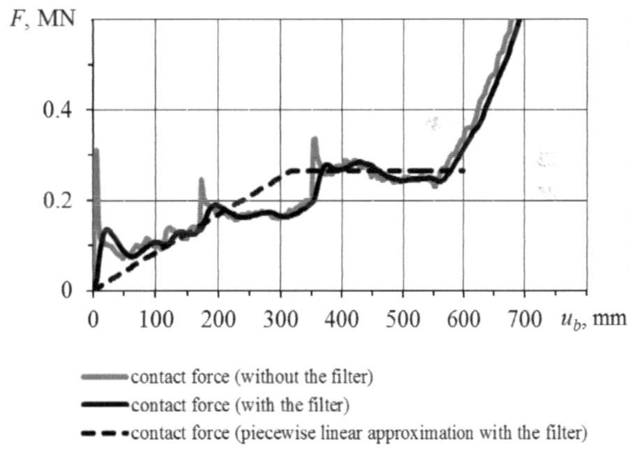

Figure 5. EAD 2 deformation diagram.

EAD 2 energy absorbing capacity is 0.12 MJ, the stroke is 600 mm. 0.24 MJ energy can be absorbed when two EAD 2 designs are deformed at a collision with 80 t impactor at 36 km/h speed. The maximum contact force transmitted to a power head body can reach 0.66 MN not exceeding the allowed value of 0.7 MN.

EAD 3geometric model is shown in Figure 6 (all dimensions are given in millimeters). EAD 3 design has been developed on the basis of EAD 1 design element 1. EAD 3 design is a parallelepiped form box with a four-layer package of hexagonal variable-height honeycombs placed on a diaphragm attached to the box. The thickness of the box walls, front and rear end plates, diaphragms is 2 mm. The length of the hexagonal cell side is 30 mm. The thickness of plates which are forming hexagonal cells is 0.7 mm. EAD 3 deformation diagram is shown in Figure 7.

IOP Conf. Series: Materials Science and Engineering **985** (2020) 012016 doi:10.1088/1757-899X/985/1/012016

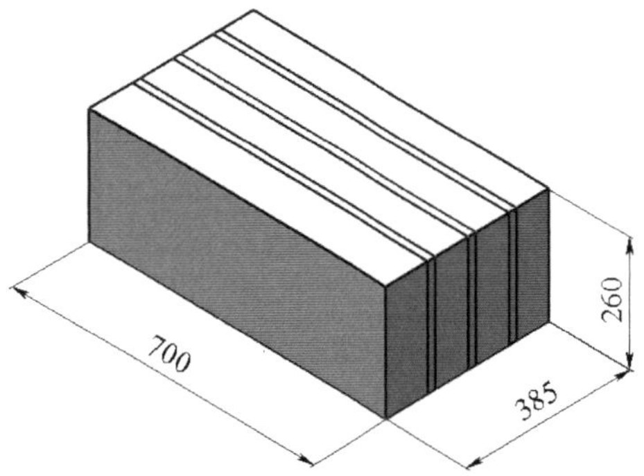

Figure 6. EAD 3 geometric model.

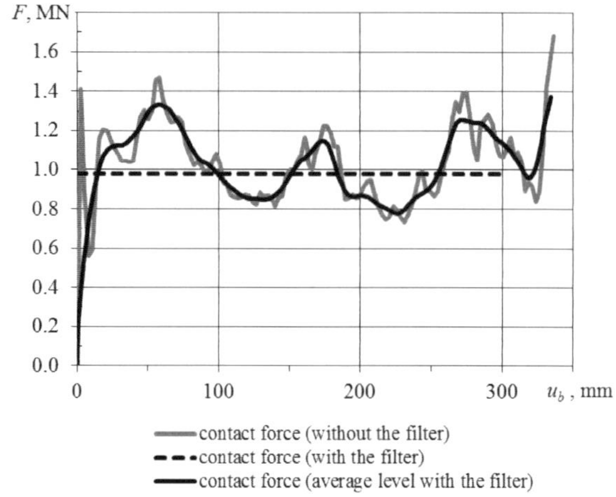

Figure 7. EAD 3 deformation diagram.

EAD 3 energy absorbing capacity is 0.3 MJ, the stroke is 300 mm. 0.6 MJ impact energy can be absorbed as a result of two EAD 3 deformations. The maximum contact force transmitted to the power head underframe is 2.6 MN not exceeding the allowed value of 3.0 MN for a coach underframe. It has been recommended to place two EAD 3 in the inter-car connection at the coupling level.

6. Conclusion

Proposals on domestic power head passive protection at collisions according to DSTU EN 15227 requirements were developed on the base of existing passive safety systems analysis. It has been recommended to install two EAD 1 at the coupling level in a power head frontal part, two EAD 2 in its window-sill part and two EAD 3 in the inter-car connection at the coupling level. Finite-element modeling of plastic deformation of EAD designs which contain cellular elements at impact has been carried out with the using of developed scientific and methodological support and the results of the crash test of the EP20 locomotive EAD prototype. As a result of it the EAD 1, EAD 2, EAD 3 designs have been developed.

EAD 1 design includes two elements located sequentially. Element 1 is a box with a single-layer package of hexagonal honeycombs inside. Element 2 is a truncated pyramid of honeycombs with triangular cells. EAD 2 design has three steps in element 2 form. EAD 3 design has been developed on the basis of element 1. Parameters of EAD 1, EAD 2, EAD 3 designs with energy absorbing capacity 0.95 MJ, 0.12 MJ and 0.3 MJ respectively are determined.

The developed scientific methodology and mathematical models can be used to select the parameters of energy absorbing devices with the required energy absorbing capacity to carry out the collision scenarios according to DSTU EN 15227 on the stage of the new-generation passenger rolling stock design.

References

[1] Railway applications – Structural requirements of railway vehicle bodies 2010 BDS EN 12663-1: 2010 p 18
[2] Railway applications. Crashworthiness requirements for railway vehicle bodies 2010 BS EN 15227:2008+A1:2010 p 37
[3] Decision of the Commission of the Customs Union on July 15, 2011 № 710 On Approval of the technical regulations of the Customs Union "On the security of railway rolling stock", "On high-speed rail safety" and "On the safety of railway infrastructure" TR-CU 001/002/003/2011 2011 p 270
[4] Railway applications – Structural requirements of railway vehicle bodies 2009 GOST R 53076-2008 p 8

[5] Emergency crash-systems railway rolling stock for passenger transportations. Technical requirements and methods of control 2014 GOST 32410-2013 p 11

[6] Zaliznychnyi transport. Strukturni vymohy orhaniv zaliznychnoho transportu. Chastyna 1. Lokomotyvy ta pasazhyrskyi rukhomyi sklad (i alternatyvnyi metod dlia vantazhnykh vahoniv) 2016 DSTU EN 12663-1: 2015 (EN 12663-1:2010, IDT) p 18 (in Ukrainian)

[7] Zaliznychnyi Transport. Vymohy do udarostiikosti reikovykh transportnykh zasobiv 2016 DSTU EN 15227: 2015 (EN 15227:2008+A1:2010, IDT) p 37 (in Ukrainian)

[8] Zaliznychnyi transport. Vymohy do konstruktsii kuzova zaliznychnoho transportnogo zasobu. Chastyna 1. Lokomotyvy ta pasazhyrskyi rukhomyi sklad (i alternatyvnyi metod dlia vantazhnykh vahoniv) 2018 DSTU EN 12663-1: 2018 (EN 12663-1:2014, IDT) p 18 (in Ukrainian)

[9] Roberts J, Fraikin B and Leveque D 2008 *Proc. Int. Symp. 20–21.11.2008 on Passive Safety of Rail Vehicles. Innovation in Passive Safety and Interior Design* (Berlin) Vol. 17 (Berlin: IFV Bahntechnik e.V.) pp 237–248

[10] Bruggemann A 2010 Elektropoezd serii 422 *Zheleznye dorogi mira* **5** pp 37–49 (in Russian)

[11] 2019 Proizvodstvo podvizhnogo sostava v stranah Vostochnoi Evropy *Zheleznye dorogi mira* **3** pp 61–65 (in Russian)

[12] Nazarov O N 2009 Tekhnicheskie osobennosti vysokoskorostnogo elektropoezda Velaro Rus *Tekhnika zheleznyh dorog* **1** pp 18–33 (in Russian)

[13] Mangler R 2012 "Desiro rus" – perspektivnyi elektropoezd dlya prigorodnykh perevozok v Rossii *Lokomotiv-inform* **10** pp 20–24 (in Russian)

[14] Bogatyrev B and Moguchev A 2016 Elektropoezd EP2D: budushchee uzhe zdes *ZAO "Transmashkholding": ZHurnal dlya partnerov* **2 (6/2016)** pp 4–11 (in Russian)

[15] Ignatov G S 2012 Mezhregionalnyi dvukhsistemnyi elektropoezd *Lokomotiv-inform* **10** pp 28–33 (in Russian)

[16] Backlash coupling device for locomotive driving passage car and railcar rolling stock of 1520 mm. Coupler contour line and dimensions 2002 OST 32.193-2002 p 14

[17] Sobolevska M and Telychko I 2013 *Proc. Int. Symp. 21–22.02.2013 on Passive Safety 2013 – Passive Safety of Rail Vehicles and Safe Interiors* **43** (Berlin: IFV Bahntechnik e.V.) pp 63–80

[18] Sobolevska M and Telychko I 2017 Passive safety of high-speed passenger trains at accident collisions on 1520 mm gauge railways *Transport problems* **12** pp 51–62

[19] Krieg R D and Key S W 1976 Implementation of a time independent plasticity theory into structural computer programs *ASME* **20** pp 125–137

[20] Cowper G R and Symonds P S 1957 *Strain Hardening and Strain Rate Effects in the Impact Loading of Cantilever Beams* (Providence: Brown University) p 28

[21] Oden J 1976 *Konechnye Elementy v Mekhanike Sploshnykh Sred* (Moscow: Mir) p 464 (in Russian)

[22] Gonorovskij I S 1986 *Radiotekhnicheskie Tsepi i Signaly* (Moscow: Radio i svyaz) p 512 (in Russian)

IOP Conf. Series: Materials Science and Engineering **985** (2020) 012017 doi:10.1088/1757-899X/985/1/012017

Track circuits adjusting calculation method under current influence traction interference and electromagnetic compatibility

S Razghonov[1,4], V Kuznetsov[2], O Zvonarova[2] and D Chernikov[3]

[1]University of Customs and Finance, Faculty of Innovative Technologies, St. V.Vernadsky, 2/4, 49000, Dnipro, Ukraine
[2]Dnipro National University of Railway Transport named after Academician V. Lazaryan, Educational and Scientific Center "Organization of Road Construction and Track Operation", St. Academic Lazaryan, 2, 49010, Dnipro, Ukraine
[3]CMA CGM Shipping Agencies Ukraine ltd, St. Pisarzhevskogo, 1A, office 513, 49000, Dnipro, Ukraine

[4]srazgonov@gmail.com

Abstract. The problem of railway transport electromagnetic compatibility automation systems in recent years has been aggravated by the use of power frequency converters in electric transport drives, which significantly increase the efficiency of traction characteristics of 3-phase electric engine. However, estimate relevant harmonic levels in the frequency range used in track circuits of railway automation and telemechanic systems. The paper provides theoretical justification for practical measurements of interference. It is shown, that the interference coefficients monotonically increase in the following cases: 1) increase in the value of the harmonic frequency; 2) increase in the propagation constant, taking into account the supports resistance of the contact network; 3) increase in the circuits track length. Operating frequencies of tonal track circuits up to 800 Hz, the interference increase rate in given length of the rail circuit slows down. The increase in interference coefficients depending on the mentioned factors is explained by the fact that with current part growth of the harmonic flows from the traction substation through the earth increases and thereby creates currents asymmetry of the track circuits.

1. Introduction

Recent years electromagnetic compatibility problem has become aggravated on Ukrainian railway network. One of the problem influence factors is use of power frequency converters in electric vehicles. These converters significantly increase traction characteristics efficiency of the 3-phase "dual power" electrical engine. However, these converters are one of sources of strong electromagnetic interference.

To change the power of electric drives traction the method of smooth sliding traction current frequency regulation in the range from infra-low frequencies to 50 Hz is used. Experience shows that with this method of power regulation a wide spectrum of harmonic components with relatively large amplitude is induced in the current traction. The induced harmonic oscillations amplitude can be comparable in level with the information tonal frequencies signals of the track circuits.

As a result, phase-sensitive track circuits receivers at frequencies of 25 Hz and 50 Hz can falsely record condition that violates railway automation safety and telemechanic system. Hence, it is advisable to create diagnostic devices in the track circuits or on-board devices of an electric locomotive, which

Content from this work may be used under the terms of the Creative Commons Attribution 3.0 licence. Any further distribution of this work must maintain attribution to the author(s) and the title of the work, journal citation and DOI.
Published under licence by IOP Publishing Ltd

IOP Conf. Series: Materials Science and Engineering **985** (2020) 012017 doi:10.1088/1757-899X/985/1/012017

will monitor the dangerous level of induced harmonics and exclude their effect on the track circuits. The current situation with interference forces developers to use more stringent requirements for the limit levels of harmonics and permissible values of the parameters of information signals. In this regard, it seems relevant to assess the levels of harmonics in the range of operating frequencies, which are used in the track circuits of railway automation and telemechanic systems. This is also important for the development of protective equipment against interference caused by the influence of current electric locomotive traction. As result, this leads to an adjustment of the calculating track circuits method. As result confirmed experimental studies theoretically obtained results which will be presented in the next article.

2. Problem Formulation

The well-known International Electrotechnical Commission decision No.77 of 1976 defines electromagnetic compatibility as the ability of devices to function correctly in a given electromagnetic environment without introducing excessive disturbances into this environment. It was also noted that the problem of higher harmonics is one of the most important problem of electricity quality components and electromagnetic compatibility.

Presence in the traction network dangerous interference higher harmonics track circuits at information frequencies creates an interfering effect on the operation of the track circuits of the railway automation system and the safety of train traffic.

The purpose of the Paper. In current article discussed methodology for calculating harmonics and compares them with the amplitudes of permissible levels in the current electric locomotive total traction.

3. Purpose and objectives

The purpose of research is to improve the methodology for calculating track circuits, taking into account the dangerous and interfering electromagnetic influence of the current traction. In the process of improving the methodology, various factors were taken into account. The schemes and track circuits configurations and reverse traction current sewerage schemes were analyzed. Additionally, other factors of signaling systems malfunctioning on certain sections of the railway during the operation of new types of rolling stock were considered.

4. Research results

It is known, that the biggest current reverse traction influence is exposed to the track circuits, to which traction substation suction feeder is connected. Figure 1 shows a diagram of the section between the electric locomotive and the traction substation with the designation of currents, voltages, conductivities, ground leakage currents and other elements that ensure the flow of traction currents. In particular, I_K is current in the contact drive and in the electric locomotive circuit; I_0, I_R, I_L – are currents at the point of connection of the feeder to the rail circuit, to the right and to the left of the traction substation; I_1, I_2 – currents, etc. Other parameters of the equivalent circuit: g_0 – supports conductivity connected to one rail; U_1, U_2 – rail-to-ground voltage, U_0, U_e – potentials at the point of drainage of electric locomotive currents and at the point of connection to the substation feeder; Z_P – rail resistance; Z_L, Z_R – input resistance of rails at the point of connection of an electric locomotive or electrical substation, respectively; M_{12} – rail mutual induction; M_{KP} – mutual induction between the electric locomotive and each rail; g_1, g_2 – rail-to-ground conductivity, respectively.

The methodology for track circuits asymmetry assessing considered in current paper is supplemented by calculations of the tonal spectrum harmonics. Additions of methodology were made taking into consideration resistance of the R65 rails and the voltage drop at these frequencies on the steel elements of the rail line (rail connectors, choke jumpers, etc.).

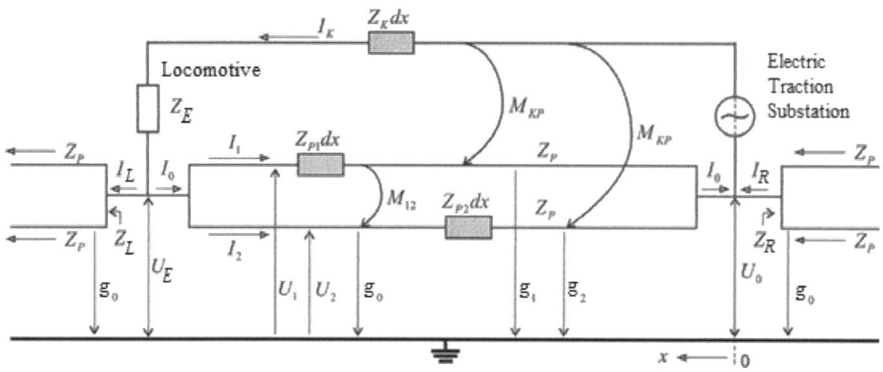

Figure 1. The equivalent circuit of the section between the electric locomotive and the traction substation.

Note that the equations given in [1] cover the main effects of the traction current and its harmonics on the equipment in the normal operation of track circuits under various climatic factors, but without taking into account the steel elements of the rail line. The equations take into account the interactions of three distributed circuits connected by mutual inductance: 'first rail-to-ground', 'second rail-to-ground' and 'contact wire-to-ground'. In this case, each "rail-to-ground" circuit has a conductivity g_1 or g_2 and a specific impedance of the rails Z_P to a current with a harmonic frequency. Both circuits are connected by mutual inductance M_{12} and have a conductivity g_{12} along the surface of the sleepers. The 'contact wire-ground' circuit is connected to the rails by mutual inductance M_{KP}. The value of the current in the contact wire depends on the load of the electric locomotive and little depends on rail circuit parameters, therefore, in the calculations, the current I_K was set to a constant value.

The technique of calculating the asymmetry algebraically by equations is rather complicated and time consuming. Practical experience shows that the greatest asymmetry of the traction current harmonics in rail circuit is observed in winter when one of the track rails is used to ground the supports of the contact network with the difference $g_1 - g_2 \approx g_0$. Therefore, without significant loss of accuracy, the conductivity of one rail can be neglected. In practical calculations, it is possible to take the "rail-to-ground" conductivity equal to g_0. Measurement statistics have shown that the value of g_0 remains very stable throughout the year. To further simplify, the calculations of two-strand track circuits, we assume that the load resistance is significantly less than the input resistance of the rail line. Therefore, it is possible to solve the problem of distribution of current in a rail line with a short circuit at the ends of the circuit (Figure 1).

It has been established that the critical insulation resistance R_{iKP} of a rail line, as shown by calculations of the control mode of tonal track circuits (which lengths are in the range from 0.25 to 1.5 km), can reach up to 1.8 Ohm·km. This value approaches the value of the insulation resistance of the rail line in winter conditions at the junction point of the contact network support and the rails. Below will be given levels assessment of harmonic currents affecting the track receivers of tone track circuits in the control mode.

From the calculated equivalent circuit (Figure 1) can be seen that the components of the current I_K, which flows in the contact wire between the electric locomotive and the substation, also flow through the grounding of the rail lines, which located to the left of the electric locomotive (I_L) and to the right of the traction substation (I_R).

Given accepted simplifications of rail circuit of length l, the boundary conditions can be expressed as follows:

$$I_K = I_0 + I_R \text{ at } x = 0; \tag{1}$$

$$I_K = I_e + I_L \text{ at } x = l, \quad I_R = \frac{U_0}{Z_R}, \quad I_L = \frac{U_1}{Z_L}; \tag{2}$$

where: Z_R and Z_L are the input impedances of the rail circuit with respect to the ground, to the right of the substation and to the left of the locomotive respectively.

The differential equations of the design circuit ('Helmholtz-type' scheme) with respect to the currents I_1 and I_2 are omitted. The integration constants in the known replacement equations, expressed through the value of the current I_K, after the transformations will be determined by the formulas:

$$A_1 = \frac{I_K}{P}\left(1 - 2\frac{Z_{KP}}{Z_P + Z_M}\right)\left(\sinh(\gamma_0 l) + \frac{z_0}{Z_L} + \frac{z_0}{Z_R}\cosh(\gamma_0 l)\right), \tag{3}$$

$$B_1 = \frac{I_K}{P}\left(1 - 2\frac{Z_{KP}}{Z_P + Z_M}\right)\left(\cosh(\gamma_0 l) - 1 + \frac{z_0}{Z_R}\sinh(\gamma_0 l)\right), \tag{4}$$

where: $\gamma_0 = \sqrt{g_0 Z_p}$ – signal propagation constant, $z_0 = \sqrt{\frac{Z_p}{g_0}}$ – characteristic impedance.

$$P = \sinh(\gamma_0 l) + 2\left(\frac{Z_P - Z_M}{Z_P + Z_M}\right)\cdot\left(\frac{\cosh(\gamma_0 l) - 1}{\gamma_0 l}\right) + \left[\cosh(\gamma_0 l) + \left(\frac{Z_P - Z_M}{Z_P + Z_M}\right)\cdot\left(\frac{\sinh(\gamma_0 l)}{\gamma_0 l}\right)\right] \times \left[\frac{z_0 + z_0}{Z_L + Z_R}\right] + \frac{z_0^2}{Z_L Z_R}\sinh(\gamma_0 l). \tag{5}$$

The input impedance of a circuit Z_R, which has a length l_R, with the signal propagation constant of the supports γ_{0R} to the right of the substation is expressed by the equation:

$$Z_R = z_0 \frac{\cosh(\gamma_{OR} l_R) + \left(\frac{Z_P - Z_M}{Z_P + Z_M}\right)\cdot\left(\frac{\sinh(\gamma_{OR} l_R)}{\gamma_{OR} l_R}\right)}{\sinh(\gamma_{OR} l_R) + 2\left(\frac{Z_P - Z_M}{Z_P + Z_M}\right)\cdot\left(\frac{\cosh(\gamma_{OR} l_R) - 1}{\gamma_{OR} l_R}\right)} \tag{6}$$

The input impedance to the left of the electric locomotive Z_L is expressed by the same equation, but taking into account the corresponding left circuit length and supports conductivity.

The asymmetry coefficient of the rail line according to the rails resistance for given harmonic at $x = l$ in the tonal track circuits between the substation and the electric locomotive, in the final form, is determined by formula:

$$K_o = \frac{1}{2P}\left(1 - 2\frac{Z_{kp}}{Z_p + Z_m}\right)\left[\sinh(\gamma_0 l) - 2\frac{\cosh(\gamma_0 l) - 1}{\gamma_0 l} + \frac{z_0}{Z_R}\left(1 - \frac{\sinh(\gamma_0 l)}{\gamma_0 l}\right) + \frac{z_0}{Z_R}\left(\cosh(\gamma_0 l) - 2\frac{\sinh(\gamma_0 l)}{\gamma_0 l}\right)\right], \tag{7}$$

the coefficient of asymmetry of the rail line near the traction substation at $x = 0$:

$$K_n = \frac{1}{2P}\left(1-2\frac{Z_k}{Z_p+Z_m}\right)\left[\begin{array}{l}\sinh(\gamma_0 l)-2\dfrac{\cosh(\gamma_0 l)-1}{\gamma_0 l}+\\ +\dfrac{z_0}{Z_R}\left(1-\dfrac{\sinh(\gamma_0 l)}{\gamma_0 l}\right)+\dfrac{z_0}{Z_R}\left(\cosh(\gamma_0 l)-2\dfrac{\sinh(\gamma_0 l)}{\gamma_0 l}\right)\end{array}\right]. \quad (8)$$

In formulas (5) and (6) the coefficients and actually determine the asymmetry of the harmonic currents. The K_0 coefficient refers to the track circuit that is connected to the traction substation feeder and is located between the electric locomotive and the traction substation. The K_0 coefficient is expressed by the formula:

$$K_o = \frac{(I_2-I_1)_o}{2I_k}. \quad (9)$$

The coefficient K_n characterizes the interference in the rail circuit, which is adjacent to the feeder on the opposite side of the electric locomotive. The K_n coefficient is determined by the formula:

$$K_n = \frac{(I_2-I_1)_{oS}}{2I_k}. \quad (10)$$

In the rail circuit to the right of the electric substation, at the connection point of the suction feeder, the current difference, in its final form, is determined by the formula:

$$(I_2-I_1)_{oS} = I_n \frac{\sinh(\gamma_{oR} l_R)-2\dfrac{Z_m}{Z_p+Z_m}\cdot\dfrac{\cosh(\gamma_{oR} l_R)-1}{\gamma_{oR} l_R-1}}{\sinh(\gamma_{oR} l_R)-2\dfrac{Z_p-Z_m}{Z_p+Z_m}\cdot\dfrac{\cosh(\gamma_{oR} l_R)-1}{\gamma_{oR} l_R-1}}. \quad (11)$$

where: I_n - total current in the track circuit to the right of the electric substation. In its final form, I_n is determined by the formula:

$$I_n = \frac{I_k}{P}\left(1-2\frac{Z_k}{Z_p+Z_m}\right)\frac{z_0}{Z_R}\left(\cosh(\gamma_0 l)+\frac{z_0}{Z_R}\sinh(\gamma_0 l)\right). \quad (12)$$

5. Asymmetry calculated data analysis and interference from each harmonic of the current traction

The calculated data, which characterize the numerical harmonics values of the current traction in the normal mode, are presented in Figure 2 as the coefficients K_0 and K_n. The calculations were carried out according to the methodology described in [1] for two-strand track circuits up to 1 km long. The maximum permissible value of the asymmetry coefficient of the reverse traction current in two-line track circuits of alternating current when using a choke-transformer DT-150 is 4% [9]. These coefficients represent the ratio of the half-difference of the harmonic currents at the ends of the rail circuit to the current of this harmonic in the contact wire.

Figure 2 shows the change in current I_K depending on the rail circuit length l, the harmonic frequency and the specific conductivity of the overhead contact network supports g_0. Thus, the coefficients show the proportion of the harmonic current in the overhead wire. The fraction of the current contained in the rail circuit at the maximum "rail-to-earth" insulation resistance, i.e. $g \to \min$, determines the interference in the rail circuit adjacent to the traction substation feeder on the side opposite to the electric locomotive.

Interference levels are calculated according to formulas (5)-(8) for each value of conductivity g_0 (0.25; 0.5; 1.0 Ohm·km), length of the rail circuit l (0.5; 0.8; 1.0 km) and conductivity $g_0 = 0.25\,S$.

The parameters of rail lines are calculated at a temperature of 20^{0}C and an earth conductivity $g_0 \to 95\,S$. In this case, the maximum interference is observed, which is created by each harmonic.

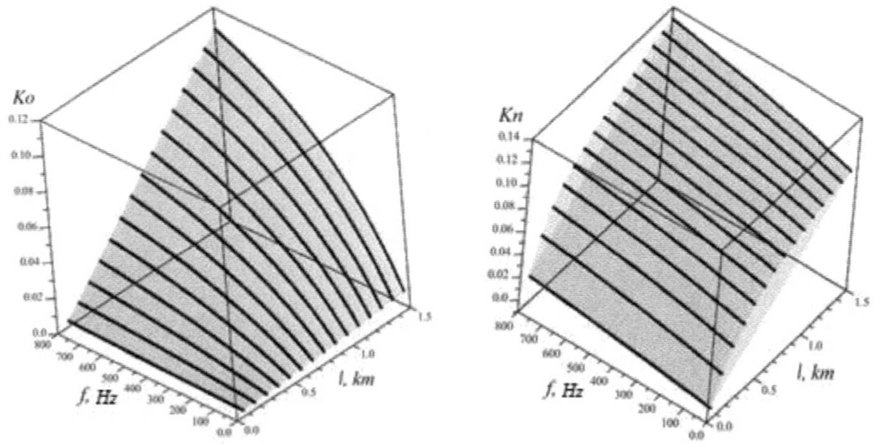

Figure 2. Variation dependences of K_o and K_n asymmetry coefficients on the frequency f and length l of the rail circuit at $g_0 = 0.25\,S$

Figure 2 shows that the asymmetry coefficients increase monotonically: with an increase in the value of the harmonic frequency, with an increase in the propagation constant γ_0, taking into account the supports of the contact network, and with an increase in the rail circuit length. At frequencies of tone track circuits up to 800 Hz, the rate of increase of the electrical interference for a given circuit length slows down. The growth of the interference coefficients depending on the above factors is explained by the fact that with an increase in γ_0, that part of the harmonic current increases, which flows to the traction substation through the ground and creates an asymmetry of currents.

Coefficient K_0, other things being equal, is less in magnitude than coefficient K_n despite the fact that the current I_0 flowing along the rails between the electric locomotive and the traction substation is greater than the current I_n entering the adjacent rail circuit from the traction substation feeder through the ground. This effect is explained by the phenomenon of current induction in the rails from the contact wire. This contributes to the equation of currents in the rail circuit.

Figure 3 shows the dependence of the change in the interference current on the frequency and length of the rail circuit for $g_0 = 0.25\,S$.

It was found that on a double-track section with a symmetrical connection of the tracks to the suction feeder, the return current of electric locomotives from the substation, flowing in each of the four track circuits adjacent to the feeder, is equal to $I_{Ke} = \dfrac{I_{mn(\text{single track section})}}{2} = \dfrac{I_{mn(\text{double track section})}}{4}$.

On a single-track section, the estimated contact network current is half the current of the substation. Thus, the currents between the rail lines in each circuit remain the same, regardless of whether there is a second path of the calculated current flow in the contact wire or there is one path with the same currents. At the same time, as measurements show, it is enough to evaluate the effect of interference with two electric locomotives moving along the same path.

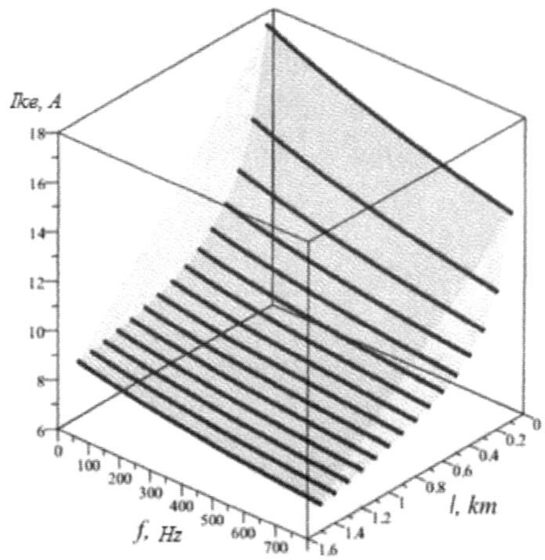

Figure 3. Dependences of the change in the interference current I_{Ke} on the frequency f and length l of the rail circuit at $g_0 = 0.25\,S$.

6. Conclusion

Analyzes main factors that characterize the electromagnetic influence of traction current on track circuits. In order to improve the methodology for determining the levels of hazardous interfering electromagnetic influence of the traction current, specific rail circuit diagrams, their configurations, various train situations under different conditions and operating modes of track circuits were analyzed.

The maximum permissible values of the traction current electromagnetic effect on the track circuits were determined both by theoretical calculation methods and by laboratory studies.

Experimental measurements of interference were carried out directly on railway sections, as well as in the power circuits of the rolling stock. As a basis for determining the maximum permissible levels of harmonic interference caused by reverse traction current, criteria of reliable operation of track circuits in all modes under the influence of the most unfavorable factors, according to safety conditions, were used.

References

[1] Bryleev A, Shishlyakov A and Kravtsov V 1966 *Devices and operation of rail chains* (Moscow: Transport press) p 264

[2] Vakhnin M, Penkin M and Pokrovsky N 1956 Signaling devices for AC electric traction *Scientific Bulletin of Railway Research Institute* (Moscow: JSC "VNIIZhT" press) p 126

[3] Kupfmüller K 1960 *Fundamentals of Theoretical Electrical Engineering* (Moscow: National Research University "Moscow Power Engineering Institute" press) p 217

[4] Razgonov A, Dyakov V, Zhuravlev Yu and Razgonov S 2009 Protection of track circuits in the docking zone of electric traction systems *Monthly scientific-theoretical and production technical journal «Automation, communication and Informatics»* **9** pp 19–22

[5] Arkatov V 2006 *Rail chains of the main railways: reference book* (Moscow: Misia press) p 496

[6] Razgonov A 1998 On the effect of the electric arc on the current collectors of an electric locomotive during ice on track circuits *Scientific and technical journal «Information and Control Systems on Railway Transport»* 6 pp 36–39

[7] Mariscotti A 2003 *Distribution of the traction return current in AC and DC electric Railway Systems* (IEEE Transactions on Power Delivery. 2003 Vol. 18 No. 4 pp1422–1432)

[8] CENELEC DS/CLC/TS 50238-2:2010. Railway Applications – Compatibility Between Rolling Stock And Train Detection Systems – Part 2: Compatibility With Track Circuits, 01.07.2010, 42 p.

[9] Instruction "On the procedure for the operation of track circuits on the Oktyabrskaya railway" // *Approved by order of the head of the railway dated April 3, 2014 No. Oct-183*

IOP Conf. Series: Materials Science and Engineering **985** (2020) 012018 doi:10.1088/1757-899X/985/1/012018

Reducing of coal dust release from train wagon with barrier

M M Biliaiev[1,6], V V Biliaieva[2], V A Kozachyna[1], O V Berlov[3], M O Oladipo[4] and P S Kirichenko[5]

[1]Department of Hydraulics and Water Supply, Dnipro National University of Railway Transport named after academician V. Lazaryan, Dnipro, Ukraine
[2]Department of AeroHydro Mechanics and Energy and Masstransfer, Oles Honchar Dnipro National University, Dnipro, Ukraine
[3]Department of Life Safety, Prydniprovska State academy of Civil Engineering and Architecture, Dnipro, Ukraine
[4]GeoLab Drilling, Winder, USA
[5]Department of Heat, Gas and Water Supply, Drainage Systems and Ventilation, Kryvyi Rih National University, Kryvyi Rih, Ukraine

[6]diit.hydro.eco@gmail.com

Abstract. Intensive environment pollution takes place during coal transportation in open wagons. Emission of coal dust from the coal wagons cause contamination of atmosphere and territory adjacent to the railway track. Different ways to reduce coal dust emission from the wagon are used in the world. Unfortunately, in Ukraine, this problem is far from solution and there is no serious research work in this field. The aim of this work was laboratory study of coal dust emission from the wagon model which had different barriers installed on the wagon. Laboratory experiments were carried for coal wagon without barrier and for coal wagon which had barriers of two types. Barrier of the first type had downwind wing. Barrier of the second type had upwind wing. The contamination zones, concentration near the model were studied. The obtained results illustrate that installation of barriers influence intensity of transport corridor contamination. Also a numerical model was developed to estimate wind flow and coal dust dispersion from the coal wagon. Equation of potential flow and equation of coal dust dispersion were used. Implicit difference schemes of splitting were used for numerical simulation of governing equations. Results of numerical experiment, which were performed, are presented.

1. Introduction

Exploitation of railway transport and development of railway infrastructure [1] has a different effect on environment. Many problems arise, which are connected with ballast layer contamination, noise and emission of harmful compounds [2, 3], waste management [4] etc.

Worthy of note, that coal trains have an intensive negative impact on environment [5-7]. Transportation of coal in open wagons results in cargo loss and intensive environment pollution. Reducing of coal dust emission from trains is a very important problem [8, 9]. Coal dust emission intensity depends on the following factors: train velocity, cargo shape in the wagon, moisture etc. There are two important tasks in the current problem. The first task is prediction of the atmosphere and underlying surface pollution from coal train. The second task is minimization of coal dust emission intensity from wagon. In order to solve the second task, various methods are used, for example, shelter

Content from this work may be used under the terms of the Creative Commons Attribution 3.0 licence. Any further distribution of this work must maintain attribution to the author(s) and the title of the work, journal citation and DOI.
Published under licence by IOP Publishing Ltd

cover systems [10, 11], canvas usage, special solutions usage, watering of coal, etc. Analysis of scientific publication showed that, in Ukraine, there is not research work connected with reducing of coal dust emission from wagons. The purpose of the work was study of barrier installation on coal wagon to reduce coal dust emission. In this paper authors show results of laboratory study which was performed in Dnipro National University of Railway Transport named after academician V. Lazaryan (Ukraine, Project 0115U007227 «Environment protection at the elements of transport infrastructure»).

2. Experimental setup

The laboratory experiments were performed for a 1:100 scale model of the open wagon. This model (scenario #1) is shown in Figure 1. Laboratory experiments to study the efficiency of barrier installation on the coal wagon to reduce the coal dust environment pollution were performed at Hydraulics department of the University.

Figure 1. Model of the coal wagon without barrier (scenario #1).

Two types of barriers on the coal wagon were studied. The first type of barrier was the barrier with downwind wing (scenario #2). This type of the barrier is shown in Figure 2.

Figure.2. Model of the coal wagon with barrier: 1 – barrier (barrier + downwind wing, scenario #2).

The second type of the barrier was the barrier with upwind wing (scenario #3). This type of the barrier is shown in Figure 3.

Figure 3. Model of the coal wagon with barrier: 1 – barrier (barrier + upwind wing, scenario #3).

Sketch of the laboratory setup is shown in Figure 4. The industrial fan was used to initiate wind flow over the wagon model (Figure 4). The Reynolds number in the experiments was $Re = 10^5$. The length of the wagon as the characteristic dimension for calculation Re number was chosen. Coal from Mezhdurechenskaya Central Enriching Factory was used in the experiments. During the experiments concentration of coal dust PM2.5 has been measured behind the model. The Air Quality Detector WP6910 was used to measure coal dust concentration near the coal wagon model. To measure air speed anemometer GM8908 was used.

To weight coal dust which was emitted from the wagon model and set near the model the balance KL168 was used.

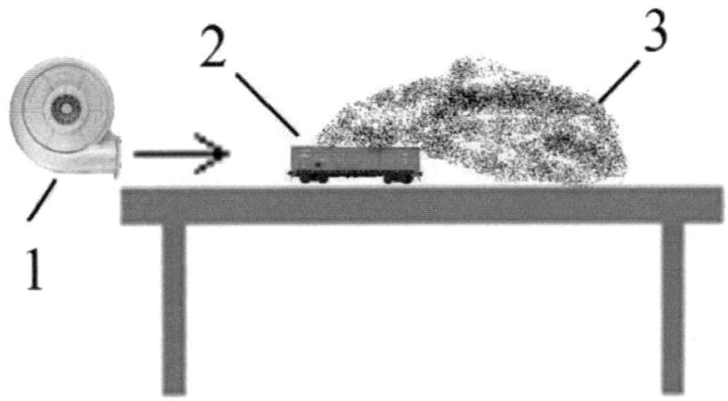

Figure 4. Sketch of laboratory setup: 1 – fan; 2 –model of the coal wagon; 3 – coal dust dispersion.

3. Results

During the laboratory experiments some aspects of the problem connected with coal dust emission from wagon were studied. First of all, form of contamination zone on the ground near the wagon model was investigated. Figures 5–7 show contamination zones for all three scenarios.

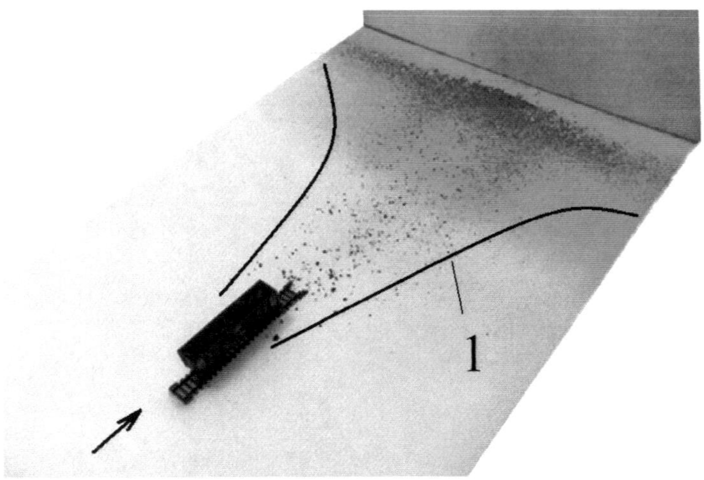

Figure 5. Contamination zone near the coal wagon (no barrier on the wagon, scenario #1).

The contamination zone, which was formed near the wagons, was quite different for each scenario (Figures 5–7). For the case when an open wagon without barrier is used (Figure 5) this contamination zone has intensive "black" color which indicates on the intensive emission of coal dust from the wagon. This contamination zone has conical form. For this scenario the ballast layer and the ground adjacent to the railway track will be hardly contaminated.

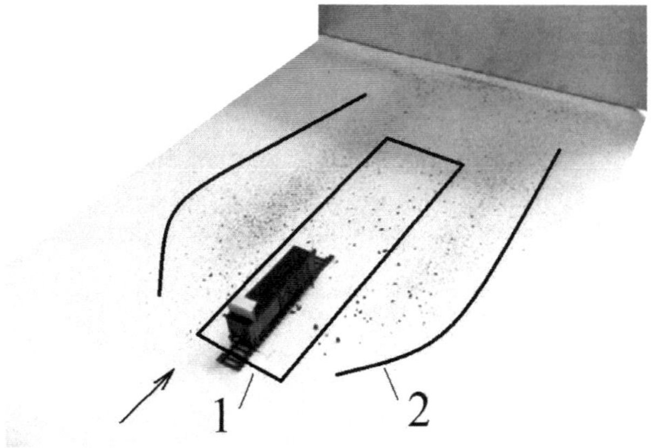

Figure 6. Contamination zone near the coal wagon with barrier (barrier + downwind wing, scenario #2): 1 – internal subregion with low contamination; 2 – boundary of second subregion.

For the case when an open wagon with "barrier+ downwind wing" is used, the contamination zone can be separated in two subregions (Figure 6). The first subregion (first contamination zone) is light colored and it is situated inside the whole contamination zone and it covers railway track. The second subregion is a boundary, like a belt, for the first subregion. It has «black» color, but less intensive as in scenario #1 (coal wagon without barrier). The first contamination zone has a rectangular form. In this case the region near the track is not contaminated hardly. The pollutant mostly set in another part of the ground, far from the track. That means that in this scenario ballast layer will be less contaminated then in scenario #1.

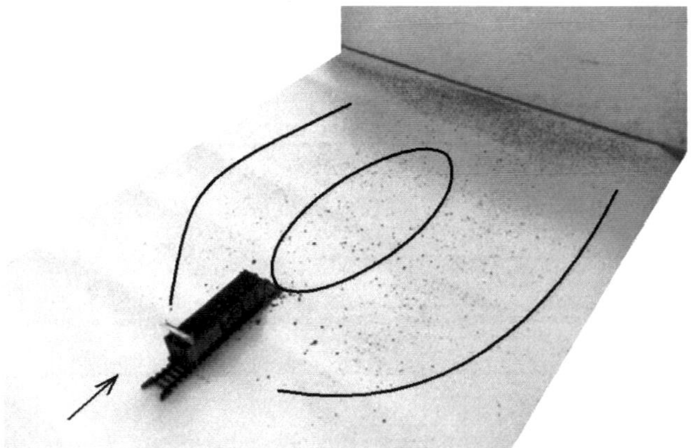

Figure 7. Contamination zone near the coal wagon with barrier (barrier + upwind wing, scenario #3).

For the case when an open wagon with «barrier + upwind wing» is used, the contamination zone can be separated in two subregions too (Figure 7). The first subregion (first contamination zone) has about an oval form.

In Table 1 coal dust concentration (PM2.5) is presented, which was measured at different distances from the model of a coal wagon. The concentration was measured at the distances $L, 2L, 3L$, where L is the length of the wagon. The concentration was measured at height of the wagon.

Table 1. Coal dust concentration for considered scenarios.

Distance	Dust concentration ($\mu g \cdot m^{-3}$)		
	scenario #1	scenario #2	scenario #3
L	500	95	23
2L	128	54	17
3L	68	36	11

It is clear (Table 1), application of barriers allows to reduce air pollution near the coal wagon. If the mass of coal dust which is emitted from the model without barrier is equal to 100%, then for scenario #2 (barrier + downwind wing) mass of emitted coal dust is about 89% and for scenario #3 (barrier + upwind wing) mass of emitted coal dust is about 84%.

4. Mathematical model

For practice, it is necessary to have mathematical models to predict air pollution near the coal wagon which allow quickly to perform numerical experiment to study the process of coal dust emission from wagons. To simulate air pollution near coal wagon the following modeling equations were used [12, 13]:

$$\frac{\partial C}{\partial t}+\frac{\partial uC}{\partial x}+\frac{\partial (\nu-w_s)C}{\partial y}=\frac{\partial}{\partial x}\left(\mu_x\frac{\partial C}{\partial x}\right)+\frac{\partial}{\partial y}\left(\mu_y\frac{\partial C}{\partial y}\right)+\sum q_i\delta(x-x_i)\delta(y-y_i), \tag{1}$$

$$\frac{\partial^2 P}{\partial^2 x}+\frac{\partial^2 P}{\partial^2 y}=0;\ u=\frac{\partial P}{\partial x},\ \nu=\frac{\partial P}{\partial y}, \tag{2}$$

where C is coal particles concentration; t is time; μ_x, μ_y are turbulent diffusion coefficients; (x_i, y_i) are coordinates of coal particles emission point source; $\delta(x-x_i)\delta(y-y_i)$ is Dirac's delta-function; q_i is coal particles emission rate form the cargo surface; w_s is coal particles fallout speed; u, v are wind velocity components; P is velocity potential.

Parameters μ_x, μ_y, w_s are computed using empirical formulae [13].

Boundary conditions and initial ones are discussed in [14]. To form cargo surface in wagon markers technique was used [14]. Emission rate q of coal dust from each part of cargo surface depends on local speed and was calculated using empirical formula [15].

5. Numerical integration

Wind flow near the coal wagon is computed on the base of equation (2). Before integration of this equation, it was written it in the following form [16]:

$$\frac{\partial P}{\partial t}=\frac{\partial^2 P}{\partial^2 x}+\frac{\partial^2 P}{\partial^2 y}, \tag{3}$$

where t is fictitious time.

To solve equation (3), the following two steps scheme of splitting was used [16]:

$$\frac{P_{i,j}^{n+\frac{1}{2}}-P_{i,j}^{n}}{\Delta t}=\left[\frac{-P_{i,j}^{n+\frac{1}{2}}+P_{i-1,j}^{n+\frac{1}{2}}}{\Delta x^2}\right]+\left[\frac{-P_{i,j}^{n+\frac{1}{2}}+P_{i,j-1}^{n+\frac{1}{2}}}{\Delta y^2}\right],\ \frac{P_{i,j}^{n+1}-P_{i,j}^{n+\frac{1}{2}}}{\Delta t}=\left[\frac{P_{i+1,j}^{n+1}-P_{i,j}^{n+1}}{\Delta x^2}\right]+\left[\frac{P_{i,j+1}^{n+1}-P_{i,j}^{n+1}}{\Delta y^2}\right]. \tag{4}$$

At each step of splitting an explicit formula is used to calculate P. The computation procedure is finished if the following condition is fulfilled:

$$\left|P_{ij}^{n+1} - P_{ij}^{n}\right| \le \varepsilon , \qquad (5)$$

where ε is a small number (e.g., $\varepsilon = 0.001$); n is an iteration number.

Air velocity vector components are computed at the sides of computational cell:

$$u_{ij} = \frac{P_{ij} - P_{i-1,j}}{\Delta x}, v_{ij} = \frac{P_{ij} - P_{i,j-1}}{\Delta y}. \qquad (6)$$

For the numerical integration of equation (1) the change–triangular difference scheme was used [14]. To create code, FORTRAN language was used.

In Figures 8, 9 results of numerical simulations which were performed on the basis of developed numerical model are presented. These figures show dimensionless concentration in per cents from the maximum concentration C_{max}. In these figures numbers are printed using format «Integer». It means, for example, if concentration is equal to 15.34%, then the printed number is 15%. Worthy of note, that computational time was 5 s.

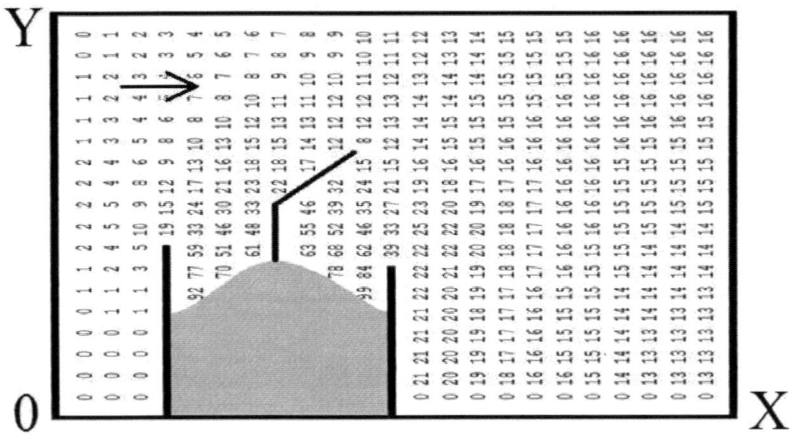

Figure 8. Computed coal dust concentration near the coal wagon with barrier (barrier + downwind wing, scenario #2), $C_{max} = 32.82$.

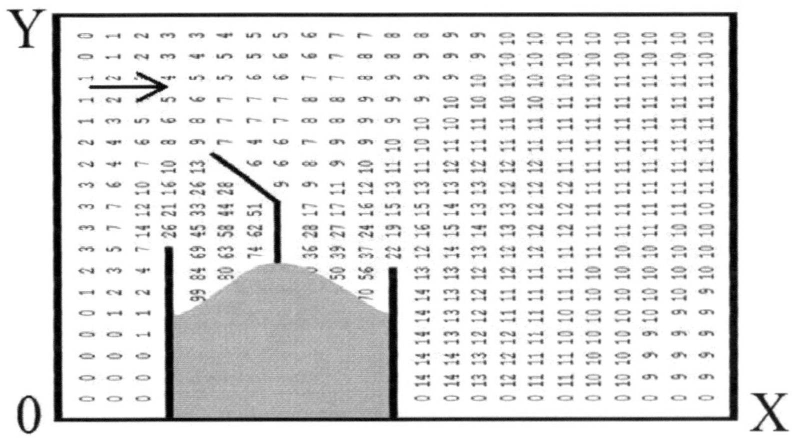

Figure 9. Computed coal dust concentration near the coal wagon with barrier (barrier + upwind wing, scenario #3), $C_{max} = 35.58$.

IOP Conf. Series: Materials Science and Engineering **985** (2020) 012018 doi:10.1088/1757-899X/985/1/012018

6. Conclusions

Coal dust emission from trains has a negative impact on environment. To minimize coal dust emission from the wagon, it was proposed to use barriers which are installed on the wagon. Results of laboratory experiments are presented. The experiments were performed to study the influence of barriers having different forms. During the laboratory experiments some aspects were studied: form of ground contamination zones, influence of barrier form on the coal dust emission rate and coal dust concentration near the wagon. Results discussed are preliminary ones and the experiments will be continue for another conditions. Numerical model was developed to compute coal dust concentration field near wagons with barriers. This model takes into account air flow velocity, diffusion, different coal dust emission rate from different parts of cargo in wagon, geometrical shape of barrier which is placed on the wagon. Model allows to obtain very quickly coal dust concentration near the coal wagon and near the railway tracks. This information can be used to evaluate level of pollution at different zones for different barriers installed on the coal wagon.

References

[1] Gołębiowski P and Kukulski J 2020 Preliminary study of shaping the railway track geometry in terms of their maintenance costs and capacity *Archives of Transport* **53** pp 115–128

[2] Żak J, Gołębiowski P and Popiela K 2018 Selected Aspects of the Environmental Friendliness Assessment of Railway Transport Using the Indicator Method *Proceedings of 22nd International Scientific Conference Transport Means* (Kowno: Kaunas University of Technology) pp 950–955

[3] Gołda I J, Gołębiowski P, Izdebski M, Kłodawski M, Jachimowski R and Szczepański E 2017 The evaluation of the sustainable transport system development with the scenario analyses procedure *Journal of Vibroengineering* **19** 7 pp 5627–5638

[4] Merkisz-Guranowska A, Merkisz J, Jacyna M, Stawecka H and Wasiak M 2014 Recycling guidelines of the rolling stock *Proceedings of The Second International Conference on Railway Technology: Research, Development and Maintenance Railways* (Stirlingshire, Scotland) pp 1–10

[5] Connell Hatch 2008 *Environmental Evaluation of Fugitive Coal Dust Emissions from Coal Trains* (Queensland: Queensland Rail Limited) p 414

[6] NSW Chief Scientist & Engineer 2016 *Final Report on the Independent Review of Rail Coal Dust Emissions Management Practices in the NSW Coal Chain* (Sydney: NSW Chief Scientist & Engineer) p 114

[7] Andrew Vernon and Lisa Smith 2014 *Literature Review of Coal Train Dust Management Practices* (Brisbane: Katestone Environmental Pty.) p 112

[8] Dukić M. and Planner J 2011 Dust Emissions from Coal Wagons: Reducing Emissions from the Surface of Coal *Bulk solids handling* **31** pp 20–24

[9] Emmitt G D 1999 Fugitive coal dust: An old problem demanding new solutions *Port Technology International* **9** pp 125–128

[10] Ferreira A D, Viegas D X and Sousa A C M 2003 Full-scale measurements for evaluation of coal dust release from train wagons with two different shelter covers *Journal of Wind Engineering and Industrial Aerodynamics* **91** 10 pp 1271–83

[11] Ferreira A D and Vaz P A 2004 Wind tunnel study of coal dust release from train wagons. *Journal of Wind Engineering and Industrial Aerodynamics* **92** 7-8 pp 565–77

[12] Marchuk G I 1982 *Mathematical modeling in the environmental problem* (Moscow: Nauka) p 320

[13] Bruyatskiy Ye V 2000 *Atmosphere diffusion theory of radioactive emissions* (Kyiv: Institufe of hydromechanics NAS of Ukraine) p 444

[14] Biliaiev M M, Oladipo M O and Kirichenko P S 2018 *Environment protection during coal transportation* (Kryvyi Rih: Kozlov publ.) p 92

[15] Biliaiev M M, Kozachyna V A and Oladipo M O 2019 Numerical analysis of atmosphere pollution from coal train *East European Scientific Journal* **3** 43 pp 10–15

[16] Samarskiy A A 1983 *Theory of differential schemes* (Moscow: Nauka) p 616

IOP Conf. Series: Materials Science and Engineering **985** (2020) 012019 doi:10.1088/1757-899X/985/1/012019

Simulation of environmental pollution from diesel locomotive

M M Biliaiev[1,6], **T I Rusakova**[2], **V A Kozachyna**[1], **O V Berlov**[3], **V M Poltoratska**[4] **and Z M Yakubovska**[5]

[1]Department of Hydraulics and Water Supply, Dnipro National University of Railway Transport named after academician V. Lazaryan, Dnipro, Ukraine
[2]Life Safety Department, Oles Honchar Dnipro National University, Dnipro, Ukraine
[3]Department of Life Safety, Prydniprovska State Academy of Civil Engineering and Architecture, Dnipro, Ukraine
[4]Department of Ecology and Environmental Protection, Prydniprovska State Academy of Civil Engineering and Architecture, Dnipro, Ukraine
[5]Department of Power Engineering, Ukrainian State University of Chemical Technology, Dnipro, Ukraine

[6]diit.hydro.eco@gmail.com

Abstract. Diesel locomotives are widely used at Ukrainian railway stations as maneuvering locomotives. Emissions from these diesel locomotives contain different toxic chemicals. It is important to understand air contamination patterns which are formed at railway stations. Understanding of these contamination patterns allows to evaluate the negative impact of diesel locomotive emissions on environment. To solve this problem it is necessary to use mathematical models which take into account the most important factors influencing formation of contamination zones. This paper introduces numerical model which allows to simulate pollutants dispersion from moving diesel locomotive. Developed numerical model is based on three dimensional equation of potential flow and three dimensional equation of pollutant dispersion. The model takes into account pollutants (NO, NO_2) chemical transformation. To solve three dimensional equation of potential flow the implicit difference scheme of splitting was used. To solve three dimensional equation of pollutant dispersion the implicit difference scheme of splitting was used. Euler method was used to solve numerically equations of pollutant chemical transformation. Developed numerical model allows to take into account influence of buildings at the railway station on the contamination zones formation. Developed model consumes not much computer time. Results of performed numerical experiment are presented.

1. Introduction

Various types of transport have an intense impact on the environment. Environment protection is a problem of great interest for exploitation of different types of transport [1, 2]. Noise pollution and chemical pollution of the environment are typical for cars and rail transport and this set a problem of their reducing [2-4]. High pollution levels are observed near roads, railways where diesel engines work. Accurate prediction of air pollution near traffic routes requires mathematical models which take into account most important factors which influence the pollutants dispersion. In assessing the negative impact of transport on the environment, two classes of problems are considered. The first class of problems is the assessment of the intensity of noise or chemical pollution from specific sources (car, locomotive, etc.). The second class of problems is predicting zones of noise and chemical

Content from this work may be used under the terms of the Creative Commons Attribution 3.0 licence. Any further distribution of this work must maintain attribution to the author(s) and the title of the work, journal citation and DOI.
Published under licence by IOP Publishing Ltd

pollution of environment both in the transport corridor and beyond its borders [5-7]. This paper discusses a numerical model for assessing the level of air pollution during the operation of a diesel locomotive.

During diesel locomotive work the intensive air pollution takes place (Figure 1). The main diesel engine pollutants are NO_x, CO. Air quality near the railway track, where diesel locomotives work, depends on a few factors. Among these factors the most important are speed of locomotive, chemical transformations of pollutants and presence of different buildings near railway track. Unfortunately, the existing predictive models in Ukraine, such as OND-86 model, do not take into account these factors. The purpose of this work is development of physically based, quick computing predictive model to evaluate air pollution near railway track where diesel locomotives work. The main feature of the model is taking into account diesel locomotive movement, chemical transformation of pollutants and influence of buildings near transport corridor on contamination zone formation.

Figure 1. Emissions from diesel locomotive.

2. Mathematical model

From the modern point of view, air quality predictive models must take into account pollutants chemical transformations. Emissions from diesel locomotives include, in particular, NO, NO_2. In this work the following transformations of these pollutants are taken into account [7, 8]:

$$NO_2 + h_\nu \xrightarrow{J} NO + O, \tag{1}$$

$$O + O_2 \rightarrow O_3, \tag{2}$$

$$NO + O_3 \xrightarrow{k_1} NO_2 + O_2. \tag{3}$$

Process of these pollutants dispersion from the moving diesel locomotive can be described by the following equations of conservation [9, 10]:

$$\frac{\partial[NO]}{\partial t} + \frac{\partial u[NO]}{\partial x} + \frac{\partial v[NO]}{\partial y} + \frac{\partial w[NO]}{\partial z} = \frac{\partial}{\partial x}\left(\mu_x \frac{\partial[NO]}{\partial x}\right) + \frac{\partial}{\partial y}\left(\mu_y \frac{\partial[NO]}{\partial y}\right) + \frac{\partial}{\partial z}\left(\mu_z \frac{\partial[NO]}{\partial z}\right) + \\ + Q_{NO}\delta(x - x_0(t))\delta(y - y_0(t))\delta(z - z_0), \tag{4}$$

$$\frac{\partial[NO_2]}{\partial t} + \frac{\partial u[NO_2]}{\partial x} + \frac{\partial v[NO_2]}{\partial y} + \frac{\partial w[NO_2]}{\partial z} = \frac{\partial}{\partial x}\left(\mu_x \frac{\partial[NO_2]}{\partial x}\right) + \frac{\partial}{\partial y}\left(\mu_y \frac{\partial[NO_2]}{\partial y}\right) + \\ + \frac{\partial}{\partial z}\left(\mu_z \frac{\partial[NO_2]}{\partial z}\right) + Q_{NO_2}\delta(x - x_0(t))(\delta(y - y_0(t))(\delta(z - z_0)), \tag{5}$$

$$\frac{\partial[O_3]}{\partial t} + \frac{\partial u[O_3]}{\partial x} + \frac{\partial v[O_3]}{\partial y} + \frac{\partial w[O_3]}{\partial z} = \frac{\partial}{\partial x}\left(\mu_x \frac{\partial[O_3]}{\partial x}\right) + \frac{\partial}{\partial y}\left(\mu_y \frac{\partial[O_3]}{\partial y}\right) + \frac{\partial}{\partial z}\left(\mu_z \frac{\partial[O_3]}{\partial z}\right), \tag{6}$$

where Q_{NO} is emission rate of NO from diesel locomotive, Q_{NO_2} is emission rate of NO_2 from diesel locomotive; u,v,w are wind speed components; μ_x, μ_y, μ_z are coefficients of turbulent diffusion; $x_0(t)$, $y_0(t)$, z_0 are coordinates of emission source; $\delta(x-x_0(t))\delta(y-y_0(t))\delta(z-z_0)$ are Dirac delta-function.

Equations (4), (5) describe dispersion of NO, NO_2 from moving diesel locomotive. Speed of locomotive is set by dependencies $x_0(t)$, $y_0(t)$ which describe position of locomotive for each time step. Equation (6) describes change of O_3 concentration in computational region due to wind speed and turbulent diffusion.

Process of chemical transformation of pollutants (1), (2), (3) is described as following [7, 8]:

$$\frac{d[NO]}{dt} = -k_1[NO][O_3] + J[NO_2], \tag{7}$$

$$\frac{d[NO_2]}{dt} = k_1[NO][O_3] - J[NO_2], \tag{8}$$

$$\frac{d[O_3]}{dt} = -k_1[NO][O_3] + J[NO_2], \tag{9}$$

where J, k_1 are empirical constants [7, 8].

To simulate wind pattern over buildings the model of potential flow was used [9]:

$$\frac{\partial^2 P}{\partial x^2} + \frac{\partial^2 P}{\partial y^2} + \frac{\partial^2 P}{\partial z^2} = 0, \tag{10}$$

where P is velocity potential.

The components of wind velocity are determined as:

$$u = \frac{\partial P}{\partial x}, \; v = \frac{\partial P}{\partial y}, \; w = \frac{\partial P}{\partial z}.$$

Boundary conditions for modeling equations are discussed in [9].

3. Numerical model

To compute wind pattern it is necessary to integrate equation (10). For the numerical integration this equation was written in a «time dependent» form:

$$\frac{\partial P}{\partial t} = \frac{\partial^2 P}{\partial x^2} + \frac{\partial^2 P}{\partial y^2} + \frac{\partial^2 P}{\partial z^2}, \tag{11}$$

where t is fictitious time.

For the numerical integration equation (11) the difference scheme of «conditional» approximation was used [11]. In this case the procedure of numerical integration of equation (11) is split in two steps. At the first step the following approximation was used [11]:

$$\frac{P_{i,j,k}^{n+\frac{1}{2}} - P_{i,j,k}^{n}}{\Delta t} = \left[\frac{-P_{i,j,k}^{n+\frac{1}{2}} + P_{i-1,j,k}^{n+\frac{1}{2}}}{\Delta x^2}\right] + \left[\frac{-P_{i,j,k}^{n+\frac{1}{2}} + P_{i,j-1,k}^{n+\frac{1}{2}}}{\Delta y^2}\right] + \left[\frac{-P_{i,j,k}^{n+\frac{1}{2}} + P_{i,j,k-1}^{n+\frac{1}{2}}}{\Delta z^2}\right]. \tag{12}$$

At the second step the following approximation was used:

$$\frac{P_{i,j,k}^{n+1} - P_{i,j,k}^{n+\frac{1}{2}}}{\Delta t} = \left[\frac{P_{i+1,j,k}^{n+1} - P_{i,j,k}^{n+1}}{\Delta x^2}\right] + \left[\frac{P_{i,j+1,k}^{n+1} - P_{i,j,k}^{n+1}}{\Delta y^2}\right] + \left[\frac{P_{i,j,k+1}^{n+1} - P_{i,j,k}^{n+1}}{\Delta z^2}\right]. \quad (13)$$

Unknown P_{ijk} in (12), (13) is computed using explicit formulae at each step. To begin computation procedure it is necessary to set field of potential at fictitious time t=0, for example: $P = 0$ at $t = 0$. The computation procedure is finished if:

$$\left|P_{i,j,k}^{n+1} - P_{i,j,k}^{n}\right| \le \varepsilon,$$

where ε is a small number; n is number of iterations.

Velocity components are computed as follows:

$$u_{ijk} = \frac{P_{i,j,k} - P_{i-1,j,k}}{\Delta x}, v_{ijk} = \frac{P_{i,j,k} - P_{i,j-1,k}}{\Delta y}, w_{ijk} = \frac{P_{i,j,k} - P_{i,j,k-1}}{\Delta z}.$$

For the numerical integration of equations (4)–(6) the implicit difference scheme of splitting was used [9, 10]. For the numerical integration equations (7)–(9) Euler method was used [11]. FORTRAN language was used to develop code.

4. Results

The developed numerical model was used to compute contamination zone formation at *Gruzovoy* Railway Station (Pridneprovsk railway, Ukraine) during the work of diesel locomotive *ChME3* (Figure 2). Arrows in Figure 2 indicate direction of the diesel locomotive movement.

Numerical experiment was performed for the following data: west wind direction; wind speed at the height 10m was 6m/s; NO emission rate was 0.32 g/s; NO_2 emission rate was 1.98 g/s; diesel locomotive speed was 10.5 km/h; O_3 background concentration was 0.15 μg/m^3, it was set that wind with this O_3 concentration entered the computational region through the inlet boundary.

Figure 2. Sketch of computational region (*Gruzovoy* Railway Station, Google Image, 2020): 1 – receptor position.

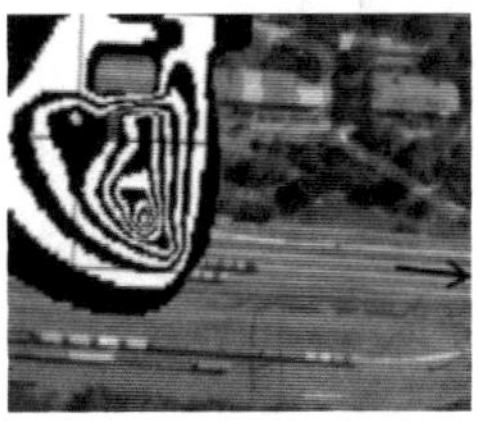

Figure 3. Predicted NO_2 concentration contours, level $z = 3m, t = 8s$.

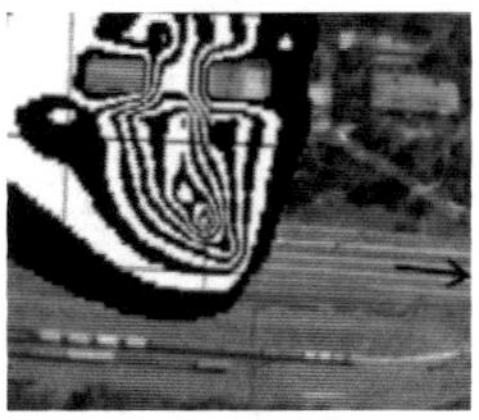

Figure 4. Predicted NO_2 concentration contours, level $z = 3m, t = 15s$.

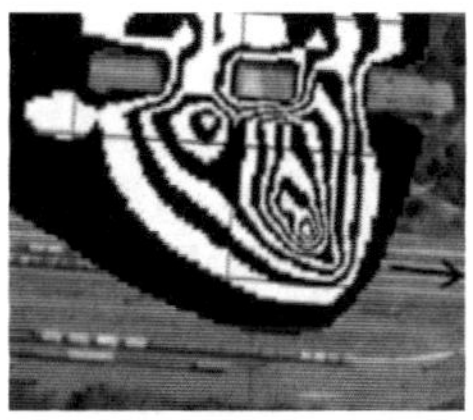

Figure 5. Predicted NO_2 concentration contours, level $z = 3m, t = 22s$.

As one can see from Figures (3)–(5), large contamination zone is formed near the railway track during diesel locomotive movement at the railway station. Station buildings deformed the contamination zone. One can see that pollutant can penetrates through windows inside the buildings near the railway tracks.

In Table 1 predicted NO_2 concentrations at receptor position (Figure 2) are presented.

Table 1. NO_2 concentration at receptor (level $z = 2m$).

time, s	12	18	29
NO_2 concentration (μg·m^{-3})	0,18	0,41	0,11

As one can see from Table 1, NO_2 concentration at receptor position is higher than the Admitted Level. So, diesel locomotive work has negative effect on environment. Worthy of note, that computational time was 10 s.

5. Conclusions

During diesel locomotives work intensive air pollution occurs. The computer model has been developed to assess the zones of chemical pollution. The model allows calculating the dynamics of the formation of pollution zones taking into account diesel locomotive movement and various obstacles (station buildings, etc.) that are located near the railways. The model is based on the numerical integration of 3D equation of potential flow and 3D equation of pollutant dispersion. Chemical transformations of NO, NO_2 are taken into account. The model can be used to understand the process of air pollution at railway stations where diesel locomotives are widely used. The developed numerical model consumes not much computational time and does not require much computational memory. This is very important in case when many numerical experiments must be performed. Developed numerical model can be used at the laboratories of environment protection at railways.

References

[1] Sobota A, Żochowska R, Szczepański E and Gołda P 2018 The influence of tram tracks on car vehicle speed and noise emission at four-approach intersections located on multilane arteries in cities *Journal of Vibroengineering* **20** 6 pp 2453–2468

[2] Gołębiowski P, Żak J and Jacyna-Gołda I 2020 Approach to the Proecological Distribution of the Traffic Flow on the Transport Network from the Point of View of Carbon Dioxide *Sustainability* **12** 17 pp 1–16

[3] Pyza D, Jacyna-Gołda I, Gołda P and Gołębiowski P 2018 Alternative Fuels and Their Impact on Reducing Pollution of the Natural Environment *Annual Set The Environment Protection* **20** pp 819–836

[4] Merkisz J, Jacyna M, Merkisz-Guranowska A and Pielecha J 2014 The parameters of passenger cars engine in terms of real drive emission test *Archives of Transport* **32** 4 pp 43–50

[5] Deborah M S M, Alexandre L B and Armando M A 2012 A numerical model for pollutant dispersion simulation in street canyons *Mecánica Computacional* **XXXI** pp 211–235
Sanchez B, Santiago J, Martilli A, Palacios M and Kirchner F 2016 CFD modeling of reactive pollutant dispersion in simplified urban configurations with different chemical mechanisms *Atmospheric Chemistry and Physics* **16** pp 12143–12157

[6] Overman H T J 2009 *Simulation model for NO_x distributions in a street canyon with air purifying pavement: Master thesis* (Netherlands: University Twente) p 69

[7] During I, Bachlin W, Ketzel M, Baum A, Friedrich U and Wurzler S 2011 A new simplified NO/NO_2 conversion model under consideration of direct NO_2-emissions *Meteorologische Zeitschrift* **20** 1 pp 67–73

[8] Zgurovski M, Scopetski V, Khrutch V and Biliaiev M 1997 *Numerical modeling pollutant dispersion in environment* (Kyiv: Naukova dumka) p 368

[9] Biliaiev M, Gunko O, Kirichenko P and Muntyan L 2017 *Risk assessment in case of toxic substances emission on railway transport* (Kryvyi Rih: Kozlov Publ.) p 127

[10] Samarskii A A and Gulin A V 1989 *Numerical methods* (Moscow: Nauka) p 432

IOP Conf. Series: Materials Science and Engineering **985** (2020) 012020 doi:10.1088/1757-899X/985/1/012020

The results of brake tests of the DPKr-3 diesel train

S Dovhaniuk[1,2], V Shaposhnyk[1], O Shykunov[1], A Shatunov[1] and V Visloguzov[1]

[1]Dnipro National University of Railway Transport named after Academician V. Lazaryan, Department of Cars, Lazaryana str, 2, 49010, Dnipro, Ukraine

[2]dovganyuk@ukr.net

Abstract. JSC "Ukrzaliznytsia" purchased the first of three planned regional three-car (two main motor cars and one intermediate) diesel train DPKr-3 manufactured by PJSC "Kryukovsky railway car building works". To confirm the declared characteristics and safety indicators of the DPKr-3 diesel train, specialists department of Cars of Dnipro National University of Railway Transport named after Academician V. Lazaryan conducted complex certification tests, which confirmed the compliance of the stated parameters. One of the points of the Terms of Reference was to determine the brake pressure of the linings per 100 weight of the train in terms of cast iron brake pads. For this purpose, according to the methodology adopted in the territory of the former CIS, force sensors, which were installed in the tick mechanisms instead of the brake linings, measured the magnitudes of the forces of pressing the brake linings on the disks. However, during the tests it was found out that the current standards for calculating the provision of train brakes cannot be applied to modern rolling stock. It is recommended for rolling stock where the use of brake pads, and even more cast iron, is not provided by the design, to develop normative documents that would determine the standards for brakes by the results of calculations and tests.

Summary

The key role in the development of the economy of our country is played by Ukrainian railways. An extensive network of railroads connects hundreds of cities, allowing not only the connection of industrial centers but also the provision of passenger transportation in all directions. The rolling stock of Ukrainian railways in 2019 was updated with modern freight and passenger cars, Tryzub diesel locomotives manufactured by General Electric (USA), modernized the rail bus produced by the Polish company PESA, purchased the first of three planned regional three-car diesel DPKr-3 manufactured by PJSC "Kryukovsky railway car building works" (Ukraine). To confirm the declared characteristics and safety indicators of the DPKr-3 diesel train, specialists of Dnipro National University of Railway Transport named after Academician V. Lazarian conducted complex certification tests, which confirmed the compliance of the declared parameters. The brake system is the most important for providing safety, so determining the brake parameters of a diesel train was a mandatory test point.

Literature review

The question of efficiency and expediency of application certain brake systems on rolling stock, and especially on high-speed rolling stock remains relevant [1, 2]. In article [3] it is stated that a modern

Content from this work may be used under the terms of the Creative Commons Attribution 3.0 licence. Any further distribution of this work must maintain attribution to the author(s) and the title of the work, journal citation and DOI.
Published under licence by IOP Publishing Ltd

rolling stock should be equipped with several types of brakes that supplement the basic pneumatic brake. The modern rolling stock should be equipped with onboard brake control and diagnostics systems. It allows you to change the input parameters: brake weight percentage, braking modes, train length and track profile. The on-board unit registers the train deceleration and determines the optimum speed value [4]. Also, a lot of attention in international scientific editions has been paid to anti-use defense systems [5, 6]. The procedure for conducting and processing data of certification and acceptance brake tests of innovative rolling stock of railways is presented in the article [7] in which the results of the braking tests of a 2EV120 trunk freight locomotive with a design speed of 140 km/h are given. The tests were carried out in accordance with GOST R 55364. In article [8], it is noted that the braking distance calculation norms set by EN 14531 may not always give fairly accurate results and require revision. This conclusion was obtained when testing the EMU V250 passenger electric train with a maximum operational speed of 250 km/h designed for high-speed lines in Belgium and the Netherlands. Taking into account the values of the friction coefficients of friction brake linings, which depend on the speed and force of pressing, it became possible to obtain reliable results of the values of the brake path that coincide with the experimental data.

In article [9] is also noted the need to revise some regulatory documents for evaluating the performance of the rolling stock braking system. The differences between the requirements of the normative documents of Western Europe and the countries of the former Soviet Union are pointed out. The brakes of wagons made in Western Europe are calculated according to the requirements of TSI, where the output parameter is the braking weight, and for the 1520 mm track wagons, the force of pushing the brake pads in terms of cast iron pads. This problem is especially acute in countries that operate wagons built according to different standards, such as Lithuania.

Analysis of the literature in the field of research confirmed the relevance of the chosen topic. It should be noted that it is necessary to revise and introduce new national regulatory documents that would establish a methodology for conducting and analyzing the results of brake tests of modern rolling stock on Ukrainian railways.

The results of the brake tests of the DPKr-3 diesel train

After 4 years of successful operation of JSC "Ukrzaliznytsia" of the three-car diesel train DPKr-2, manufactured by PJSC "Kryukovsky railway car building works" [10]. On order of JSC "Ukrzaliznytsia" for regional passenger transportation, it was made with some changes to three wagon diesel train DPKr-3. The composition of the diesel train includes two main motor cars and one intermediate. The train diagram for stationary tests is shown in Figure 1. Figure 1 shows the wagon numbers and their weight in the loaded state, the designation of the trolleys and the total estimated pressure of the brake linings on the wheels of the trolley.

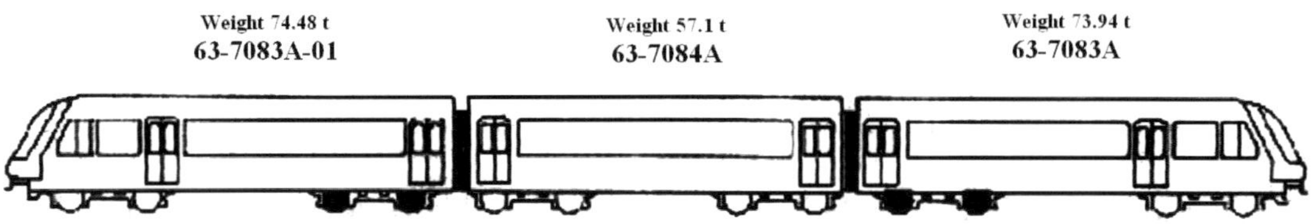

Figure 1. Scheme of DPKr-3 diesel train during stationary brake tests.

The diesel train is equipped with a system for regulating the pressure in the brake cylinders, depending on the load (population) of each individual cart. This makes it possible to realize the same specific braking force of each cart of a diesel train wagon, which provides smooth braking. One of the points of the Terms of Reference (TOR) [11] was to determine the brake pressure of the linings per

100 weight of the train in terms of cast iron brake pads. For this purpose, according to the methodology adopted in the territory of the former CIS, force sensors, which were installed in the tick mechanisms instead of the brake linings (Figure 2), measured the magnitudes of the forces of pressing the brake linings on the disks. Wheeled brakes are mounted on a diesel train. The pressing forces were measured in the empty and loaded states three times with pneumatic emergency, electro pneumatic and parking brake operation.

Figure 2. Installation of force sensors in determining the force of pressing the pad.

There were no "emissions" when determining the force of pressing the plates on the disks and the average values of the three measurements were taken into account. Table 1 shows the average values of the actual presses of the brake pads on discs in the empty and loaded states under different braking modes.

Knowing the magnitudes of the actual pressures, the calculated formulas [13] for the composite brake pads have been calculated using the known formula.

$$K_P = 1.22K\frac{0.1K+20}{0.4K+20}, \quad (1)$$

where K is the value of the actual pressing of the brake lining of the brake disc, kN.

To assess the braking performance of the rolling stock, use the force factor of the brake pad lining [13].

$$\delta_P = \frac{\sum K_P}{Q}\frac{r}{R}, \quad (2)$$

where $\sum K_P$ is the total estimated compression of the brake linings of the wagon, kN;

Q – weight of the wagon, kN;

r – is the average friction radius of the brake disc, 322 mm;

R – is the wheel radius of the rolling circle, 467.5 mm.

The results of the calculations of the calculated pressures and the coefficients of the pressing force in terms of the composite and cast iron brake pads of the individual wagons and the brake coefficient of the diesel train as a whole are shown in Table 2.

Table 1. The average values of the actual presses of the brake pads on discs in the empty and loaded states under different braking modes.

Kind of braking, others	State	Pressing, kgF (63-7083A)			Pressing, kgF (63-7084A)		
		Car Truck TB1	Car Truck MB1	Total	Car Truck TB2	Car Truck TB3	Total
Pneumatic	Loaded	15203	15244	30447	10468	10610	21078
	Empty	15318	14280	29598	10223	10225	20448
Electro pneumatic	Loaded	15763	15483	31246	11509	11598	23107
	Empty	15078	14662	29740	11175	11266	22441
Parking brake	Loaded	13104	11987	25091	8085	8139	16224
	Empty	12980	11983	24963	8046	8149	16195
Weight, kgF	Loaded		73940			57100	
	Empty		69193			52126	

Kind of braking, others	State	Pressing, kgF (63-7083A-01)			Diesel-train
		Car Truck MB2	Car Truck TB4	Total	
Pneumatic	Loaded	15088	15069	30157	81682
	Empty	14317	14389	28706	78752
Electro pneumatic	Loaded	15337	15625	30962	85315
	Empty	14443	15056	29499	81680
Parking brake	Loaded	12200	13220	25420	66735
	Empty	12191	13053	25244	66402
Weight, kgF	Loaded		74480		205520
	Empty		69706		191025

Table 2. The results of the calculations of the calculated pressures and the coefficients of the pressing force in terms of the composite and cast iron brake pads of the individual wagons and the brake coefficient of the diesel train.

Wagon	Weight, kN	Compositional		Cast iron	
		$\sum K_P$, kN	δ_P	$\sum K_P$, kN	δ_P
63-7083A	725.35	289.06	0.27	342.92	0.33
63-7084A	560.15	212.81	0.26	270.56	0.33
63-7083A-01	730.65	286.82	0.27	340.86	0.32
Diesel train	2016.15	788.69	0.269	954.32	0.326

At a speed of 140 km/h. according to the chart given in the guide to traction calculations [12], the effectiveness of composite brake pads (pads) is increased by 1.22 times compared to cast iron. Pressing the composite pads on the axle 342.92/4=85.73 kN of the motor car 63-7083A will correspond in terms of cast iron pads to the pressure of 104.59 kN.

In this case, the force coefficient of pressing δ_P in terms of cast iron brake pads will be:

$$\delta_P = \frac{104.59 \times 4}{725.35} \frac{322}{467.5} = 0.397 . \quad (3)$$

According to the Terms of Reference the single minimum brake pressure, in terms of cast iron brake pads, must be at least 78 tons for every 100 tons of diesel train weight.

Taking into account the coefficient of efficiency at a speed of 140 km/h. brake pressure, in terms of cast iron brake pads, for every 100 tons of diesel train weight is 39.7 tons. The condition of the TOR according to the above calculations is not fulfilled.

According to the Standards for non-self-propelled rolling stock [13, Table 9.2] the minimum permissible value of the coefficient of pressure of the brake pads (composite) or lining of disc brakes of passenger cars at maximum speeds up to 160 km/h and pneumatic brakes is 0.3. There is no value for speeds up to 140 km/h.

According to the Standards for non-self-propelled rolling stock [14, Table 9.2] the minimum permissible value of the compression ratio of composite brake pads for wagons weighing 64-75 t (diesel wagon motor cars) and speeds of 140 km/h is 0.25, which is less than calculated and satisfies the condition. For wagons weighing 53-64 tones (towed diesel train car), the minimum permissible value is 0.27, which is more than calculated and does not satisfy the condition. These regulations do not apply to diesel trains, and the requirements of the existing braking regulations [15, 16, 17] cannot be applied to DPKr-3 diesel trains equipped with KNORR-BREMSE wheel-disc brakes.

The task of determining the single minimum brake pressure in terms of cast iron brake pads, for every 100 tons of diesel train weight, was decided to solve from the opposite, applying the nomogram of the brake path of the passenger train with cast iron brake pads during emergency braking.

During running tests for pneumatic emergency braking in the loaded state on a straight section with a value of ±2 ppm (thousandths), the average value of the braking distance according to the results of three brakes was: at a speed of 140 km/h – 868 m, at a speed of 120 km/h – 634 m at a speed of 100 km/h – 473 m, at a speed of 80 km/h – 290 m. If found the amount of braking pressure per 100 tons of train weight by nomogram (Figure 3) [18, Figure 317a] with such values of the braking distance, its value is more than 100 tons per 100 tons of train weight (Figure 3).

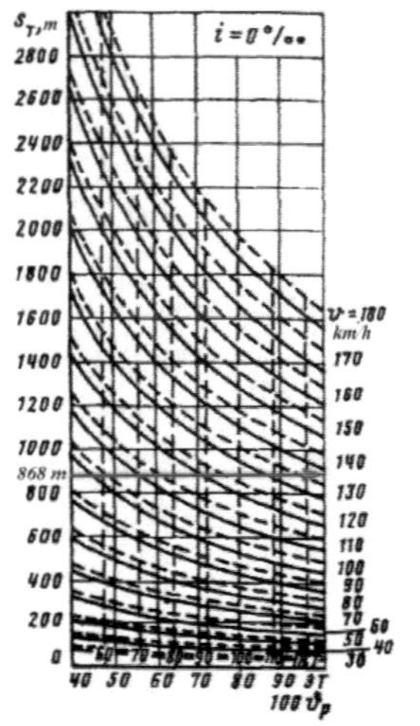

Figure 3. Nomogram of the brake path of the passenger train with cast iron brake pads.

Conclusions

For such diesel trains (DPKr-2, DPKr-3), where the use of brake pads, and even more cast iron, is not provided by design, it is necessary to develop normative documents that would determine the brake standards for this type of rolling stock based on the results of calculations and tests. Otherwise, the VU-45 brake certificate data does not reflect the actual values. Taking into account the coefficient of efficiency at a speed of 140 km/h. brake pressure, in terms of cast iron brake pads, for every 100 tons of diesel train weight is 39.7 tons. Found the amount of braking pressure of train by nomogram its value is more than 100 tons per 100 tons of train weight.

References

[1] Sharma R C, Dhingra M and Pathak R K 2015 Braking systems in railway vehicles *Int. J. Engineering Research & Technology* **4** 1 pp 206–211
[2] Reidemeister O H and Shaposhnyk V Yu 2015 Determination of stopping distance of unpowered rolling stock by the method of sequential braking *Science and Transport Progress* 6 60 pp 127-133
[3] Hasegawa I and Uchida S Braking systems 1999 *Japan Railway Transp* **20** pp 52–59
[4] Presciani P, Malvezzi M, Bonacci G and Balli M 2001 Development of a braking model for speed control systems *World Congress on Railways Research* (Cologne, Germany November 2001) pp 25–29
[5] Mousavi A, Markazi A and Masoudi S. 2017 Adaptive Fuzzy Sliding-Mode Wheel Control Slide Protection Device for ER24PC Locomotive *Latin American Journal of Solids and Structures* **14** pp 2019–2045
[6] Ahmad H A 2013 Dynamic braking control for accurate train braking distance estimation under different operating conditions *Mech. Eng., Virginia Polytech. Instit. State Univ.* (Blacksburg, VA, USA) p 119
[7] Domanov K I 2019 Determination of the brake parameters of a double-powered electric locomotive according to the results of certification tests *Bulletin of the UrGUPS* **1** 41 pp 89–95
[8] Pugi L, Malvezzi M, Papini S and Vettori G 2013 Design and preliminary validation of a tool for the simulation of train braking performance. *J Mod Transp* **21** pp 247–257
[9] Bureika G, Mikaliūnas Š 2008 Research on the Compatibility of the Calculation Methods of Rolling Stock Brakes *Transport* **23** 4 pp 351–355
[10] Savchenko K, Ryzhov V, Shykunov A, Ryzhov S and Shaposhnyk V 2015 Research of brake system trains ДПКр-2 *Abstracts of the 75 international scientific & practical conference «The problems and prospects of railway transport development»* (Dnepropetrovsk, May 14-15) pp 72–73
[11] Specification "Diesel train DPKr-3 for regional passenger transportation" 63 DPKr-3.TZ
[12] Grebenyuk P T, Dolganov A N and Skvortsova A I 1987 *Directory. Traction calculations* (Moscow: Transport) p 272
[13] 1996 Norms of calculation and design of railway cars of the gauge of the gauge of 1520 mm (not self-propelled) (Moscow: GosNIIV – VNIIZhT) p 317
[14] 1991 Standards for the calculation and design of new and upgraded rail cars of 1520 mm gauge (non-self-propelled) (Moscow: VNIIV – VNIIZhT) p 260
[15] 2002 Operating instructions for rolling stock brakes on the railways of Ukraine **TST-TSV-TSL-0015** *Kyiv* p 145
[16] 2002 Instruction on maintenance, repair and testing of brake equipment of locomotives and motor-car rolling stock **TST-0058** *Kyiv* p 223
[17] 1998 Braking standards **TSV-0011** *Kyiv* p 18
[18] Krylov V I and Krylov V V 1989 Directory. Railway rolling stock braking equipment (Moscow: Transport) p 487

IOP Conf. Series: Materials Science and Engineering **985** (2020) 012021 doi:10.1088/1757-899X/985/1/012021

Optimal Control Method of High-Voltage Frequency Converters with Damaged Cells

V Busher[1,6], O Chornyi[2,7], O Glazeva[1,6], V G Kuznetsov[3,8], V Tytiuk[4,9] and M Tryputen[5,10]

[1]Department of Ship`s Electromechanics and Electrotechnics, National University "Odessa Maritime Academy", 8, Didrikhson str., Odessa, 65029, Ukraine
[2]Institute of Electromechanics, Energy Saving and Automatic Control Systems, Kremenchuk Mykhailo Ostrohradskyi National University, 20, Pershotravneva Street, Kremenchuk, Ukraine
[3]Electric Power Department, Railway Research Institute, 50, Chlopickiego str., Warsaw, Poland
[4]Department of electromechanics, Kryvyj Rih National University, 11, Vitaly Matusevich str, Kryvyi Rih, Ukraine
[5]Department of Automation and Instrumentation, Dnipro University of Technology, 19, D. Yavornitsky ave., Dnipro, Ukraine

[6]victor.v.bousher@gmail.com, [7]alekseii.chornyi@gmail.com, [8]VKuznetsov@ikolej.pl, [9]dinalt2006@gmail.com, [10]nikolay.triputen@gmail.com

Abstract. Application of high-voltage frequency converters is one of the most prospective trends in the development of current powerful electric drives, i.e. in high-speed trains. In terms of such converters, a multilevel cascaded H-bridge converter is becoming the most widespread one owing to its increased reliability, survival in terms of one or more damaged cells, and quick repairability due to its modular structure. The research proposes a control method, which provides maximum possible linear voltage in case of one or several damaged cells owing to the shift of zero point and turning of phase voltages. It is also demonstrated how turning angles of phase voltages should be set to preserve spatial position of the linear voltage vectors which provides the least complicated electromagnetic transients in terms of nominal rotation frequency; in case of low or middle frequencies, damage of the cells may have no effect on the motor operation. Fast algorithm to calculate turning points of phase voltages is proposed; the algorithm may be applied in the industrial microprocessor control systems. Algorithm of the operations, which provides implementation of the method, is described.

1. Introduction

High-voltage frequency converters are becoming more and more popular in different fields of technology where powerful electric drives are required. These are metallurgical enterprises, metal rolling mills, ship power systems, rail transport and much more. Along with the common benefits – smooth and efficient velocity control of the alternating-current machines – their implementation provides multiple reduction of both consumption of electrotechnical materials (first of all, that is copper) and THD (Total Harmonic Distortion), increase of EEDI (Energy Efficiency Design Index) and EEOI (Energy Efficiency Operational Index) giving the incontestable benefits while designing up-to-date electromechanical systems, which include transportation systems, i.e. high-speed trains [1]. It is in the

Content from this work may be used under the terms of the Creative Commons Attribution 3.0 licence. Any further distribution of this work must maintain attribution to the author(s) and the title of the work, journal citation and DOI.
Published under licence by IOP Publishing Ltd

latter systems that high-voltage multilevel frequency converters can provide energy recovery to the common network, solving the problem of increasing energy efficiency [2]. However, a number of problems also arise, since high-voltage semiconductor power devices are still less reliable. Therefore, it is necessary to solve the problems of both improving the reliability of electronic components and finding methods to reduce damage in the event of failure of some of them.

The aim of the work is to find such a method for controlling high-voltage multilevel frequency converters that would provide the least voltage drop on the motor and the least ripple of the electromagnetic torque in case of damage to one or more cells in the inverter phases.

2. Analysis of the problem

High-voltage frequency converters (HVFC) with the following circuitries are the most widespread ones [3]:

1. Two-transformer FC with low-voltage inverter;
2. Multilevel FC with input sectional transformer.

Systems of the first types are being gradually ousted from the market as powerful high-voltage switches are entering the market; among the multilevel converters, several types are the most widespread ones [3]:

- diode-clamped converter;
- neutral point clamped converter;
- floating-capacitor converter;
- cascaded H-bridge converter.

Cascaded H-bridge converters are considered to be the most advanced ones from the viewpoint of reliability, survivability, and maintainability. Such circuit helps obtain any amount of voltage using standard low-voltage technologically independent modules (Figure 1) [4]. Moreover, the design modularity makes it possible to preserve operating capacity, if any or even several modules may get out of order.

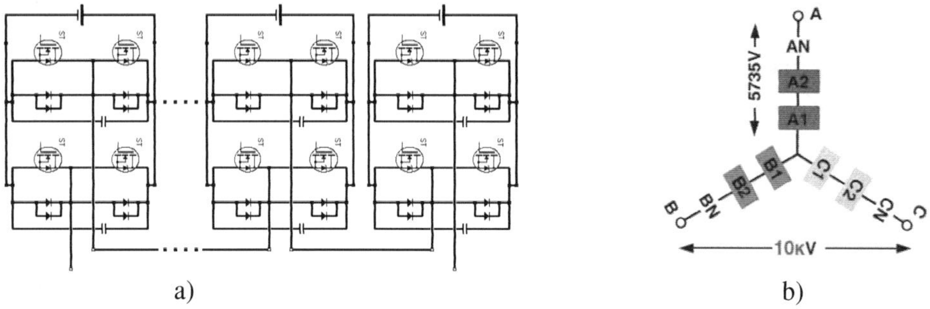

a) b)

Figure 1. The principle of building a cascaded H-bridge converter: a) one phase of the cascaded H-bridge converter; b) the formation of a three-phase high voltage system.

Layout and principles of the inverter control are determined by the power semiconductor components (IGBT, GTO, IGCT, SGCT) [5] and Pulse Width Modulation (PWM) methods [6–8]:

- sinusoidal PWM;
- modified algorithms of sinusoidal PWM;
- PWM with variable parameters;
- Space Vector PWM (SVPWM);
- algorithm of elimination of specific harmonics;
- wavelet-modulation etc.

In terms of multilevel cascaded H-bridge converter, response to the breakdowns of one or more modules is the most critical action [9, 10]. Common short circuit of the damaged module results in the side-to-side linear voltage tilts, disbalance of currents, consequently, electric and electromechanical shock loads and a threat to the safety of the operation of the mechanism and personnel [11–13]. Thus, different correction algorithms are applied. For instance, in terms of Allen Bradley HV FC 6000, when defect is detected in one of the phases, the "like" cells in two other phases are closed [14]. In this case, voltage drops by $(N-1)/N$ times. In some cases, the method of linear voltage balancing is applied by changing the interphase angles and the shift of zero point, as in terms of FC SIEMENS of Robicon PERFECT HARMONY series [15, 16]. However, solutions of the same problem in case of breakdown of several cells and reduced shock load at the moment of transition from standard to emergent operating mode as well as the evaluation of possible application of SVPWM in the emergency modes are still problematic.

Objective of the research is to develop a universal method to correct linear voltages in terms of damaged cells of multilevel cascaded H-bridge converter with the minimization of shock loads during transition from standard to emergency operating modes and to estimate efficiency of the use of Space Vector Pulse Width Modulation in case of the emergency modes.

3. Materials of the research

Preliminary analysis has demonstrated that in terms of the multiphase electric machines, considerable oscillations of electromagnetic moment take place not only due to voltage changes but also because of the turning of linear voltage vectors at that moment.

The proposed correction method is based on the shift of zero point and turning of FC phase voltages to provide symmetry of the amplitudes and preserve spatial position of the linear voltages in terms of damaged cells (Figure 2).

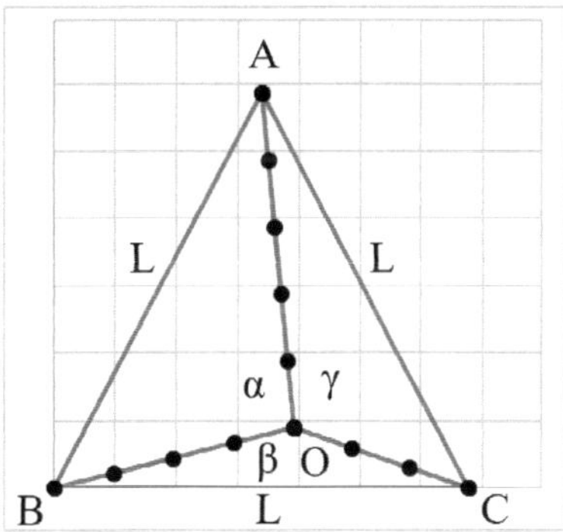

Figure 2. Linear and phase voltage vectors after damage of cells.

Since all the cells generate certain similar voltage U, it is possible to move on to the equations in relative units having substituted absolute values of phase voltages with the number of cells:

$$\begin{cases} N_A = U_A / U, \\ N_B = U_B / U, \\ N_C = U_C / U \end{cases} \quad (1)$$

and having expressed linear voltages in their relation to U. Then, in terms of standard operating mode at $N_A = N_B = N_C = N$, we obtain $L_n = \sqrt{3}N$.

To determine interphase angles α, β, γ in the emergency mode, it is required to meet the equality of the linear voltages to some unknown value L:

$$\begin{cases} N_A^2 + N_B^2 - 2N_A N_B \cos(\alpha) = L^2, \\ N_B^2 + N_C^2 - 2N_B N_C \cos(\beta) = L^2, \\ N_C^2 + N_A^2 - 2N_C N_A \cos(\gamma) = L^2, \\ \alpha + \beta + \gamma = 2\pi, \end{cases} \tag{2}$$

where N_A, N_B, N_C are relative values of voltages of each phase, $\alpha = \angle(U_A, U_B) = \angle AOB$, $\beta = \angle(U_B, U_C) = \angle BOC$, $\gamma = \angle(U_C, U_A) = \angle COA$.

In case of breakdown of one or several cells in two phases, two of the three angles become obviously equal; then, system (1) is reduced to a square equation being solved analytically. Thus, if $N_A = N_B \neq N_C$, we obtain:

$$\begin{cases} a = 4N_A^2, b = -2N_A N_C, c = -3N_A^2 + N_C^2, \\ D = b^2 - 4ac, \\ \beta = \gamma = \arccos\dfrac{-b - \sqrt{D}}{2a}, \\ \alpha = \pi - 2\beta, \\ L = 2N_A \sin(\pi - \alpha). \end{cases} \tag{3}$$

If $N_A \neq N_B \neq N_C$, then equation system (2) may be solved only by iteration methods or by searching through possible solutions within the admissible ranges of value changes; the latter method is simpler to implement in microprocessor control systems.

As the analysis has shown, a calculation method based on the angle's selection turns to be rather slow as it is necessary to use two nested loops for two of the three angles with the searching through all the possible combinations.

Much faster solution is possible owing to a method based on Heron theorem. The idea is in the fact that the area of equilateral triangle $S_{\Delta ABC}$, in terms of point O located within it, is equal to the total of the three internal triangles:

$$S_{\Delta ABC} = S_{\Delta AOB} + S_{\Delta BOC} + S_{\Delta COA}, \tag{4}$$

where:

$$\begin{cases} S_{\Delta AOB}^2 = \dfrac{2N_A^2 N_B^2 + 2N_A^2 L^2 + 2N_B^2 L^2 - N_A^4 - N_B^4 - L^4}{16}, \\ S_{\Delta BOC}^2 = \dfrac{2N_B^2 N_C^2 + 2N_B^2 L^2 + 2N_C^2 L^2 - N_B^4 - N_C^4 - L^4}{16}, \\ S_{\Delta COA}^2 = \dfrac{2N_C^2 N_A^2 + 2N_C^2 L^2 + 2N_A^2 L^2 - N_C^4 - N_A^4 - L^4}{16}, \\ S_{\Delta ABC} = \sqrt{3}L^2 / 4. \end{cases} \tag{5}$$

In case of damage of one or several cells, $S_{\Delta ABC}$ becomes less than the initial one, and solution is possible if $L_{\min} <= L <= L_{\max}$, where:

$$\begin{cases} N_{\max} = \max(N_A, N_B, N_C), \\ L_{\min} = \min(N_{\max}, |N_A - N_B|, |N_B - N_C|, |N_C - N_A|),. \\ L_{\max} = \min(\sqrt{3}N_{\max}, N_A + N_B, N_B + N_C, N_C + N_A) \end{cases} \tag{6}$$

Consequently, while reducing (with some step) L from L_{max} to L_{min}, it is possible to find a value meeting condition (4). Such calculation is over, if:

$$S_{\Delta ABC} - S_{\Delta AOB} - S_{\Delta BOC} - S_{\Delta COA} <= 0. \quad (7)$$

Further, use a cosine theorem to determine angles:

$$\alpha = \arccos\frac{N_A^2 + N_B^2 - L^2}{2N_A N_B},\ \beta = \arccos\frac{N_B^2 + N_C^2 - L^2}{2N_B N_C},\ \gamma = \arccos\frac{N_C^2 + N_A^2 - L^2}{2N_C N_A}. \quad (8)$$

Below, Table 1 represents several solutions for 3-7-level inverters. The table shows the amount of undamaged cells N_A, N_B, N_C, corresponding angles α, β, γ, value of linear voltage L and ratio L to L_n, and ratio L_o / L_n during the balancing by preserving similar number of cells in each phase. Rows in bold correspond to the standard operating mode. The table data may be entered in the controller database which helps simplify considerably the software.

Table 1. Several solutions for 3-7-level inverters.

Na	*Nb*	*Nc*	*α*	*β*	*γ*	*L*	*L/L$_n$*	*L$_o$/L$_n$*
7	**7**	**7**	**120.0**	**120.0**	**120.0**	**12.1**	**1.00**	**1.00**
7	7	6	110.8	124.6	124.6	11.5	0.95	0.86
7	6	6	114.3	131.4	114.3	10.9	0.90	0.86
7	6	5	104.4	138.5	117.1	10.2	0.85	0.71
7	5	5	105.6	148.9	105.6	9.63	0.79	0.71
7	5	4	94.0	161.5	104.4	8.88	0.73	0.57
7	4	4	89.0	182.1	89.0	8.00	0.66	0.57
6	**6**	**6**	**120.0**	**120.0**	**120.0**	**10.3**	**1.00**	**1.00**
6	6	5	109.2	125.4	125.4	9.78	0.94	0.83
6	5	5	113.1	133.7	113.1	9.20	0.88	0.83
6	5	4	101.4	142.8	115.8	8.54	0.82	0.67
6	4	4	101.4	157.2	101.4	7.84	0.75	0.67
6	4	3	86.4	177.3	96.3	7.00	0.67	0.50
5	**5**	**5**	**120.0**	**120.0**	**120.0**	**8.66**	**1.00**	**1.00**
5	5	4	107.2	126.4	126.4	8.05	0.93	0.80
5	4	4	111.3	137.4	111.3	7.45	0.86	0.80
5	4	3	96.9	150.0	113.1	6.77	0.78	0.60
5	3	3	93.6	172.9	93.6	5.99	0.69	0.60
4	**4**	**4**	**120.0**	**120.0**	**120.0**	**6.93**	**1.00**	**1.00**
4	4	3	104.0	128.0	128.0	6.31	0.91	0.75
4	3	3	108.2	143.6	108.2	5.70	0.82	0.75
4	3	2	89.0	164.5	106.6	4.96	0.72	0.50
3	**3**	**3**	**120.0**	**120.0**	**120.0**	**5.20**	**1.00**	**1.00**
3	3	2	98.9	130.5	130.5	4.56	0.88	0.67
3	2	2	101.4	157.2	101.4	3.92	0.75	0.67

It is clear that in all cases correction of zero-point position and phase voltage directions helps obtain better result. For instance, Figure 3 shows vectors of phase and linear voltages for a case when one and two cells break down in phases B and C respectively in a 5-level inverter.

Blue color is for the vectors in standard operation mode; dashed line is for standard solution in that situation (three cells are preserved in all the phases). It is obvious that use of the calculation results from Table 1 (red lines with black highlights) makes it possible to reduce almost by half the drop of linear voltages.

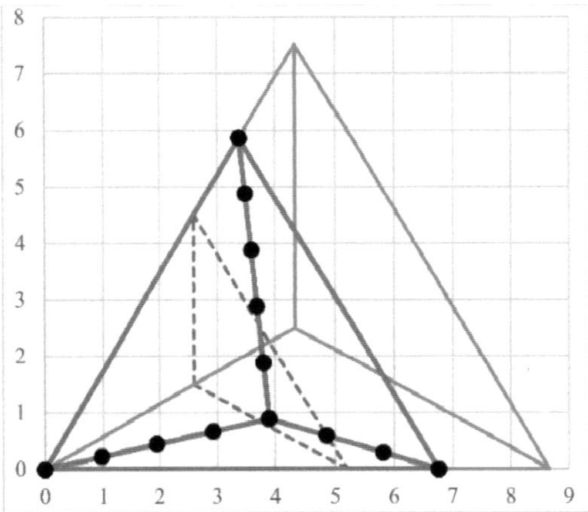

Figure 3. Vectors of linear and phase voltages in a 5-level inverter.

Next stage involves application of those results while controlling the inverter. It is required to complete several operations in the system of control at the moment of emergency diagnosis.

First, change the phase shift by the difference between the position of vector of the corresponding phase in standard and emergency modes not interrupting the calculation of the turning angle of the field vector. Here we can address to the classic geometry to help us. Assuming that in terms of standard operation, vector of phase A is located vertically, we obtain new positions of vectors $\alpha_0, \beta_0, \gamma_0$:

$$\alpha_0 = \frac{\pi}{2} + \arcsin\left(\frac{N_B}{L}\sin(\alpha)\right), \quad \beta_0 = \alpha_0 + \alpha, \quad \gamma_0 = \beta_0 + \beta. \tag{9}$$

Secondly, multiply the generated sinusoidal signals in each phase by value L_n / L which will help preserve linear voltage without any changes during the low and medium speed operation.

Sequence of those operations is represented by the model fragment in MATLAB / SIMULINK shown in Figure 4. Further, in terms of the system generating control signals of transistors in the phases with damaged cells, preserve and correlate the shifts of the number of the comparison triangular signals being similar to the number of the undamaged cells left; multiply the amplitudes of sinusoidal signals by the number of undamaged cells.

It is important to note that the proposed algorithm does not exclude the possibility of applying Space Vector Pulse Width Modulation. However, if linear voltage increases by 16% in terms of standard operation, the efficiency will decrease slightly in the emergency modes. The cause is in the fact that if the operation is standard, then the 3rd harmonic with relative amplitudes of 1/6, which is added to the phase voltages, forms symmetric curves with the saddle-like peak [17–19].

When cells are damaged, shifts of phases of the first harmonics "deform" the peaks. Thus, having calculated preliminarily one of the periods of phase voltages, it is necessary to define such a boundary admissible value of the main harmonic so that the amplitude of sum signal will not exceed N_A, N_B, N_C [20–22]. It is important that in terms of the 3rd harmonic, optimal shift will be the one of the undamaged phase in the standard operation mode. Then, linear voltages may be increased by 2…10 % [23].

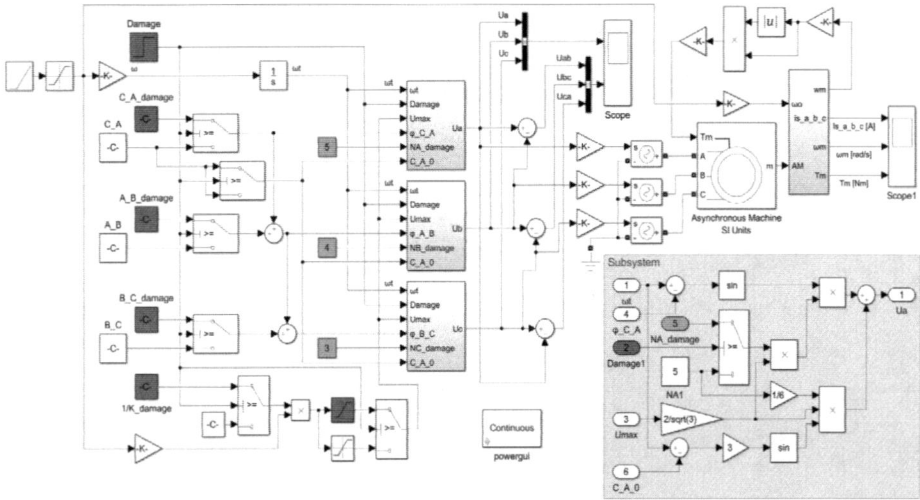

Figure 4. Model of electric drive "Frequency converter – asynchronous machine with ventilating load" with the correction system.

The results of the system operation are represented in Figure 5 (in the context of reduced frequency of the generated voltage) and Figure 6 (at nominal frequency) while using the following law: $U / f = const$.

It is seen that in case of underfrequency, changes in an inverter do not affect the motor operation; in terms of nominal frequency, moment fluctuations experience the minimal possible amplitude owing to both minimized voltage changes and preservation of the spatial position of the linear voltage vectors.

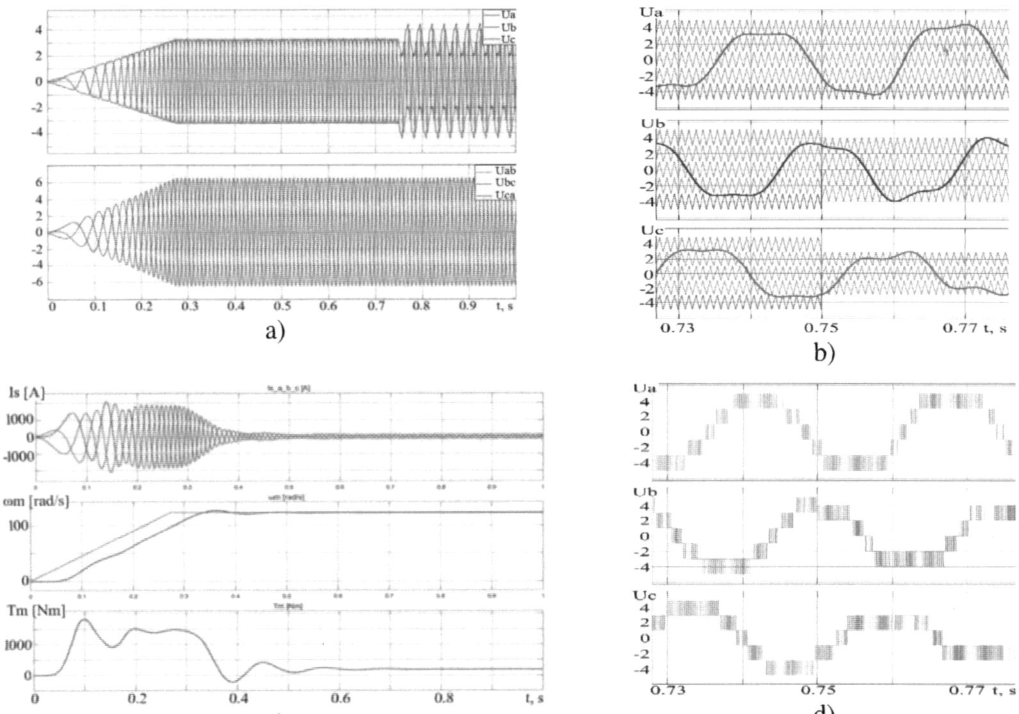

Figure 5. Transients at 39 Hz: a) phase and linear voltages; b) signals of SVPWM; c) stator currents, speed, and torque; d) phase voltages.

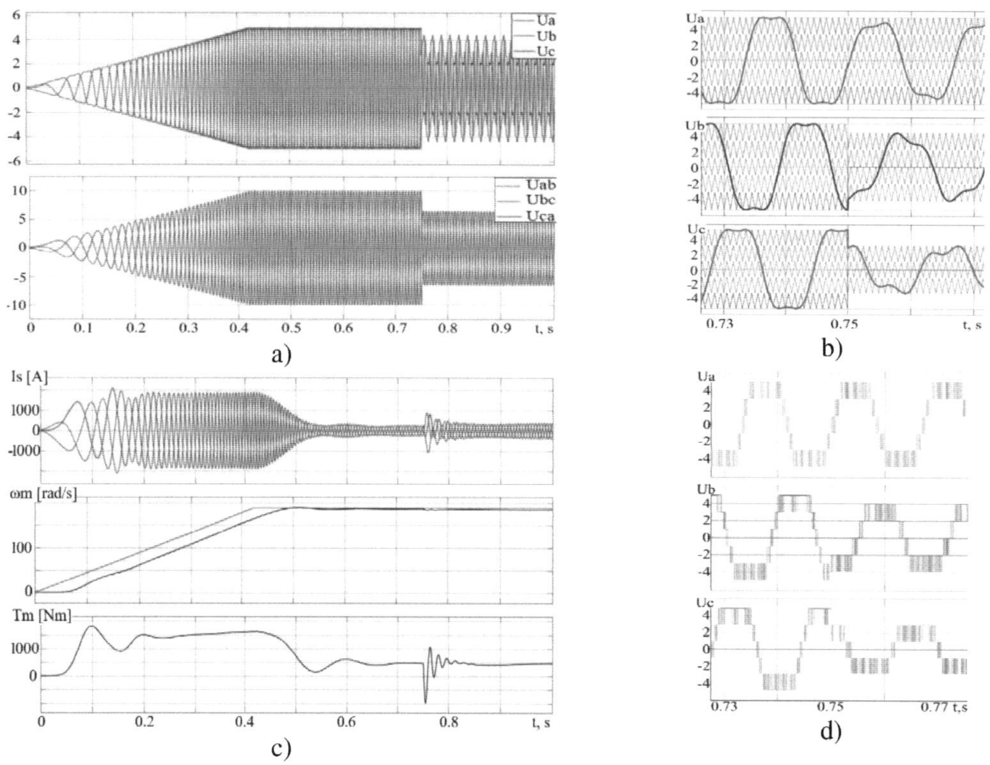

Figure 6. Transients at 60 Hz: a) phase and linear voltages; b) signals of SVPWM; c) stator currents, speed, and torque; d) phase voltages.

4. Conclusions

The paper proposes a method to control multilevel cascaded H-bridge converter in case of damage of one or several cells in different phases. The method provides minimization of shock loads in terms of transitions from the standard to emergency operating modes owing to such shift of zero point and turning of phase voltages, at which amplitude of linear voltages decreases by the minimal possible value and spatial position of the linear voltage vectors remains the same. Fast algorithm to calculate turning points of the phase voltages is proposed; the algorithm may be applied in the industrial microprocessor control systems. Algorithm of the operations, which provides implementation of the method, is described. It is demonstrated that in terms of nominal frequency, application of the method provides shock load minimization; in case of low and medium frequencies, damage of cells may not affect the motor operation (under conditions of fast reaction and disconnection of the damaged cells). Conditions of the application of Space Vector Pulse Width Modulation in the emergency modes have been identified which help increase considerably the operating efficiency of the frequency converters.

Acknowledgments: This paper is elaborated in the framework of the project co-financed by the Polish National Agency for Academic Exchange.

References

[1] Glazeva O V 2018 The use of high-voltage frequency converters as a method of increasing the energy efficiency index at the marine industry *Materials of the scientific-methodical conference "Ship engineering, electronics and automation"* pp 70–78

[2] Urbaniak M, Kardas-Cinal E and Jacyna M 2019 Optimization of Energetic Train Cooperation *Symmetry* **11** 9 1175 https://doi.org/10.3390/sym11091175

[3] Burdasov B K, Nesterov S A and Fedotov Yu B 2015 Frequency converters for high-voltage AC drives *APRIORI Series Natural and technical sciences* **4** pp 2–15

[4] Malinowski M, Gopakumar K, Rodriguez J and Pérez M A 2010 A Survey on Cascaded Multilevel Inverters *IEEE Transactions on Industrial Electronics* **57** 7 pp 2197–2206
[5] Manimala V, Geetha N and Renuga P 2011 Design and simulation of five level cascaded inverter using multilevel sinusoidal pulse width modulation strategies *IEEE 3rd International Conference on Electronics Computer Technology* **2** pp 280–283
[6] McGrath B P and Holmes D G 2002 Multicarrier PWM strategies for multilevel inverters *IEEE Transactions on Industrial Electronics* **49** 4 pp 858–867
[7] Grahame D 2003 *Pulse width modulation for power converters. Principles and practice* (New Jersey: John Wiley & Sons, Inc.) p 724
[8] Kolpakov A 2009 Multilevel Converter Control Algorithms *Silovaya Elektron.* **2** pp 57–65
[9] High voltage frequency converters series N5000 // http://www.tekhar.com/ Programma/ HYUNDAI/ index_V_V_invert.html
[10] Frequency converters VEDADRIVE 315–25000 kVA // www.danfoss.ru/VLT
[11] Busher V V 2020 Research of high-voltage frequency converters in ship electric power systems *9th Conference on Ships' Electrical Engineering, Electronics and Automation NU «OMA» Publ* pp 231–237
[12] Krishnapriya S and Unnikrishnan L 2015 Multilevel Inverter Fed Induction Motor Drives, *International Journal of Research in Engineering and Technology (IJRET)* **04** 09 pp 60–64
[13] Rudyk T, Szczepański E, Jacyna M 2019 Safety factor in the sustainable fleet management model. *Archives of Transport* **49** 1 pp 103–114
[14] PowerFlex 6000 Medium Voltage Variable Frequency Drive Firmware, Parameters, and Troubleshooting Manual Catalog Number 6000G//Publication 6000-TD004E-EN-P - September 2019. http://www.rockwellautomation.com/support
[15] High voltage frequency converters Robicon PERFECT HARMONY 225 kW–120 MW // www.siemens.com/robicon-perfect-harmony
[16] Puchkov A P and Osipov O I 2018 Correction of the Control System of a Multilevel Converter with a Defect in the Power cell of its Inverter *Electrotechnical Systems and Complexes* **3** 40 pp 42–46
[17] Kuznetsov V, Tryputen N and Kuznetsova Y 2019 Evaluating the effect of electric power quality upon the efficiency of electric power consumption *IEEE 2nd Ukraine Conference on Electrical and Computer Engineering* pp 556–561
[18] Dymerets A V, Yershov R D, Gorodny A N, Denisov Y O, Boiko S, Kuznetsov V 2020 Dynamic Characteristics of Zero-Current-Switching Quasi-Resonant Buck Converter under Variation of Resonant Circuit and Load Parameters *IEEE 40th International Conference on Electronics and Nanotechnology* pp 848–853
[19] Denisov Y, Gordienko V, Gorodny A, Stepenko S, Yershov R, Prokhorova A, Kostyrieva O 2016 Power losses in MOSFET switch of quasi-resonant pulse converter with series resonant circuit *2nd International Conference on Intelligent Energy and Power Systems* pp 1–6
[20] Kuzenkov O, Kuznetsov V, Tryputen N 2019 Analysis of phase trajectories of the third-order dynamic objects *IEEE 2nd Ukraine Conference on Electrical and Computer Engineering* pp 1235–1243
[21] Gorodny A, Dymerets A, Kut Y, Denisov Y, Natalia D 2020 Generalized method of commutation processes calculation in high-frequency switched-mode power converters *Advances in Intelligent Systems and Computing* **1019** pp 71–80
[22] Kuznetsova Y, Kuznetsov V, Tryputen M, Kuznetsova A, Tryputen M and Babyak M 2019 Development and Verification of Dynamic Electromagnetic Model of Asynchronous Motor Operating in Terms of Poor-Quality Electric Power *Proceedings of the International Conference on Modern Electrical and Energy Systems* pp 350–353
[23] Perekrest A, Chornyi O, Mur O, Nikolenko A, Kuznetsov V, Kuznetsova Y 2018 Preparation and preliminary analysis of data on energy consumption by municipal buildings *Eastern-European Journal of Enterprise Technologies* **6** 8 96 pp 32–42

Choosing of Asynchronous Motor Protection Equipment in Production Environment

V V Kuznetsov[1,6], M M Tryputen[2,7], V G Kuznetsov[3,8], M Tryputen[4,9], A Kuznetsova[4,10] and Y Kuznetsova[5,11]

[1]Department of the electrical engineering and electromechanic, National metallurgical academy of Ukraine, 4, Gagarina ave., Dnipro, Ukraine
[2]Department of Automation and Instrumentation, Dnipro University of Technology, 19, D. Yavornitsky ave., Dnipro, Ukraine
[3]Electric Power Department, Railway Research Institute, 50, Chlopickiego str., Warsaw, Poland
[4]Department of Calculating Mathematics and Mathematical Cybernetics Oles Honchar Dnipro National University Dnipro, Ukraine, 35, D. Yavornitsky ave., Dnipro, Ukraine,
[5]Department of humanitarian, fundamental and general engineering disciplines Institute of Integrated Education National metallurgical academy of Ukraine, 4, Gagarina ave., Dnipro, Ukraine

[6]wit1975@i.ua, [7]nikolay.triputen@gmail.com, [8]VKuznetsov@ikolej.pl, [9]triputen2014@i.ua, [10]alisa20002014@i.ua, [11]wit_jane2000@i.ua

Abstract. The current article is devoted to the topic problem of decision making concerning the choice of the protective means for asynchronous motors operated within industrial shop electric circuits under challenging conditions of improper electric supply. In this paper we show how energy economizing model for the asynchronous motor can be presented in the form of predicate disjunction and be applied with the pattern recognition algorithm for making solutions. The major advantage of a new model is its open character and the possibility of knowledge accumulation with respect to electromechanical equipment. We submit the information on the software and hardware complex applied for the research on the characteristics of the circuit voltage and the asynchronous motors in the real time mode directly within the enterprise industrial shop. The publication also reports on the searching for the best protective solution for asynchronous motors. This work is based on the known recognition algorithm of statistical optimization for non-linear objects given as the aggregates of predicates. The algorithm fruitfully applies the local selection principle. The approach proposed in this publication has been approbated at the protective means selection procedure for the asynchronous motor of 7.5 kW capacity, which performance is 80% of the working cycle under conditions of improper electric energy supply in the experimental shop of Ukrspetsservis.

1. Introduction

As known [1], the electric network mode parameters often do not meet the requirements of GOST 32144-2013 in Russia, of GOST 13109-97 in Ukraine and the guidelines No.39/2015/TT-BCT and No.25/2016/TT-BCT in Vietnam. In real operating conditions, there is very often a non-sinusoidal mode in electrical networks, the consequences of which are voltage and current harmonics. The problem of the presence of low-quality electricity in electrical networks is described in the articles [2-4]. The

Content from this work may be used under the terms of the Creative Commons Attribution 3.0 licence. Any further distribution of this work must maintain attribution to the author(s) and the title of the work, journal citation and DOI.
Published under licence by IOP Publishing Ltd

problem of the negative influence of voltage and current harmonics on electrical equipment, on the efficiency of electric energy use, has recently been increasingly represented in international publications and conferences [5-7]. Even in the countries of Southeast Asia, scientists pay attention to this problem. In paper [8] the authors note that, the parameters of electrical network modes do not meet the requirements of Russian GOST 32144-2013 and the guidelines of Vietnam.

The principal ways of decrease of poor-quality electricity negative impact on electric motor operation in production environment and consequently on the efficiency of production in general are as follows: application of "individual" LC-filters for protection of principal electric drives [9]; application of "sectional" poor-quality supply voltage compensating devices on a workshop level [10]; suppressing of supply voltage distortion in the points of its origin. Rejection of any measures is also considered acceptable despite insignificant motor lifetime reduction. Each of the aforesaid options incurs certain integration cost and expected economic effect.

The known methodology for choosing of protection equipment to secure asynchronous motor (AM) [10] operating under the conditions of poor-quality electric energy is based on its energy-efficient pattern. The above methodology implements computing algorithms involving stochastic model of linear voltage within workshop power supply network, nonlinear electromagnetic and thermal model of AM and economic model as well [11, 12]. However, problems related to practical realization of computing procedures in each particular case prevents its implementation in production.

The goal of this article is justification of the possibility of the above methodology implementation in production environment based on SCADA of Zenon system software installed on PC; and application of predicate models and non-relational data model-oriented recognition algorithms.

2. Research methods and results

Taking a decision on economic viability of the choice (or refusal) of a particular protection equipment depends on the value of several variables (input technical and economic): total harmonic distortion K_U, coefficients of specific harmonic components $K_{U(m)}$ (m=7), negative sequence ratio K_{2U}, zero-sequence index K_{20}, protection equipment cost C_j $(i=\overline{1,r})$, where r – is the number of different types of protection devices. Herewith, indexes K_U, $K_{U(m)}$, K_{2U} and K_{20} depend on objective laws of linear voltage variation within electric network and asynchronous motor operation pattern [13].

For the purposes of determination of characteristics of linear voltage within workshop electric network and asynchronous motors in real time there was a hardware and software suite developed, see Figure 1.

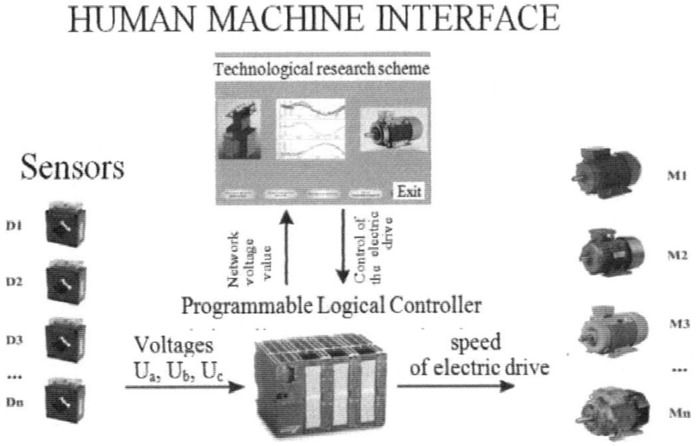

Figure 1. Schematic structure of the system for study of electric network and asynchronous motors.

Hardware package of the suite has been developed on the basis of VIPA System 200 V programmable logical controller (PLC). PLC is a remote analogue signal input module. Software provides computation process and human-computer interface on HMI/SCADA zenon Supervisor 7.0 based PC [14].

Interaction between programmable logical controller and PC with software suite is implemented by means of Ethernet interface. Current values of linear voltage and motors parameters are reflected on a PC screen and are saved for further processing. The suggested suite helps perform simultaneous study of all engines operating in a workshop.

Technical and economic values have some deviations conditioned by either measuring precision (for technical values) or economic situation (for costs) and are measured within certain range. This makes possible to represent energy-efficient model of AM by a sum-of-predicates form (discrete form) [15]:

$$Z_{em}\left[\vec{X},\vec{C}\right]=V_{p=1}^{q}\,V_{l}^{\lambda_p}Z_{p,l}\left[\vec{X},\vec{C}\right], \tag{1}$$

where:

$$\begin{aligned} Z_{p,l}\left[\vec{X},\vec{C}\right] &= 2^{-n}\prod\nolimits_{j=1}^{n}\left\{1+\operatorname{sgn}\left[(X_j-X_{j\min}^{pl})(X_{j\max}^{pl}-X_j)\right]\right\}+ \\ &+2^{-r}\prod\nolimits_{j=1}^{r}\left\{1+\operatorname{sgn}\left[(C_j-C_{j\min}^{pl})(C_{j\max}^{pl}-C_j)\right]\right\}, \end{aligned} \tag{2}$$

here: V – logic operation of disjunction, q – number of loss experience categories resulting from integration of protection equipment or their clusters, λ_p – number of predicates determining p – range, n and r – number if technical and cost values respectively, $X_{j\min}^{pl}, X_{j\max}^{pl}, C_{j\min}^{pl}, C_{j\max}^{pl}$ – model constants.

Formula (1) defines the relationship between technical $\vec{X}$ and economic $\vec{C}$ variables in discrete form – as the sum of predicates $Z_{p,l}\left[\vec{X},\vec{C}\right]$.

In this case, the element $Z_{p,l}\left[\vec{X},\vec{C}\right]$ defines a "hyperparallelepiped" in a multidimensional feature space and, according to (2), takes the value "1" if the current technical and economic situation falls inside the "hyperparallelepiped", and the value "0" – otherwise.

Generation of predicates parameters and their consolidation in categories may be commenced in the course of teaching the model according to the criterion of minimal economic losses resulting from the availability of AM protection equipment (or their unavailability):

$$E_p \to \min. \tag{3}$$

In this case in the course of input values sampling population recognition learning it's requisite by setting different criteria E_p within the interval $E_{p.\max} \div E_{p.\min}$ to split factor space into two categories: M_1 if $E_S < E_p$ and M_2, if $E_S > E_p$. Provided that the criteria values changes within the range $\Delta E_p = (E_{p.\max} - E_{p.\min})/q$, the q splitting the categories of hypersurfaces will be received, which pursuing the methodology of analytical description by means of methods admitting splitting of the factor space into elementary subfields may be represented by predicate equation (1). Here: ΔE_p – permissible deviation of economic losses from estimated value.

Teaching the model is performed on the basis of computing experiment, structural pattern of which is shown on Figure 2. In the course of experiment, implementation control unit (ICU) generates random sequence of input values within the prescribed limit.

In section "Energy Saving Pattern" calculation of economic losses incurred by application (abandoning) of protection equipment for electric drives in electric networks with poor-quality electrical energy is performed. See design formula for their determination pursuing [16] in Table 1.

Generation of predicate pattern components is commenced in section "Education and Adaptation" pursuing [17-19]. Herewith the number of predicates of completely defined predicate pattern depends upon parameters of input variables and defined with the help of the following formula [15]:

$$K_q = \prod_{i=1}^{n} \frac{d_i}{\Delta x_i}, \tag{4}$$

here $d_i;\Delta x_i$ – turndown and sample spacing of input value. The ratio $\frac{d_i}{\Delta x_i}$ in formula (4) determines the number of intervals in which the current value of the input variable i^{th} (technical or economic) can fall. Then, the total number of intervals, and hence the total number of predicates in formula (1) will be equal to the product of the number of intervals for all n – input variables.

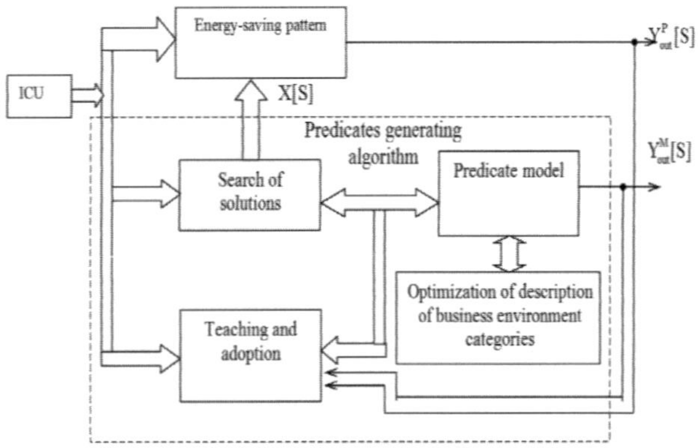

Figure 2. Schematic structure of predicates generation model.

Table 1. Phase of economic losses computation.

Phase no.	Computing subsection	Output parameter		
		Symbol	**Design formula**	**Name**
1	2	3	4	5
1	**Voltage pattern in workshop networks**	K_U	$K_U = \sqrt{\sum_{n=2}^{40} U_n^2} \cdot \frac{100}{U_{nom}}$	Harmonic distortion ratio
		$K_{U(n)}$	$K_{U(n)} = \frac{U_n}{U_{nom}} \cdot 100$	Specific harmonical component ratio
		K_{2U}	$K_{2U} = A_2 / A_1$	Reverse sequence ratio
		K_{20}	$K_{20} = A_2 / A_1$	Zero-sequence ratio
2	**Electric magnetic pattern of AM**	I_{Aeq}	$I_{Aeq} = \sqrt{\frac{1}{N}\sum_{n=0}^{N}(i_{An})^2}$	Equivalent stator current value (calculated for each phase)
		I_{Req}	The same for rotor current	The same for rotor
		$\underline{I}_{M_n}$	$\underline{I}_{M_n} = \underline{I}_{statorn} + \underline{I}_{rotn}$	Excitation current
		ΔP_{m1}	$\Delta P_{m1} = \left(I_{Aeq}^2 + I_{Beq}^2 + I_{Ceq}^2\right) R_{stator}$	Losses in stator copper
		ΔP_{m2}	The same for rotor	The same for rotor

Continuation of Table 1

1	2	3	4	5
2	**Electric magnetic pattern of AM**	ΔP_s	$\Delta P_s = 3 \cdot I_M^2 \cdot R_S$	Losses in steel
		ΔP_Σ	$\Delta P_\Sigma = \Delta P_{m1} + \Delta P_{m2} + \Delta P_s$	Losses in steel
		P_1	$P_1 = U_A I_A + U_B I_B + U_C I_C$	Consumed actual power
		Q_1	$Q_1 = \sqrt{S_1^2 - P_1^2}$	Consumed reactive power
		S_1	$S_1 = U_{Aeq} I_{Aeq} + U_{Beq} I_{Beq} + \ldots + U_{Ceq} I_{Ceq}$	Consumed total power
		P_2	$P_2 = \omega_{av} \cdot M_{av}$	Shaft power
		η	$\eta = P_2 / P_1$	Efficiency factor
		$\cos\varphi$	$\cos\varphi = P_1 / S_1$	Coefficient of performance (with the regard of distortions)
		THD_I	$THD_I = \frac{1}{I_1}\sqrt{\sum_{n=2}^{N}(I_{avn})^2}$	Harmonic current distortions ratio
		THD_T	The same for the torque	The same for the torque
3	**AM thermal model**	$\tau(t)$	$\tau_k = \tau_{(k-1)} + 1/C\left(\Delta P - A\tau_{(k-1)}\right)$	Excessive temperature-time relationship
		τ_{av}	$\tau_{av} = \frac{1}{M}\sum_k \tau_k$	Average temperature of isolation
		α'	$\alpha' = \frac{1}{T_{cycle}}\sum_n(\Delta\tau_n \cdot t_n)$	Equivalent duration of AM operation involving overheating
		T	$T = T_H \cdot e^{-\beta\cdot\alpha'}$	Isolation life time
4	**Economic damage pattern**	$E_{generalized}$	$E_{generalized} = \Delta P_{\Sigma allowable} \cdot C \cdot T_{work}$	Collateral damage
		E_{year}	$E_{year} = E_{summ1} - E_{summ2} - e \cdot K$	Annual economic damage
5	**Determination of selected protection equipment parameters**	C_1	assigned 1 μF	Choke filter capacity
		L_1	$L_1 = \frac{1}{\omega_r^2 C}$	Choke filter coefficient of induction
		C_2	Evaluated iteratively	Capacity in the part of a "star" of a compound filter
		E_{TC}	Evaluated separately	Protection equipment cost

See Table 2 for data on input values parameters during the study of 7.5 kW AM operation under the conditions of poor-quality electrical energy. As it appears from Table 2 and (4) $K_q = 1.664 \cdot 10^{13}$. Computation of such number of predicates within reasonable timeframes is rather difficult.

Table 2. AM input values parameters.

No.	Input value	Turndown	Variation range	Note
1	Total harmonic distortion	2–15%	0.5%	
2	Specific harmonic components ratio	0 – 10%	0.5%	First 7 harmonic components
3	Reverse sequence ratio	0 – 5%	0.1%	
4	Zero sequence ratio	0 – 5%	0.1%	
5	Protection equipment cost	UAH 0…200000	UAH 2000	10 options of technical solution

To overcome the above problem called "curse of dimensionality" in the course of teaching the predicate model, the algorithm of accelerated education has been applied [15]. This algorithm allows to include untaught fields of factor space into predicative pattern once simple criteria for two predicates of a certain class are met:

$$\begin{cases} X^1_{u\min} \le X^2_{u\min} \\ X^1_{u\max} \ge X^2_{u\max}, \quad when \quad u = \overline{1,n}; u \ne 1' \end{cases} \tag{5}$$

where $X^1_{u\min}, X^1_{u\max}, X^2_{u\min}, X^2_{u\max}$ – parameters of the merged fields projections, u – number of factor space feature axis towards which subfields are combined.

The system of inequalities (5) defines an area in the n-dimensional feature space, which is not defined in (1). To describe the specified area, it is necessary to have the corresponding technical and economic situations in the training sequence. However, taking into account (4), the expectation of their appearance can go beyond reasonable time intervals.

This problem can be overcome by the invariance of the mathematical structure of the predicate equation (2) to the size of the domain specified by it. Therefore, a strict fulfillment of one of the inequalities (5) is sufficient to form the parameters of the corresponding predicate and include it in (1).

"Predicate Model" module generates economic environment in the form of a predicate and assigns it to p-class based on defined values of technical and economic parameters and economic damages from application of protection equipment computed with energy saving pattern involved. The number of the class is defined using the following formula:

$$p = entier\left|E_p \times \Delta E_p^{-1}\right| + 1. \tag{6}$$

In (6), E_p is the range of values of economic losses when choosing various methods of protecting an asynchronous motor, and ΔE_p is the possible deviations of economic losses from the calculated ones caused by the forecast of fluctuations in exchange rates, inflation and other economic macroindicators in the considered time interval.

It is also worth noting that an adaptation algorithm has been developed for predicate model which makes possible its updating to reflect expansion of hardware park and its cost changes:

$$\begin{aligned} &Z_p\left[\vec{X}, \vec{C}\right] = \\ &= \left[\mathrm{V}_{l=1}^{L_1+\lambda_p} Z_{p,l}\left[\vec{X}, \vec{C}\right]\right] \Lambda \left[V_{v=1}^{L_2} Z_{t,v}\left[\vec{X}, \vec{C}\right]\right], \end{aligned} \tag{7}$$

where L_1 and L_2 – is the value obtained as a result of recognition of the first and second order controversies (respectively), Λ – logical operation of the conjunction.

The first order controversy should be thought of as affiliation of a predicate with P class though the given predicate should be referred to T class pursuing economic losses value (as a result of technical and economic conditions) and the second order controversy should be thought of as affiliation of a predicate with T class though the given predicate should be referred to P class.

Realization of adaptation algorithm pursuing (7) leads to gradual structural complication of predicate model and difficulties in its real-world application. To overcome structural complication of the model is possible by means of application of algorithms used in "Reduction of Economic Conditions Categories Description". Reduction of categories description provides enlargement of subareas by way of their merging with the following encoding of the parameters of predicate equation which determines enlarged area [20, 21].

Merging of subareas is commenced upon the fulfillment of equality conditions in the right and left parts of the formulas in (6). Hence, the resultant subarea has the minimal and maximum value of $u-$ attribute determined as $X^{12}_{u\min} = \min\left\{X^{1}_{u\min}, X^{1}_{u\min}\right\}; X^{12}_{u\max} = \max\left\{X^{1}_{u\max}, X^{2}_{u\max}\right\}$.

It's obvious that in the context of enlargement of subareas the transition of predicates from one category to another and consequently resolution of the first and second order controversies is possible.

Encoding of predicate equations parameters involves determination of their numbers on feature axis in the form of some vector $\vec{B}$ and its collapse to some scalar by formula [15]:

$$K = \sum\nolimits_{u=1}^{2(n+r)} b_j q^{(2(n+r)-j)}, \tag{8}$$

where b_j – collapsing vector $\vec{B}$ component, matching with j – feature axis; q – system base.

Moving from encoding figures K_γ to vector $\overrightarrow{B_\gamma}$ is done by the formula:

$$b_{j\gamma} = \mathrm{mod}(\left[K_\gamma / q^{(2(n+r)-j)}\right], q), j = \overline{1,(n+r)}. \tag{9}$$

Transformations (8) and (9) make it possible to reduce the recording of the mathematical model given by formulas (1), (2), and, hence, to reduce the requirements for computing and information resources of the software and hardware that implement them.

Determination of the best technical option of AM protection according to predicate model is based on algorithm of recognition static optimization in "Solutions Search" unit as follows. For current technical values $z_{em}\left[\vec{X}, \vec{C}\right]$ is computed starting from the first category of economic $p = 1$, which is an equivalent of minimal value of economic losses. If $z_{1l}\left[\vec{X}, \vec{C}\right] = 0$ for all $l = \overline{1, \lambda_1}$,, then the second category of economic conditions should be analyzed, etc. This procedure is performed unless some $p = c$ and $l = z_{cg}\left[\vec{X}, \vec{C}\right] = 1$. Then according to the values of the chosen predicate constants financial expenditures and consequently the chosen technical option of protection are determined.

In [15] mentions that predicate equations (1) may be represented by relational data model. It helps describe the processes of teaching, adaptation, minimization and search of the optimum solutions based on single mathematical apparatus – α – algebra.

Taking into consideration that relational model is supported by Data base control system this approach towards determination of optimum protection equipment for AM with regard to its operation under poor-quality electricity conditions is easy to implement in production environment [22, 23] and railway transportation systems [24-26].

The proposed approach has been tested while choosing asynchronous motor protection equipment with 7.5 kW capacity operating during 80% of operation cycle under the conditions of poor-quality electrical energy in research-and-development shop of "Ukespetsservice" Ltd. The results of the research are listed in the Table 3:

It follows from Table 3 that for the purposes of protection of this induction motor it's enough to use passive LC-filter at the cost of UAH 2.000.00. With regard to the above mentioned annual economic losses will be E_p= UAH 3.700 per year.

Table 3. Results of the research.

No.	Input variable	Technical parameter	Cost	Note
1	Total harmonic distortion coefficient	2.5%	-	
2	Specific harmonic components coefficients	1.5%	-	
		0.5%		
		0.5%		
		2.5%		
		0.5%		
		2.0%		
		2.0%		
3	Reverse sequence coefficient	2.3%	-	
4	Zero-sequence coefficient	1.5%	-	
5	The cost of protection equipment	-	UAH 2.000.00	Passive filter

3. Conclusions

The analyzed approach towards determination of optimum option of protection of electrical equipment operating under conditions of poor-quality electricity networks is possible to implement in production environment. The proposed algorithms make it possible to derive mathematical models of electrical units under research in the form of logical sum of predicates standing for specific economic losses. Based on predicate models it's easy to find a range of optimal solutions for different conditions of electric devices operation. Obtained solutions may be saved on data store electronic component.

For practical application of the obtained results it's enough to estimate the quality of electric energy on specific enterprise and engine`s health and afterwards, by means of data base control system, to choose the most economically reasonable protection equipment.

Acknowledgments: This paper is elaborated in the framework of the project co-financed by the Polish National Agency for Academic Exchange.

References

[1] Kovernikova L I, Sudnova V V, Shamonov R G and Voropay N I 2017 *The quality of electric energy: current status, problems and suggestions for solving them* (Novosibirsk: Nauka Publ.) p 219

[2] Ded A V, Sikorskiy S P and Danyukov I B 2018 Data processing of experimental measurements of indicators of the quality of electric energy by the example of levels of voltage deviations *Electrical engineering. Energy Omsk Scientific Herald* **2** 158 pp 55–59

[3] Guzhov S V, Yanchenko S A 2016 Analysis of the operability of the functioning of electrical complexes and systems in non-sinusoidal current modes *Bulletin of the Kuzbass State Technical University* **6** 118 pp 145–152

[4] IEEE Recommended Practice for Establishing Liquid-Filled and Dry-Type Power and Distribution Transformer Capability When Supplying Nonsinusoidal Load Currents *IEEE Std C57.110-2008* pp 1–52

[5] Dao Thinh, Abdull Halim H, Liu Z and Phung B T 2016 Voltage Harmonic Effect on Losses in Distribution Transformers *2016 International Conference on Smart Green Technology in Electrical and Information Systems* pp 27–32

[6] Singh J, Singh S, Singh A 2017 Effect of Harmonics on Distribution Transformer Losses and Capacity *International Journal of Engineering Technology Science and Research* **4** 6 pp 48–55

[7] Abbas A, El Zahab E A and Elbendary A 2015 Thermal modeling and ageing of transformer under harmonic currents *Proceedings of 23-rd International Conference on Electricity Distribution* pp 1–5
[8] Ngo Van Kyong and Kovernikova L I 2019 Prediction of the influence of the non-sinousoidal network mode on power transformers *Modern Technologies. System Analysis. Modeling* **64** 4 pp 36–43
[9] Kovalenko D V, Kiselev B Yu and Plotnikov D I 2017 Methodology for calculating passive filters designed to compensate for higher current harmonics in power supply systems of industrial enterprises *International Research Journal* **1** 55 pp 82–86
[10] Tryputen M, Kuznetsov V, Kuznetsova A, Tryputen M, Kuznetsova Y and Serdiuk T 2021 Improving the Reliability of Simulating the Operation of an Induction Motor in Solving the Technical and Economic Problem *Advances in Computer Science for Engineering and Education III Advances in Intelligent Systems and Computing* **1247** pp 143–152
[11] Kachan U G and Dyachenko V V 2016 *Means of efficient use of electrical energy. Monograph* (Zaporizhia) p 156
[12] Kuznetsova Y, Kuznetsov V, Tryputen M, Kuznetsova A, Tryputen M and Babyak M 2019 Development and Verification of Dynamic Electromagnetic Model of Asynchronous Motor Operating in Terms of Poor-Quality Electric Power *Proceedings of the International Conference on Modern Electrical and Energy Systems* pp 350–353
[13] Kuznetsov V, Tryputen N and Kuznetsova Y 2019 Evaluating the effect of electric power quality upon the efficiency of electric power consumption *2019 IEEE 2nd Ukraine Conference on Electrical and Computer Engineering* pp 556–561
[14] Boiko O A 2014 About the possibility of interaction of zenon SCADA system with external software *http://www.svaltera.ua/press-center/articles/8773.php* p 28
[15] Tryputen M M, Kuznetsov V V α-algebra in problems of increasing the energy efficiency of induction motors operating under low-quality electricity *Mining Electromechanics and Automation, Scientific and Technical Collection* **101** pp 110–114
[16] Kuznetsov V V 2014 Methodological bases of choice economically and expedient facilities of defence of asynchronous motors, working in the conditions of off-grade electric power *Electrification of vehicles* **7** pp 92–97
[17] Kusko A and Thompson M 2008 *Power Quality in Electrical Systems* (Moscow) p 336
[18] Akagi H 2005Active Harmonic Filters *Proceedings of the IEEE* **93** 12 pp 2128–2141
[19] Burman A P, Rozanov Iu K and Shakarian Iu G 2012 Managing the flows of electricity and improving the efficiency of electric power systems: textbook for high schools (Moscow) p 336
[20] Kachan Yu G 1984 Image adaptation in the recognition algorithm of production situations *Mechanization and automation of control* **1** pp 15–17
[21] Kachan Yu G 1986 Minimization of the description of images in recognition problems of production situations *News of universities. Mountain Journal* **7** pp 119–122
[22] González De La Rosa J J, Pérez A A, Palomares-Salas J C, Fernández J M S, Ramiro Leo J G, Sedeño D A and Moreno-Muñoz A 2012 Power quality analysis using higher-order statistical estimators: Characterization of electrical sags and swells *Lecture Notes of the Institute for Computer Sciences, Social-Informatics and Telecommunications Engineering* **82** pp 22–29
[23] Perekrest A, Chornyi O, Mur O, Nikolenko A, Kuznetsov V, Kuznetsova Y 2018 Preparation and preliminary analysis of data on energy consumption by municipal buildings *Eastern-European Journal of Enterprise Technologies* **6** 8 96 pp 32–42
[24] Zhelieznov K I, Akulov A S, Zabolotnyi O M, Ursulyak L V, Chabanuk E V, Shvets A O, Kuznetsov V G, Radkevych A V 2019 The revised method for calculating of the optimal train control mode *Archives of Transport* **51** 3 pp 21–34
[25] Kisilowski J, Zalewski J 2019 Selected aspects of motor vehicle dynamics on the example of a power-off straight line maneuver *Archives of Transport* **50** 2 pp 57–76
[26] Pyza D, Golda P 2011 Transport Cargo Handling Shipments in Air Transport in the Aspect of Supply Chains *21st International Conference on Systems Engineering* Las Vegas NV pp 442–445

IOP Conf. Series: Materials Science and Engineering **985** (2020) 012023 doi:10.1088/1757-899X/985/1/012023

Mathematical modeling of shock wave interaction with wagon

V K Khrutch[1], **M M Biliaiev**[2,5], **V A Kozachyna**[2], **O V Berlov**[3], **P S Kirichenko**[4] **and V V Biliaieva**[1]

[1]Department of AeroHydro Mechanics and Energy and Masstransfer, Oles Honchar Dnipro National University, Dnipro, Ukraine
[2]Department of Hydraulics and Water Supply, Dnipro National University of Railway Transport named after Academician V. Lazaryan, Dnipro, Ukraine
[3]Department of Life Safety, Prydniprovska State Academy of Civil Engineering and Architecture, Dnipro, Ukraine
[4]Department of Heat and gas supply, drainage and ventilation, Kryvyi Rih National University, Kryvyi Rih, Ukraine

[5]diit.hydro.eco@gmail.com

Abstract. In case of some accidents on railways there may be situations when the shock wave appears and interacts with different objects on the railway (wagons, cargo, buildings etc.). In these cases it is necessary to predict the possible effect of shock wave diffraction on the different objects. Study of these problems on the basis of physical experiments (laboratory experiment or field experiment) demand expensive and unique experimental facility. In some cases physical experiment can't be set. That is why mathematical simulation plays the important role in solving problem connected with shock wave propagation. For practice it is necessary to have predictive quick computing mathematical models which allow to perform numerical experiment on the basis of non-powerful computers. Now, in Ukraine, there is a real deficit of mathematical models which allow to compute quickly shock wave interaction with different objects. The aim of this work was development of quick computing numerical model to simulate shock wave propagation and its interaction with the wagons. The model is based on the numerical integration of Euler equations which are written in integral form. To solve modeling equations difference scheme of splitting was used. Results of numerical modeling are presented.

1. Introduction

Explosion of railway tanks may cause shock wave appearance. Also, use of explosives at railway stations results in shock wave appearance and wave interaction with objects. That is why, shock wave interaction with different objects on the railway stations, shock wave interaction with wagons are very important problems in the field of mechanics [1-5]. Worthy of note, that mathematical modeling is the main tool to solve this problem because of physical experiment can't be used for every day study this complicated problem. Study diffraction processes of shock wave propagation needs very expensive equipment, special experimental fields and sometimes it can't be organized for the problems which are under interest.

Two approaches are used in the field of mathematical modeling of shock wave interaction with different objects. The first way is based on the numerical integration of Euler equations. Modeling

Content from this work may be used under the terms of the Creative Commons Attribution 3.0 licence. Any further distribution of this work must maintain attribution to the author(s) and the title of the work, journal citation and DOI.
Published under licence by IOP Publishing Ltd

shock wave interaction on the basis Euler equations does not take into account viscous effects. Computation time when this approach is used is small. That is very convenient to perform series of numerical experiments during one working day. The second way is based on the numerical integration Navier-Stokes equations. This approach needs much computational time. On practice, different commercial codes are used to model shock wave interaction with different objects.

The goal of this work is development numerical model on the basis of Euler equations for modeling shock wave interaction with objects on the railway transport.

2. Mathematical model

We consider shock wave propagation on the railway station. We model this shock wave interaction with the cargo which is situated on the railway platform (figure 1). Near this railway platform there is another track with cargo train. It means that, in common, we model the situation of shock wave diffraction on two objects.

Figure 1. Sketch of computational region: 1 – shock wave front (http://www.fpk.ru/perevozka_tehniki.html).

To simulate the shock wave diffraction on two objects we used Euler equations which were written in the form of integral laws of conservation. These equations were written for the non-orthogonal difference cell [6]:

$$\frac{\partial}{\partial t}\oint_{\Delta G} \rho d\tau + \oint_{\Delta\Gamma} \rho V_n ds = 0 \quad (1)$$

$$\frac{\partial}{\partial t}\oint_{\Delta G} \rho\vec{V} d\tau + \oint_{\Delta\Gamma} \rho\vec{V} V_n ds + \oint_{\Delta\Gamma} P\vec{n} ds = 0 \quad (2)$$

$$\frac{\partial}{\partial t}\oint_{\Delta G} \rho E d\tau + \oint_{\Delta\Gamma} \rho I V_n ds = 0 \quad (3)$$

where ΔG is square of the computational cell; $\Delta\Gamma$ is the boundary of this computational cell; P is pressure; u, v are components of air velocity vector; ρ is density; $E = \frac{u^2}{2} + \frac{v^2}{2} + \frac{P}{\rho(1-\gamma)}$ is internal energy; $I = E + \frac{P}{\rho}$ is enthalpy.

Initial and boundary conditions are considered in [6, 7]. Gas parameters behind the shock wave front are determined using Renkin-Gugonoio relationships [7].

3. Numerical model

The computational region is covered by non-orthogonal difference grid. Sketch of non-orthogonal computational cell is shown in figure 2 [7].

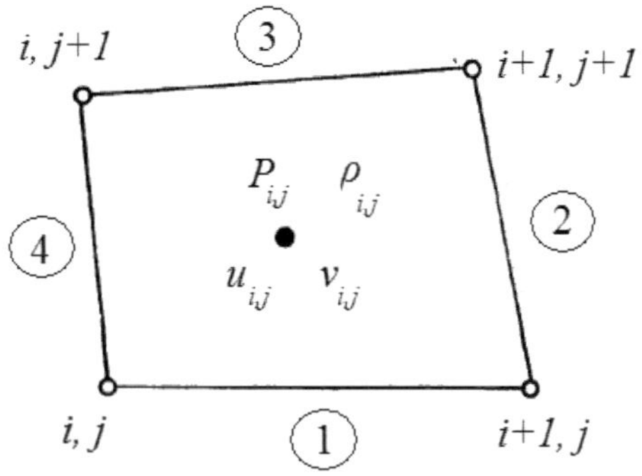

Figure 2. Sketch of non-orthogonal computation cell.

The numerical procedure to solve equations (1)-(3) was as following [7].

At first step we approximated the modeling equations as follows:

$$\frac{\rho_{ij}^{n+1}-\rho_{ij}^{n}}{\Delta t}\Delta G_{ij}+\sum_{k=1}^{4}(\rho_k^+W_k^+ + \rho_k^-W_k^-)^{n+1}\Delta s_k = 0, \quad (4)$$

$$\frac{\rho\vec{V}_{ij}^{n+1}-\rho\vec{V}_{ij}^{n}}{\Delta t}\Delta G_{ij}+\sum_{k=1}^{4}((\rho\vec{V})_k^+W_k^+ + (\rho\vec{V})_k^-W_k^-)^{n+1}\Delta s_k + \\ +\sum_{k=1}^{4}P_k^{n+1}\vec{N}_k\Delta s_k = 0, \quad (5)$$

$$\frac{\rho E_{ij}^{n+1}-\rho E_{ij}^{n}}{\Delta t}\Delta G_{ij}+\sum_{k=1}^{4}((\rho I)_k^+W_k^+ + (\rho I)_k^-W_k^-)\Delta s_k = 0, \quad (6)$$

where $\rho_k^+W_k^+ = \rho_{ij}^{n+1}\frac{1}{2}\left(V_{N,k}^{n+1}+\left|V_{N,k}^{n+1}\right|\right)$,

$$\rho_k^-W_k^- = \begin{Bmatrix} \rho_{i,j-1}^{n+1}\ for\ k=1 \\ \rho_{i+1,j}^{n+1}\ for\ k=2 \\ \rho_{i,j+1}^{n+1}\ for\ k=3 \\ \rho_{i-1,j}^{n+1}\ for\ k=4 \end{Bmatrix}\frac{1}{2}\left(V_{N,k}^{n+1}-\left|V_{N,k}^{n+1}\right|\right), \quad (7)$$

$$V_{N,k}^{n+1} = \frac{1}{2}(\vec{V}_{ij} \cdot \vec{N}_k)^{n+1} + \frac{1}{2}(\begin{Bmatrix} \vec{V}_{i,j-1} \ for\ k=1 \\ \vec{V}_{i+1,j} \ for\ k=2 \\ \vec{V}_{i,j+1} \ for\ k=3 \\ \vec{V}_{i-1,j} \ for\ k=4 \end{Bmatrix} \cdot \vec{N}_k) \ , \tag{8}$$

where $\vec{N}_k$ – is external normal to the «k» part of the cell boundary.

The pressure on the cell boundary is determined as:

$$P_k^{n+1} = \frac{1}{2} P_{ij}^{n+1} + \frac{1}{2} \begin{Bmatrix} P_{i,j-1} \ for\ k=1 \\ P_{i+1,j} \ for\ k=2 \\ P_{i,j+1} \ for\ k=3 \\ P_{i-1,j} \ for\ k=4 \end{Bmatrix} . \tag{9}$$

If we use the following designations:

$$T_k^+ \rho_{ij}^{n+1} = \frac{1}{\Delta G_{ij}} \rho_k^+ W_k^+ \Delta s_k \ , \tag{10}$$

$$T_k^- \rho_{ij}^{n+1} = \frac{1}{\Delta G_{ij}} \rho_k^- W_k^- \Delta s_k \ , \tag{11}$$

$$L_k P_{ij}^{n+1} = P_k^{n+1} \Delta s_k \frac{1}{\Delta G_{ij}} , \tag{12}$$

Using expression (7)-(12) we can write the difference equations as:

$$(E + \Delta t \sum_{k=1}^{4} (T_k^+ + T_k^-)) \rho_{ij}^{n+1} = E \rho_{ij}^n , \tag{13}$$

$$(E + \Delta t \sum_{k=1}^{4} (T_k^+ + T_k^-)) \rho \vec{V}_{ij}^{n+1} + \Delta t \sum_{k=1}^{4} L_k P_{ij}^{n+1} = E \rho \vec{V}_{ij}^{n} , \tag{14}$$

$$E \rho E_{ij}^{n+1} + \Delta t \sum_{k=1}^{4} (T_k^+ + T_k^-) \rho I_{ij}^{n+1} = E \rho E_{ij}^n . \tag{15}$$

At the second step we write the difference scheme of splitting:

$$\left(E + \frac{\Delta t}{2} (T_4^- + T_2^+ + T_1^- + T_3^+) \right) \rho_{ij}^{n+\frac{1}{4}} = E \rho_{ij}^n , \tag{16}$$

$$\left(E + \frac{\Delta t}{2} (T_2^- + T_4^+ + T_3^- + T_1^+) \right) \rho_{ij}^{n+\frac{1}{2}} = E \rho_{ij}^{n+\frac{1}{4}} , \tag{17}$$

$$\left(E + \frac{\Delta t}{2} (T_4^- + T_2^+ + T_3^- + T_1^+) \right) \rho_{ij}^{n+\frac{3}{4}} = E \rho_{ij}^{n+\frac{1}{2}} , \tag{18}$$

$$\left(E + \frac{\Delta t}{2} (T_2^- + T_4^+ + T_1^- + T_3^+) \right) \rho_{ij}^{n+1} = E \rho_{ij}^{n+\frac{3}{4}} , \tag{19}$$

$$\left(E+\frac{\Delta t}{2}T^{++}\right)\rho\vec{V}_{ij}^{n+\frac{1}{4}}+\frac{\Delta t}{4}\sum_{k=1}^{4}L_k P_{ij}^{n+\frac{1}{4}}\vec{N}_k = E\rho\vec{V}_{ij}^{n}a\,, \tag{20}$$

$$\left(E+\frac{\Delta t}{2}T^{--}\right)\rho\vec{V}_{ij}^{n+\frac{1}{2}}+\frac{\Delta t}{4}\sum_{k=1}^{4}L_k P_{ij}^{n+\frac{1}{2}}\vec{N}_k = E\rho\vec{V}_{ij}^{n+\frac{1}{4}}, \tag{21}$$

$$\left(E+\frac{\Delta t}{2}T^{+-}\right)\rho\vec{V}_{ij}^{n+\frac{3}{4}}+\frac{\Delta t}{4}\sum_{k=1}^{4}L_k P_{ij}^{n+\frac{3}{4}}\vec{N}_k = E\rho\vec{V}_{ij}^{n+\frac{1}{2}}, \tag{22}$$

$$\left(E+\frac{\Delta t}{2}T^{-+}\right)\rho\vec{V}_{ij}^{n+1}+\frac{\Delta t}{4}\sum_{k=1}^{4}L_k P_{ij}^{n+1}\vec{N}_k = E\rho\vec{V}_{ij}^{n+\frac{3}{4}}, \tag{23}$$

$$E\rho E_{ij}^{n+\frac{1}{4}}+\frac{\Delta t}{2}T^{++}\rho I_{ij}^{n+\frac{1}{4}} = E\rho E_{ij}^{n}, \tag{24}$$

$$E\rho E_{ij}^{n+\frac{1}{2}}+\frac{\Delta t}{2}T^{--}\rho I_{ij}^{n+\frac{1}{2}} = E\rho E_{ij}^{n+\frac{1}{4}}, \tag{25}$$

$$E\rho E_{ij}^{n+\frac{3}{4}}+\frac{\Delta t}{2}T^{+-}\rho I_{ij}^{n+\frac{3}{4}} = E\rho E_{ij}^{n+\frac{1}{2}}, \tag{26}$$

$$E\rho E_{ij}^{n+1}+\frac{\Delta t}{2}T^{-+}\rho I_{ij}^{n+1} = E\rho E_{ij}^{n+\frac{3}{4}}, \tag{27}$$

In expressions (24)-(27) we use the following designations:

$$T^{++} = T_4^- + T_2^+ + T_1^- + T_3^+\,, \tag{28}$$

$$T^{--} = T_2^- + T_4^+ + T_3^- + T_1^+\,, \tag{29}$$

$$T^{+-} = T_4^- + T_2^+ + T_3^- + T_1^+\,, \tag{30}$$

$$T^{-+} = T_2^- + T_4^+ + T_1^- + T_3^+\,. \tag{31}$$

If we eliminate the fractional steps we obtain the following scheme [7]:

$$\left(E+\frac{\Delta t}{2}T^{+-}\right)\left(E+\frac{\Delta t}{2}T^{-+}\right)\rho_{ij}^{n+1} = E\rho_{ij}^{n+\frac{1}{2}}, \tag{32}$$

$$\left(E+\frac{\Delta t}{2}T^{--}\right)\left(E+\frac{\Delta t}{2}T^{+-}\right)\left(E+\frac{\Delta t}{2}T^{-+}\right)\rho_{ij}^{n+1} = E\rho_{ij}^{n+\frac{1}{4}}, \tag{33}$$

$$\left(E+\frac{\Delta t}{2}T^{++}\right)\left(E+\frac{\Delta t}{2}T^{--}\right)\left(E+\frac{\Delta t}{2}T^{+-}\right)\left(E+\frac{\Delta t}{2}T^{-+}\right)\rho_{ij}^{n+1} = E\rho_{ij}^{n}. \tag{34}$$

As this numerical scheme has the first order of accuracy the shock wave front is «smeared» on some computational cells.

FORTRAN was used to develop code on the basis of described numerical model. Developed code includes the following SUBPROGRAMs:

1. «DENS» – subprogram to compute density field in computational domain.
2. «VEL» – subprogram to compute air velocity components field in computational domain.
3. «PRES» – subprogram to compute pressure field in computational domain.

IOP Conf. Series: Materials Science and Engineering **985** (2020) 012023 doi:10.1088/1757-899X/985/1/012023

4. «P5GF» – subprogram to illustrate parameter distribution in computational domain.
Coordination of all subprograms is performed by main program.

4. Results

Numerical experiment was performed for scenario when the pressure behind the shock wave front was 15 atm. Dimensionless pressure field for different time of shock wave diffraction on two objects is shown in figures 3-5. Time is also dimensionless.

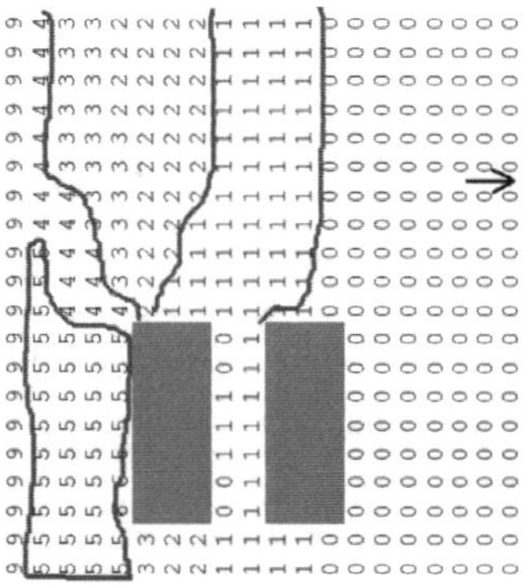

Figure 3. Pressure field for time t=0.43.

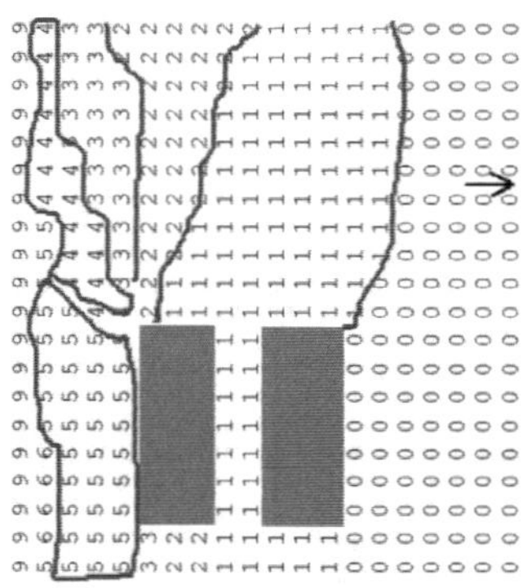

Figure 4. Pressure field for time t=0. 56.

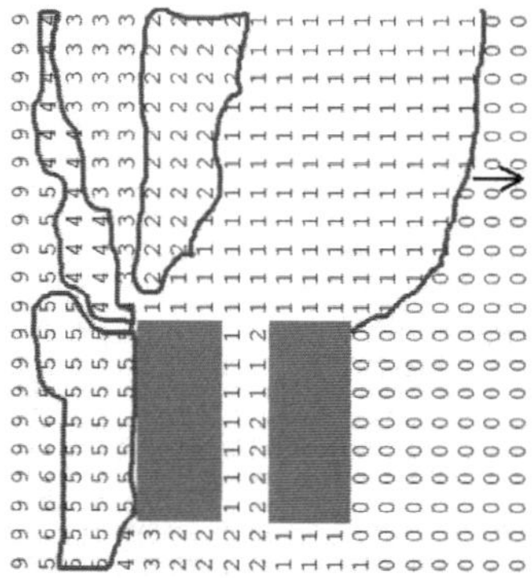

Figure 5. Pressure field for time t=0.69.

Each «number» in figures 3-5 shows the range of pressure value. For example, number «1» means that in this region pressure is in the range of about 10% of the maximum pressure at this time. As we see from the figures 3-5, the pressure field near the objects has the complicated form which is caused by shock wave interaction with objects. We see that pressure rises near the first obstacle caused by gas flow breaking on the surface of this object. Worthy of note that computational time was 3s.

5. Conclusions

Briefly, numerical model was developed to simulate shock wave interaction with objects on the railway station. To model shock wave diffraction Euler equations are used. The model can be used for preliminary evaluation of shock wave impact on different objects. Further, we shall develop model to solve conjugate problem: «reaction of object + load from shock wave».

References

[1] Zhou A, Wang K, Wang L, Du F and Li Z 2015 *International Journal of Mining Science and Technology* **25** 1 pp 107–12

[2] Zeitoun D E, Burtschell Y, Graur I A, Ivanov M S., Kudryavtsev A N and Bondar Y A 2009 *Shock Waves* **19** 4 pp 307–16

[3] Mortazawy S M 2018 *Numerical Simulation of Shock Wave Propagation in Ducts with Grooves: Dissertation Thesis* (Florida: Embry-Riddle Aeronautical Univeristy) p 389

[4] Klimenko I, Černiauskaite L, Neduzha L and Ochkasov O 2018. Mathematical Simulation of Spatial Oscillations of the "Underframe-Track" System Interaction *Proc. of 12th Int. Conf. ITELMS'2018* (Panevėžysm, Lithuania: Kaunas Univ. of Techn.) pp 105–114

[5] Myamlin S, Lunys O, Neduzha L and Kyryl'chuk O 2017 Mathematical Modeling of Dynamic Loading of Cassette Bearings for Freight Cars *Proc. of 21st Int. Sci. Conf. Transport Means 2017* pt. III (Juodkrante, Lithuania: Kaunas Univ. of Technology) pp 973–976

[6] Loytsyanski L 1978 *Fluid Mechanics* (Moscow: Nauka) p 736

[7] Biliaiev M and Khrutch V 1984 *Numerical simulation of supersonic flows* (Kyiv: Visha shkola) p 224

IOP Conf. Series: Materials Science and Engineering **985** (2020) 012024 doi:10.1088/1757-899X/985/1/012024

Formulation of the mathematical model for the planning system in the carriage of dangerous goods by rail

S Bibik[1], O Strelko[2], H Nesterenko[1,4], M Muzykin[1,5] and A Kuzmenko[3,6]

[1]Dnipro National University of Railway Transport named after academician V Lazaryan, Lazaryan Str., 2, 49010, Dnipro, Ukraine
[2]State University of Infrastructure and Technology, Kyrylivska Str., 9, 040071, Kyiv, Ukraine
[3]University of Customs and Finance, St. Volodymyra Vernadskoho, 2/4, 49010, Dnipro, Ukraine

[4]galinamuzykina@rambler.ru, [5]mihailmuzykin@gmail.com, [6]alia1971@i.ua

Abstract. The paper aims to the topical issue of improving the system in the planning of carriage of dangerous goods by rail subject to ensuring a high level of safety of the transport process and minimal costs for its implementation. Formalization of this process leads to the formation of an optimization task of the two-stage mathematical model. To implement the first stage, a mathematical model in searching the best route based on risk minimization is developed. In order to exclude the unacceptably complex and expensive options for the carriage of dangerous goods from the many possible routes, the authors abided by the condition not to exceed the critical operating costs. The second stage in the process of planning routes for dangerous goods takes into account the hazard identification study in the technological process of car movement with dangerous goods. Considering that the probability of a potential traffic accident is a complex random variable, which is due to a set of elementary previously unknown events, its estimation was made on the basis of the Bayesian approach. Based on the simulation results, the following reliability values are determined at which a transport accident should not emerge.

1. Introduction

The analysis of reporting data shows that the volumes of carriage of dangerous goods (DG) by Ukrainian railways is constantly increasing. Depending on the basic properties and type of hazard, these goods are transported in covered cars and containers by carloads and part load consignments, and liquid dangerous goods are in tanks. But in case of non-compliance with safe conditions of carriage or storage, such goods can cause disease, poisoning, burns of people and animals, explosion, fire, as well as cause damage to other goods, rolling stock, structures and devices, environmental pollution. Therefore, it is much-needed to pay special attention to the rational organization in the planning of rational routes of trains with dangerous goods and control of their timely and safe movement.

It should be noted that more than 70% of the total turn-around time, freight cars among which there are ones with DG, are located at the stations under technical operations, during which the risk of accidents increases significantly. Thus, railway stations are objects of increased danger associated not only with a high degree in the probability of emergence of accidents but also with the scope of their potential consequences. This determines the relevance of this work, which aims to improve the system in the planning of carriage of dangerous goods by rail subject to ensuring a high level of safety of the transport process and minimal costs for its implementation.

Content from this work may be used under the terms of the Creative Commons Attribution 3.0 licence. Any further distribution of this work must maintain attribution to the author(s) and the title of the work, journal citation and DOI.
Published under licence by IOP Publishing Ltd

2. Formulation of the problem

Railway transport is a complex logistics system, in which a man always takes part in the management structure. Therefore, achieving absolute security is impossible. But any technology for the carriage of dangerous goods should be based on the use of a criterion of improving traffic safety. The question accordingly arises of the system improvement in the planning of carriage of DG by rail with ensuring the high level in the safety of the transport process and the minimum costs for its implementation by choosing the optimal functioning criteria at the tactical level.

3. Analysis of literary data

Many scientists and practitioners have made a significant contribution to the development in the research on the development of planning and operational management systems for railway transport, technology in the organization of carriage of DG, determination of optimal patterns for the car movement, application of information technology in operational management. For example, authors of article [1] aim to develop a mathematical model that allows to construct the intelligent control system for the operating technique of the station and will be able to minimize the total operating costs, provide simultaneously an acceptable level of risk while the breaking up/making-up of trains including the cars with DG, as well as when performing technological operations with them at the station.

Article [2] highlights a method for determining the rational traffic intensity of car traffic volume on railway transport on the basis of a mathematical model. The mathematical model based on stochasticity and allows optimizing, under risks, the interval between the train movement and the transfer of cars at stations. The simulation results make it possible to find a balance between the costs of railways and cargo owners, including the carriage of DG.

Authors of article [3] discuss the substantial point of non-destructive monitoring and control of mechanical and geometric parameters for the roadbed, which are one of the most important in the practice of safe operation of railways during the organizing of the carriage of DG. It is established that the uncertainty in the motion process of trains over long distances significantly complicates the procedure in the planning of the transport process and ensuring traffic safety. This is due to risk factors in the organization of train traffic. In order to ensure the high reliability of all railway subsystems, the method is proposed for a more accurate analysis of the diagnostic parameters for the roadbed to predict possible deformations and destructions.

Scientific research in the field of the carriage of DG in international traffic is primarily aimed at solving the problems of increasing the safety of transportation and reducing the delivery time of cargo. Paper [4] argues that the issues of improving the technology of passing the trains through the border crossings, where the railways with different track gauge connect, are of great importance. The traditional technology of cars crossing through joints of railways of different standards by means of transshipment or handling, which is connected with the change of track gauge, does not meet the current carriage safety requirements. One of the most effective measures to speed up the carriage and improve safety is to launch the continuous operation of international trains with the use of technologies in the automatic crossing the break-of-gauge with different track width – 1520 mm and 1435 mm. At the carriage of DG, it is proposed to use bogies with a variable gauge SUW-2000 system.

Notably, the concept of risk has been introduced and is widely used in the world science and practice of technogenic safety [5, 6] which made it possible to subsume the danger under the position of categories that are measured. This means that in determining the safety of functioning, it is necessary to scientifically justify from the economic point of view the level of risk (in natural exponents) that can be achieved while ensuring the optimal functioning of transport, taking into account the need to maintain this level of risk with resource facilities The implementation of the transport process associated with dangerous goods is accompanied by the risk of unfavorable events [7]. Therefore, it is important to consider risks when solving the task to justify the process of managing the movement of cars with DG.

The first works to solve the routing task in the carriage of DG appeared in the eighties of the twentieth century. In 1991, George F. List, Pitu B. Mirchandani, Mark A. Turnquist and Konstantinos G. Zografos [8] proposed several mathematical models to optimize the carriage of dangerous goods. In them as a criterion for choosing a route, there were respectively: minimizing the number of people affected by the accident and minimizing the length of the route. In 1985, a model was proposed [9], which allowed to precisely and accurately formulate the consequences of an accident and its impact on the population. A model is constructed on the basis of two criteria. They are the criterion of minimum risk and the criterion of minimum probabilities of an accident. The above models have a different concept of risk, the risk is often interpreted as a subjective value that cannot be measured. In addition, the formulation of multicriteria models at that time did not allow solving large-dimensional problems for choosing a route on a real transport network.

Since the 1990s, with the advent of the Value-At-Risk concept [6], risk began to be identified through the probability of losses. This approach has led to structure the statement of the task in finding the optimal route for the carriage of dangerous goods on condition of risks treatment. This task, as a ruler, boils down to making a choice of the shortest tracks in the transport network, that is, to choose one low-risk route or to find a Pareto-optimal solution where different methods, models and criteria can be used [10].

The model for optimization at the carriage of DG with the use of mathematical programming to calculate the optimal routes is of scientific interest [11]. The optimization task is set in terms of finding the minimum cost of the traffic in the network, where the purpose of seeking is to minimize the total cost of the carriage along the route. Work [12] proposes a distributed decision support system (DDSS) to choose the route for the carriage of dangerous goods. The user's interface was designed involving a geo-information system (GIS) to visualize the optimal route, while in the model the main criterion was to minimize travel time between departure and destination points. But other objectives of the task were presented in the form of a large number of restrictions, in particular a traveled distance, traffic accident probability, population density, and population risk.

4. Research results

One of the most difficult stages in the organization of the carriage of DG by rail is the planning process at the tactical level. Planning at the tactical level involves decisions concerning the choice of the route for cars with dangerous goods, including the direction of movement and train class, station stops and time-schedule, with regard to the current plan in the formation of the freight trains. Fail-safeness in the process of carriage of DG and the efficiency in the implementation of the quality of the declared services by rail depends on the planning level, therefore it is necessary to improve the existing information and control system for managing the carriage of dangerous goods by rail.

Given the level of danger that arises at the carriage of DG, the most important criterion for route planning is to ensure a higher level of traffic safety. Within the transport planning at the tactical level in compliance with the clearly defined regulations and rules for the carriage of DG, various organizational decisions are possible on the procedure for sending cars with such cargoes. This can further reduce the risk of traffic accidents during carriage. The application of enhanced safety measures for the carriage of goods is accompanied by an increase in the service time of cars at stations. They include regulatory compliance in holding and special rules for shunting operations, the need in some cases to increase the car-mileage when choosing a route other than the shortest one, for reasons of avoiding the large transport nodes and so forth. Implementation of such measures leads to an increase in operating costs, which is not always acceptable in the economic expediency at the carriage of DG. In such circumstances, in addition to ensuring a high level of safety in the rail transport process, the criterion of minimizing operating costs is required.

Therefore, in the conditions of the transport market development in the segment for the organization of carriage of DG, the railway is faced with the task of improving the planning system of carriage of DG at the tactical level, provided that a high level of safety of the transport process and minimal costs for carriage. Formalization of this process leads to the formation of an optimization task. With regard to the contradictions between the proposed criteria in choosing the optimal route of carriages, it is advisable to solve the task on the basis of the two-stage mathematical model.

In order to implement the planning system at the tactical level, in the research authors propose to formulate such a mathematical model, the idea of which is to choose a set of possible routes in cars movement with dangerous goods, its power is limited by the established critical operating costs and finding the optimal route based on minimizing the risks [13].

The criterion of the effectiveness in choosing the route for the car movement with DG is to minimize the total cost concerning the car movement in trains and handling at the stations

$$F = \sum_{i=1}^{M}\sum_{j=1}^{M} c_{ij}x_{ij} \to \min_{\{x_{ij}\}}. \tag{1}$$

Restrictions take into account the continuity of the route, and each intermediate point along the route can be visited only once:

to enumerate all k -th arcs, incoming to i -th vertex of a route:

$$\sum_{k} x_{ki} = 1, i = \overline{2, M}; \tag{2}$$

to enumerate j -th arcs, outgoing from i -th vertex of a route

$$\sum_{j} x_{ij} = 1, i = \overline{1, M-1}. \tag{3}$$

If the i -th vertex is not included in the shortest route, the corresponding sum for both the incoming and outgoing arcs of the graph vertex must be equal to zero. Then, for any point in the network, except origin and destination, the condition must be satisfied:

$$\sum_{k} x_{ki} - \sum_{j} x_{ij} = 0. \tag{4}$$

Origin point is $\sum_{j} x_{sj} = 1$, destination is $\sum_{k} x_{kt} = 1$, $x_{ij} \geq 0$ for all i, j.

The transport system in the task is an oriented graph, it is a two-pole net, S is the origin station number, t is the destination number. All way-points of the route can be divided into origin, intermediate and destination. Obviously, there should be one incoming and one outgoing arc at each intermediate point, and for the origin and destination points, it can be only one outgoing or incoming arc, respectively. Thus constraints (2-3) reflect the requirement that at each vertex along the route there is only one arc exiting the input and one arc enters the input. Constraint (4) ensures equality in the number of arcs incoming and outgoing into each intermediate vertex.

Using criterion (1) and constraints (2-4) allows to find the shortest route for cars movement with dangerous goods from vertex S to t one, but to form an option in moving the cars under the condition of high reliability it is necessary to determine on the graph G' several k -rather short simple tracks (chains) from the origin vertex to the destination vertex.

To solve this task it is proposed to use the Yen algorithm [14], the scheme of which involves the application of the Floyd-Worschell algorithm [15] to find one optimal route between two vertices on the network.

The essence of a logical-functional scheme of the algorithm is to find with Floyd-Worschell technique [16] one of the shortest routes at k -bundle graph of the railway network. After this, searches are made for the next suboptimal k routes, every of which has to differ from already found in at least by one arc e'_{ij}. As the suboptimal route, one should be interpreted the route that leads from the origin to the destination vertex but differs from the optimum route by at least one arc.

The model is launched to search for many routes in the carriage of DG with minimizing operating costs is performed in the m-file init_train_route.m. The results of this program to choose routes for car movement with dangerous goods based on minimizing operating costs and risks for the arbitrary rail network graph $G' = (V', E')$ are shown in Figure 1 and Figure 2.

According to the second stage in the process of planning the routes for dangerous goods, except minimizing the operating costs of carriage, hazard identification is the most important in the technological process of car movement with dangerous goods. Given the technological specifics of rail transport, we use the risk concept of the traffic accident. It means the probability of a potentially possible traffic accident and its related losses when performing train or shunting operation or while stopping/ trains standing/cars carrying dangerous goods. It is established that the decision-making process regarding the choice of the least accident-free route for cars carrying the dangerous goods is accompanied by uncertainty. Its main source is information on the reliability of railway infrastructure and rolling stock, impossibility in the reliable prediction of personnel errors in traffic management, intervention of unauthorized persons into the railways activity, or the emergence of natural disasters.

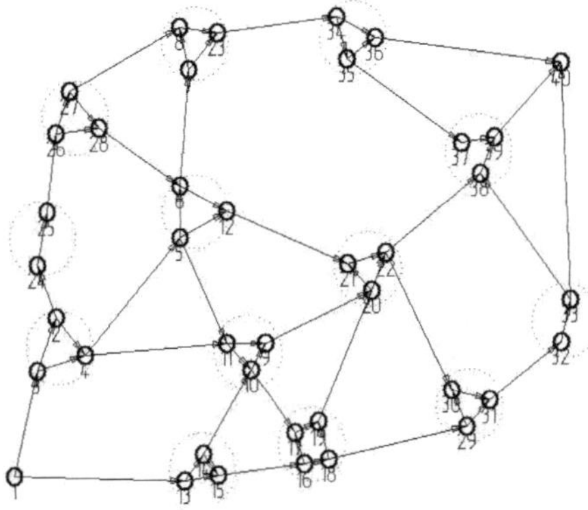

Figure 1. Visualization of an arbitrary graph for the railway network $G'=(V',E')$.

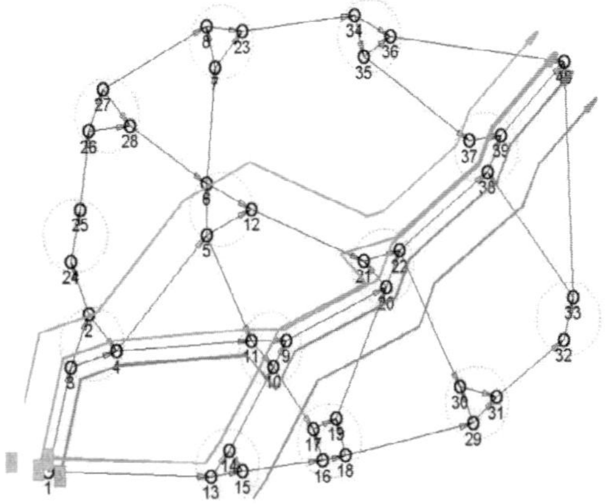

Figure 2. Optimal (number one) and suboptimal routes (No. 2-4) at cars movement with DG on an arbitrary rail network graph $G'=(V',E')$.

Assuming that reducing the risk of danger on the route under uncertainty is the main criterion for choosing the best route from the k -found shortest simple tracks on the graph $G'=(V',E')$, it is advisable to use a mathematical model based on a theory of conflict, which reproduces the independence of players' actions or a game theory [13].

It should be noted that the uncertainty of the game regarding the route of car movement with dangerous goods can be reduced by conducting a statistical experiment that will allow the "statistics" to obtain additional probabilistic information about the strategies of "nature".

To determine strategies of nature at the emergence of a traffic accident, it is necessary to calculate the function of risk values $R(\Theta,d)$ by expression:

$$R(\Theta,d)=M\left[L(\Theta,\mu)\right]=\int_{x} L(\Theta_q,\mu_k)dF(x\,|\,\Theta_q), \tag{5}$$

where $M\left[L(\Theta,\mu)\right]$ – expectation operator,

Θ_q – individual state of nature.

It is proposed to use the Bayesian Belief Network [16, 17] to find the probabilities in the emergence of a traffic accident on a route (probability in state of nature). Given that the probability in the emergence of a traffic accident is a complex random variable, which is caused by a set of elementary that is not known in advance. Reliability indices are determined on the basis of probabilistic and statistical assumptions, and therefore, if there is information concerning the traffic safety along the route, quantitative evaluation in the reliability of subsystems is a simple procedure and can be solved in accordance with [18]. Construction of the Bayesian Belief Network that describes route safety in the HUGIN system is to create vertices of the network graph and establish cause-and-effect relationships between them using special BBN network software. The system interface is shown in Figure 3. After plotting the network graph for each vertex, their state is established.

The next step in designing is the correct table compiling (matrices) of conditional probabilities (CPT) of each vertex. The list of conditional probabilities may include "Incident", "Collision", "No fence", "Incorrect actions", "Wrong route" and others. After the network design is completed, HUGIN system moves to the compute mode. For this purpose, the procedure of network compilation is carried out at which the correctness of the introduced probabilities is checked. The sum of opposite events should not exceed unity.

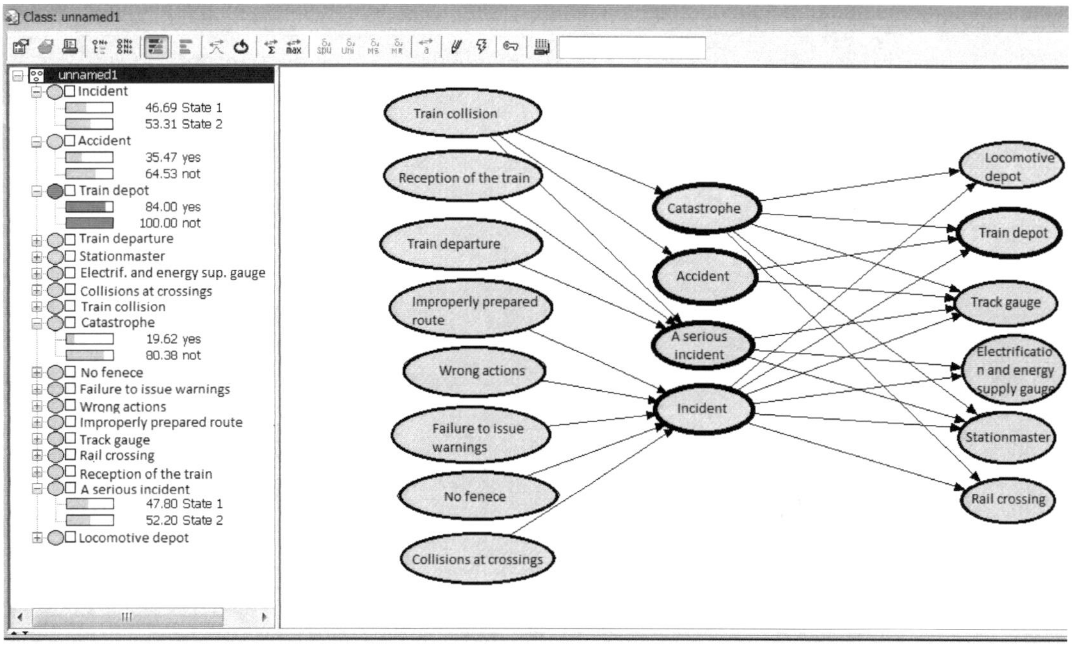

Figure 3. System interface while determining the probabilities for the state of all vertices of the network, subject to establishing the reliability in "Car facilities" subsystem is 84%.

Based on the receipt of values for the reliability of the subsystems into the BBN network, in accordance with the current situation on the route, the process of distribution through the network of newly received evidence of nodes state is carried out. Since the graph is acyclic, the equilibrium is set in the recalculation of estimations: the messages cease to spread and the probability values cease to change. According to the results of the simulation, the following reliability values were determined at which no traffic accident should occur. For example, in the case of the permissible state in the "Car facilities" subsystem, see Figure 3) it is found by using the Max propagation tool that the most probable state of all vertices is 84% in reliability. According to the developed network, the most probable fact is that with such a value of the subsystem reliability, no traffic accident should emerge.

5. Conclusion

Considering the level of danger that the process of carriage of DG entails, the most important requirement for planning the appropriate route is to ensure the high level of traffic safety. Therefore, the task of improving the system of planning for the carriage of DG at the tactical level can be solved, provided that the high level of safety of the transport process and minimum costs for the implementation of carriage. Given the contradictions between the proposed criteria when choosing the optimal route of carriage, it is advisable to solve the task on the basis of the two-stage mathematical model. To this end, a mathematical model of finding the optimal route based on risk minimization was proposed and a probabilistic model based on the Bayesian Belief Network was constructed. This model is a set of variables and their cause-effect relationships, that is, their probabilistic dependencies. The implementation of models in to the practice of organizing carriage of dangerous goods by rail allows the logistics operator to make the decision on the planning of optimal routes with minimal risk of traffic accidents. Also, the model can be easily customized to other case studies and adapted to different routing problems.

6. References

[1] Butko T, Prokhorov V and Chekhunov D 2018 Intelligent control of marshalling yards at transportation of dangerous goods based on multiobjective optimization *Science and Transport Progress* **5** 77 pp 41–52

[2] Butko T, Muzykin M, Prokhorchenko A, Nesterenko H and Prokhorchenko H 2019 Determining the rational motion intensity of train traffic flows on the railway corridors with account for balance of expenses on traction resources and cargo owners *Transport and Telecommunication* **20** 3 pp 215–228

[3] Trofimov A, Kuzmenko A, Nesterenko H, Avramenko S, Muzykin M, Mormul N and Sokhatsky A 2019 Non-destructive control data analysis of railroad foundation constructions *MATEC Web of Conferences* **294** 03012

[4] Nesterenko G, Kuzmenko A and Muzykina S 2012 Investigation of performance problems of implementation of SUW-2000 for international freight transportation *Bulletin of Volodymyr Dahl East Ukrainian National University* **6** 177 part-1 pp 49–54

[5] Barlow R E, Scheuer E M 1969 An Introduction to Reliabiliti Theory *CEIR* pp 118–123

[6] Crouhy M, Galai D, Mark R 2000 A comparative analysis of current credit risk models *Journal of Banking & Finance* **24** pp 59–117

[7] Jovanović D, Živković N 2010 Routing problems in transportation of hazardous materials: *Working and Living Environmental Protection* **7** 1 pp 43–51

[8] List G, Mirchandani P, Turnquist M, Zografos K 1991 Modeling and Analysis for Hazardous Materials Transportation: Risk Analysis, Routing/Scheduling and Facility Location *Transportation Science. Transportation of Hazardous Materials* **25** 2 pp 100–114

[9] Saccomanno F Chan A 1985 Economic Evaluation of routing strategies for hazardous road shipments *Transportation Research Record* **1020** pp 12–18

[10] Barilla D, Leonardi G, Puglisi A 2009 Risk Assessment for Hazardous Materials Transportation *Applied Mathematical Sciences* **3** 46 pp 2295–2309

[11] Leonelli P, Bonvicini S, Spadoni G 2000 Hazardous materials transportation: a risk-analysis based routing methodology *Journal of Hazardous Materials* **71** 1-3 pp 283–300

[12] Frank W C, Thill J C, Batta R 2000 Spatial decision support system for hazardous material truck routing *Transportation Research Part C: Emerging Technologies* **8** 1-6 pp 337–359

[13] Butko T, Prokhorchenko A and Muzykina S 2012 Forming prediction models consequences of appearance of emergencies at railway transport in carriage of dangerous goods *Bulletin of Volodymyr Dahl East Ukrainian National University* **3** 174 pp 18–23

[14] Yen J Y 1971 Finding the K shortest loopless paths in a network *Management Science* **17** pp 712–716

[15] Kormen T, Leiserson C, Rivest R and Stein C 2006 Algorithms: construction and analysis *Introduction to Algorithms* (Moscow: Williams) p 1296

[16] Druzdzel M J 1996 Qualitative verbal ex planati ons in Bayesian b elief networks *Artificial Intelligence and Simulation of Behaviour Quarterly* **94** pp 43–54

[17] Cheng J, Druzdel M J 2000 Latin hypercube sampling in Bayesian networks In: Proc. *Of the Thirteenth International Florida Articial Intelligence Research Symposium (FLAIRS-2000)* (Orlando: AAAI Publishers) pp 287–292

[18] Gruntov P 1986 Operational reliability of stations (Moscow: Transport) pp 192–239

Coefficient of stability against lift by longitudinal forces of freight cars in trains

A O Shvets[1,3,4], O V Shatunov[2,3], S S Dovhaniuk[2,3], L A Muradian[2,3], A L Pularyia[2,3] and V V Kalashnik[2,3]

[1]Department «Theoretical and Structural Mechanics»
[2]Department «Cars and Car Facilities»
[3]Dnipro National University of Railway Transport named after Academician V. Lazaryan, Lazaryan St., 2, Dnipro, Ukraine, 49010

[4]angela_Shvets@ua.fm

Abstract. The paper is devoted to a theoretical study of the car movement stability when exposed to longitudinal forces of a quasistatic nature. The authors obtained the expressions for calculating the resistance coefficient of the car lift by the longitudinal compressive force acting on the car as part of a freight train. The influence of some factor's combination on the longitudinal forces, at which the car movement stability is still maintained, was analyzed using the analytical dependencies to assess the longitudinal loading of cars in trains. The study was carried out by analytical method for assessing the freight car stability when moving at different speeds along curved track sections. The calculations were made in a curve of small radius taking into account the inertia forces from the unbalanced acceleration. In a theoretical study, the influence of quasistatic longitudinal compressive forces depending on changes in speed and force value, as well as the influence of friction in the wheel flange and rail contact and the eccentricity of fastening the automatic coupler shank on the stability were considered. When applying the results obtained, the stability of freight rolling stock can be increased, which in turn will remove some existing restrictions on the permissible speeds and increase the technical speed of train movement. Using the described methodology for determining the lift resistance coefficient will make it possible to justify the cause of derailment, as well as to develop and put into practice technical measures to prevent the lift of carriages, thrusts and shears of the track.

1. Introduction

Ensuring the movement stability of cars in long freight trains with increased carrying capacity remains an urgent problem for a long period of time. The values of the permissible longitudinal forces are included in the regulatory and technical documents, according to which all newly designed and constructed cars must satisfy the conditions of lift resistance and rolling off the track [1-3]. These documents also serve as guidance when organizing the movement of long freight trains with increased carrying capacity. However, during operation, there is a need to adjust regulatory documents, especially when increasing movement speed and cancelling too stiff restrictions on the conditions of trains driving [4-6].

The longitudinal quasistatic compressive force in a train is the main operational parameter. Therefore, one of the priority tasks of improving the technology of trains driving is to reduce this power factor in the process of operational work. In addition, the transfer of longitudinal forces of a

Content from this work may be used under the terms of the Creative Commons Attribution 3.0 licence. Any further distribution of this work must maintain attribution to the author(s) and the title of the work, journal citation and DOI.
Published under licence by IOP Publishing Ltd

quasistatic nature to the elements and units of cars, as well as to the railway track structure, has not yet been sufficiently studied, especially in theoretical terms [7-9].

2. Methodology

The main criterion for assessing the dynamic qualities of the rolling stock is the criterion of the carriage derailment resistance, which is the maximum permissible ratio of lateral horizontal and vertical forces acting on the climbing wheel [10-13]. The choice of design schemes for cars and trains is determined by the set task and the criterion for assessing the movement stability of the car. The position of car is set by the rotation angle of the central axle and/or the value of the lateral swaying of the body relative the track axle [4]. In addition, the design scheme should take into account the possibility of placing a separate car on a rail track, depending on the compressive or tensile forces in a curve or a straight track section [12, 13]. The design scheme should also reflect the characteristics of the transmission of longitudinal forces to the elements of bogies in the vertical and horizontal (transverse) directions. Each case of forces application to the car in a vertical plane should be considered together with various cases of car placing in a horizontal plane and separately for tensile and compressive longitudinal forces in straight and curved track sections [4, 13].

In regenerative braking mode, the cars occupy the position of maximum skew within the gap in the rail track, and the rotation angles of the experimental and neighboring cars are of the same sign (Figure 1, Table 1). When driving along a curve in traction mode, cars occupy a predominantly chordal position. With an increase in the quasistatic compression of the train, the skews increase, causing an increase in the lateral and vertical effects of the wheels of car undercarriage on the track structure. At the same time, the wheels of neighboring bogies act on opposite rail tracks, that is, they tend to thrust the track or shift it in different directions [4, 7].

The main reasons of the additional group influence of the bogie wheels on the track caused by quasistatic stretching and compression of the train are as follows: eccentricities of fastening the automatic coupler shank in the vertical Δ and horizontal plane e; vertical skews of freight cars during the train compression. The automatic coupler shank of SA-3 (SA-3M) model transfers the compressive force to the base plate at a point which moves in a horizontal plane across the carriage during the skew-symmetrical skews of cars (according to the herringbone pattern). Reasonably practicable value for cars is $e = 20\,\text{mm}$. Eccentricity e is possible as a result of deviation from the design position of the automatic coupler shank during the carriage manufacture at the factory, and the value Δ is caused by the car loading, the vertical stiffness of the central spring suspension and the wheels wear. When an empty car is placed on a train with fully loaded cars, the value can reach 80 mm [7, 10].

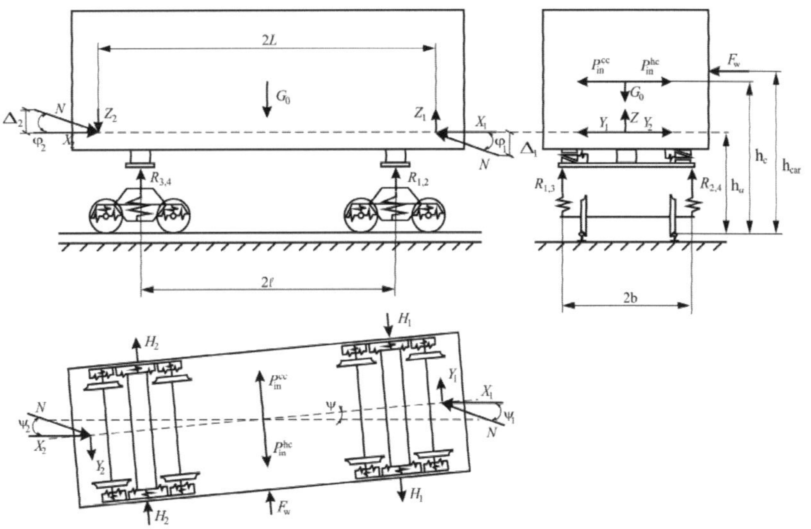

Figure 1. Scheme of forces acting on the car at skew-symmetrical skew.

IOP Conf. Series: Materials Science and Engineering **985** (2020) 012025 doi:10.1088/1757-899X/985/1/012025

Table 1. Parameters used in determining the resistance coefficient of car lift by longitudinal forces.

Designation	Parameter	Dimension
N	longitudinal quasistatic force in an automatic coupler under the action of compressive forces on the car	kN
G_0	car body weight	kN
H_f	frame force acting on the wheel set	kN
G_{bog}	bogie weight	kN
C_h	horizontal stiffness of spring suspension of one bogie	kN/m
C_v	vertical stiffness of spring suspension of one bogie	kN/m
φ_1, φ_2	angles in the vertical plane due to the difference in levels of the automatic coupler axles in the connection of two cars	rad
Δ_1, Δ_2	difference in levels of the automatic coupler axles in front and behind the car	m
e	eccentricity of the automatic coupler shank as a result of deviation from the design position during the manufacture	m
$2b$	the distance between the middle of the wheel set axle necks	m
$2\delta_0$	total lateral acceleration of the car body frame relative to the track axle in the guiding section along the center pin	m
$2L_c$	length over coupler pulling faces	m
2ℓ	wheel base	m
$2L$	distance between coupler followers	m
R	curve radius	m
a	automatic coupler body length from pulling face to the shank end	m
$2S$	distance between wheel rolling circles	m
h_a	automatic coupler axle height above the rail heads level	m
h_{hs}	height above the rail heads plane to the upper plane of the central spring group	m
r	radius of a medium-worn wheel	m
h_c c	the height of the car gravity center above the rail heads level	m
h_{car}	height above the rail heads plane to the gravity center of the side surface of the car body	m
a_{un}	unbalanced acceleration	m/s^2
g	acceleration of gravity	m/s^2
μ	wheel-rail friction coefficient	
μ_2	wheel-rail friction coefficient on non-climbing wheel	
β	inclination angle formed by the conical surface of the wheel flange to the horizontal axle	

Currently, in Ukraine there are significant differences in determining the stability of freight cars under the action of quasistatic compressive longitudinal forces in the regulatory documentation DSTU 7598:2014 "Freight cars. General requirements for the calculations and design of new and modernized 1520 mm gauge cars (unpowered cars)" and the procedure for calculating wheel unloading during forensic inquiries [14]. The document DSTU 7598:2014 regulates calculations when cars are placed according to the maximum skew-symmetric skew scheme ("herringbone" position). At the same time, during the technical examinations on the rolling stock derailments, the necessary calculations are performed for the chordal arrangement of cars on a railway track [14]. The methodology presented in this study will make it possible to carry out necessary calculations for unloading the wheels when the cars are skewed according to the scheme of the maximum skew-symmetric skew ("herringbone" position).

The basis of the methodology for determining the lift resistance coefficient by longitudinal forces is the research presented in the works [9, 15, 16]. The carriage movement in the curve is considered at different angles on both sides of the car: the tilt angles of the bodies to the track plane φ_1 and φ_2; automatic coupler rotation angles relative to the track axle ψ_1 and ψ_2; the car body rotation within the gap in the track by an angle ψ is taken into account. The car (Figure 1) is subject to compressive longitudinal forces, and the carriage is placed on the rail track according to the maximum skew-symmetrical skew scheme. The longitudinal forces N_1 and N_2, acting in automatic couplings in front and behind the car, are assumed to be the same.

The forces acting on the car are projected on the track plane, as well as on the planes perpendicular to it – longitudinal and transverse relative to the track axle. The designations given in Figure 1, as well as the parameters used to determine the resistance coefficient of car lift by longitudinal forces, are summarized in Table 1. The efforts depicted in Figure 1 are determined according to the methodology given in the works [9, 16].

Angles in the vertical plane due to the difference in levels of the automatic coupler axles in the connection of two cars:

$$\varphi_1 = \frac{\Delta_1}{2a}, \; \varphi_2 = \frac{\Delta_2}{2a}. \tag{1}$$

Consider the definition of angles ψ, ψ_1 and ψ_2 in the horizontal plane during compression of the car in the case of installation in a rut with the deviation of the body center plate across the path (in different directions from the longitudinal axis of the car) by the value δ_0 (Figure 2).

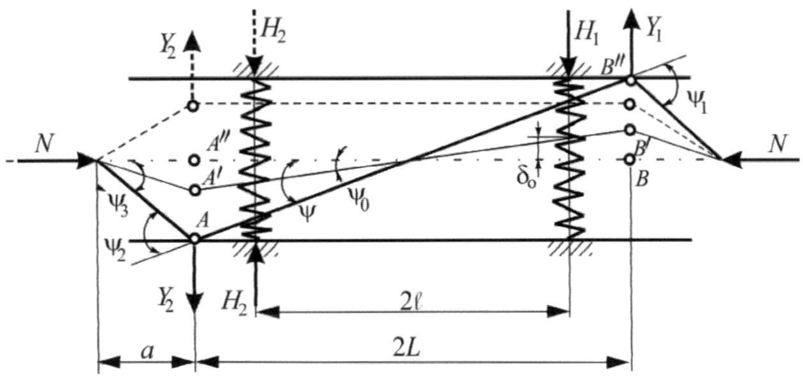

Figure 2. The scheme of the car to determine the components of the longitudinal forces in the horizontal plane.

In Figure 2, the dashed line shows the interaction forces and the position of the central axis with the chordal arrangement of the car under study under the action of compressive forces.

When the car axis is skewed (Figure 1) due to a lateral deflection from the track axis of the center plate section of the car frame due to the wheel set acceleration in the rail track, bearings are skewed along the axle necks, axle boxes – relative to the bogie frame ($BB' = A'A'' = \delta_0$), the car axis will additionally deflect due to spring deformation under the action of transverse components of longitudinal force ($B'B'' = A'A$). Let us find the rotation angles of the car body in a curve in the horizontal plane:

$$\psi = \frac{\delta}{\ell}, \tag{2}$$

$$\psi_1 = \frac{\delta}{\ell}\left(1 + \frac{L}{a}\right) + \frac{L_c}{R}, \tag{3}$$

$$\psi_2 = \frac{\delta}{\ell}\left(1 + \frac{L}{a}\right) - \frac{L_c}{R}. \tag{4}$$

Since the angles ψ, ψ_1 and ψ_2 depend on the argument δ, which in turn depends on H_f, the calculation by formulas (2)-(4) is possible by the method of successive approximations. With sufficient accuracy for calculation, it is possible to limit ourselves to the first approximation and formulas (2), (3), (4), taking into account the elastic transverse deformation of the spring sets of the bogies, will take the following form:

$$\psi = \frac{\delta_0 + e}{\ell} \cdot \left[1 + \frac{N \cdot L}{2\ell^2 \cdot C_h}\left(2 + \frac{L}{a}\right)\right], \tag{5}$$

$$\psi_1 = \frac{\delta_0 + e}{\ell} \cdot \left[1 + \frac{N \cdot L}{2\ell^2 \cdot C_h}\left(2 + \frac{L}{a}\right)\right] \cdot \left(1 + \frac{L}{a}\right) + \frac{L_c}{R}, \tag{6}$$

$$\psi_2 = \frac{\delta_0 + e}{\ell} \cdot \left[1 + \frac{N \cdot L}{2\ell^2 \cdot C_h}\left(2 + \frac{L}{a}\right)\right] \cdot \left(1 + \frac{L}{a}\right) - \frac{L_c}{R}. \tag{7}$$

From expressions (5), (6), (7) it is possible to obtain dependencies for determining the angles in the case of a "clear" skew of the car in a straight track section. If we neglect the elastic transverse deformation of the spring sets of bogies and the eccentricity of the automatic coupler shank location, these formulas take the form obtained in the works [4, 15] and DSTU 7598:2014.

The vertical and horizontal forces H_f acting on the wheel set are determined from the balance of the unsprung part of the bogie. As a result of solving a system of linear equations, we determine the vertical and lateral transverse reaction of the rail to the climbing wheel, as well as the lift resistance coefficient by longitudinal forces.

$$V = \frac{G_0}{8} + \frac{G_{bog}}{4} + \frac{N}{8} \cdot \left[(\psi_1 - \psi_2) \cdot \frac{h_a}{S} - \left[\varphi_1 \cdot \left(\frac{L}{\ell} + 1\right) + \varphi_2 \cdot \left(\frac{L}{\ell} - 1\right)\right] \cdot \frac{b}{S} + (\psi_1 + \psi_2 + 2\psi) \cdot \frac{L \cdot h_{hs}}{\ell \cdot S}\right], \tag{8}$$

$$L = \mu \cdot \left(\frac{G_0}{8} + \frac{G_{bog}}{4}\right) + \frac{N}{8} \cdot \left[\begin{array}{l} (\psi_1 - \psi_2) \cdot \left(2 - \mu \cdot \frac{h_a}{S}\right) - \mu \cdot \left[\varphi_1 \cdot \left(\frac{L}{\ell} + 1\right) + \varphi_2 \cdot \left(\frac{L}{\ell} - 1\right)\right] \cdot \frac{b}{S} + \\ + (\psi_1 + \psi_2 + 2\psi) \cdot \frac{L}{\ell} \cdot \left(2 - \mu \cdot \frac{h_{hs}}{S}\right) \end{array}\right], \tag{9}$$

$$C_{lr}=\frac{tg\beta-\mu}{1+\mu\cdot tg\beta}\cdot\frac{P_{car}+N\cdot\left[(\psi_1-\psi_2)\cdot\frac{h_a}{S}-\varphi_e\cdot\frac{b}{S}+(\psi_1+\psi_2+2\psi)\cdot\frac{L\cdot h_{hs}}{\ell\cdot S}\right]}{\mu\cdot P_{car}+N\cdot\left[(\psi_1-\psi_2)\cdot\left(2-\mu\cdot\frac{h_a}{S}\right)-\mu\cdot\varphi_e\cdot\frac{b}{S}+(\psi_1+\psi_2+2\psi)\cdot\frac{L}{\ell}\cdot\left(2-\mu\cdot\frac{h_{hs}}{S}\right)\right]}. \quad (10)$$

If it is necessary to take into account inertia forces in the curve, the vertical (8) and the lateral (9) reactions of the rail on the climbing wheel are added to the inertia forces caused by unbalanced acceleration and wind pressure on the side surface of the car body.

The dependency for determining the resistance coefficient of car lift by longitudinal forces, taking into account the inertia forces and wind pressure on the car body`s side surface in the curve for the climbing wheel set of the front carriage, has the following form:

$$C_{lr}=\frac{tg\beta-\mu}{1+\mu\cdot tg\beta}\cdot$$
$$\cdot\frac{P_{car}^{st}+\frac{N^2}{C_h}\cdot\frac{\psi_a^2\cdot h_{hs}}{(\delta_0+e)\cdot S}+N\cdot\left[2\psi_a\cdot\frac{h_{hs}}{S}+\psi_{cur}\cdot\frac{h_a}{S}\right]+P_{in}\cdot\frac{h_c}{S}+F_w\cdot\frac{h_{car}}{S}}{\left[\mu\cdot P_{car}^{st}+\frac{N^2}{C_h}\cdot\frac{\psi_a^2}{(\delta_0+e)}\cdot\left(2-\mu\cdot\frac{h_{hs}}{S}\right)+N\cdot\left[2\psi_a\cdot\left(2-\mu\cdot\frac{h_{hs}}{S}\right)+\psi_{cur}\cdot\left(2-\mu\cdot\frac{h_a}{S}\right)\right]+ +P_{in}\cdot\left(2-\mu\cdot\frac{h_c}{S}\right)+F_w\cdot\left(2-\mu\cdot\frac{h_{car}}{S}\right)\right]}. \quad (11)$$

Where P_{car}^{st} – is the static pressure of the car, taking into account unloading from longitudinal force, kN.

In dependency (11), it is recalculated for the climbing wheel set of the front carriage;

$$P_{car}^{st}=P_{car}-N\cdot\varphi_e\cdot\frac{b}{S}, \quad (12)$$

$P_{car}=G_o+2G_{bog}$ – car weight, kN;

φ_e – the rotation angle of the car body in a vertical plane, caused by the presence of a difference in the levels of automatic coupler axles in front and behind the car, rad;

$$\varphi_e=\varphi_1\cdot\left(\frac{L}{\ell}+1\right)+\varphi_2\cdot\left(\frac{L}{\ell}-1\right), \quad (13)$$

ψ_a – the angle formed by the longitudinal axle of the automatic coupler body and the axle of the central sill of the car frame in a horizontal plane, rad;

$$\psi_a=\frac{(\delta_0+e)\cdot L}{\ell^2}\cdot\left(2+\frac{L}{a}\right), \quad (14)$$

ψ_{cur} – car rotation angle, depending on the location in the curve, rad;

$$\psi_{cur}=\frac{2L_c}{R}. \quad (15)$$

$h_{hs}=r+r_n\approx r$ – height above the level of the rail heads plane to the upper plane of the central spring set, m. Often this parameter in a number of studies is taken equal to the radius of the average worn wheel.

P_{in} – inertia force from unbalanced acceleration, kN;

$$P_{in} = \left(G_o + 2 \cdot G_{bog}\right) \cdot \frac{a_{un}}{g} = P_{car} \cdot \frac{a_{un}}{g}. \tag{16}$$

Dependency (15) is calculated and cannot be used to determine stability in the experiments. The calculation is based on a static calculation scheme. However, the car is a complex mechanical system and during movement, the interaction between its individual parts and between the car and the track structure is dynamic [9].

In the work [17], a dependency is given for assessing the movement safety; on the basis of the experiment, it takes into account the wheel set speed influence on the friction coefficient value in the contact between the wheel flange and the rail. The dependency for determining the lift resistance coefficient by longitudinal forces, taking into account the inertia forces in the curve and the component of the wind loading action on the side surface of the car body, will take the following form:

$$C_{lr} = \frac{tg\beta - \mu_0 \cdot (1 - 0.002v)}{1 + \mu_0 \cdot (1 - 0.002v) \cdot tg\beta} \cdot \frac{P_{car}^{st} + \frac{N^2}{C_h} \cdot \frac{\psi_a^2 \cdot h_{hs}}{(\delta_0 + e) \cdot S} + N \cdot \left[2\psi_a \cdot \frac{h_{hs}}{S} + \psi_{cur} \cdot \frac{h_a}{S}\right] + P_{in} \cdot \frac{h_c}{S} + F_w \cdot \frac{h_{car}}{S}}{\left[\begin{array}{l} \mu_2 \cdot P_{car}^{st} + \frac{N^2}{C_r} \cdot \frac{\psi_a^2}{(\delta_0 + e)} \cdot \left(2 - \mu_2 \cdot \frac{h_{hs}}{S}\right) + N \cdot \left[2\psi_a \cdot \left(2 - \mu_2 \cdot \frac{h_{hs}}{S}\right) + \psi_{cur} \cdot \left(2 - \mu_2 \cdot \frac{h_a}{S}\right)\right] + \\ + P_{in} \cdot \left(2 - \mu_2 \cdot \frac{h_c}{S}\right) + F_w \cdot \left(2 - \mu_2 \cdot \frac{h_{car}}{S}\right) \end{array}\right]}. \tag{17}$$

Where v – speed of movement in km/h;

μ_0 – is the friction coefficient at the adhesion limit at $v = 0$. For railway rolling stock, in the calculations it is recommended to take this parameter equal to $\mu_0 = 0.3 \div 0.33$.

3. Calculation results

To assess the influence of the longitudinal forces on the lift resistance coefficient, Figures 3, 4 show the results of movement calculations of empty and loaded open car of 12-532 model in a curve with a radius of 250 m and elevation of 150 mm with a transverse gap of the car body frame relative to the track axle in directing section of 50 mm. The calculations were performed without taking into account the friction forces and taking into account the friction in the contact of the wheel flange and the rail at a speed from 10 to 120 km/h.

It was assumed that the longitudinal forces acting on the cars are static or quasistatic external forces that slowly change in time. During the calculations, the difference in height between the longitudinal axles of automatic coupler of a freight train was taken into account. It is permitted no more than $\Delta_1 = 80$ mm in front of a group of empty cars standing behind the loaded cars in a train. Behind the car under study, the difference in the levels of the automatic coupler axles is taken to be $\Delta_2 = 0$ mm, in addition, the presence an eccentricity of fastening the automatic coupler shank $e = 20$ mm was taken into account. During calculation of the influence of wind loading on the side surface of the body was taken into account. This force is applied to the geometric center of the side surface, directed outward of the curve and taken equal to 10 kN.

The resistance coefficient of an empty open car lift in Figure 3 is given as follows: (a), (c) excluding the friction forces in the contact of the wheel flange and the rail and (b) taking into account these factors; (a), (b) with and (c), (d) without the eccentricity of fastening the automatic coupler shank. Analysis of the obtained values of the resistance coefficient of an empty open car lift (Figure 3) shows that dependency (17) gives stiffer results as compared to dependency (11).

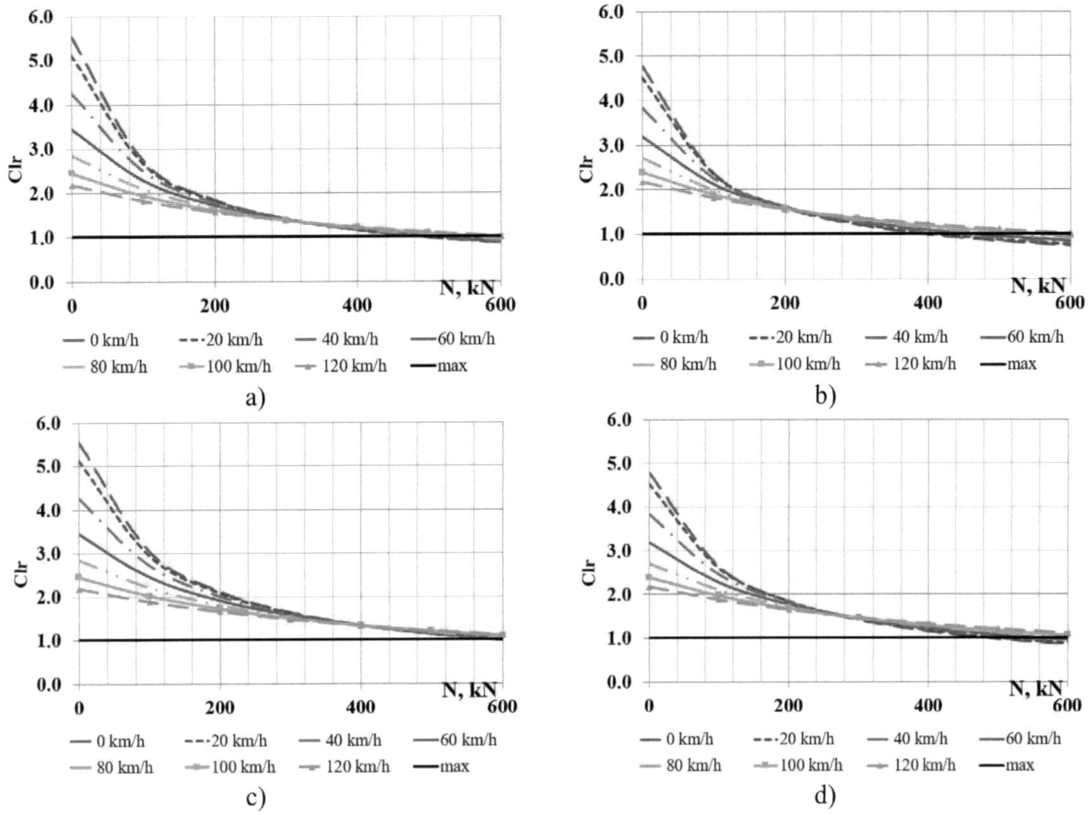

Figure 3. Resistance coefficient of an empty open car lift.

Thus, accounting the different friction in the contact between the wheel and the rail with an eccentricity of fastening the automatic coupler shank according to (17) shows that there is no lift resistance even at 400 kN, and (11) with a compressive force of 480 kN. The eccentricity of fastening the automatic coupler shank has a significant influence on the value of longitudinal force that can lift an empty freight car. The absence of eccentricity leads to an increase in the longitudinal force by 100 kN when calculating for both dependencies. Excluding wind power, the longitudinal force decreases by an average of 30-50 kN. If the wind pressure is directed inside the curve, then the force of the wind loading will contribute to the additional unloading of the climbing wheel, which will significantly affect the conditions of car lift.

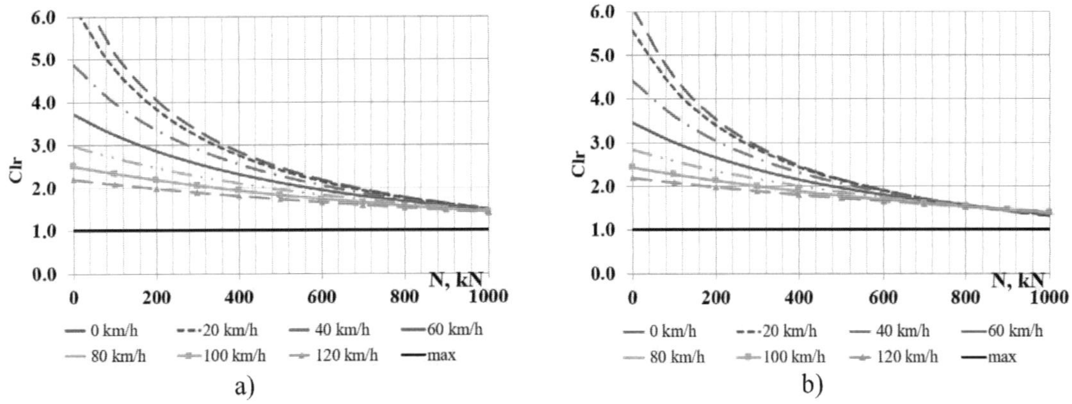

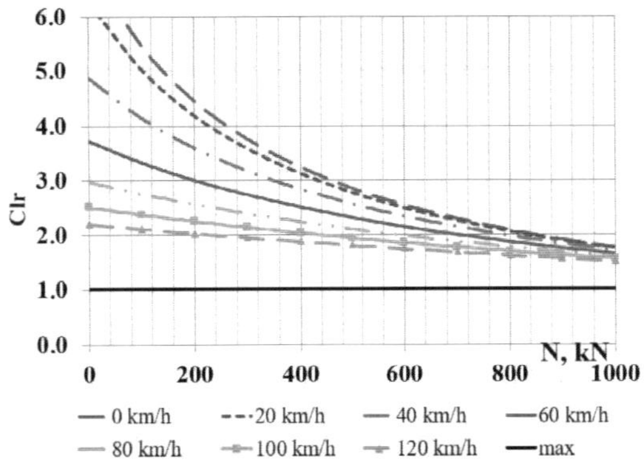

c)

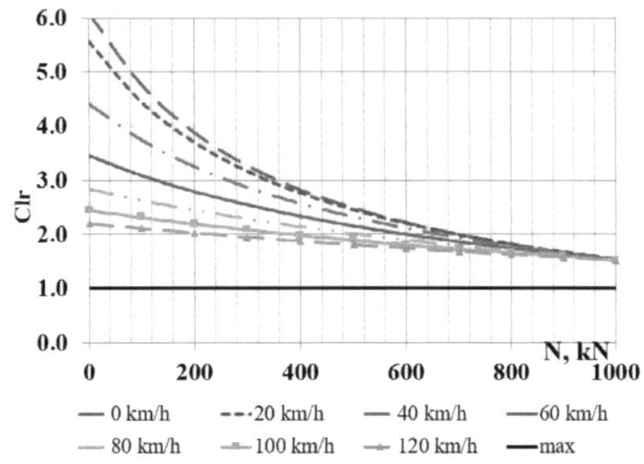

d)

Figure 4. Resistance coefficient of a loaded open car lift.

The resistance coefficient of a loaded open car lift in Figure 3 is given in the same way: excluding the friction forces in the contact of the wheel flange and the rail and (b) taking into account these factors; (a), (b) with and (c), (d) without the eccentricity of fastening the automatic coupler shank. In the loaded mode (Figure 4), the open car under study maintains the established resistance coefficient for freight cars under the action of a quasistatic compressive force of 1 MN permissible by regulatory documentation.

4. Conclusions

As a result of theoretical studies, the dependencies of the resistance coefficient of car lift by longitudinal forces are obtained taking into account the eccentricity of fastening the automatic coupler shank and the friction value in the contact of the wheel and rail. The influence of longitudinal compressive forces on the stability of freight rolling stock when moving in a curve of small radius with speeds up to a design value of 120 km/h is investigated.

Based on the theoretical studies of the stability of the freight carriage using the example of open car of the model 12-532, it is possible to draw the following conclusions:

- the presence of eccentricity of fastening the automatic coupler shank for empty cars significantly affects the value of the car lift resistance coefficient;
- the influence of freight cars on the track is possible by reducing the total transverse gap of the car body frame δ by reducing the eccentricity of fastening the automatic coupler shank e and increasing the horizontal stiffness of the spring suspension C_h, as well as by reducing the gaps in axle boxes, center plate arrangements and other units of rolling stock, affecting lateral movements (skews) of the body;
- in the loaded state, the open car under study maintains the set lift resistance coefficient for freight cars under the action of quasistatic compressive force of 1 MN permissible by the regulatory documentation.

The investigated dependence for determining the lift resistance coefficient by longitudinal forces was compiled for a freight rolling stock on the 18-100 bogie models and similar ones. For the running gears with structural differences from the 18-100 three-piece bogie, dependencies (11) and (17) can have significant differences.

Compared to the existing methods for determining the stability of freight rolling stock under the action of compressive longitudinal forces, the proposed method allows assessing the influence of longitudinal force and movement speed of a freight car on the value of the stability coefficient. It also allows taking into account the eccentricity of fastening the automatic coupler shank, the pressure of wind loading on the side surface of the body, as well as calculating the unloading or additional loading of each wheel separately during the technical examination.

Using the described methodology for determining the lift resistance coefficient will allow justifying the cause of derailment, as well as to develop and put into practice the technical measures to prevent car lifts, thrusts and track shears. When applying the results obtained, the stability of freight rolling stock can be increased, which in turn will remove some existing restrictions on permissible speeds and increase the technical speed of train movement.

References

[1] Pshinko O, Ursulyak L, Kostrytsia S, Fedorov Ye and Shvets A 2019 The influence of the «train-track» system parameters on the maximum longitudinal forces' level *Transport Problems* **4** 14 pp 161–172
[2] Fomin O V 2015 Increase of the freight wagons ideality degree and prognostication of their evolution stages *Scientific Bulletin of National Mining University* **2** pp 68–76
[3] Kelrykh M and Fomin O 2014 Perspective directions of planning carrying systems of gondolas *Scientific and technical journal «Metallurgical and Mining Industry»* **6** pp 64–67
[4] Cherkashin Yu M and Kostin G V 1992 Determination of permissible longitudinal forces in a train under the condition of ensuring the stability of the movement of cars *Study of strength, stability, impact on the track and maintenance of wagons in trains of increased mass and length. Collection of works VNIIZhT* (Moscow: Publisher Transport) pp 4–31
[5] Shimanovsky A O, Sakharau P A and Kovalenko A V 2018 Modeling of train longitudinal dynamics in MSC.ADAMS software *Aktualnye voprosy mashinovedeniya* 7 pp 75–78
[6] Shimanovsky A O and Sakharau P A 2019 Effect of gap clearances in automatic coupling devices on longitudinal forces in intercar connections of homogeneous train. *Mekhanika mashin, mekhanizmov i materialov* **2** 47 pp 42–50
[7] Lysyuk V S 2002 *The causes and mechanisms of the vanishing wheel from the rail. The problem of wear of wheels and rails* (Moscow: Publisher Transport) p 215
[8] Zhelieznov K, Akulov A, Zabolotnyi O, Ursulyak L, Chabanuk Ye, Shvets A, Kuznetsov V and Radkevych A 2019 The revised method for calculating of the optimal train control mode *Archives of Transport* **3** 51 pp 21–34
[9] Shvets A A, Zheleznov K I, Akulov A S, Zabolotnyi A N and Chabanyuk Ye V 2016 Determination the permissible forces in assessing the lift resistant factor of freight cars in trains *Science and Transport Progress* **1** 61 pp 189–192
[10] Fomin O, Shvets A, Hauser V and Prokopenko P 2019 Transversal displacement of freight wagons bogies *AIP Conference Proceedings* **2198** 020002
[11] Shvets A A, Zhelieznov K I, Akulov A S, Zabolotnyi A N and Chabaniuk Ye V 2015 Some aspects of the definition of empty cars stability from squeezing their longitudinal forces in the freight train *Science and Transport Progress* **4** 58 pp 175–189
[12] Shvets A A, Zhelieznov K I, Akulov A S, Zabolotnyi A N and Chabaniuk Ye V 2015 Determination of the issue concerning the lift resistance factor of lightweight car *Science and Transport Progress* **6** 60 pp 134–148
[13] Danovich V D and Malysheva A A 1998 Mathematical model of spatial oscillations of the coupling of five cars moving along a rectilinear section of the track *Transport. Stress loading and durability of a rolling stock* (Dnepropetrovsk: Dnipro National University of Railway Transport named after Academician V. Lazaryan) pp 62–69
[14] Sokol E N 2002 *Derailment and collision of rolling stock* (Kiev: Transport of Ukraine) p 364
[15] Vershynskiy S V 1970 Dynamics, durability and the stability to heavy wagons and high-speed trains *Collection of works VNIIZhT* (Moscow: Publisher Transport) **425** p 208
[16] Shvets A O 2020 Stability of freight cars under the action of compressive longitudinal forces *Science and Transport Progress* (Preprint gr-qc/10.15802/stp2020/199485)
[17] Andrievsky S M and Krylov V A 1969 Wheel derailment *Research in the dynamics and strength of locomotives* (Moscow: Publisher Transport) **393** pp 20–41

Energy efficiency of heat tests for traction electric machines

A M Afanasov[1,3], **O S Shapovalov**[2,3,4], **S N Holik**[1,3], **S V Arpul**[1,3] **and D S Bilukhin**[1,3]

[1]Department «Electric Rolling Stock of Railways»
[2]Department «Electrical Engineering and Electromechanics»
[3]Dnipro National University of Railway Transport named after Academician V. Lazaryan, Lazaryan St., 2, Dnipro, Ukraine, 49010

[4]shapovalov93as@gmail.com

Abstract. The article discusses one of the options to reduce the electricity consumption for the carry out acceptance of post-repair tests of traction electric machines. Heating tests of traction electric machines at the mutual load stand are the most energy-intensive part of the entire test program. Energy consumption for this type of test can be reduced, for this, an analysis of thermal processes in the windings of traction electric machines during their heating test was given. The energy efficiency coefficient of the mutual load system is formulated. The energy efficiency coefficient for the traction electric machine NB-406 is calculated. The calculation results showed that with an increase in the test current by 30%, the total losses in the motor windings decrease by about one and a half times, while the test time decreases by almost three times. An increase in test current and a decrease in test time do not reduce the quality of the test, and the results obtained. A methodology for assessing the energy efficiency of heating tests is proposed. It can be applied at test sites of depots, as well as repair plants. This will reduce the material costs of testing, without compromising the quality of the tests themselves.

1. Introduction

The reduction of energy consumption for acceptance tests after the repair of traction electric machines is one of the urgent problems at the repair plants of traction rolling stock. The most energy-intensive part of the entire test program of traction electric machines at a back-to-back stand are heat tests. Performing this kind of test, energy consumption can be reduced both by increasing the energy efficiency of the loading back test, and by optimizing the loading mode of traction electric machines.

2. Research problem statement

In accordance with IEC 60349-1: 2010, NEQ [1], traction electric machines are tested during acceptance tests for one hour at a rated voltage and current giving a temperature rise corresponding to a temperature rise in nominal mode. In fact, this current value corresponds to the one-hour mode.

Obviously, part of the thermal energy released in the windings of the tested electric machine is emitted into the environment and does not participate in the formation of the thermal potential of the windings, during the time of the heating test [2, 3]. The goal of these tests is to heat the windings of the traction electric machine to a certain temperature rise in certain time at a given heat transfer power. From this point of view, that part of the thermal energy released in the windings that is used to increase their temperature is useful. It is determined by the equivalent heat capacity and the specified

Content from this work may be used under the terms of the Creative Commons Attribution 3.0 licence. Any further distribution of this work must maintain attribution to the author(s) and the title of the work, journal citation and DOI.
Published under licence by IOP Publishing Ltd

IOP Conf. Series: Materials Science and Engineering **985** (2020) 012026 doi:10.1088/1757-899X/985/1/012026

temperature rise, and does not depend on the test mode. Therefore, a reduction in the energy consumption for thermal tests can be achieved by minimizing the total heat transfer energy during heating to a predetermined temperature rise [4-6].

A preliminary qualitative analysis of the thermal processes occurring during these tests shows that the heat transfer energy during heating to a predetermined temperature rise of the windings of the electric machine decreases with increasing current. An increase in current leads to a decrease in heating time as well. Therefore, the ceteris paribus, the greater the load current value of the tested electric machine, the less the energy consumption for its heating.

Let's take a closer look at the thermal processes when heating the armature windings of a traction electric machine during its test with a constant current load for a more accurate quantitative assessment of the influence of the test value of current on the power consumption. Anchor winding is a heating limit [7] for most traction electric motors of electric rolling stock.

3. Determination of the energy efficiency of heating the armature winding

The differential equation for heating the armature winding is [8]:

$$\Delta P_{rd}\mathrm{d}t = C_{rd}\mathrm{d}\tau + B_{rd}\tau \cdot \mathrm{d}t\ , \tag{1}$$

where ΔP_{rd} – reduced power loss;

C_{rd} – reduced heat capacity;

B_{rd} – reduced heat transfer;

τ – temperature rise;

t – time.

Dividing the left and right sides of expression (1) by $\mathrm{d}t$, we obtain the power balance equation:

$$\Delta P_{rd} = C_{rd}\frac{\mathrm{d}\tau}{\mathrm{d}t} + B_{rd}\tau\ . \tag{2}$$

Let us introduce the notation for the terms on the right side of the equation (2):

$$P_c = C_{rd}\frac{\mathrm{d}\tau}{\mathrm{d}t}, \tag{3}$$

$$P_в = B_{rd}\tau\ , \tag{4}$$

where P_c – power losses for heating the windings of an electric machine;

P_e – heat transfer power loss.

Then equation (2) takes the form:

$$\Delta P_{rd} = P_c + P_в\ . \tag{5}$$

Integrating the right and left sides of the expression (1) we obtain the equation of thermal energy balance:

$$\int_0^{t_1}\Delta P_{rd}\mathrm{d}t = C_{rd}\tau_1 + \int_0^{t_1} B_{rd}\tau\mathrm{d}t\ , \tag{6}$$

where t_1 – test time;

τ_1 – temperature rise at time t_1.

Definite integral $\int_0^{t_1} \Delta P_{rd} dt$ represents the total thermal energy released in the armature winding. At time t_1, one part of this energy, equal to $C_{rd}\tau_1$, was spent on raising the temperature of the winding, and the other, equal to $B_{rd}\int_0^{t_1} \tau dt$, was scattered into the environment.

Reduced power loss ΔP_{rd} can be represented as [8]:

$$\Delta P_{rd} = I^2 R_0 + I^2 R_0 \alpha \tau + k_c \Delta P_c, \quad (7)$$

where R_0 – armature winding resistance at ambient temperature;

α – temperature coefficient of resistance;

k_c – coefficient taking into account the influence of iron loss on heating of the armature winding;

ΔP_c – iron loss.

In general, the losses ΔP_{rd} can be represented as:

$$\Delta P_{rd} = \Delta P_e + \Delta P_\tau, \quad (8)$$

where $\Delta P_e = I^2 R_0 + k_c \Delta P_c$ – equivalent losses provided $\tau = 0$;

$\Delta P_\tau = I^2 R_0 \alpha \tau$ – additional temperature rise-dependent losses.

To keep it simple, let's introduce the following notation for the terms of the equation (6) the balance of thermal energy:

$$Q_p = \int_0^{t_1} \Delta P_{rd} dt; \quad (9)$$

$$Q_c = C_{rd}\tau_1; \quad (10)$$

$$Q_в = B_{rd}\int_0^{t_1} \tau dt. \quad (11)$$

where Q_p – energy expended in testing;

Q_c – energy spent on heating the motor windings;

$Q_в$ – energy dissipated in the environment.

Then equation (1) takes the form:

$$Q_p = Q_c + Q_в, \quad (12)$$

$$k_{ef} = \frac{Q_c}{Q_p} = \frac{Q_c}{Q_c + Q_в}, \quad (13)$$

where k_{ef} – coefficient of energy efficiency.

The purpose of thermal tests is to heat the armature winding at a given current I and heat transfer B_{rd} to temperature rise τ_1 during time t_1, then the energy Q_c is useful, that is, spent only on increasing τ. The ratio of this energy to total energy Q_p will characterize the energy efficiency of the heating test.

The coefficient of energy efficiency will shows how much of the thermal energy released in the armature winding is spent on raising its temperature.

To maximize k_{ef} provided $Q_c = \text{const}$, compliance with condition $Q_в \to \min$ is necessary.

Figure 1 shows the temperature rise τ and power ΔP_{rd}, P_c, $P_в$ included in equation (5) curves × time t.

The area of the figure ABE0 represents the total thermal energy Q_p for time t_1. The area of the figure 0CE is the energy $Q_в$, which is scattered into the environment. The area of the figure ADE0 represents the energy Q_c, which is spent on heating until the temperature rises τ_1.

A preliminary analysis of the curves in Figure 1 shows that with an increase in the heating rate $\frac{d\tau}{dt}$ at a fixed value of τ_1, the energy $Q_в$ will decrease. Therefore, with increasing current I, the coefficient k_{ef} will increase.

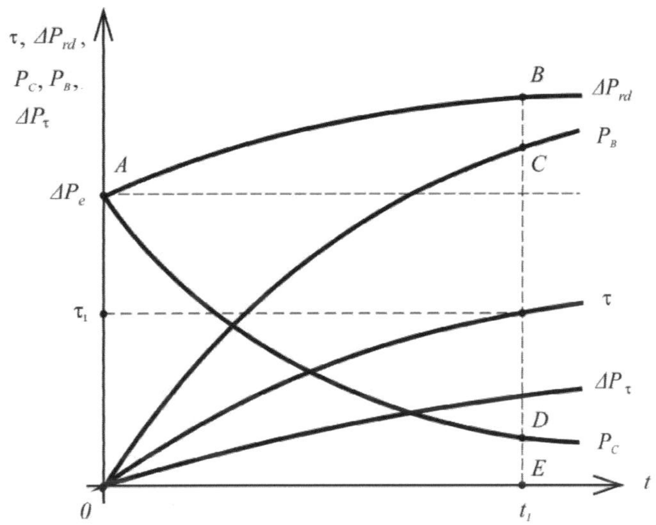

Figure 1. Curves of changes in motor parameters during tests.

Let's find the dependence $k_{ef} = f(I)$ in an analytical form according to formula (6) for the possibility of a quantitative analysis of the effect of the current value I on energy consumption during the test.

The dependence of the temperature rise τ on time t can be represented in the form [8]:

$$\tau = \tau_\infty \left(1 - e^{-\frac{t}{T_e}}\right), \tag{14}$$

where τ_∞ – steady-state temperature rise;

T_e – equivalent heating time constant.

The steady-state temperature rise and the equivalent heating time constant can be calculated by the formulas [4]:

$$\tau_\infty = \frac{\Delta P_e}{B_{rd} - I^2 R_0 \alpha}, \tag{15}$$

$$T_e = \frac{C_{rd}}{B_{rd} - I^2 R_0 \alpha}. \tag{16}$$

With a known value of τ_1, the test time t_1 can be found from equation (14) in the form:

$$t_1 = T_e \cdot \ln \frac{\tau_\infty}{\tau_\infty - \tau_1}. \tag{17}$$

Substituting expressions (14) into equation (11), we obtain:

$$Q_в = B_{rd} \int_0^{t_1} \tau \mathrm{d}t = B_{rd} \int_0^{t_1} \tau_\infty \left(1 - e^{-\frac{t}{T_e}}\right) \mathrm{d}t = B_{rd} \left(\tau_\infty \cdot t_1 - \tau_1 \cdot T_e\right). \tag{18}$$

Substituting expressions (9-11) and (18) for Q_c and $Q_в$ into equation (13), after the transformations, we obtain:

$$k_{ef} = \frac{\tau_1}{\tau_1 + \left(\tau_\infty \cdot t_1 - \tau_1 \cdot T_e\right) \cdot \frac{B_{rd}}{C_{rd}}}. \tag{19}$$

The ratio B_{rd}/C_{rd} will be found from equation (16):

$$\frac{B_{rd}}{C_{rd}} = \frac{1}{T_e} + \frac{\alpha \Delta P_e}{C_{rd}}. \tag{20}$$

The ratio $\Delta P_e/C_{rd}$ can be found from an analysis of equation (2) of the power balance at time $t = 0$. For this point in time, the expressions:

$$\Delta P_{rd} = \Delta P_e, \tag{21}$$

$$B_{rd} \cdot \tau = 0. \tag{22}$$

The expression (2) can be written as equation:

$$\Delta P_{rd} = C_{rd} \left.\frac{\mathrm{d}\tau}{\mathrm{d}t}\right|_{t=0}. \tag{23}$$

The derivative $\frac{\mathrm{d}\tau}{\mathrm{d}t}$ at time $t = 0$ can be found from expression (14) as equation:

$$\left.\frac{\mathrm{d}\tau}{\mathrm{d}t}\right|_{t=0} = \frac{\tau_\infty}{T_e}. \tag{24}$$

Substituting expressions (24) into equation (23) for ΔP_e, we obtain:

$$\Delta P_e = \frac{C_{rd} \tau_\infty}{T_e}. \tag{25}$$

Therefore, the ratio $\Delta P_e/C_{rd}$ can be found in the form equation:

$$\frac{\Delta P_e}{C_{rd}} = \frac{\tau_\infty}{T_e}. \tag{26}$$

After appropriate transformations, expression (20) can be written as equation:

$$\frac{B_{rd}}{C_{rd}} = \frac{1+\tau_\infty \alpha}{T_e}. \tag{27}$$

By substituting (27) into expression (19) after transformations, we obtain:

$$k_{ef} = \frac{T_e \cdot \tau_1}{\tau_\infty \left(t_1 + \alpha\left(\tau_\infty t_1 - \tau_1 T_e\right)\right)}. \tag{28}$$

The reciprocal of k_{ef} is the ratio as equation:

$$\frac{1}{k_{ef}} = \frac{Q_p}{Q_c}. \tag{29}$$

Which shows how many times the total energy of losses in the armature of the traction motor when testing it is more than the energy spent on heating the armature winding.

Table 1 shows the results of calculating the dependencies $t_1 = f(I)$, $k_{ef} = f(I)$, and $1/k_{ef} = f(I)$ obtained for the traction motor NB-406 using the thermal characteristics of its armature winding $(\tau_1 = 120°C)$ [7].

Table 1. Results of calculating the dependencies $t_1 = f(I)$, $k_{ef} = f(I)$, and $1/k_{ef} = f(I)$.

I, A	t_1, min	k_{ef}	$1/k_{ef}$
380	56.5	0.43	2.32
440	28.9	0.57	1.75
485	20.4	0.63	1.59
530	14.9	0.67	1.5
600	10.4	0.7	1.42

Graphs of functions $t_1 = f(I)$ and $k_{ef} = f(I)$ are presented in Figure 2. It can be seen from the graphs in Figure 2 that with an increase in the test load current of the traction motor, the energy efficiency of the heating tests increases, and the test time decreases significantly.

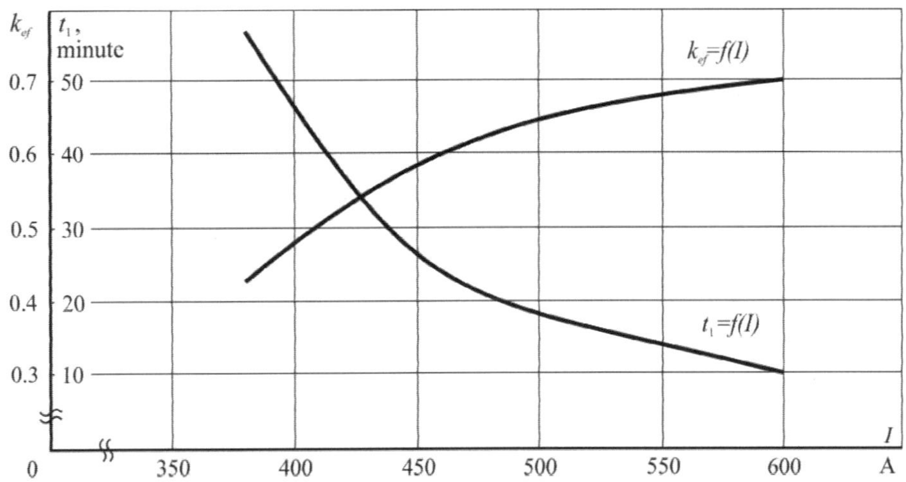

Figure 2. Graphs of functions $t_1 = f(I)$ and $k_{ef} = f(I)$.

4. Conclusions

Table 1 shows that during heating tests of the NB-406 traction motor with a load current close to the starting value (485 A), the total energy loss in the armature is almost one and a half times lower than in the one-hour mode (380 A). At the same time, the time of thermal testing is reduced by almost three times. It should be noted that the electric losses in other windings of the tested electric machine with an increase in the test current value decrease no less than the losses in the limiting winding (armature winding). These conclusions will be true for other types of traction electric machines of traction rolling stock.

References

[1] IEC 60349-1: 2010, NEQ 2010 Traction electric rotating machines. General specifications
[2] Zakharchenko D D, Rotanov N A and Gorchakov E V 1979 *Traction electric machines and transformers* (Moscow: Transport) p 303
[3] El Hayek J, Sobczyk T and Skarpetowski G 2010 Experiences with a traction drive laboratory model *Electromotion* **17** pp 30–36
[4] Afanasov A M 2012 *The systems of mutual loading of traction electrical machines of the constant and pulsing current* (Dnipropetrovsk: Makovetsky) p 248
[5] Liu Y and Eberle W 2009 Developments in switching mode supply technologies *IEEE Canadian Review. Switching Mode Power Supplies* **61** pp 9–14
[6] Loza P O 2009 Improvement of power and other indicators of acceptance tests of traction engines of electric locomotives *Science and Transport Progress* **27** pp 81–83
[7] Grebenyuk P T, Dolganov A N, Nekrasov O A and Lisitsyn A L 1985 *Rules of traction calculations for train work* (Moscow: Transport) p 287
[8] Rosenfeld V E, Isaev I P and Sidorov N N 1983 *Theory of electric traction* (Moscow: Transport) p 328

IOP Conf. Series: Materials Science and Engineering **985** (2020) 012027 doi:10.1088/1757-899X/985/1/012027

Research of the modern absorbing apparatus power characteristics influence on the freight train inter-car forces

A O Shimanovsky[1,2], **P A Sakharau**[1] **and M G Kuzniatsova**[1]

[1]Belarusian State University of Transport, 34 Kirova Str., Gomel, 246653, Belarus

[2]tm.belsut@gmail.com

Abstract. A mathematical model of the inter-car connections used in the long and heavy train is presented. It takes into account the elastic-frictional properties of absorbing devices and gaps in automatic coupler devices connecting the train cars. There are demonstrated the calculation results of a homogeneous train motion along a straight horizontal track section at such non-steady modes as at starting, braking by the locomotive electrodynamic brakes and passing along a track profile section of a complex shape. A comparative assessment of the influence of the kind of the absorbing devices power characteristic on the maximal longitudinal forces arising in the train inter-car connections is carried out using the MSC ADAMS engineering software package. Transients with a smooth and sharp change in traction and braking forces are considered. Based on the obtained results, the main positive and negative properties of rigid, linear and soft power characteristics of the automatic couplers' absorbing devices used on the railway rolling stock are determined. The results of performed computer simulation can be applied to the sphere of design and modernization of absorbing devices of automatic couplers and to the process of composition of various types of cars in trains of any length and weight.

1. Analysis of recent research

The general trend of the railway freight transportation is the maintenance of long and heavy trains, aimed at fuel and energy resources saving. In the inter-car connections of such trains, large longitudinal forces arise and have a significant effect on the cars motion stability and safety. These forces reach their maximal values at unsteady driving conditions: at starting, braking and passing along a track profile section of a complex shape. An increase in train weight standards requires a detailed study of trains longitudinal loading and a development of recommendations for their formation and operation with traffic safety improvement.

An analysis of modern absorbing devices [1] showed that the perspective elastomeric shock absorbers are the most effective from the point of view of longitudinal forces reducing in automatic couplings. The least effective are friction ones. A comparative evaluation of the operational qualities of absorbing apparatuses of various classes was carried out by the authors of [2, 3]. The presented statistics indicates that the introduction of elastomeric devices led to a decrease in the number of railcar injuries during shunting collisions by 3.8 times, incidents with dangerous goods due to damage to the boiler – by 2.1 times, and breakdowns of automatic couplings – by 2.7 times.

To evaluate the absorbing devices operation, a computer simulation is widely used, for example, in [4, 5], so it is required the accurate mathematical description of the objects to obtain reliable results. Articles [6, 7] present mathematical models and force characteristics of absorbing devices, a considered list of indicators and operating conditions of shock absorbers. In [8] it is investigated the possibility of creating a mathematical model based on the dependences of the change in stiffness, the

Content from this work may be used under the terms of the Creative Commons Attribution 3.0 licence. Any further distribution of this work must maintain attribution to the author(s) and the title of the work, journal citation and DOI.
Published under licence by IOP Publishing Ltd

coefficients of dry and viscous friction on the elastomeric absorbing devices' deformation. A comparison of the experimental data with the mathematical modeling results for the elastomer flow regime is given in [9]. It was shown that the laminar model has a smaller error in comparison with the turbulent one. Friction-polymer shock absorbers are considered in [10, 11]. In [10] an experimentally proved mathematical model of polymer elements is based on the concepts of hyperelasticity. The article [11] is devoted to the analysis of the influence of ambient temperature on the polymer shock absorbers' force characteristics. It is shown that the temperature decrease from +15 °C to – 60°C leads to a fivefold increase in the dynamic forces in the absorbing devices. The influence of gaps in the automatic couplers on longitudinal forces in homogeneous train inter-car connections is analyzed in the paper [12]. The attention of the article [13] is focused on the efficiency and relative simplicity of the machines' and mechanisms' dynamics modeling using the NX Motion Software, which allows to solve complex problems without possessing in-depth knowledge in the field of mechanics.

The final goal of the operation modeling of the automatic couplers' absorbing devices is the ability of correct determination of the longitudinal dynamics of the train as a whole. The analysis of longitudinal forces using the Matlab/Simulink program was described in [14]. In [15] it is shown that an increase in deceleration from 0.65 to 1.39 $\mathrm{m/s^2}$ leads to an increase in longitudinal forces maximum values by more than 2.3 times. At the same time, when the braking force growth time changes from 2 to 5 s, the maximal forces decrease by about 20%.

The paper [16] presents the simulation results of the homogeneous and heterogeneous train motion at starting and emergency braking. A comparative assessment of the arising longitudinal forces at unsteady movement modes is given for the train equipped with the spring-friction and with the friction-polymer absorbing devices. It is pointed that when using the last mentioned ones, the forces in the inter-car joints decreases by 10-20% on the average. In [17], the influence of the absorbing device initial tightening force on the train longitudinal accelerations at its transient motion is considered.

The purpose of the presented work is to investigate the influence of the absorbing device force characteristic shape on the value of the maximal longitudinal forces arising in the train inter-car joints.

2. Computer modelling

To solve the problem, there was created a scheme of a train motion model, shown in Figure 1. In the scheme the train is a chain of absolutely solid bodies – train cars moving rectilinearly without vertical vibrations and angular movements.

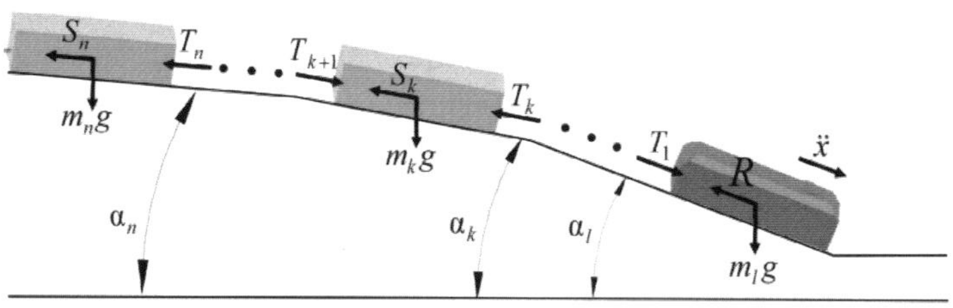

Figure 1. A scheme of a train motion model.

A system of differential equations of train motion is [18]:

$$\left.\begin{aligned} & m_l\ddot{x}_l - T_1 - m_l g\sin\alpha_l + R = 0; \\ & m_k\ddot{x}_k + T_k - T_{k+1} + S_k - m_k g\sin\alpha_k = 0;\ k = 1, 2, \ldots, n-1; \\ & m_n\ddot{x}_n + T_n + S_n - m_n g\sin\alpha_n = 0, \end{aligned}\right\} \quad (1)$$

where m_l – a locomotive mass; $\ddot{x}_l$, $\ddot{x}_k$ – longitudinal acceleration of the locomotive and the k-th car, respectively $(k = 1, 2, \ldots, n)$; T_k – forces, acting from the side of the inter-car couplings; g – acceleration of gravity; α_l, α_k – the slope of the way which the locomotive and the k-th car move along; R – an external force acting on the locomotive (at traction or electrodynamic braking), including the resistant to movement forces; S_k – force of resistance to movement of the k -th car (the calculated formulas of the resistant to the cars movement forces are given in [19]).

The longitudinal force between the cars simulates the operation of the couplers with absorbing devices. The force characteristic of the elastic-frictional absorbing device was adopted as a model of an automatic coupler device ad it can be determined in accordance with the expression [20]:

$$|T| = \begin{cases} T_L(q), \text{ if } q\dot{q} \geq 0; \\ T_U(q), \text{ if } q\dot{q} < 0; \\ |T| \in [T_L(q), T_U(q)], \text{ if } \dot{q} = 0, \end{cases} \tag{2}$$

where q, $\dot{q}$ – the device compression and compression rate correspondingly; $T_L(q)$, $T_U(q)$ – force characteristics during loading and unloading.

The last line in the expression (2) shows that for $\dot{q}=0$ the force T is controversially determined and it can have any value in the specified interval. The presented mathematical description simulates the operation of only one shock absorber. A model of an automatic coupler with gaps in the inter-car joints can be represented as a force defined by the expression:

$$T = \begin{cases} \left(C_L\left|q-\frac{\delta}{2}\right|^{n_L} + T_{0L}\right)\operatorname{sgn}(\dot{q}),\ q \in \left[-h-\frac{\delta}{2};-\frac{\delta}{2}\right) \cup \left(\frac{\delta}{2};h+\frac{\delta}{2}\right],\text{ if } q\dot{q} \geq 0; \\ 0,\ q \in \left[-\frac{\delta}{2};\frac{\delta}{2}\right]; \\ \left(C_U\left|q-\frac{\delta}{2}\right|^{n_U} + T_{0U}\right)\operatorname{sgn}(\dot{q}),\ q \in \left(-h-\frac{\delta}{2};-\frac{\delta}{2}\right) \cup \left(\frac{\delta}{2};h+\frac{\delta}{2}\right),\text{ if } q\dot{q} < 0, \end{cases} \tag{3}$$

where C_L and C_U – coefficient of the device rigidity at loading and unloading, respectively; n_L, n_U – exponent at loading and unloading; h – maximal stroke of the inter-car connection device; δ – total clearance in automatic couplings; T_{0L} and T_{0U} – initial protraction force of the absorbing device at loading and unloading.

A computer model corresponding to the presented mathematical description is implemented in the MSC.ADAMS software package. It allows to set the most important characteristics of the train's movement for research: the number and weight of cars, the gap and the elastic-frictional characteristics of the inter-car joints, the track section profile. To estimate the influence of the absorbing device force characteristic's type on the maximum longitudinal forces in the train three variants of characteristics were considered: rigid, linear and soft one.

Lines of loading and unloading are defined from the equation:

$$T(q) = Cq^n + T_0, \tag{4}$$

where C – stiffness coefficient of the device; n – exponent depending on the design of the device; T_0 – force of the initial protraction.

The unloading line is assumed to be the same for the three variants of absorbing devices, and the load lines are selected in accordance with (3) so that the maximum energy intensity of the inter-car joint absorbing devices is the same and equal to 207 kJ. Line equations have the following form:

– for the soft characteristic $T(q) = 4260q^{0.6} + 200$;
– for the linear characteristic $T(q) = 10600q + 200$;
– for the rigid characteristic $T(q) = 88000q^2 + 200$;
– for the unloading line $T(q) = 23000q^2 + 50$.

In the presented equations the stroke of the devices q is assumed in meters.

The maximum values of the inter-car joint devices stroke and the coefficient of irreversible energy absorption are 180 mm and 0.74, respectively. Initial protraction forces are $T_{0L} = 200$ kN, $T_{0U} = 50$ kN. All parameters correspond to the absorbing device class T1. It should be noted that ceteris paribus, the force reaches 3 MN, for linear – 2.1 MN, and for soft – 1.7 MN for the case of full compression of the device with a rigid characteristic.

In this work, it is considered the movement of a homogeneous freight train of weight equal to 8000 tons (100 train cars, 80 tons each) along a straight horizontal section of the rail track. The investigated transient modes are: starting, traction force disengaging, braking by the locomotive electrodynamic brakes, braking by the locomotive after disengaging the traction. In all cases, the gaps in the inter-car joints are taken equal to 65 mm, this corresponds to the values for cars after their long operation [21]. In the calculations, the values of external forces (traction or braking one), acting from the locomotive are taken equal to 500 kN, and they change from zero to maximum in 20 seconds.

3. Simulation results

The process of starting off is considered for the case of a pre-compressed train. Graphical dependencies of the maximum longitudinal forces in the train and their distribution along the train length are shown in Figure 2 (in Figure 2 and further, k is the serial number of the train's inter-car joint).

With a smooth increase in traction force, the tendency for the dynamic forces development and growth occurred in the inter-car joints with soft characteristics. This fact is probably caused by a high rate of force growth at the compression of the absorbing device, which leads to a decrease in the flexibility of the inter-car connection as a whole and an increase in the impact forces propagated along the train length under tension.

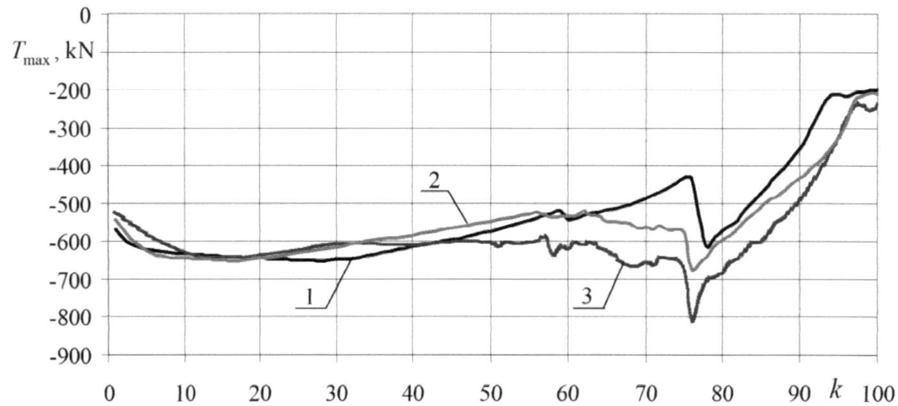

Figure 2. Distribution of maximum forces along the train length at smooth starting off and rigid (1), linear (2) and soft (3) force characteristic of the inter-car joints.

The greatest impact forces with soft characteristics arose in the region of the train free end and exceeded the friction force by 60%. In the train with rigid characteristics, the maximum forces are elastic and distributed in the first half of the train, while the dynamic forces do not reach large values. For the three cases, the elastic forces in the inter-car connections exceeded the starting force by 15%.

The process of the locomotive electric braking is considered for two cases: the compressed and the stretched train. The initial braking velocity is 30 km/h. The results of the stretched train braking (Figure 3) are similar to the results obtained for case pre-compressed train starting off. Smooth braking leads to the impact forces appearance in the inter-car joints with soft characteristics, the largest of these forces exceed the braking force by 30%. It should be noted that the soft characteristic leads to the growth of not only shock compressive forces, but also tensile ones arising in the free end of the train at its pulling.

At smooth braking, the elastic damped oscillations occur in the train and the maximal compressive forces at it do not exceed the external influence value. The compressive forces of shock nature are not observed. All types of shock absorbers are characterized by the appearance of insignificant tensile forces in the inter-car joints of the first ten cars.

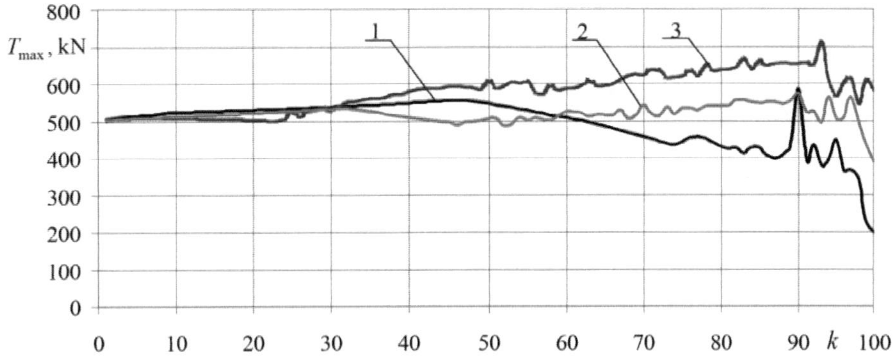

Figure 3. Maximal compressive forces along the stretched train length at its smooth braking for the cases of the rigid (1), linear (2) and soft (3) force characteristic of the inter-car joint.

There is investigated the process of a smooth transition to the locomotive braking mode after a sharp traction disabling. The train velocity at the beginning of the transition process is assumed equal to 30 km/h, the forces of traction and braking – 500 kN. The calculation results are presented in Figure 4.

As in previous cases, the greatest impact loads occurred in the train equipped with shock absorbers with soft force characteristics. At the same time, the maximal forces in automatic couplings of the train second half exceed the braking force by 30%. In trains with rigid and linear characteristics, the longitudinal forces exceed the braking force by less than 20%.

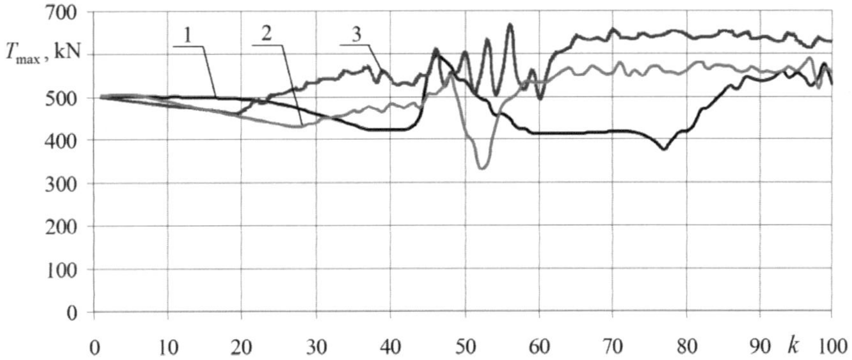

Figure 4. Distribution of the maximal compressive forces along the length of the train at its smooth braking after disabling the traction force for the cases of the rigid (1), linear (2) and soft (3) force characteristic of the inter-car joint.

The presented calculation results showed that a smooth change in the external force over 20 seconds at various transient motion regimes leads to an increase in dynamic impact forces, appearing due to the presence of the not eliminated gaps at the initial moment of force application. The maximal values arise in the trains equipped with absorbing devices with soft force characteristics. For the further analysis of the influence of the force characteristic type on the longitudinal forces in the train, there are considered the transient movement modes of the train at sharp changes in the external forces.

The transition from the traction to braking mode is investigated. The initial data is the same as for the case of a smooth force change. The only difference is that the locomotive braking power is applied sharply. The calculation results are shown in Figure 5. The highest values of forces, reaching 1.75 MN, arise in the train with rigid characteristics. The least compressive (about 1 MN) and the highest tensile (up to 0.5 MN) forces are observed in the case of soft characteristics of the inter-car connection.

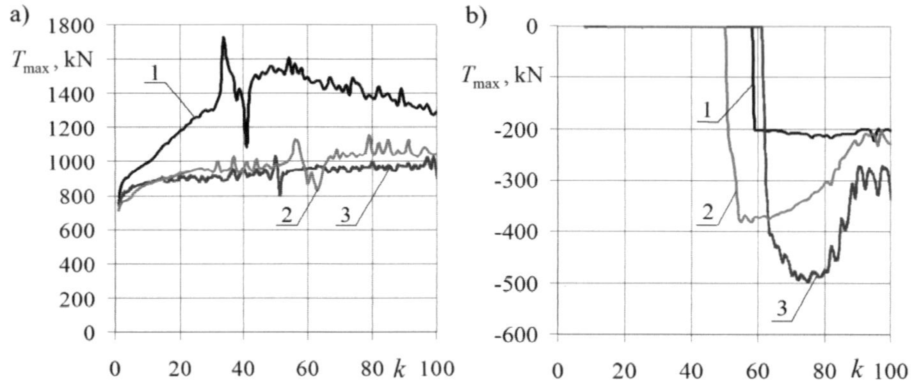

Figure 5. Distribution of the maximal compressive (a) and tensile (b) forces along the train length at the sharp traction force disabling for the cases of the rigid (1), linear (2) and soft (3) characteristics of the inter-car joints.

The similar results are obtained for the cases of the sharp starting off of the compressed train and the stretched train emergency braking:

- at a sharp starting off there is a significant increase in impact forces, exceeding the traction force by two times or more for the all types of the inter-car connections. The maximal values of these forces arise in the free train end. On average, the forces in the inter-car joints with a rigid characteristic are greater than the others by 100 kN;
- at an emergency (sharp) braking, the dynamic forces are 1.8-2 times higher than the braking force; the greatest ones arise in the trains with rigid force characteristics.

The longitudinal loading of the compressed train at its emergency braking is investigated. It should be noted that the dynamic forces occur in the first two thirds of the train length and the forces exceed the braking force by up to 42% for the case of shock absorbers with rigid power characteristics. With linear and soft characteristics, these forces almost do not exceed the applied external force.

The transient process is also caused by switching to the idle mode when the driver controller is turned off. After the traction force disabling, the stretched elastic elements begin to compress, this leads to the appearance of the elastic vibrations in the train. The traction force disabling is shifted by 30 seconds for each case. The smallest compressive forces are observed between cars with soft force characteristic, but in this case, tensile stresses arise in the last cars and reach 450 kN.

4. Conclusions

Based on the presented results, it can be concluded that there are both positive and negative features of all the considered force characteristics of the automatic couplers' absorbing devices. The positive features of the shock absorber with a rigid force characteristic include the smoothness of the forces transfer between the cars with a slow change in the external force (traction or braking one) and a low tendency to the tensile forces appearing and increasing at the free train end pulling. With a sharp increase in the control action in the train equipped with such shock absorbers, the maximal longitudinal forces arise in comparison with other absorbing devices.

The soft characteristic of the absorbing device leads to a significant impact forces reduction with a sharp change in external influence, leading to the rapid longitudinal elastic vibrations damping in the train. At the same time, the main disadvantages of its operation are: the growth of dynamic forces in the train with a smooth change in the external influence, as well as the facilitation of the tensile forces appearing and increasing at the free train end pulling during transient movement modes.

According to the results of the performed investigation, the linear form of the force characteristic reflected the combination of the properties of hard and soft ones.

The presented results can be implemented in the design and modernization of absorbing devices of automatic couplers and in the composition of various types of cars in trains.

References

[1] Solonenko V G, Sekerova Sh A and Adilhanov E G 2010 Analysis of modern absorbing devices' design *Bulletin of the Kazakh Academy of Transport and Communications named after M. Tynyshpaev* **6** 67 pp 37–42 [in Russian]
[2] Boldyrev A P and Gurov A M 2014 Efficiency of using high-energy draft gears on freight cars *Transport of Russian Federation* **3** pp 43–45 [in Russian]
[3] Boldyrev A P, Stupin D A and Gurov A M 2018 The main trends in the development and implementation of new designs of draft gears for automatic couplers of freight cars *Bulletin of the Petersburg University of Railways* **1** pp 30–35 [in Russian]
[4] Pugi L, Malvezzi M, Papini S and Vettori G 2013 Design and preliminary validation of a tool for the simulation of train braking performance *J Mod Transp* **21** 4 pp 247–257
[5] Sakharau P A and Shimanovsky A O 2019 Evaluation of the influence of the characteristics of intercar connections in a train on the value of longitudinal forces at electric braking *Mechanics. Research and innovation* **12** pp 171–181 [in Russian]
[6] Wu Q, Spiryagin M and Cole C 2015 Advanced dynamic modelling for friction draft gears *Vehicle System Dynamics* **53** 4 pp 475–492
[7] Wu Q, Cole C, Luo S and Spryagin M A 2014 Review of dynamics modelling of friction draft gear *Vehicle System Dynamics* **52** 6 pp 733–758
[8] Andriyanov S S 2015 Mathematical model of the elastomeric draft gear *World of transport* **13** 2 pp 214–218 [in Russian]
[9] Vojnovsky M G 2015 Mathematical modeling of the power characteristics of an elastomeric shock absorber *Transport of Russian Federation* **3** pp 38–40 [in Russian]
[10] Govorov V V, Halaev A A and Boldyrev A P 2009 Modeling and calculation of a perspective automatic coupler shock absorber with elastic elements made of modern polymer materials *New materials and technologies in mechanical engineering* **10** pp 16–20 [in Russian]
[11] Mickoski H, Mickoski I, Djidrov M and Zdraveski F 2018 Mathematical Model of New Type of Train Buffer Made of Polymer Absorber – Determination of Dynamic Impact Curve for Different Temperatures *Machines* **6** 4 47 pp 1–12
[12] Shimanovsky A O and Sakharau P A 2019 Influence of clearances in automatic couplers on longitudinal forces in intercar connections of a homogeneous train *Mechanics of machines, mechanisms and materials* **2** pp 42–50 [in Russian]
[13] Nikitchenko A, Artiukh V, Shevchenko D and Murgul V 2017 Modeling of Operation of Elastic-Frictional Draft Gear by NX Motion Software *Procedia Engineering* **187** pp 790–796
[14] Jha P K and Gokhale S S 2018 Validation of longitudinal dynamics of Gatimaan Express model in Matlab/Simulink *Mechanics & Industry* **19** 3 307 pp 1–15
[15] Fang K, Wu P and Wang Q 2015 Research on Dynamic Performance of Coupler for Rescue EMUs. *Advances in Computer Science Research* **12** pp 1173–1178
[16] Naumenko N E, Hizha I Yu and Nikitchenko A A 2009 Assessment of the influence of the power characteristics of perspective draft gears on the dynamics of a freight train under non-stationary modes of movement *Technical Mechanics* **2** pp 27–31 [in Russian]
[17] Belyaev V I 2013 Standards and new design of absorbing devices for automatic SA-3 couplers *Rail Technology Review* **53** 3-4 pp 43–46
[18] Shimanovsky A and Sakharau P. 2019. Investigation of the Longitudinal Track Profile Influence on the Forces Acting in the Train Inter-car Connections Using the MSC.ADAMS Software *CEUR Workshop Proceedings* **2353** pp 555–569
[19] *Rules for traction calculations for train operation*: approved by decree of JSC «RRW» N 867p from 12.05.2016 (Moscow: JSC «RRW») [in Russian]
[20] Manashkin L A, Myamlin S V and Prihod'ko V I 2007 *Vibration dampers and shock absorbers for rail vehicles (mathematical models)* (Dnepropetrovsk: Art-press) [in Russian]
[21] Blochinas E, Dailydka S, Lingaitis L and Ursuliak L 2016 *Nestacionarieji ir kvazistatiniai geležinkelio traukinių judėjimo režimai* (Vilnius: Technika)

IOP Conf. Series: Materials Science and Engineering **985** (2020) 012028 doi:10.1088/1757-899X/985/1/012028

Methods and results of evaluating the dual-power electric train crew elements service life

V Horobets[1,3], O Sablin[1], E Fedorov[1], E Horobets[2], O Bolotov[1] and O Jangulova[1]

[1]Dnipro National University of Railway Transport named Academician V. Lazaryan, 49010, Ukraine, Dnipro, str. Lazarian, 2
[2]SPA «SPI Ukrtransacad», 49107, Ukraine, Dnipro, Zvezdny boulevard, 1, building 1, of. 3

[3]vgor5650@gmail.com

Abstract. The article assesses the long-term strength of the vibration damper assembly of a double power transmission of electric trains manufactured by «Hyundai-Rotem Company». The technique is described and the analysis of the service life of the elements of the chassis of the electric train is made. The presence and nature of micro cracks in the loaded elements of the vibration damper assembly was evaluated after a full cycle of vibration bench tests. The influence of vertical vibrations of the electric train body on the level of dynamic load and the service life of the bearing elements of the vibration damper assembly is considered. The service life of the vibration damper at the end of the service life has been estimated by comparing working hours and "statistical playback". The calculation of the resource by the method of "statistical reproduction" confirms the possibility of operating the vibration damping module (second modernization) for 50 years from the date of construction. The influence of the vibration damper position on its dynamic loading is checked. The main stages and sequence of tests for estimation the resource of the longitudinal vibration damper module of a double-fed electric train are also considered, including methodological, calculated and experimental confirmation of the reliability and durability of the unit, which was studied as part of a comprehensive work.

1. Introduction

This article is devoted to a review of the work, carried out by specialists of the Dnipro National University of Railway Transport named Academician V. Lazaryan (DNURT) in assessing the long-term strength of the vibration damper assembly of the dual electric power train.

The purpose of the research was:

- assessment of the presence and nature of microcracks in the loaded elements of the vibration damper assembly after a full cycle of bench vibration testing;
- evaluation and consideration of the influence of the vertical oscillations of the body of the electric train on the level of dynamic loading and the life of the bearing elements of the vibration damper assembly;
- estimation of the end-of-life service life of the vibration damper by methods of comparing the operating hours and "statistical playback";
- formation of conclusions regarding the quality of construction and the durability of the vibration damper assembly.

Content from this work may be used under the terms of the Creative Commons Attribution 3.0 licence. Any further distribution of this work must maintain attribution to the author(s) and the title of the work, journal citation and DOI.
Published under licence by IOP Publishing Ltd

- verification of the conformity of the materials [12,13] of the units of increased reliability of the pivot beam of the interregional electric train (hereinafter referred to as EP) of the Hyundai-Rotem double feeder (hereinafter referred to as the vibration damper assembly) used in conducting bench vibration tests and in operation

The vibration damping unit of the bogie of the electric train (Figure 1) is intended to improve the dynamic performance of the crew, namely to prevent the development of resonance oscillations of wagging when traveling at critical speeds. The hydraulic damper 1 02R-2217-001 [1] (Figure 1, a) is connected to the frame of the bogie and the body bolster. The bracket of the damper 2 on the frame of the body is connected to it by bolted connections and is located on the tide at the side of the pivot beam. The tide is strengthened with kerchiefs 3 (Figure 1, b).

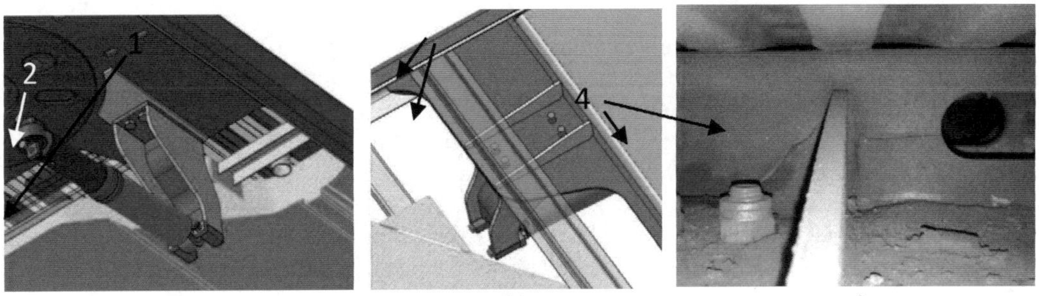

a) b) c)

Figure 1. Unit of vibration damping device:
a) view from damper side; b) top view; c) crack of the side wall of the body bolster.

After a short (about 2 years) operation period, fatigue cracks 4 (Figure 1, c) were found [3] on the side wall of the pivot beam. The appearance of cracks, in addition to the intense dynamic loading of the node, could be due to the presence of a stress concentrator near the kerchiefs and residual welding stresses.

In accordance with the above, the first unit upgrade was proposed (Figure 2).

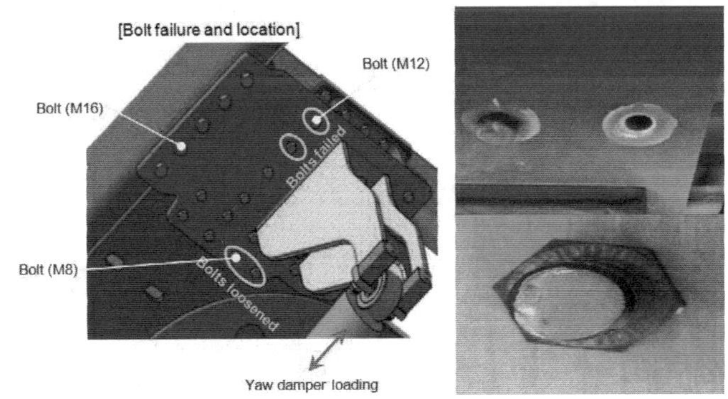

a) b)

Figure 2. Unit vibration damper wobble, upgrade 1:
a) view from the damper side; b) failure of bolted connections.

This modernization (according to the results of the strength tests carried out by the DNURT [2]) has improved the dynamic loading of the unit. However, there was a problem associated with loosening the bolted joints (Figure 1.2, b) and the subsequent cutting of the bolts.

At the moment, the second upgrade of the high reliability unit has been proposed (Figure 3).

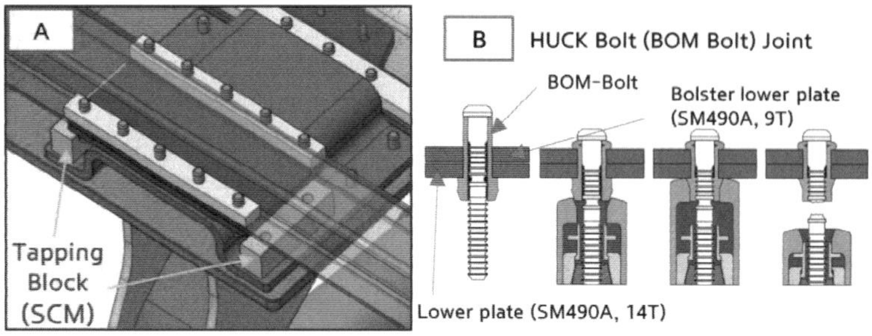

a) b)

Figure 3. Unit of vibration damping oscillator, modernization 2:
a) top view; b) bolted connections.

Its meaning lies in the use of self-locking bolted joints in conjunction with the replacement of nuts with rigid centering blocks (tapping block) and bolts with a crimping ring (huck bolt).

Qualitative indicators of this modernization are the object of research of this work.

Conclusions: The vibration damping unit of the bogie turned out to be a weak place for the crew part of the EP design, which required its two-stage modernization.

2. Methods and composition of research objects strength and resource studies

The resource estimates are obtained by calculation and compared with the unit's calculated resource as indicated below.

The ratio of the resource of the investigated design to the resource of the analogue (when performing vibration tests) d is found as the inverse relation of the developments:

$$d = \frac{D_d}{D_a} \quad . \tag{1}$$

In the indicated expression: D_d is the specific (per unit of distance) operating time of the design in use; D_a – specific (per unit of time) operating time of the module – analog in the testing process. Then the resource of the structure under study will be determined as follows:

$$R_d = \frac{d \cdot T_a}{L_d \cdot R_a}, \tag{2}$$

where L_d is the average annual mileage of the structure under investigation during operation; T_a – time tests of analog; R_d – resource, expressed in terms of the lifetime of the structure under study; R_a – the resource gained during the test.

To carry out "statistical playback" a model of endurance is used, generally corresponding to the model adopted in the Norms [4]. In this case, as a cyclic load, a harmonic perturbation is adopted with the most unfavorable statistical parameters of the Gaussian distribution (m is the average value of the perturbation, σ is its standard deviation) obtained during strength tests [2].

3. Baseline data for research. Visual inspection of the presence of visible cracks

Bench vibratory tests of the damper damping unit EP [7] are divided into two stages: the first stage – cyclic loading tests from 0 to 9 000 000 cycles; the second stage is cyclic loading tests from 9 000 000 to 10 000 000 cycles.

Before the second stage of the test, a visual PT control of the presence of visible cracks was performed. The order of this and subsequent (final) control did not contradict the requirements of national regulatory requirements [6]. The control was carried out with the participation of the representative of the HRC Testing Center, Mr. Do Yoon Kim (certificate DY-451/G, valid until Nov.6 2020). The control points are shown in Figure 4.

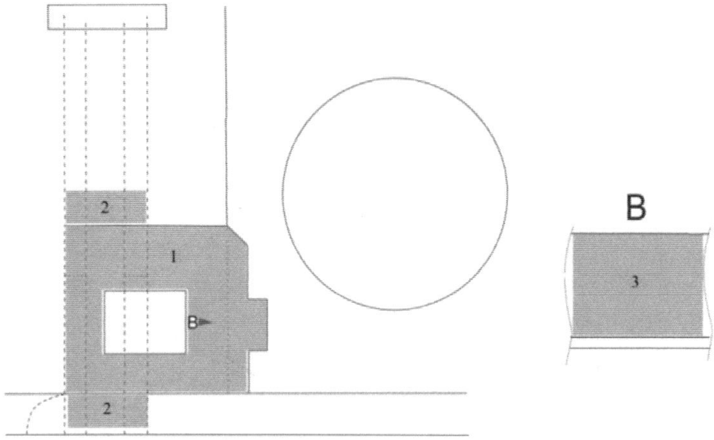

Figure 4. The places of penetrant control.

In Figure 4: zone 1 – all areas of bolted connections of the support plate of the wobbling damper support bracket; 2 – transverse beam in places near the support plate; 3 – side wall of the pivot beam in the vicinity of the support plate of the vibration damper support.

After the procedures for the preparation of control zones were completed, an intermediate visual control of the presence of cracks in the load-bearing elements of the structure was made (Figures 5, 6).

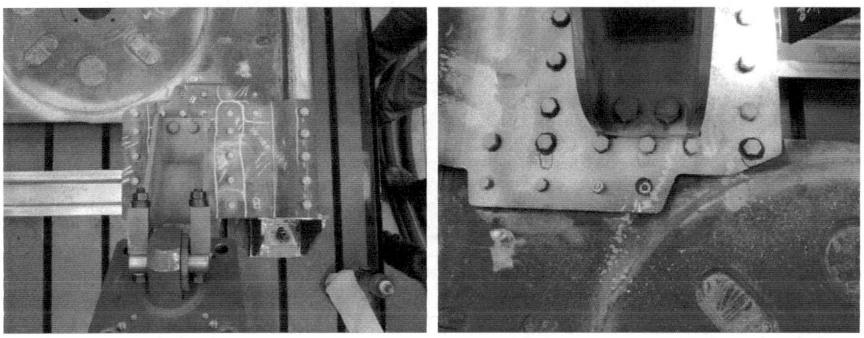

Figure 5. Marking of places and results of intermediate penetrant control zone 1.

Figure 6. Marking of places and results of intermediate penetrant control of zones 2 and 3.

4. Evaluation of the operating times of the load-bearing elements of the vibration damping assembly of the bogie during strength and bench testing

During the dynamic strength tests [2] dynamic processes were registered according to the scheme of location of the sensors, shown in Figure 7. This sensor placement scheme was also used when performing wall-vibration (resource) tests at the "Hyundai-Rotem Company" Test Center with the participation of representatives of the DNURT.

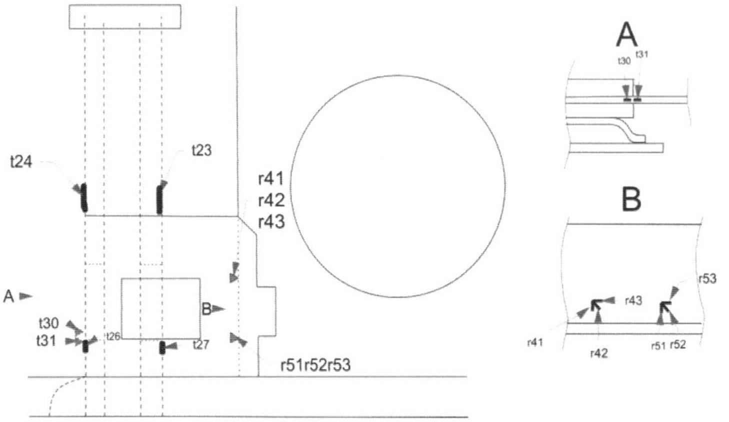

Figure 7. Scheme of the distribution of strain gauges during dynamic strength and bench vibration (resource) tests.

The most loaded element is the transverse beam in the area of connection with the support plate of the vibration damper bracket on the body frame. This makes it possible to use its stress state (sensor t30) as an indicator of the operating time for running strength and bench vibration testing.

A typical example of the dynamic loading of this point, determined during dynamic strength tests, is shown in Figure 8.

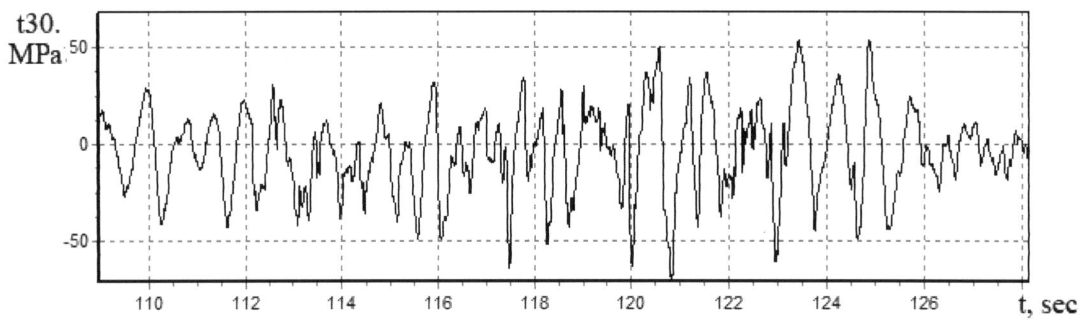

Figure 8. A typical example of the process of dynamic loading of the transverse beam of the body frame (experiment 6, entry 29).

To evaluate the level of dynamic loading, three records were selected which are characterized by the highest level of process variance: experiment 4, record 26 and 29 and experiment 6, record 29. These processes were schematized according to the rain method according to [9]. The results of determining the operating time D_d – the specific (per unit distance) operating time of the structure under investigation during the operation are given in Table 1.

Table 1. Work during realization of different experiments.

Experiment / Record	6 / 29	4 / 29	4 / 26
Work on 1 km, MPa9	1.16^{14}	1.22^{13}	6.3^{13}

The coefficient of the power-law curve of endurance m was assumed according to [10] as $m = 16 / K_{\sigma}$, where $K_{\sigma} = 1.72$ was taken from [7] and amounted to $m = 9.3$.

5. Research results. Analysis of the results of intermediate and final visual inspection of the presence of visible cracks

After the work on an intermediate visual control of the presence of visible cracks, no contrasting penetrate residues were found in the regions of the individual bolts of the base plate, which were additionally examined (Figure 9) using a 25-500x portable optical computer microscope U500X.

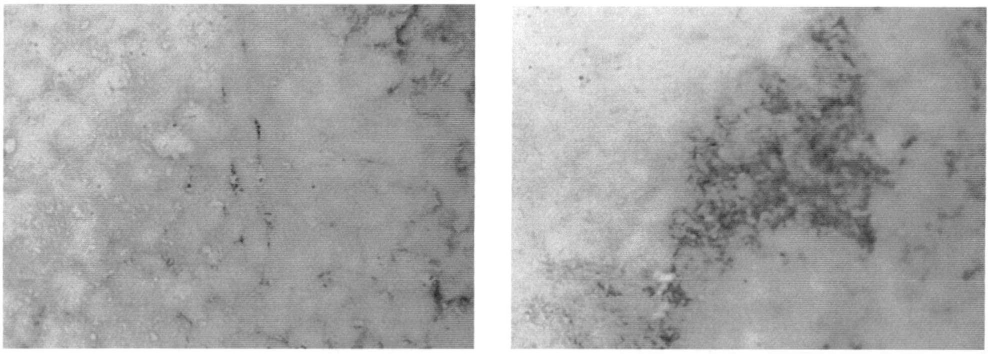

Figure 9. Additional investigation of possible locations of cracks with the use of an optical computer microscope U500X.

A study of all the selected control zones showed no visible cracks. This allowed us to proceed to the second stage of the cyclic loading of the structure without any observations. After the completion of the second test phase, the final PT control of the presence of visual cracks was repeated. After performing all the necessary operations, no cracks were found (examples – Figure 10).

So, after two stages of cyclic loading, visible cracks in the loaded structural elements are not revealed. The vibration damping unit of the wagon of the second modernization truck sustained a cyclic loading with a maximum operational amplitude of the load (9 kN) in the volume $N0 = 10\ 000\ 000$ cycles. This volume of tests is taken into account for the total operating time when assessing its final service life.

Figure 10. The results of the final penetrant control.

6. Evaluation of the influence of vertical dynamics on the loading of the vibration dampener assembly of oscillations of a bogie of an electric train bogie

Below, carried out special calculations of the stress-strain state of the vibration damper assembly [8], namely for cases of horizontal load application, as well as with a deviation of 10 in the positive and negative directions.

The results of the calculations are given in Table 2 and in graphical form – in Figure 11.

Table 2. Maximum stresses in the body structure of the vibration damper assembly.

Cases of load	Description	Max. Stresses, MPa
Case 1	Load of damper 9 kH with angle –10° for normal condition	66
Case 2	Load of damper 9 kH with angle +10° for normal condition	60
Case 3	Load of damper 9 kH with angle 0° for normal condition	63

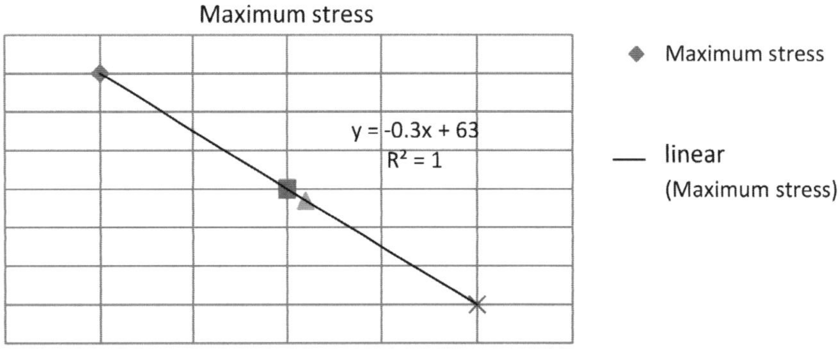

Figure 11. Dependence of the maximum voltage in the construction on the angle of inclination of the damper.

Taking into account the installation of a damper with a positive inclination angle of 1 [1] and a linear dependence of the maximum stresses in the unit at small angles of the damper's inclination, we determine the coefficient of influence of the vertical dynamics on the stress level in the structure.

The value of $\Delta_{\sigma\ \max}$ (according to the data in [5]) is 3.3 MPa. The coefficient K_{vi} of the influence of vertical dynamics is taken as:

$$K_{vi} = \frac{1 - \Delta_{\sigma\ \max}}{\sigma_{\text{mid}}} = 0.947, \tag{3}$$

where $\sigma_{mid} = 62.7$ MPa is the average voltage (in the absence of vertical displacements of the frame of the car body with respect to the frame of the telegraph).

As can be seen from Figure 10, the vertical movement of the carriage frames relative to the body frame can both increase and decrease the maximum stresses in the structure. This depends on the ratio of the frequencies and phases of oscillation of wobble and vertical vibrations. Given the complexity of this analysis, we will assume that the frame of the bogie is constantly in an unfavorable position from the standpoint of stresses in the construction.

The effect of vertical dynamics on the increase in stresses in the unit of the vibration damping oscillator is taken into account by a constant coefficient $K_{vi} = 0.947$, which assumes a permanent finding of the frame of the bogie in an unfavorable position with respect to the body frame. This correction factor will be used in calculating the design life by artificially reducing the stress level during the bench test.

7. Evaluation of the end-life of the load-bearing elements of the vibration damping unit of the bogie

The method of direct comparison of the developments [11] is based on the direct use of the power endurance curve and on the fact that the total operating time C of the $\sigma^m N = C$ form [14] is constant under the condition of linear accumulation of damages in the structure.

When carrying out bench vibration tests, the damper assembly was loaded with the maximum operating force from the vibration damping side ($F = 9$ kN) with a frequency $f = 5$ Hz. The voltage at point $t30$ (Figure 7) during the bench vibration testing was [7] $\sigma_{t30} = 65.5$ MPa. At the same time, the total operating time, obtained by the design, is:

$$D_a \cdot T_a = \frac{D_a \cdot N_0}{f} = \frac{(\sigma_{t30} K_{vi})^m N_0}{f}, \tag{4}$$

where: D_a – working out of the power form (see previous paragraph) per unit time; T_a – time of testing; N_0 – according to the conclusions of previous paragraph; f – according to previous paragraph; σ_{t30} – according to [7]; K_{vi} – according to the formula (3); m – according in accordance with the discussion of test results.

The results of estimating the site resource by comparing the developments are given in Table 3. The design of the wobble damper module (the second upgrade of enhanced reliability) has a significant $n_R = 210$ run time relative to the design life of 50 years.

The main idea of the method of "statistical playback" is the mathematical simulation of the process of accumulation of damages. At the same time, the input of the mathematical model receives a harmonic signal of the fundamental oscillation frequency f_e with the normal distribution of the amplitudes, while the statistical characteristics of the distribution coincide with the operational loading.

When using steel similar to SM 490A for a crossbeam, we will assume that its endurance characteristics have the same proportion [15] compared with SUS301L steel as their yield strength, i.e. 0.793. Conditionally, we will reduce the operating time when testing 0.7939 times. For this case, table 3 takes the form shown below.

Table 3. Evaluation of site resource by direct comparison of developments.

Experiment No.– Record No.	6-29
Operating time per 1 km D_a, MPa^{9}km^{-1}	$1.16 \cdot 10^{14}$
Operating time, tests $D_a \cdot T_a$, MPa^9	$4.38 \cdot 10^{23}$
Ratio of operating time $D_a \cdot T_a / D_d$, (residual mileage, thousand km.)	3778080
Yearly mileage L_d, residual mileage, thousand km.	360
Resource R_1 (R_d), years	10494
Life length 50 years D_{d50}, MPa^9	$2.085 \cdot 10^{21}$
Run-up reserve $n_R = (D_a \cdot T_a) D_{d50}$	210
Reserve reduced to linear dependence on stresses $n_\sigma = n_R^{1/m}$	1.78
Experiment No. – No. of recording	6-29
Operating time, tests $D_a \cdot T_a$, MPa^9	$4.38 \cdot 10^{23}$
Ratio of operating time $D_a \cdot T_a / D_d$, (residual mileage, thousand km.)	468522
Resource R_1 (R_d), years	1300
Life length 50 years D_{d50}, MPa^9	$2.085 \cdot 10^{21}$
Run-up reserve $n_R = (D_a \cdot T_a) D_{d50}$	26
Reserve reduced to linear dependence on stresses $n_\sigma = n_R^{1/m}$	1.44

Thus, when using steel SM 490A, the design can be operated for 50 years.

The "rain" treatment [9] of the chosen most stress-laden implementation of the process of dynamic stresses at the point $t30$ gave the distribution of their amplitudes and probabilities, shown in Table 4.

The average amplitude of the dynamic stresses [4] is $M_{t30} = 12.68$ MPa, the standard deviation is $s_{t30} = 11.31$ MPa. Thus, the calculated amplitude according to the Norms [4] is $M_{t30} + 2s_{t30} = 35.3$ MPa.

IOP Conf. Series: Materials Science and Engineering **985** (2020) 012028 doi:10.1088/1757-899X/985/1/012028

To determine the dominant oscillation frequency, a spectral analysis of the loading process of the damper assembly was carried out, and the spectrum of the oscillations is shown in Figure 12. The dominant frequency of the oscillations in this case was $f_e = 1.4$ Hz.

Table 4. Distribution of amplitudes of dynamic stresses.

Amplitude of the discharge of strain, MPa	Probability
5.0	0.503086
10.0	0.159122
15.0	0.108368
20.0	0.065158
25.0	0.049383
30.0	0.038409
35.0	0.024691
40.0	0.021262
45.0	0.011660
50.0	0.007545
55.0	0.006859
60.0	0.001372
65.0	0.002058
70.0	0.000686

The design response to a single overload is defined as:

$$\sigma_{-1}^{+} = \sqrt[m]{\sigma_{-1}^{m} - \frac{\sigma_v^{m}}{N_0}} \uparrow \sigma_v > \sigma_{-1},, \qquad (5)$$

where σ_{-1}^{+} – the endurance limit of the structure after the action of a single overload; σ_{-1} – the endurance of the structure to the action of a single overload; σ_v – current amplitude of the dynamic voltage cycle: m – an indicator of the degree of the endurance curve; N_0 is the test base, on the basis of which the initial endurance limit of the structure is estimated.

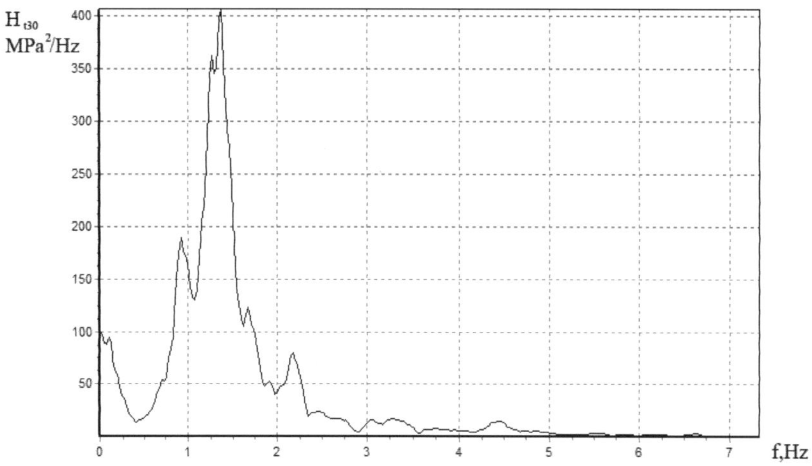

Figure 12. The spectrum of the dynamic loading process.

The initial data for performing "statistical playback" is given in Table 5.

Table 5. Initial data for estimating the final resource by the "statistical playback" method.

№	Indicator name	Value
1	Estimated amplitude of dynamic stresses, MPa	35.3
2	Assessment of endurance limit on the basis of tests N_0, MPa	65.5
3	Base of tests N_0	10^7
4	Endurance factor N_0	1.85
5	Estimated average annual mileage, thousand km	360
6	Average speed, km/h*	100
7	Estimated operating time per year, thousand h	3.6
8	Dominant frequency of oscillation f_e, Hz	1.4

*Note: the average speed estimation was obtained for InterCity + trains by analyzing the data of the official Internet resources http://intercity.uz.gov.ua/ and http://booking.uz.gov.ua/ru/

The results of estimating the resource of the estimated "statistical playback" are shown in Figure 13 and in Table 6. It is believed that the construction begins to break down when the endurance limit is reduced to $\sigma_{-1}^{+} = 0$.

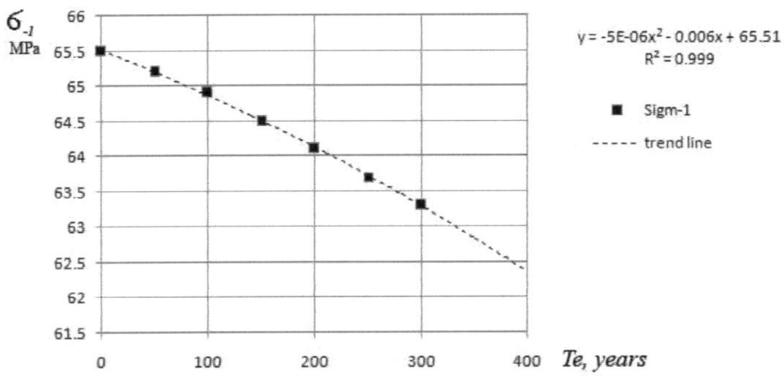

Figure 13. Dependence of lowering the endurance limit on life.

Table 6. Estimate the site resource by the method of "statistical playback".

№	Indicator name	Value
1	Initial value of the endurance limit, MPa	65.5
2	Estimated service life for 50 years, thousand h	180
4	Maximum fixed endurance limit, MPa	110.3
5	Duration to the first exceedance of the current endurance limit, h	0.3
6	The value of the endurance limit after the estimated service life, MPa	65.2
7	Decrease in endurance limit after estimated service life, %	0.45

Thus, the calculation of the resource by the method of "statistical playback" confirms the possibility of operating the vibration damping module (second modernization) for a period of 50 years from the construction.

It is necessary to pay special attention to the fact that all the results given are obtained for normal operating conditions of EP and can not be extended to cases of emergency situations and other incidents leading to a critical disruption of the strength of the elements of its crew part.

Main conclusions

The complex of the conducted researches allows to draw the following conclusions:

1. After carrying out two stages stand cyclic loading of the bogie wagging oscillation damper (the second modernization), visible cracks in the structure loaded elements are not revealed.

2. The influence of vertical dynamics on the increase of stresses in the unit of the vibration damper is about 5 percent when the frame of the ET carriage is constantly in an unfavorable position relatively to the main frame of carriage ET.
3. Calculation of the finite value of the modernized vibration damping module resource (second modernization with increased reliability) by two methods confirms the possibility of its exploitation during 50 years from building. This conclusion does not depend of of cross-beams material (SUS301L or SM 490A steel).
4. The obtained results can not be extended to cases of emergency situations and other incidents leading to a critical disruption in strength of ET crew part elements.

References

[1] Service and Maintenance Manual of KONII Dampers for Rotem 2014 Ukraine EMU 90R project
[2] DIIT-80.210-106-16 / ITs Protocol of acceptance running tests of the modernized unit of the interregional electric double-feed train "Hyundai-Rotem" 2017
[3] DSTU ISO 3057: 2016 Non-destructive testing. Metallographic replica method for surface inspection (ISO 3057: 1998, IDT)
[4] Standards for calculating and assessing the strength of load-bearing elements and dynamic qualities of the undercarriage of a multi-unit rolling stock of the railways of the Ministry of Railways of the Russian Federation of 1520 mm gauge 1997
[5] PM 36.001-17 / ONIL TSBT Methodology for researching the quality indicators of the upgraded unit of increased reliability of the pivot beam of the car of the interregional electric train of double power supply "Hyundai-Rotem" 2017
[6] TsT-0215 Instructions for color flaw detection of critical parts and assemblies during repairs: VND UZ 32.2.04.009-2012CT 2012 Ukrainian Railway
[7] DIIT -91.375-17/ONILDPPS Protocol of acceptance running strength tests of modernized unit of interregional dual-system electric train «HYUNDAI-ROTEM» 2017
[8] REDE208202 Finite Element Analysis for evaluation of vertical damping effect Ukraine CS2 EMU Hyundai-Rotem Co
[9] GOST 25.101-83 Methods for schematization of random processes of loading machine elements and structures and statistical presentation of results 1983
[10] Standards for the calculation and design of railroad cars of the Ministry of Railways with a gauge of 1520 mm (non-self-propelled) VNIIV-VNIIZhT 1983
[11] Gorobets V, Pilipenko S and Batyushin I 2011 Scientific substantiation of the assessment of the resource of load-bearing structures of traction rolling stock *Bulletin of the Dnepropetrovsk certification body for railway transport* **4** 1 pp 56–68
[12] JIS G 3106:2004 Rolled Steel for Welded Construction: Japanese Industry Standard, 2004
[13] JIS G 4305: 1999 Cold rolled stainless steel plates, sheets and strips: Japanese Industry Standard, 1999
[14] Stussi F. Die teorie der Dauertesigkeit und die Versuche von August Wöhler/Mitt TKVSB **13** pp 15–25
[15] Heywood R 1962 *Designing against fatigue* (Chapman and Hall ltd)

IOP Conf. Series: Materials Science and Engineering **985** (2020) 012029 doi:10.1088/1757-899X/985/1/012029

Study of strength characteristics of the long wheelbase flat cars

D V Fedosov-Nikonov[1], A O Sulym[1,3], V V Ilchyshyn[1], O M Safronov[1] and M B Kelrikh[2]

[1]State enterprise "Ukrainian Scientific Railway Car Building Research Institute", 33, I. Prykhodko St., Kremenchuk, 39621, Ukraine

[2]State University of infrastructure and technologies, 19, Ivana Ohiienka Str., Kyiv, 02000, Ukraine

[3]sulim1.ua@gmail.com

Abstract. Transportation of goods in large containers with loading length of 80 feet is one of the promising directions of railway transport development. However, despite considerable experience in construction of long-wheelbase flat cars, occasionally problems with the strength of load-bearing elements arise at the calculation, design and manufacture of such products. The article presents the technical characteristics of a long wheelbase flat car, as well as the results of experimental studies of fatigue tests of its load-bearing elements. The optimization of the design of the long wheelbase flat car was achieved both by increasing the sizes and shapes, and by using materials of increased strength. An analysis of the study's results of an improved design showed the compliance of the values of the safety factor of fatigue strength and the service life of the flat car with regulatory and technical documentation. The conducted studies allow us to make recommendations regarding the design, rebuilding and testing of long wheelbase flat cars. The primary results of fatigue tests of the main load-bearing structural elements within the scope of experimental studies turned out to be mostly below acceptable values. This article contains facts confirming it. The results of experimental studies to determine the fatigue strength proved that the design elements require improvement and strengthening.

1. The first paragraph after a heading is Analysis of research and publications

Problems of the dynamics and strength of the rolling stock, as well as the possibility of using mathematical modeling to solve this problem were considered in the works of M.B. Kelrich [1-3], S.V. Myamlin [4-7], O.V. Fomin [8-11], A.V. Donchenko [12], V.M. Ishchenko [8, 13], L. Neduzha [6, 7, 14] and other scientists. The research papers of M.B Kelrich, A.V. Donchenko and V.M Ishchenko address the issues of experimental and computational studies of the strength qualities of long wheelbase flat car designs. The studies of O.V. Fomin focus on scientific and practical basis for construction and improvement of long wheelbase flat cars design and also technical solutions to improvement of existing flat car models and other type of freight rolling stock. Research and simulation of spatial oscillations of the railway rolling stock are dealt with in papers by S.V. Myamlin

Content from this work may be used under the terms of the Creative Commons Attribution 3.0 licence. Any further distribution of this work must maintain attribution to the author(s) and the title of the work, journal citation and DOI.
Published under licence by IOP Publishing Ltd

IOP Conf. Series: Materials Science and Engineering **985** (2020) 012029 doi:10.1088/1757-899X/985/1/012029

and L. O. Neduzha. However, nowadays, design and experimental study of the structural strength of new models of long wheelbase flat cars raise a lot of questions regarding their dynamic and strength properties.

2. The purpose of the work

Is to study the stress-strain state of the long wheelbase flat car and increase its strength characteristics. One of the railcar-building factories in Ukraine completed works on the development, conducting a set of relevant tests and serial production set-up of a new model of a long wheelbase flat car. The frame of the long wheelbase flat car is shown in Figure 1, the main parameters and dimensions of the flat car are shown in Table 1.

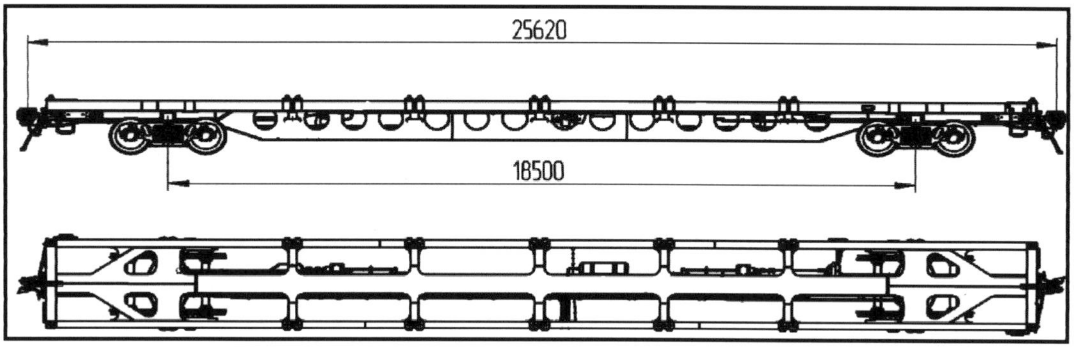

Figure 1. The frame of the long wheelbase flat car.

Table 1. The main parameters and dimensions of the long wheelbase flat car.

Name and dimension of the parameter	Normal value
1 Payload, t, not more than	69.6
2 Weight (tare), t	23.9 ± 0.5
3 Design static load of the wheelset on the rails, kN (tf)	230.5 (23.5)
4 Length along the automatic coupling axle, mm	25620^{+64}_{-11}
5 Design speed, km/h	120
6 Dimension in accordance with GOST 9238 (DSTU B V.2.3-29 *)	1-T
7 Height above top of the rail up to automatic coupler axle (empty platform), mm	1040-1080
8 The distance between the internal axles of the wheelsets (long-wheelbase), mm	16650
9 Two two-axial bogies:	model 18-7055
10 Minimum weight (tare) during operation, t	22.124

These products are intended for the transportation of large capacity universal containers according to DSTU ISO 668 [15] or tank containers (for non-hazardous goods) according to DSTU ISO 668 types 1A (1AA, 1AX), 1B (1BB, 1BX), 1C (1CC, 1CX), 1EE on the 1520 mm gauge railways.

The loading scheme shown in Figure 2 was selected for the tests. Such a loading scheme is considered less favorable and, when operating long wheelbase flat cars, causes the most damaging effect to the structural members, which is confirmed by the results of calculations at the design stage.

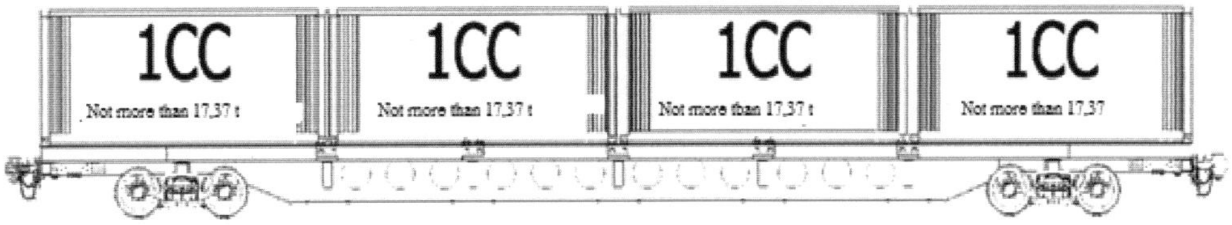

Figure 2. Disposition of four containers or tank containers of standard size 1C (1CC, 1CX).

However, at the stage of the full complex of preliminary experimental studies, negative results of fatigue tests were obtained. The obtained values of fatigue strength coefficient in several sections of the frame are below the permissible value.

For testing, a prototype flat car was loaded with four container simulators up to a gross weight of 101.26 tons, with an overload of 8.32 tons, which corresponds to 12% of the maximum wagon loading capacity. To increase the load in dangerous sections, according to the results of static tests, 8.32 tons of cargos were loaded into two containers located closer to the central section. An increase in cargo weight was necessary to increase the stress level in the central section up to the level of operational stresses obtained from the results of running strength tests. Permissible increase in static load is not more than 30% of the payload capacity. While performing a set of tests for a long wheelbase flat car the calculation of the main indicators of fatigue strength is calculated from equations (1) - (4).

The calculation of the fatigue safety factor was carried out as follows:

$$n = \frac{\sigma_{a,N}}{\sigma_{a,e}}, \tag{1}$$

where:

- $\sigma_{a,N}$ – is the endurance limit (in amplitude) for the check area with a symmetric cycle and steady-state loading mode with the base number of cycles $N_0 = 10^7$.
- $\sigma_{a,e}$ – is the equivalent reduced amplitude of dynamic stresses in the studied section of the frame during the movement of the flat car in operation, is determined by the results of running strength tests or by calculation.

The endurance limit was determined as follows:

$$\sigma_{a,N} = \sqrt[m]{\frac{N_{test}}{N_0} \cdot \sigma^m_{a,test}}, \tag{2}$$

- N_{test} – is the number of cycles passed during testing before the crack initiation.
- N_0 – is the base number of cycles, $N_0 = 10^7$.
- m – is the exponent in the equation of the fatigue curve $m = 4$
- $\sigma^m_{a,test}$ – is the stress amplitude during fatigue strength tests, MPa.

The equivalent reduced amplitude is determined by calculation, based on the results of throwing off of the flat car from the blocks.

$$\sigma_{a,e} = \frac{\sigma_{a,b,\ell}}{\sigma_{a,b,\ell}^{cent}} \cdot \sigma_{a,e}^{cent}, \tag{3}$$

- $\sigma_{a,b,\ell}$ – the amplitude of the dynamic voltage recorded in the studied area of the frame during testing to throw off the flat car from the blocks.
- $\sigma_{a,b,\ell}^{cent}$ – the amplitude of the dynamic voltage recorded in the central section of the frame during testing to throw off the flat car from the blocks.
- $\sigma_{a,e}^{cent}$ – the calculated reduced amplitude of the dynamic voltage in the central section of the frame when the car is running in operation:

$$\sigma_{a,e}^{cent} = \sigma_{st}^{cent} \cdot K_{d,e}, \tag{4}$$

- σ_{st}^{cent} – static voltage in the central section of the frame.
- $K_{d,e}$ – is the equivalent reduced coefficient of vertical dynamics of the car.

Figure 3-4 shows the installation locations of strain gages and the appearance of cracks in the central section.

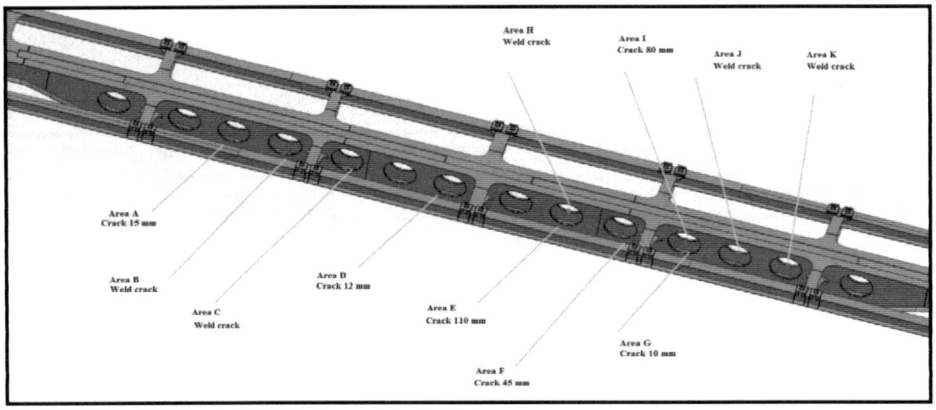

Figure 3. Locations of strain gauges and the cracks initiation during fatigue tests.

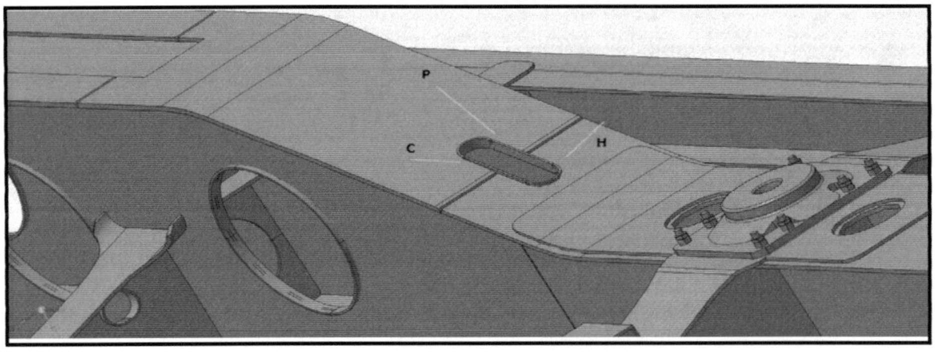

Figure 4. Zones C, H, P of the crack's initiation during fatigue tests.

The cracks started to appear on the flat car frame with the number of test cycles N_{test} =1198000. The appearance of such a number of cracks indicates insufficient structural strength. The safety factor of fatigue resistance in these zones is n =0.57-0.59 with an acceptable value $[n]\geq 1.15$.

Figure 5 and 6 show the characteristic fatigue cracks in the studied areas of the frame of a long wheelbase flat car.

Figure 5. A crack with a length of 80 mm in zone D.

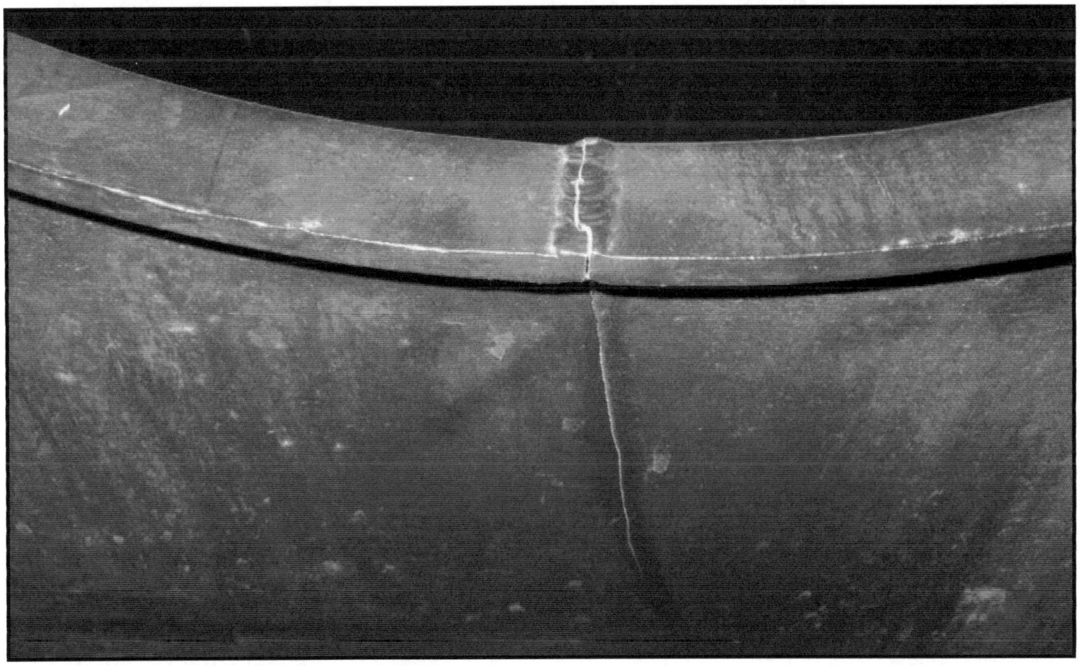

Figure 6. Crack 110 mm long in zone D.

IOP Conf. Series: Materials Science and Engineering **985** (2020) 012029 doi:10.1088/1757-899X/985/1/012029

Various force application schemes were used to simulate vertical disturbances (one and two-sided force application schemes) during the tests. Realization of the same values of dynamic stresses by one and two-sided force application schemes allowed to obtain satisfactory convergence of the achieved values of the fatigue strength coefficient. Thus, the methodology for fatigue testing of long wheelbase flat cars structure concerning the improvement of research methods was further developed.

An analysis of the fatigue tests results showed the need to strengthen and modify certain structural elements. An additional strength calculation of the structure was performed, and adjustments were made taking into account the test results.

The main changes concern the following elements:

- center sill web and upper and lower plates;
- stub sill bottom and vertical plates;
- draft sill bottom plate.

The thickness of the centre sill web plate in the central part is increased from 8 to 10 mm. The upper and lower plates of the center sill are made solid, without 10-mm upper and lower inserts, of 09G2S material with a thickness of 16 mm. In addition, the diameter of the holes on the center sill was reduced from 600 to 500 mm, and the holes of a 600 mm diameter at the beginning of a variable section were removed.

After manufacturing of the advanced prototype, fatigue tests were continued. Before crack initiation in the new design of a long wheelbase flat car in the central section area, the number of cycles was 2128000, which corresponds to a fatigue strength coefficient of 1.196 with an allowable value of $[n] \geq 1.15$.

Before the loss of bearing capacity (structural survivability), the number of cycles was 2267000. In the remaining studied zones, the values of the fatigue strength coefficient exceed the value of the fatigue strength coefficient of the central section zone. Table 2 shows the data for calculating the coefficient of fatigue strength in the central section.

Table 2. The source data for calculating the coefficient of fatigue strength.

Section	$\sigma_{a,e}$, MPa	$\sigma_{a,N}$, MPa	N_{test}	$\sigma_{a,test}$, MPa	$\sigma_{a,Ntest}$, MPa	$[n]$, MPa	n, MPa
Crack in the central section area	23.0	26.5	2128000	40.5	27.5	1.15	1.196

3. Conclusions

1. The original design of the long wheelbase flat car had insufficient strength properties, which is confirmed by experimental studies. The weakest zones are sections located in the middle of the flat car frame. An analysis of the experimental studies results of fatigue strength showed that the coefficient of fatigue strength in these zones is 49 - 52% of the permissible value.

2. The methodology of fatigue testing of long wheelbase flat cars concerning the improvement of test methods has been developed.

3. After improvement of the long wheelbase flat car design and performance of experimental studies, the minimum value of the fatigue strength coefficient in the central zones proved to be 1.196, that is higher than the allowable value $[n] \geq 1.15$.

References

[1] Fedosov-Nikonov D, Kelrikh M 2016 Research on the Structural Strength of a Long Wheelbase Flat Car *Bulletin of Volodymyr Dahl East-Ukrain. Nat Univ. Sci. Journal* **1** pp 90–94

[2] Fedosov-Nikonov D, Kelrikh M 2016 Calculative experimental estimation method of the structural strength of a long wheelbase flat car *Proc. of the Int. Sci. and Pract. Conf. Infrastructure innovations of transport and logistics systems Problems, experience, prospects* (Truskavets: V. Dahl EUNU) pp 101

[3] Kelrikh M 2014 Perspective directions of planning carrying systems of gondolas *Metallurgical and Mining Industry «Machine building»* **6** pp 64–67

[4] Myamlin S 2001 Methodology for modeling spatial vibrations of a railway crew *Railway transport of Ukraine* **2** pp 2-5

[5] Myamlin S 2000 The program for modeling spatial vibrations of the rolling stock *Railway transport of Ukraine* **3** pp 52–54

[6] Myamlin S, Lunys O, Neduzha L and Kyryl'chuk O 2017 Mathematical Modeling of Dynamic Loading of Cassette Bearings for Freight Cars *Proc. of 21st Int. Sci. Conf. Transport Means 2017 pt. III* (Juodkrante, Lithuania: Kaunas Univ. of Technology) pp 973–976

[7] Myamlin S, Neduzha L and Urbutis Ž 2016 Research of Innovations of Diesel Locomotives and Bogies *Proc. Eng.* **134** (Vilnius, Lithuania: Vilnius Gediminas Techn. Univ.) pp 469–474

[8] Fomin O, Ishchenko V and Fedosov-Nikonov D 2018 Long wheelbase flat cars Problems of structural strength *Metallurgical and mining industry* (Dnipro: MPA) **2** pp 84–90

[9] Fomin A, Fedosov-Nikonov D 2018 Scientific and practical research of the structural strength of the long wheelbase flat car frame *Scientific methodical Journal Bulletin of Science and Education* **10** 46 pp 8–15

[10] Fomin O, Fedosov-Nikonov D 2017 Research of the structural strength of the long wheelbase flat car *Proc. of the State Univ. of Infr.and Techn. Transport Systems and Technologies* **31** (Kyiv: DUIT) pp 140147

[11] Fomin O 2014 Modern requirements to carrying systems of railway general-purpose gondola cars *Metallurgical and Mining Industry «Machine building»* **5** pp 31–43

[12] Donchenko A, Fedosov-Nikonov D 2016 Methods of design and experimental studies of the long wheelbase flat car *Col. of sci. papers of the State Econ.-Tech. Univ. of Transport Systems and Technologies* **28** (Kyiv: DETUT) pp 53–60

[13] Ishchenko V, Fedosov-Nikonov D 2017 Long wheelbase flat cars: structural strength *Metallurgical and Mining Industry «Machine building»* **8** pp 18–23

[14] Lunys O, Neduzha L and Tatarinova V 2019 Stability research of the main-line locomotive movement *Proc. of the 23rd Int. Sci. Conf. Transport Means 2019 pt III* (Palanga, Lithuania: Kaunas Univ. of Techn.) pp 1341–1345

[15] Freight containers, series 1 Classification, sizes and nominal characteristics 2015 DSTU ISO 668: 2015 (ISO 668: 2013, IDT) p 25

Stand for accelerated tests of rail vehicles wheelsets

S Raksha[1,2], **P Anofriev**[1,2,3] **and O Kuropiatnyk**[1,2]

[1]Department of Applied Mechanics and Materials Science
[2]Dnipro National University of Railway Transport named after Academician V. Lazaryan, Dnipro, Ukraine

[3]anofrievp@ukr.net

Abstract. Tests and research can be greatly accelerated by applying a multi-point load to the test wheel. The aim of this work is to substantiate rational parameters of the stand for accelerated life tests of railway wheelsets, in which a multi-point load applies to the rolling surface of the test wheel. On the basis of the laws of the theory of elasticity, we performed analytical calculations of the contact stresses created by the rolling bodies of the load mechanism of the test stand. Research has shown that the dependence of the contact stresses on the wheel rolling surface on the ball's diameter is nonlinear. The decrease in stress is due to an increase in the contact area. We obtained the analytical and graphical dependencies, which make it possible to evaluate the influence of the parameters of the stand for accelerated life tests of railway wheelsets on contact stresses on the wheel rolling surface, taking into account the shape and dimensions of this surface. The authors have proposed a stand design for accelerated life tests of railway wheelsets. The use of a stand with the parameters justified in this work will reduce the test duration by 10-15 times.

1. Introduction

Wheelsets are one of the main assembly units that determine the reliability and safety of industrial, mine and railway transport. The technology of the manufacture of parts and assembly wheelsets continuously improved. In this regard, it becomes necessary to check the performance characteristics of wheelsets, such as geometric dimensions, stresses in the wheel-axle connection, static and dynamic strength, imbalance, wear resistance and electrical characteristics. Obtaining and registration of these parameters is performed out at special stands. Tolerances for geometric dimensions, electrical characteristics [1], values of cyclic loads, schemes for their application and test methods for axles are regulated by sectoral state documents [2, 3]. Axle fatigue tests are described in articles [4, 5]. The results of fatigue tests and the probability of destruction of the wheelsets axles, depending on the duration of operation, are due to many random factors [6].

2. Purpose and problem statement

The aim of this work is to substantiate rational parameters of the stand for accelerated life tests of railway wheelsets, in which a multi-point load applies to the rolling surface of the test wheel.

The interaction of the railway wheel with the rail forms a highest kinematic pair with an initial contact along the line. During operation, the wheel rolling surface is subjected to pulsating loads with large amplitude, which cause significant contact stresses. Over time, the wheels working surfaces begin to wear out and collapse; the wheels profile along the rolling circle changes. The greatest depth of the wheel surface wear is observed along the rolling circle placed on distance of 70 mm from the wheel side

Content from this work may be used under the terms of the Creative Commons Attribution 3.0 licence. Any further distribution of this work must maintain attribution to the author(s) and the title of the work, journal citation and DOI.
Published under licence by IOP Publishing Ltd

surface (Figure 1). Such a feature of changing the wheels profile should be taken into account when designing a test bench.

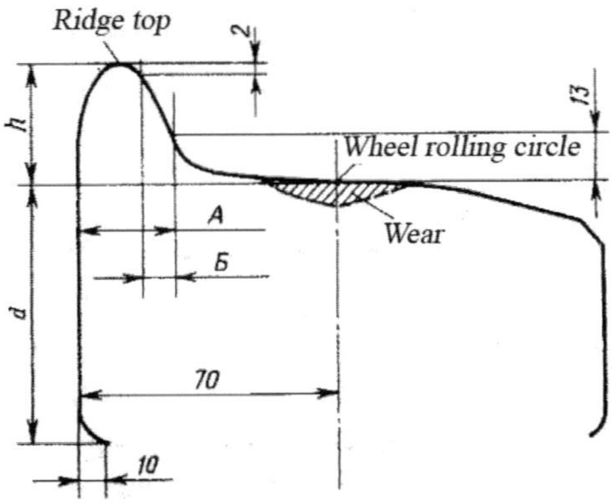

Figure 1. Wheel surface profile wear.

The load on the wheelsets can be realized using the stand rollers, simulating an endless rail. This is the easiest way to select the method and place of load application to the wheelset. The disadvantage of this test method is their considerable duration and the need to use high pressure hydraulic equipment.

The initial profile of the wheel rolling surface has a tapered shape (Figure 2). This feature allows designing the load mechanism of stand resembling to a ball or roller bearing, in which the wheel working surface plays the role of the raceway of inner ring.

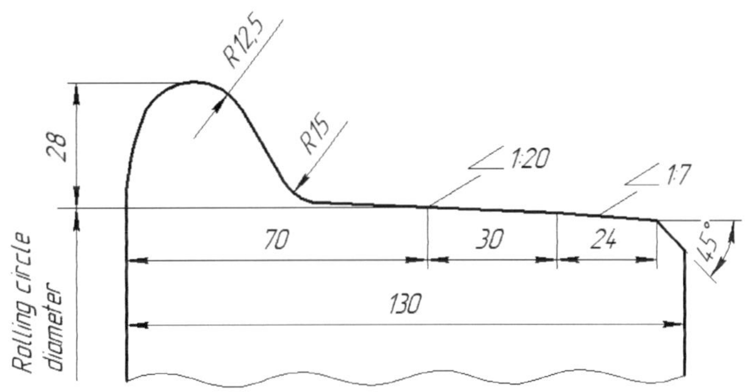

Figure 2. Wheel surface profile.

It should be established how the contact stresses on the rolling surface change depending on the shape and number of rolling elements of the test stand.

3. Designs of stands for wheelsets endurance test

The stand (Figure 3) with a pair of rollers simulating an infinite rail can be used to obtain the characteristics of endurance and wear resistance [7]. This stand contains two rollers 3 interacting with the wheels rolling surfaces, a device for rotation and braking of the wheelset and a loading device. The loading device includes independent oil stations 6 with hydraulic distributors, hydraulic cylinders with rods 5, at the ends of which wedge stops 4 are installed, resting on the wedge surfaces of the roller holders 3. The wheelset rotation device 2 includes axle boxes 1 and a coupling with an engine 11. The braking device consists of an oil station 15 with hydraulic distributors, a hydraulic cylinder 14 and brake shoes 13. The imitation of the operational misalignment of the axle box unit relative to the wheelset is carried out by a drive consisting of an engine 10, a gearbox 9 and a gear 8.

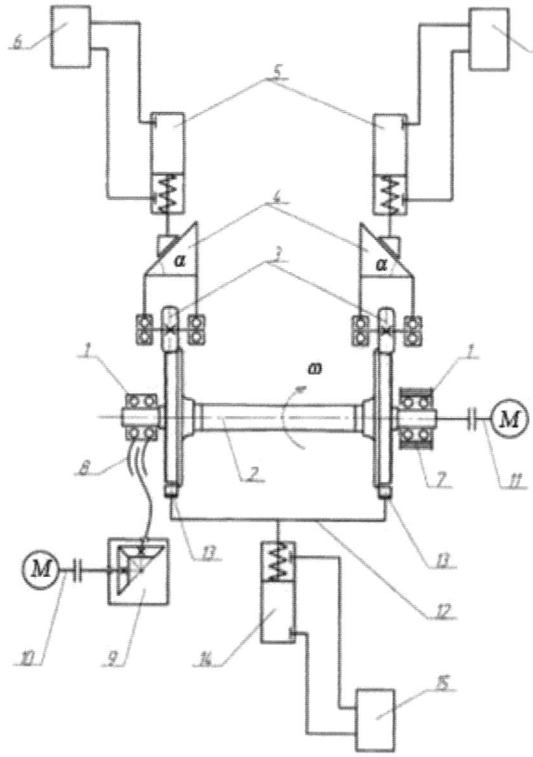

Figure 3. Wheelset test bench.

To test wheel steels for wear and contact strength, friction machine and small-sized samples made of the investigated wheel steels are used [8]. Analysis of the characteristics of small-sized samples using similarity criteria is much cheaper and easier. The test scheme assumes contact interaction of two rollers (Figure 4). One roller is cylindrical in wheel steel and the other is toroidal in carbide WC - Co. When assessing the wear, the Archald model was used, which assumes wear is proportional to the work of friction forces. This is a classic method, which gives an idea of the properties of materials; however, it does not take into account the peculiarities of the interaction between the wheel and the rail.

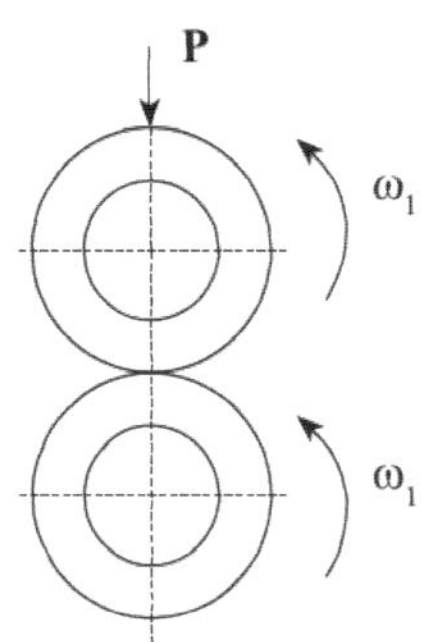

Figure 4. Scheme of rollers contact.

4. Force calculation of the interaction of a wheelset and rolling bodies of the test bench

The stand for accelerated tests of wheelsets proposed by the authors (Figure 5) contains an electric motor 1, a gear coupling 2, bearing assemblies 4 and a load mechanism 5. Wheels of the tested wheelset 3 play the role of inner bearing rings in units 4.

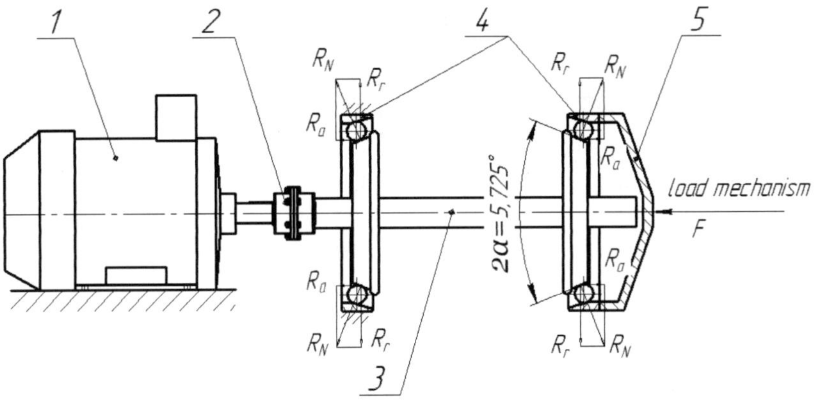

Figure 5. Scheme of wheelset load.

The calculation scheme of the stand for calculating the forces and contact stresses on the wheels rolling surfaces is similar to the calculation scheme of a ball bearing. We assume that the rolling bodies are of the same diameter and are equally loaded. For such a case, the dependence of the wheel normal reaction R_N on the magnitude of the force F developed by the load mechanism:

$$R_N = \frac{F}{n \cdot \sin\alpha}, \tag{1}$$

where F is the axial force from the loading mechanism; n is the number of rolling bodies; α is a half of the angle at the top of the generatrix of the wheel working surface.

Under the action of R_N forces, contact stresses should be created not lower than the maximum stresses on contact surface of "wheel-rail" pair during the operation of railway rolling stock.

During research the interaction of a wheel with a rail, the authors of work [9] established the range of values of contact stresses on the rolling surface of railway freight cars wheels: 1300–1700 MPa. And with the appearance of trough wear, the stress values can reach 6000 MPa. These data are the starting point for determining the load mechanism characteristics and the rolling elements dimensions.

Contact areas of the rolling bodies and the wheel rolling surface are ellipses with semi-axles [10]:

$$a = \alpha \sqrt[3]{\frac{3R_N(1-\mu^2)}{\frac{1}{\rho_{11}}+\frac{1}{\rho_{12}}+\frac{1}{\rho_{21}}+\frac{1}{\rho_{22}}}}; \tag{2}$$

$$b = \beta \sqrt[3]{\frac{3R_N(1-\mu^2)}{\frac{1}{\rho_{11}}+\frac{1}{\rho_{12}}+\frac{1}{\rho_{21}}+\frac{1}{\rho_{22}}}}, \tag{3}$$

where $\rho_{11} = \rho_{12}$ are the radii of the balls; $\rho_{21} = 0.479$ m and $\rho_{22} = \infty$ are the radii of curvature of the wheel working surface; $\mu = 0.3$ is the Poisson's ratio; $E = 210$ GPa is the elastic modulus of steel; α, β are coefficients determined from the table as functions of the auxiliary angle ψ :

$$\cos\psi = \frac{\pm\sqrt{\left(\frac{1}{\rho_{11}}+\frac{1}{\rho_{12}}\right)^2+\left(\frac{1}{\rho_{21}}+\frac{1}{\rho_{22}}\right)^2+2\left(\frac{1}{\rho_{11}}+\frac{1}{\rho_{12}}\right)\left(\frac{1}{\rho_{21}}+\frac{1}{\rho_{22}}\right)\cos 2\varphi}}{\frac{1}{\rho_{11}}+\frac{1}{\rho_{12}}+\frac{1}{\rho_{21}}+\frac{1}{\rho_{22}}}, \tag{4}$$

where $\varphi = 0$ is the angle between the main planes of curvature of the ball and the wheel, in which there are smaller radii ρ_{11} and ρ_{21}; the sign of the numerator is determined so that $\cos\psi > 0$.

The maximum contact stresses are determined by the formula:

$$\sigma_{\max} = 1.5\frac{R_N}{\pi ab}. \tag{5}$$

In order to determine the relationship of contact stresses, the wheel-ball interaction forces, the ball diameter, according to the formulas (2)-(5), variant calculations of contact stresses were performed. Studies have shown that the dependence of contact stresses on the wheel rolling surface on the ball diameter is nonlinear (Figure 6). The decrease in stress is due to an increase in the contact area.

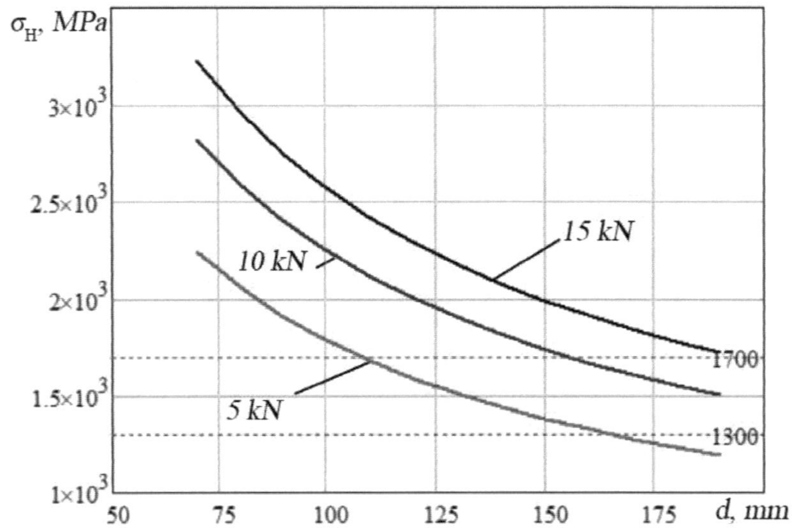

Figure 6. Dependence of contact stresses on the wheel rolling surface on the ball diameter and the load on the contact area.

Having given the value of contact stresses, according to the graphs, you can select the balls diameter and determine the load on them. Analysis of the graphs allows us to recommend a rational range of balls diameter under load:

- 5 kN – from 110 to 160 mm;
- 10 kN – from 160 to 180 mm.

The radius of the arc along which the balls can contact each other is several millimeters less than the radius of the arc passing through the ball's centers (Figure 7). The lengths of these arcs differ slightly from the ball diameter. Therefore, the length of these arcs can be approximately taken equal to the ball diameter.

The number of balls in the load mechanism of the stand, taking into account the size of the separator, can be determined by the formula:

$$n = k\pi\left(\frac{D}{d}+1\right), \tag{6}$$

where k is coefficient taking into account the dimensions required to install the separator (in the first approximation we take $k = 0.5$); D is wheel rolling circle diameter (we take $D = 957$ mm for new railway wheel); d is ball diameter.

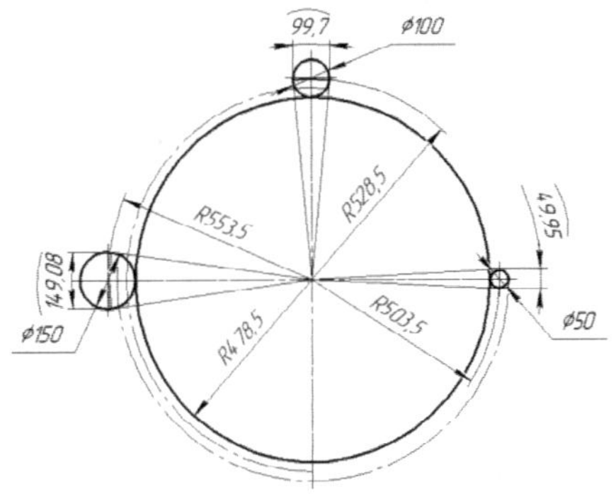

Figure 7. Geometric characteristics of balls with a diameter of 50, 100 and 150 mm in the load mechanism of the stand.

The fractional part of the number of balls obtained from (5) is removed, so the graph of the dependence of the balls number on their diameter takes on a stepped form (Figure 8). From formula (1) we determine the value of the axial force of the load mechanism of the stand.

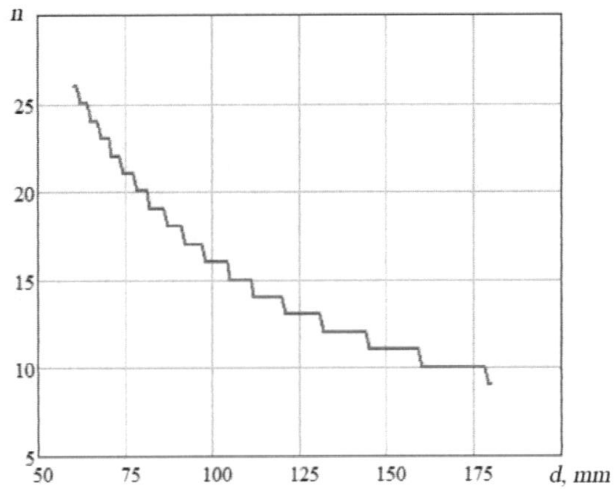

Figure 8. Dependence of the number of balls in the load mechanism of the stand on their diameter.

The test stand allows realizing contact stresses of 1300-1700 MPa with 10 balls in the load mechanism with a force of 10 kN on each ball. With a force of 5 kN per ball, there should be 11-15 balls in the load mechanism.

Thus, the authors have proposed a stand design for accelerated life tests of railway wheelsets. The use of a stand with the parameters justified in this work will reduce the test duration by 10-15 times.

5. Conclusions

1) The authors performed a review and analysis of existing designs of stands for testing axles and methods of applying cyclic loads to the wheels of wheelsets of rail vehicles.
2) The dependence of the contact stresses on the wheel rolling surface on the ball's diameter and the load on the contact area was obtained.
3) An approximate formula for calculating the number of balls in the load mechanism of the test stand is proposed.
4) We recommend a rational range of parameters of the load mechanism of the stand: at a load of 5 kN, the diameter of the balls is from 110 to 160 mm; at a load of 10 kN, diameter of the balls is from 160 to 180 mm.

5) Due to the large number of balls in the load mechanism of the stand, the tests will be significantly accelerated. The use of a stand with the parameters justified in this work will reduce the test duration by 10-15 times.

References

[1] Wheelsets of railway rolling stock. Methods for determining strength indicators 2016 GOST 33783-2016 (Moscow: FGUP «STANDARTINFORM») p 57
[2] Wheelsets of locomotives and multiple units rolling stock. Calculations and strength tests 2009 GOST 31373-2008 (Moscow: FGUP «STANDARTINFORM») p 17
[3] Wheelsets of railway cars. Technical conditions 2013 GOST 4835-2013 (Moscow: FGUP «STANDARTINFORM») p 44
[4] Raksha S, Anofriev P and Kuropiatnyk O 2019 Simulation modelling of the rolling stock axle test-bench *E3S Web of Conferences Ukrainian School of Mining Engineering* **123** Article Number 01032. https://doi.org/10.1051/e3sconf/201912301032
[5] Raksha S, Anofriev P and Kuropiatnyk O 2019 Justification of parameters of wheelset axle fatigue strength test-bench for railway rolling stock *MATEC Web Conf. Volume 294 2nd International Scientific and Practical Conference "Energy-Optimal Technologies, Logistic and Safety on Transport" (EOT-2019)* Article Number 03008 https://doi.org/10.1051/matecconf/201929403008
[6] Volokhov G M and Oguenko V V 2015 Wheelset axle resource assessment *Transport Rossiyskoy Federatsii* **3** 58 pp 71–73
[7] Yeremeev V K, Tsvik L B, Kulishov A V and Zapolskiy D V 2010 Patent 2436061 Russian Federation. Test bench for wheelsets and their elements
[8] Vorobev A A, Kerentsev D Ye and Fedorov I V 2017 Wheel steels wear and contact fatigue tests *Izvestiya Peterburgskogo universiteta putey soobshcheniya* **4** pp 628–636
[9] Vorobev A A 2009 Contact interaction of wheel and rail *Vestnik IrGTU* 3 pp 42–47
[10] Pisarenko G S, Agarev V A, Kvitka A L, Popkov V G and Umanskiy E S 1979 *Strength of materials* (Kiev: Naukova dumka) p 696

IOP Conf. Series: Materials Science and Engineering **985** (2020) 012031 doi:10.1088/1757-899X/985/1/012031

Improvement of diesel injector nozzle test techniques

D Aulin[1], O Klymenko[1], A Falendysh[2], O Kletska[2,4] and J Dizo[3]

[1]Department of Heat Engineering and Heat Engines, Ukrainian state university of railway transport, Kharkov, Ukraine
[2]Department of Industrial transportation technologies, Pryazovskyi State Technical University
[3]Department of Transport and Handling Machines, University of Zilina, Zilina, Slovakia

[4]dimmo@ex.ua

Abstract Due to changes in the characteristics of the fuel equipment during operation, a considerable part of the life cycle of locomotive engines operate under increased fuel consumption, and thermal voltage, which often causes degraded environmental performance and reduced engine capacity and, as a result, low efficiency of the locomotive. A reserve for improving the performance of the diesel plant and the diesel locomotive in general is the implementation of measures to prevent changes in the characteristics of the fuel system, for example, the development of clean-in-place technique of fuel systems from carbon deposits. In order to determine the effectiveness of such measures, it is necessary to apply new quality control methods of fuel atomization. The article offers a more informative method for evaluating the quality of fuel atomization, which can be used in locomotive depots during research and development work. The method consists of microphotography of the drops of sprayed fuel followed by microscopic analysis and processing of the results using special software. It is proposed to process the results of injectors testing using a statistical method, taking into account the parameters according to the Rosin-Rammler dependence for the relative total curve. As a result of the work, it became possible to develop requirements and implement technical improvements to the test stands for fuel equipment of diesel locomotives.

1. Introduction

In accordance with the target program and the National Transport Strategy of Ukraine until 2030, approved by the Decree of Cabinet of Ministers of Ukraine No. 430-p of May 30, 2018 [1], one of the main directions of scientific and technical development of Ukrzaliznytsya JSC is to increase the economy and the service life of its technical means.

Better reliability and service life of locomotives can be achieved through the introduction of new control methods of the technical condition and detection of faults or failures of the injector nozzles of diesel locomotives, as well as the angle of advance of fuel injection from changing the course of the spray needle. Therefore, the improvement of the test technique of diesel injector nozzles is an important component of technical measures aimed at improving the efficiency and longevity of locomotives.

This problem should be solved through comprehensive measures that include a wide range of issues related to the durability of traction rolling stock. The efficiency of locomotives operation largely depends on the reliability and durability of the fuel equipment. It is determined by the accuracy

Content from this work may be used under the terms of the Creative Commons Attribution 3.0 licence. Any further distribution of this work must maintain attribution to the author(s) and the title of the work, journal citation and DOI.
Published under licence by IOP Publishing Ltd

of diagnosis and the quality of maintenance and repair. The share of failures of locomotives on the railway network due to failure of diesel plants reaches 41 % of the total number of failures of the main locomotive assemblies, including 12-13 % of the total number of failures of diesel because of failure of the fuel system. Some elements of the fuel system operate under high thermal and mechanical load. This, together with the structural and regulatory factors, as well as the physical and chemical properties of the fuel, leads to a significant formation and accumulation of carbon deposits [2, 3, 4]. This leads to the maldistribution of fuel on the cylinders and inside the combustion chamber, which degrades the performance of the diesel. This condition of the locomotive fleet and the fuel equipment of diesel locomotives is due to the low quality of current repairs and inaccurate diagnosis, which indicates the need for further work in this area and the relevance of this topic.

Therefore, improving diesel injector nozzle test techniques subject to the conditions and modes of operation of locomotives is a relevant scientific problem.

2. Literature review and problem statement

The objective of this research is to improve both an important stage of the technological process of testing fuel equipment, and the reliability and economy of diesel locomotives.

The fuel spraying process in the combustion chamber of the internal combustion engine was considered in the works by AS. Lyshevskyi, V.A. Kutovyi; then, the proposed approaches were modified by N.F. Razleitsev and other authors [2-10]. As the processes of fuel spraying have been studied, requirements were made for the fuel equipment of the internal combustion engine, the authors used some simplifications and assumptions that lead to distortion of the object and the results of the research, but made a significant contribution necessary for the construction of the physical picture and understanding of the essence of the phenomena and establishing their relationships. Experimental studies show that a sprayed jet of fluid consists of a very large number of droplets whose sizes vary within fairly wide limits. Spray characteristics are used to evaluate the dispersion of the sprayed liquid or the spray quality. The review of the works showed that the definition of the quality of spraying is relevant in current research to test alternative fuels and new and promising fuel systems.

Obtaining a reliable analytical expression is very important because it allows calculating the number of droplets, their volume and surface, choosing reasonably the average dispersion of the sprayed liquid, and making calculations related to the evaporation of the droplets and the torch combustion in general.

3. Objective and tasks

The objective of this research is to improve diesel injector nozzle test techniques by improving the quality assessment method of fuel spraying. This will increase the efficiency and reliability of the fuel equipment and reduce the cost of their maintenance and repair.

To achieve the set objective, the below-mentioned task had to be solved:

- to develop a control method for the dispersed composition of the sprayed fuel jet;
- to search for and identify statistically conditioned interrelations of the controlled parameters obtained at the spraying quality control bench using high-speed shooting and image processing of fuel jets;
- to develop recommendations for further improvement of test stands in the characteristics of the fuel spraying flow.

4. Main part

The assessment of the impact of resource-saving measures and means (for example, the impact of in-place cleaning on the technical condition of the fuel equipment) was carried out by testing the injectors of diesel locomotive on A106 bench before and after in-place cleaning. The tests checked injection pressure, hydraulic density of the nozzles, tightness, as well as visual inspection of the fuel spraying quality by nozzles. In-place cleaning did not affect the injection pressure and the hydraulic density. During the nozzle tests on A106 bench, the spray quality was visually assessed, which allowed detecting only serious malfunctions of the nozzle. Therefore, to assess the impact of the use of in-place

IOP Conf. Series: Materials Science and Engineering **985** (2020) 012031 doi:10.1088/1757-899X/985/1/012031

cleaning on the quality of fuel spraying, a more accurate and reliable method of assessment was required.

Dispersion composition is a mandatory method of controlling the quality of fuel spraying [11]. The application of dispersion analysis methods is necessary for scientific research, for the development of new and promising fuel supply systems.

Experimental studies show that the torch of sprayed fuel consists of a very large number of droplets whose sizes vary within wide enough limits. Spraying characteristics are used to estimate the dispersion of the sprayed fuel or the quality of the spray. When estimating the dispersion of sprayed fuel, in calculations of motion, evaporation and combustion of droplets in the torch we have to use average data on the size of droplets. The average size of the droplets gives a simplified summary of the degree of fuel fineness; it serves as one of the characteristics of the dispersed composition of the droplets in the torch and allows simplifying mathematical operations in the analysis of the composition of the droplets: perform calculations related to the change of the torch of the sprayed liquid in relation to the number of droplets and their magnitude during evaporation; approximately determine the path of the torch vertex; calculate the time required to burn the torch of the sprayed fuel of a certain dispersed composition, etc. Studying fuel spraying, and especially the size and quantity of droplets, is a difficult and time-consuming task. There are various methods for its solution: micro-filming of drops in a torch, photometric method, injection into molten paraffin with subsequent freezing of drops and their screening, sedimentometric method, etc. All of these methods require sophisticated equipment and considerable cost and may not always be applicable in locomotive research. To determine the effect of indiscriminate cleaning technology on the quality of fuel spraying, the test method used was glycerol droplet trapping, followed by microscopic analysis and processing of results on special software [12-14]. The layout of the automated plant for determining the dispersion composition and size of the droplets is shown in Figure 1.

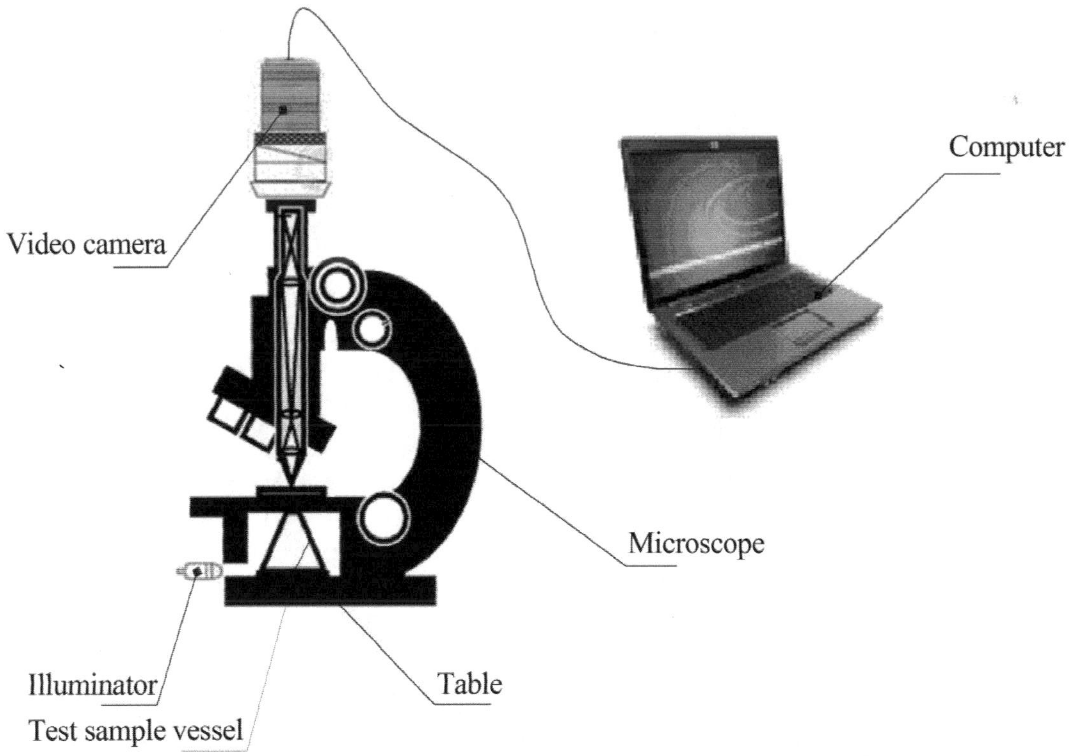

Figure 1. The automated dispersed composition determination plant layout.

The image through the microscope lens gets is converted into digital video signal in the video camera. This signal is sent to the computer and converted into a bitmap. Obtaining and further processing of the graphic image is carried out by means of the special software which allows for:

- calibration and entering of the scale factor;
- image filtering;
- image conversion to negative;
- horizontal and vertical image display;
- image splitting by shades;
- binary image conversion (black and white);
- processing of the edges of the image (removal of unwanted objects along the edges of the image);
- manual linear measurements;
- manual measurements of areas;
- automatic image analysis (calculation of linear sizes and areas);
- grouping of the data obtained from image analysis.

Statistical analysis of the nozzle test results was performed, taking the Rosin-Rammler relation parameters for a relative total curve:

$$S(z) = 1 - \exp\left[-\left(\frac{z}{B}\right)^A\right], \quad (1)$$

where A and B are constants determined from experimental data.

The parameter B is also called the size constant. It is the size of a droplet that divides the relative total curve so that 63.2% of all the sprayed fuel consists of smaller droplets, and 36.8% - of larger droplets.

Parameter A characterizes the width of the distribution, i.e. the degree of uniformity of the droplets of size z. Images of sprayed nozzle droplets after a 300-day period both with and without application of the proposed technique are presented in Figure 2.

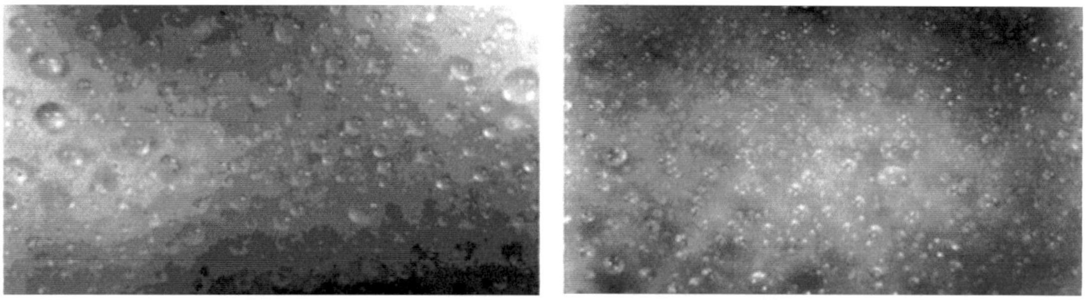

(a) (b)

Figure 2. An image of sprayed fuel droplets from a nozzle, (a) an image of sprayed fuel droplets from a nozzle with service hours $T = 300$ days (150× magnified), (b) an image of sprayed fuel droplets from a nozzle with service hours $T = 300$ days after in-place cleaning of the fuel system (150× magnified).

In addition, to determine the effect of in-place cleaning on the quality of fuel spraying, a high-speed image shooting of the sprayed fuel torch was performed (Figure 3).

As a result of the experimental studies, the differential and integral curves of the size distribution of droplets were obtained, which are shown in Figure 4 and Figure 5 respectively.

IOP Conf. Series: Materials Science and Engineering **985** (2020) 012031 doi:10.1088/1757-899X/985/1/012031

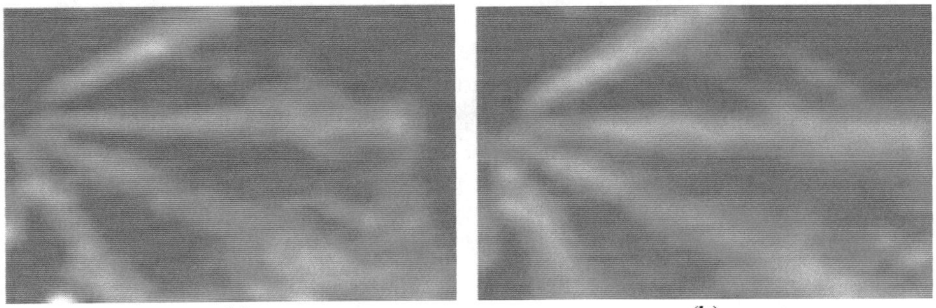

(a) (b)

Figure 3. Spraying of fuel with K6S310DR diesel nozzle, (a) spraying of fuel with a nozzle with service hours $T = 300$ days, (b)s of fuel with a nozzle with service hours $T = 300$ days after in-place cleaning of the fuel system.

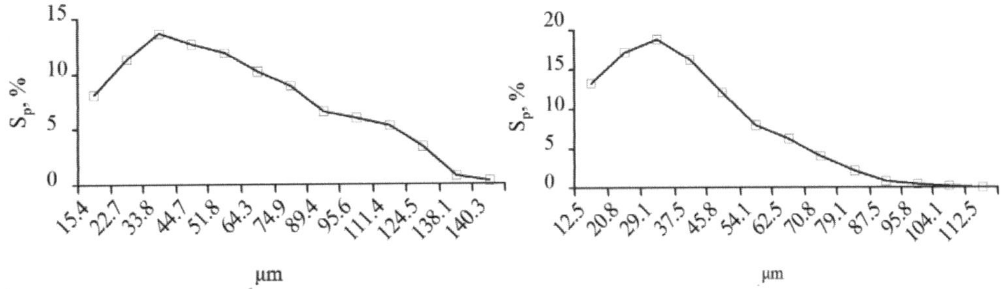

(a) (b)

Figure 4. Differential distribution curves of droplets by size, (a) nozzle service hours $T = 300$ days, (b) nozzle service hours $T = 300$ days after in-place cleaning of the fuel system.

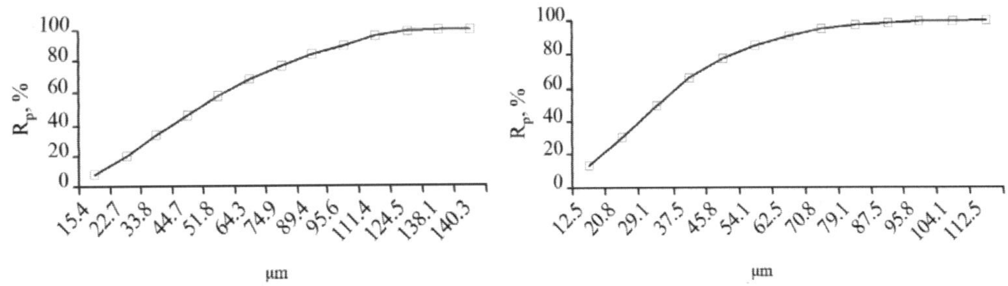

(a) (b)

Figure 5. Integral distribution curves of droplets by size, (a) nozzle service hours $T = 300$ days, (b) nozzle service hours $T = 300$ days after in-place cleaning of the fuel system.

The distribution parameters of droplets by size based on the Rosin-Rammler relation with the nozzle service hours $T = 300$ days, were: $A = 31.06$; $B = 85.21$ µm, after removal of carbon deposits: $A = 3.62$; $B = 66.06$ µm.

Thus, while maintaining the fuel supply pressure, the use of in-place cleaning reduces the size of the sprayed fuel droplets. This increases the specific surface area, accelerates fuel evaporation of the fuel, and, as a consequence, improves the quality of blending and, as a result, the combustion process and the technical and economic performance of the locomotives [15, 16].

Summary

The article presents scientifically grounded technical solutions in the development of quality control methods of fuel spaying by nozzles. The implementation of these solutions will increase the efficiency of control over the operation of fuel equipment during their maintenance and repair, as well as during research and development activities.

To estimate the dispersed composition of the sprayed fuel, glycerol droplet trapping with subsequent software-based analysis was proposed. Spraying quality was evaluated using the Rosin-Rammler equation, which helps estimate the dispersed composition of the fuel sprayed with the nozzle.

The developed guidelines for the production of a test bench based on the proposed method of determining the fuel spraying quality of nozzles were developed.

References

[1] 2018 Pro skhvalennia Natsionalnoi transportnoi stratehii Ukrainy na period do 2030 roku vid 30 travnia 2018 Kyiv. Ofitsiinyi visnyk Ukrainy vid 10.07.2018. 2018. №52. stor. 533. stattia 1848. kod akta 90720/2018 p 430

[2] Lushevskyi A S 1981 *Systemy pytanyia dyzeliv* (Moskva: Mashynostroenye) p 216

[3] Lushevskyi A S 1963 *Protsessu raspulyvanyia toplyva dyzelnumy forsunkamy* (Moscow: Mashhyz) p 180

[4] Lushevskyi A S 1971 *Raspulyvanye toplyva v sudovukh dyzeliakh* (Lenynhrad: Sudostroenye) p 248

[5] Kutovoi V A 1971 *Vprusk toplyva v dyzeliakh* (Moskva: Mashynostroenye) p 119

[6] Markov V A, Syrotyn E A and Pavlov V A 2001 Toplyvnaia apparatura dyzeliv *Hruzovyk* **7** pp 17–24

[7] Razleitsev N F 1980 *Modelyrovanye y optymyzatsyia protsessa shoranyia v dyzeliakh (Kharkiv: Vyshcha shkola)* p 169

[8] Hrekhov L V, Yvashchenko N A and Markov V A 2005 Toplyvnaia apparatura y systemy upravlenyia dyzelei (Moskva: Lehyon-Avtodata) p 344

[9] Babadi M N, Kheradmand S and Bae C 2020 Experimental and Computational Investigation of Diesel and Gasoline Injection in a Direct Injection Compression Ignition Engine *International Journal of Automotive Technology* **21** 1 pp 23–32

[10] Lee C H 2019 Entire-spray mixing characteristics over time after the start of the injection using an image-processing technique for high-speed captured images Of Cr-Di fuel sprays *ARPN. Journal of Engineering and Applied Sciences* **14** 3 pp 587–595

[11] Kouzov P A 1974 *Osnovэ analyza dyspersnoho sostava promэshlennэkh pэlei y yzmelchennыkh materyalov 2-e yzd. yspr* (Lenynhrad: Khymyia) p 279

[12] Galante-Fox J and Bennett J 2012 Diesel injector internal deposits in High Pressure Common Rail diesel engines Fuel Systems for IC Engines *Woodhead Publishing Limited* pp 157–166

[13] Karelyn B A and Lutskyi V K 1966 *Metodэ y apparatura dlia yzmerenyia razmerov chastyts* (Moskva: TsVETMET ynformatsyia) p 155

[14] Hradus L Ya 1979 *Rukovodstvo po dyspersyonnomu analyzu metodom mykroskopyy* (Moscow: Yzdatelstvo "Khymyia") p 232

[15] Aulin D, Zinkivskyi A, Anatskyi O and Kovalenko D 2018 Measures for Resource Saving for Diesel Locomotives *International Journal of Engineering & Technology* pp 152–156

[16] Kagramanian A, Stankevich P, Aulin D and Basov A 2019 Efficiency improvement of locomotive-type diesel engine operation due to introduction of resource-saving technologies for cleaning diesel and diesel locomotive systems *Procedia Computer Science* **149** pp 264–273

IOP Conf. Series: Materials Science and Engineering **985** (2020) 012032 doi:10.1088/1757-899X/985/1/012032

Theory development for determination of rational parameters of the capacitive energy storage for a metro train

A O Sulym

State enterprise "Scientific Railcar Building Research Institute", 33 I. Prykhodka St., 39621, Kremenchuk, Ukraine

sulim1.ua@gmail.com

Abstract. In the article a comparative analysis of existing methodologies and approaches on the parameters determination of the capacitive energy storages for a metro train with energy recovery systems was performed, disadvantages of each of them were determined. A complex approach for estimation of the on-board capacitive energy storage parameters was proposed; the point of approach consists in determination of the rational power and energy capacity for two parameters of the storage system at once – mass and payback period. The complex approach is based on theoretical researches, which require simulation of the metro train equipped with recuperation system operation by mean of developed software. The objective function was represented and boundary values for the power and energy capacity during determination of the rational parameters of the on-board capacitive energy storage using a complex approach were specified. The recuperation system with rational parameters of the on-board capacitive energy storage for specified service conditions in Public Utility Company "Kyiv Metropolitan" and chosen research metro train was specified. Amount of energy stored due to the installation of the energy storage system with rational parameters was estimated.

1. Introduction

At the moment, one of the priorities and urgent tasks in the subway is to reduce electricity consumption by its rolling stock. To this end, scientists have focused their work on increasing the use of regenerative braking electricity in the following areas: optimizing the train schedule, streamlining the train mode, the introduction of inverters and energy storage. In works [1,3] the organizational method on reduction of energy consumption and increase of efficiency of use of the electric power of regenerative braking which is based on optimization of movement of the subway trains is offered. Also, in these works the technical measures of reduction of electricity consumption by trains - introduction of inverters on traction substations and energy storage devices are considered in detail. In work [2] the automatic system of control of movement of the subway train is offered. Rational mode of train operation allows to significantly reduce electricity costs. Over the last years companies of the USA, Japan, Korea, European Union countries and domestic manufacturers have greatly increased electric power characteristics of the capacitive energy storages (CES), which are also called super capacitors, ultra capacitors, ionistors, electric double-layer capacitors. It led to performance of a great number of researches concerning their practical use as power supply sources, including underground railway systems for saving electrical energy of the regenerative braking and its recovery by the electric

Content from this work may be used under the terms of the Creative Commons Attribution 3.0 licence. Any further distribution of this work must maintain attribution to the author(s) and the title of the work, journal citation and DOI.
Published under licence by IOP Publishing Ltd

traction drive during train acceleration. However, determination of necessary parameters of the on-board CES for a metro train, remains one of the insufficiently studied and current problems.

2. Literature review

Researches [4-6], which concern an issue of determination of the on-board CES parameters, particularly its power and energy capacity. In works [7-9] determination of the on-board CES parameters is recommended to be performed using theoretical researches as well as date analysis results of experimental researches of energy processes between the overhead line and the electric traction drive under standard operational conditions of the rolling stock. These researches and research are presented in the works [10, 11] are based on the use of such methods and key points: a point of the electric traction theory and numerous methods of integration techniques, methods of the analysis of density characteristics for power division and amount of recuperated energy, points of the probability theory, theoretical basis of the electrical engineering etc.

However, according to the results of works [12-14] it is known that, these methods and points don't let to take into account a number of factors of the actual operational conditions completely. Key factors of these are: gauge profile, cars load, change of the train schedule during within 24 hours, composting braking. The main disadvantage of methods and approaches proposed in mentioned works is impossibility of the determination of rational parameters of the on-board CES. The analysis of advantages and disadvantages of each method is given in work [14] in more detail.

In existing works [12, 13] the choice of the on-board CES parameters is recommended to be done according to the criterion of the minimum payback period of the storage system. The storage system consists of the on-board CES, the reversible static converter and the control system of the energy-exchange processes between the on-board CES and the electric traction drive. The principle of the approach if in the parameters determination according to the result of the analysis of the payback period diagrams for chosen storage systems. The advantage of the approach use is in the determination of the rational parameters of the on-board CES. For this purpose, actual operating conditions of the metro train equipped with recuperation systems are taken into account. The disadvantage of this approach is the determination of the rational parameters of the on-board CES according to one criterion – the payback period of the storage system.

In work [14] a complex approach for the rational parameters estimation of the on-board CES was proposed, which allows choosing it according to two important factors of the storage system – weight and payback period. This approach requires performance of the experimental researches in the standard actual operating conditions. In addition to this the complex approach lets to take into account those factors, which were not considered in the above-mentioned works.

However, the complex approach, proposed in work [14], which is based on the analysis of the experimental researches data also has disadvantages. Main disadvantages of the experimental research's performance are great time consumption and financial expenses. One more important disadvantage that does not allow getting a truthful overview is the impossibility to consider load change of the train during its operation, because performance of such researches in conditions of the regular traffic is rather problematic. As a rule, during experimental researches it is assumed that a train is operated with minimum, nominal and maximum load per day. And accordingly, experimental researches of the energy processes at the best case are performed during standard train operation for three loading modes. Therefore, during performance of researches for determination of the rational power and energy capacity of the on-board CES by using data of the experimental researches there is an inaccuracy, which rises because the change of the train load during its operation was not performed. That is why in this paper in order to decrease financial expenses, time for researches and increase the accuracy of obtained results we proposed to perform researches on the determination of the rational parameters of the on-board CES using a developed mathematical model of the metro train operation equipped with recuperation systems.

Purpose – to determine rational power and energy capacity of the on-board CES using a complex approach of the parameter's estimation, which based on methods of theoretical researches.

IOP Conf. Series: Materials Science and Engineering **985** (2020) 012032 doi:10.1088/1757-899X/985/1/012032

3. Matter and methods of research

The complex approach for determination of the rational parameters of the on-board CES assumes a stage-by-stage procedure of research (Figure 1). As it was mentioned above, the point of the approach is in the determination of the rational parameters by two parameters of the storage system at once – weight and payback period.

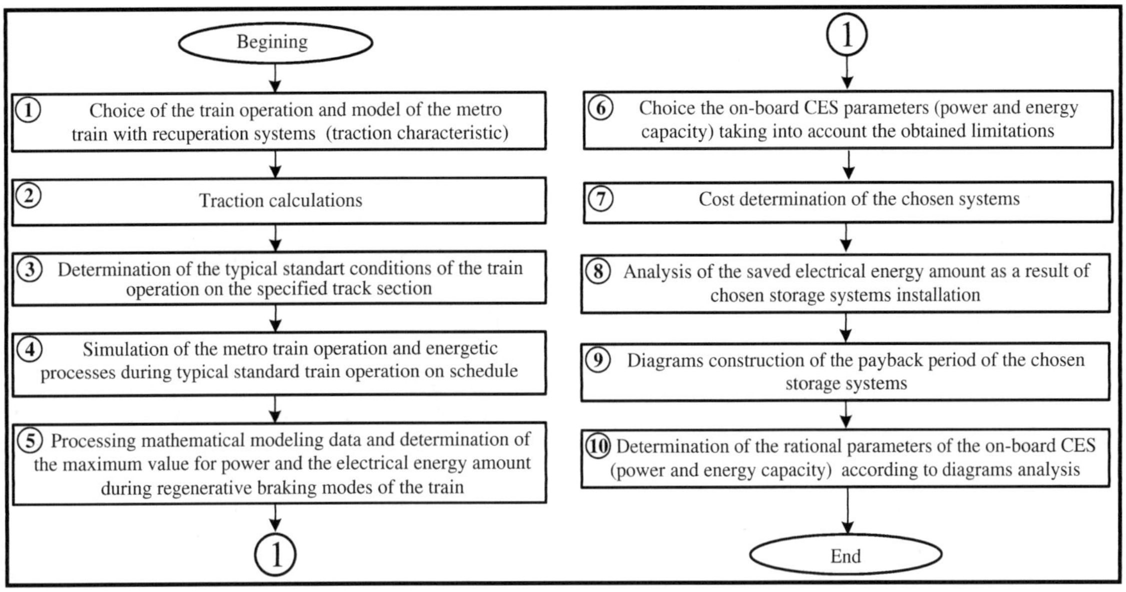

Figure 1. Procedure of the rational parameter's determination of the on-board CES using the complex approach.

While the main difference from the approach described in paper [14] is that this procedure is based on theoretical researches, the point of which is in simulation of the metro train operation with recuperation systems using the software.

A generalized mathematical description of the objective function of the complex approach with determination of the rational parameters of the on-board CES can be represented the following way:

$$F(\overline{X}) \rightarrow \text{extremum}, \ \overline{X} \in D_1 ... D_k \tag{1}$$

Where, $F(\overline{X})$ is the main criterion of the optimality, D is area of feasible solutions, which is determined by boundaries of corresponding values (parametric limitation), k is amount of accepted limitations for a search of the optimum solution.

In our case the payback period of the storage system was chosen as the main optimization criterion. Taking into account the chosen criterion definition of the optimization task: to find such values of the power and energy capacity of the on-board $CES(P_{CES}, A_{CES})$, at which:

$$T_{pp} = F(P_{CES}, A_{CES}) \rightarrow \min. \tag{2}$$

In this case the following boundary values of the parameters were specified for the objective function: $P_{CES} \in [0...P_{\Delta m}]$, $A_{CES} \in [0...A_{\Delta m}]$, $P_{CES} \in [0...P_{max}]$, $A_{CES} \in [0...A_{recmax}]$, where $P_{\Delta m}$, $A_{\Delta m}$ is maximum values of power and energy capacity by weight, which depend on the metro train type with recuperation systems and condenser modules; P_{max}, A_{recmax} is maximum values for power and energy recuperation amount for specified standard train operation.

Provided that the objective function has several minimum values, the storage system which can save maximum amount of the electrical energy must be chosen, that is:

$$\alpha = \begin{pmatrix} T_{pp1} \\ T_{pp2} \\ \ldots\ldots \\ T_{ppj} \end{pmatrix} \rightarrow \max, \quad (3)$$

where j is a number of storage systems with the same minimum payback period; α is amount of the saved electrical energy due to storage system installation.

Further using this approach, we analyzed determination of the rational power and energy capacity of the on-board CES for specified operation conditions of the metro rolling stock with energy recovery systems.

First stage. A section between terminal depots Sviatoshynsko-Brovarska line of PU "Kyiv Metropoliten" was chosen as a test section. The test rolling stock is a train of 5 five cars with asynchronous electric traction drive and energy recovery systems in which engine cars – trailer cars, middle – motor cars (train, which consists of cars models 81-7080, 81-7081, 81-7081-01).

Second stage. Performance of the traction calculations, which are based on principles of the traction effort choice (deceleration) taking into account limitations on the maximum motor torque, wheel – rail adhesion and providing specified train operation dynamics by traction motors.

Thus, according to the results of the traction calculations for specified conditions it was determined that weight of the storage system for a chosen type of the train should not exceed 7.86 t.

Forth stage involves metro train simulation. Developed and certified software Motion Simulation (further – SW "Motion Simulation") was used to perform modeling of the dynamic and energy processes under standard specified metro train operation conditions. Input data, values of dynamic and energy processes, and resulting output values, which are determined during simulation, calculation formulas for these values are reviewed and reported in paper [15] in details, so there is no need to stop on this issue in more detail. That is why it is not necessary to dwell on this issue. It should be noted that the input data implies setting: energy characteristic (dependence of the efficiency factor of the traction motor on running speed), a number of set train operation modes on a line, gauge profile parameters, static and dynamic parameters for each train operation mode.

At the *fifth stage*, mathematical modeling data is processed. Data processing was performed using SW "Motion Simulation". The main result is the determination of the following values: the amount of electric energy consumed for traction $(A_{traction})$, the amount of electric energy generated by the train during regenerative braking (A_{reg}), the maximum power in recovery mode (P_{max}).

Therefore, the first boundary values of the on-board CES parameters based on the obtained data results were established: $P_{CES} \in [0...3.88]$, $A_{CES} \in [0...45.09]$. Other boundary values of power and energy capacity are determined taking into account weight limitations and depend on chosen types of CES.

In the *sixth stage* parameters of CES are chosen (power and energy capacity. In our case area of power and energy capacity boundary values of the on-board CES is determined under conditions of using storage systems, assembled on the basis of condenser modules produced by such well-known manufacturers as EKOND (Russia), Nesscap (Korea), Maxwell (USA), Epcos (Germany). General appearance of chosen condenser modules is given in Figure 2. Storage systems of the necessary value of the service voltage, power and energy capacity were formed by means of series-parallel connection of specified condenser modules. Weight of storage systems were determined by summing up weight of condenser modules (on-board CES), operated transducers, metallic structures, connecting wires (bars), stress gauges and current transducers, cooling systems components, control systems and other additional materials.

IOP Conf. Series: Materials Science and Engineering **985** (2020) 012032 doi:10.1088/1757-899X/985/1/012032

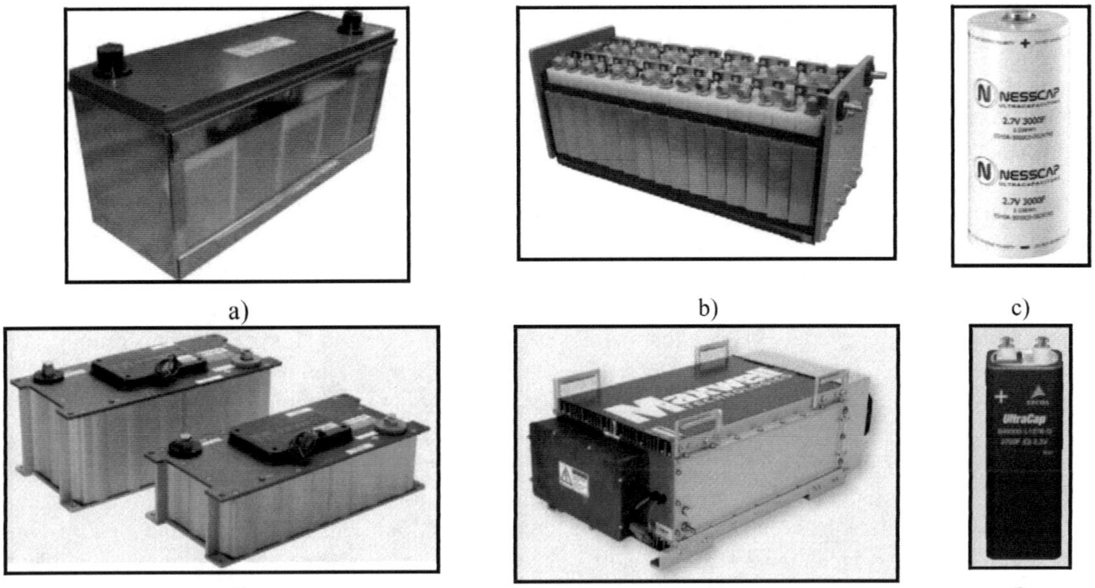

Figure 2. General appearance of the condenser modules 10EK303 (a), 30EK404 (b), ESHSR-3000C0-002R7A5 (c), BMOD0063P125 B08 (d), BMOD0165P048 B01 (e), B49300 L1276 Q (f).

Therefore, other boundary values of the on-board CES parameters were set, taking into account weight limitations. Taking into account determined limitations after processing data of the mathematical simulation and weight, the following boundary values of the parameters of the storage systems with operating voltage of 450-900 V were set:

- for storage systems of condenser modules type 10EK303 – $P_{CES} \in [0...3.88]$, $A_{CES} \in [0...15.3]$;
- for storage systems of condenser modules type 30EK404 – $P_{CES} \in [0...3.88]$, $A_{CES} \in [0...10.9]$;
- for storage systems of condenser modules type ESHSR – 3000C0-002R7A5 – $P_{CES} \in [0...3.88]$, $A_{CES} \in [0...26.9]$;
- for storage systems of condenser modules type BMOD0063P125 B08 – $P_{CES} \in [0...3.88]$, $A_{CES} \in [0...13.0]$;
- for storage systems of condenser modules type BMOD0165P048 BXX – $P_{CES} \in [0...3.88]$, $A_{CES} \in [0...21.7]$;
- for storage systems of condenser modules type B49300L1276Q – $P_{CES} \in [0...3.88]$, $A_{CES} \in [0...13.5]$.

Based on the obtained boundary values of the on-board CES parameters, for further calculations, depending on the type of condenser modules, on-board CES with the level of power and operating energy capacity were chosen according to Table. 1.

Table 1. Chosen parameters of the on-boards CESs.

Type of the condenser module	Power, MW	Energy capacity, kW·h
10EK303	1.7; 3.4	6.4; 12.7
30EK404	0.8; 1.7; 2.5; 3.4	1.7; 3.4; 5.0; 6.7
ESHSR – 3000C0 – 002R7A5	1.1; 2.2; 3.3	1.0; 1.9; 2.9
BMOD0063P125 B08	1.5; 3.1	1.1; 2.2
BMOD0165P048 BXX	1.7; 3.3	1.1; 2.2
B49300L1276Q	0.7; 1.5; 2.2; 2.9; 3.6	0.7; 1.3; 2.0; 2.6; 3.3

In this case, the total number of chosen storage systems with different levels of power and energy capacity of the on-board CES that satisfy the conditions (defined by limitations) is 18.

Seventh stage. Estimation of the cost of storage systems is performed according to the results of a technical and economic analysis of the selected on-board CESs cost, reversible converters and other ancillary equipment produced by manufacturing companies of this production. So, the cost of each chosen storage system was determined. In addition, the cost factor of chosen storage systems, depending on their parameters and the type of condenser modules used comprises 1.1-3.9 mln. UAH for 1 t.

The eighth stage is to research the amount of saved energy due to implementation of the chosen storage systems. Certain research is performed for each type of a chosen storage system. Initially, for each typical operational condition and a chosen storage system, a power limitation check is performed, the results of which, if necessary, are used to recalculate the amount of recuperated electric energy to the capacitive storage. Next, an estimation of the amount of stored electric energy is performed, taking into account the energy capacity limitations using the sub-program "Energy Recovery". The operation algorithm of the above-mentioned sub-program is particularly described in paper [16].

The following values have been determined during research: amount of stored electric energy for one cycle of storage (regenerative braking and its accumulation during the train acceleration), amount of the stored electric energy for each standard operating condition, the amount of stored electric energy over a year. Thus, using the mentioned sub-program we determined the amount of stored electric energy as a result of chosen energy storage systems implementation.

The ninth stage involves creating characteristic (diagrams) for the payback period of the storage systems depending on the operating power and energy capacity of the chosen CES.

The payback period value is determined by relation of the cost of the storage system implementation and the cost of energy stored. Calculation results of the payback period for chosen storage systems are represented by diagrams in Figure 3.

The *tenth stage* (determination of the system with rational parameters). According to the results of the analysis of the payback period diagrams for storage systems it is obvious that for specified operating conditions of the metro train the most suitable is the system with operating energy capacity of 1.7 kW·h and maximum power of 0.8 MW assembled from the condenser modules type 30EK 404. The payback period of this system is minimal and comprises 3.2 years and its weight is about 1.2 tons.

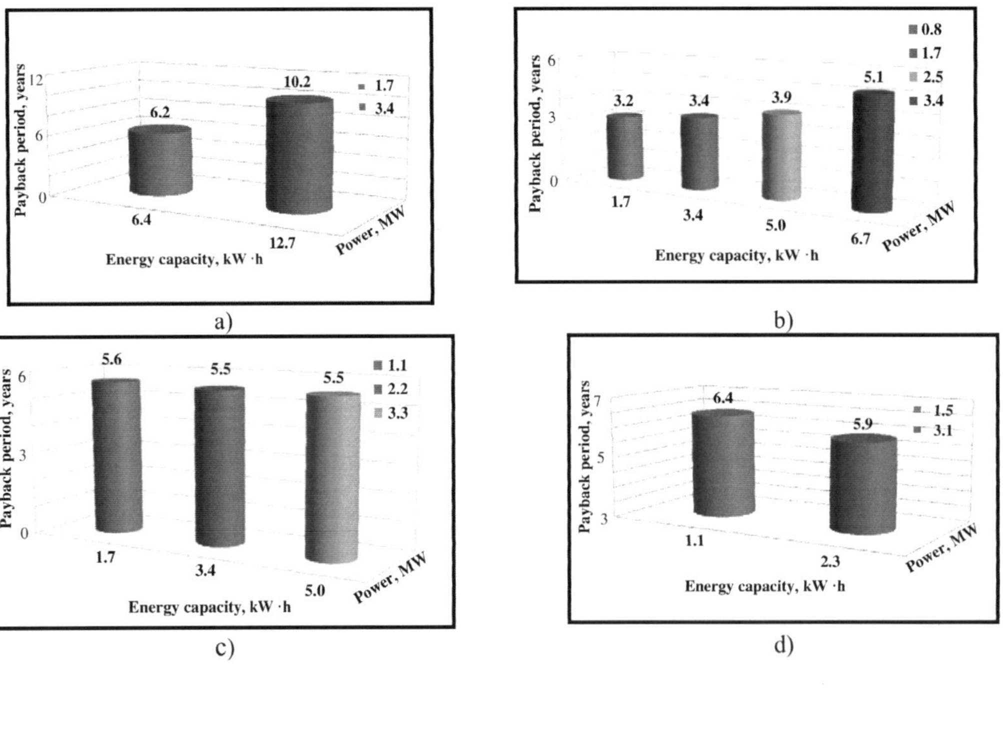

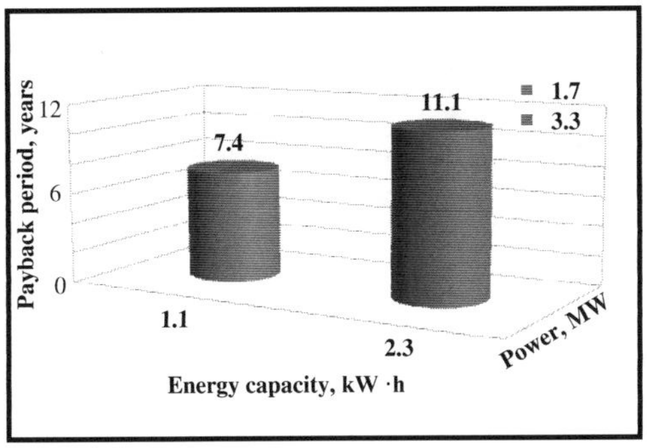

e)

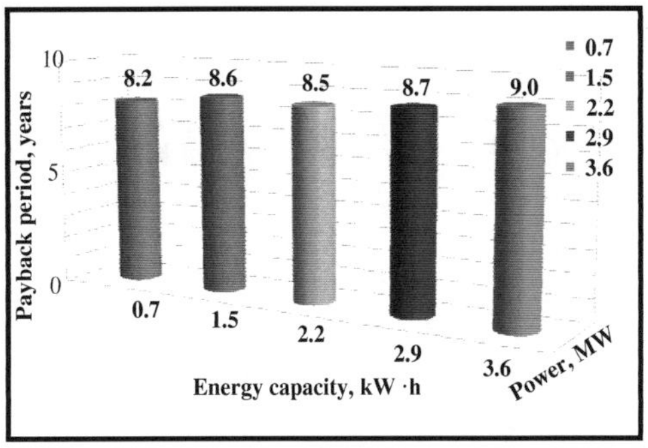

f)

Figure 3. Payback period diagrams of the chosen storage systems assembled of condenser modules type 10EK303 (a), 30EK404 (b), ESHSR-3000C0-002R7A5 (c), BMOD0063P125 B08 (d), BMOD0165P048 B01 (e), B49300 L1276 Q (f).

On condition of implementation of the storage systems with rational parameters we have calculated the amount of stored energy in relation to consumed electric energy according to formula [16]:

$$\alpha = \frac{E_r}{A_{traction(year)}} \cdot 100, \quad (4)$$

where $A_{traction(year)}$ is amount of consumed electric energy per year, kW h.

According to formula 4 it was determined that for specified operational conditions the implementation of storage systems with rational parameters on the on-board CES allows to store 11.4% of the consumed electric energy volume of the traction. In this case the weight of the storage system with rational parameters is about 0.5% and 0.8% of the metro train weight at maximum load and when empty respectively. As a result of this storage system implementation dynamics of the train acceleration will be reduced by 0.5%.

4. Conclusions

1. Using a complex approach for parameters estimation, which is based on methods of theoretical researches, rational parameters of the on-board CES were determined (maximum power and operating energy capacity) for specified operational conditions of the metro train with energy recovery systems.

2. Methodology of the complex approach for determination of the rational parameters of the on-board CES during theoretical research was improved; the object of the research is a simulation of the metro train operation with energy recovery systems using software which helps to reduce financial expenses, research time and improves their accuracy.

3. According to results of performed researches it was determined that for specified train operation modes it is rational to use storage systems with the on-board CES with operational energy capacity of 1.7 kW·h, and maximum power of 0.8MW. It was determined that the payback period of this system is 3.2 years, and its weight is 1.2 tons.

4. Implementation of the storage system with rational parameters for specified operation conditions allows to save up to 11.4 % of the electrical energy amount consumed by traction. In addition, the weight of the storage system with rational parameters is about 0.5% and 0.8% of the metro train weight at maximum load and when empty respectively. As a result of this storage system implementation dynamics of the train acceleration will be reduced by 0.5%.

5. It is proposed to develop a theory for determination of rational parameters of CES for a metro train in case of using theoretical research, which allow modeling of the metro train operation with energy recovery systems instead of experimental research. Further research should be focused on the development of theory for determination of rational power and energy capacity values of the on-board capacitive energy storages taking into account energy-optimal operating conditions of the metro rolling stock equipped with the energy recovery systems.

References

[1] Urbaniak M Kardas-Cinal E and Jacuna M 2019 Optimization of energetic traine *Symmetry* **11** 1175 pp 1–19

[2] Toruń A Sokolowska L and Jagyna M 2019 Communications-based train control system–concept based on WiFi LAN network *Transport means* 2019 – October pp 911–915

[3] Jacyna M Golebiowski P and Urboniak M 2016 Multi-option model of railway traffic organization including the energy recuperation *16-th International conference on transport systems telematics* pp 199–210

[4] Shevlyugin M V and Zheltov K S 2008 Reduction of energy consumption for the movement of trains in the Moscow metro when using capacitive energy storage *Science and technology of transport* **1** pp 15–20

[5] Vasiliev V A 2012 The energy efficiency of electric trains, DC: *author. dis. Cand. Tech Sciences: 05.22.07 Petersburg state University of Railways* – 16

[6] Shchurov N I Shcheglov K V and Shtang A A 2008 The use of energy storage systems, electric traction *Collection of scientific papers of NSTU* **1** pp 99–104

[7] Ryabov E S 2015 Determination of parameters of energy storage for electric rolling stock with asynchronous traction drive is in current limitation mode traction network *Visnyk of the National University technonogy "Kharkivskyy poltechnic Institute"* **6** 1115 pp 132–137

[8] Rybalko A Y and Dybrin S V 2008 The Choice of capacity of energy storage for reducing maximum ower consumption *Mining information-analytical Bulletin. Scientific and technical journal* **8** pp 356–361

[9] Kostin N I and Nikitenko A V 2014 Recuperative braking autonomy is a basis of reliable and efficient energy recuperation in DC electric rolling stock *Zaliznychnyi transport Ukrainy* **3** pp 15–23

[10] Sulym A, Fomin O, Khozia P and Mastepan A 2018 Theoretical and practical determination of parameters of on-board capacitive energy storage of the rolling stock *Naukovyi Visnyk NHU* **5** pp 79–87

[11] Mukha A M, Kostin M O, Kurylenko O Y and Tsyplia G V 2017 Enhancing the operational efficiency of direct current drive based on use of supercondenser power storage units *Nauka and progress transport. Herald of DUST im. V. Lazaryan* **5** 71 pp 48–60

[12] Myatezh A V and Yaroslavtsev M V 2013 Determination of the intensity of the on-Board buffer capacitor energy storage for an urban electric transport *Transport of the Russian Federation. Electricity and electrical engineering* **4** 47 pp 62–65

[13] Fomin O, Sulym A, Kulbovskiy P, Khozia P and Ishchenko V 2018 Determining rational parameters of the capacitive energy storage system for the underground railway rolling stock *Eastern-European Journal of Enterprise Technologies* **2** 92 pp 63–71

[14] Sulym A, Fomin O, Khozia P, Palant O and Stamatin V 2019 Development of a comprehensive approach to determining the rational parameters of an onboard capacitive energy accumulator for a subway train *Eastern-European Journal of Enterprise Technologies* **3** 102 pp 28–38

[15] Sulym A O, Siora O S, Khozia P O and Melnyk O O 2019 Software for metro train movement simulation. *Zbirnik naukovikh naukov praz DUIT: Seriya "Transport systems and technologies"* **34** pp 182–198

[16] Sulim A A and Lomonos A I 2013 Calculation of energy recovery when installing the drive on the subway *Zbirnik naukovikh naukov praz DETUT: Seriya "Transport systems and technologies"* **23** pp 22–29

Analytical method of determining the movement resistance of a tip for forming rectangular technological hole in the lower structure tracks

O Posmitiukha[1,5], K Hlavatskyi[1,6], S Kravets[2,7], V Suponyev[3,8] and A Koval[4,9]

[1]Dnipro National University of Railway Transport named after Academician V. Lazaryan, 2, Lazaryan St., Dnipro, 49010, Ukraine
[2]National University of Water and Environmental Engineering, 77 A. Novaka Street, building No.3, 33002, Rovno City, Ukraine
[3]Kharkiv National Automobile and Highway University, 25 Yaroslava Mudrogo st., Kharkiv, 61002, Ukraine
[4]National Transport University, 1, Mykhaila Omelianovycha - Pavlenka Str., 01010, Kyiv, Ukraine

[5]AleksandrP@3g.ua, [6]kazimir.glavatskij@gmail.com, [7]s.v.kravets@nuwm.edu.ua, [8]v-suponev@ukr.net, [9]kandr@i.ua

Abstract. The aim of this work is to theoretically determine the total resistance to soil pricking with a wedge-shaped tip to create a rectangular cavity and to determine the influence of the size of the working equipment and soil environment in the repair of earthenware or areas where trench technology can-not be used. To determine the forces of the recess of the replaceable pilot, the law of change of soil pressure on the working surfaces of the working equipment was first determined, which is based on the idea of the change of the soil elastic state at its compaction was determined, which is determined by the compression module of deformation of the soil. The processes occurring during the formation of wells and new approaches to the improvement of calculations have been described. On the basis of the pro-posed model of the piercing process, the laws of change of the normal pressure of soil resistance on the surface of the wedge working equipment are established. Based on its value, the results of theoretical studies have received a rational form and parameters of the tip of the work equipment.

1. Introduction

Static perforation of the soil for trenchless laying of underground distribution communications, when repairing the ground or adjacent territories where it is impossible to use trench technology, is the most effective method of forming a horizontal well. In this paper, a method of determining the soil resistance at its perforation is proposed with a tip to form a well of a rectangular cavity.

Installation of underground communications under railway objects and in compressed urban conditions requires simultaneous laying of several linear-extended objects (LPOs), or their protective flares, through one common cavity. Substantiation of rational design of soil-boring working equipment for simultaneous laying of two or more communications and establishment of its parameters is an urgent task.

The purpose of the work is to determine the soil resistance during its perforation with a wedge-shaped tip for forming a rectangular cavity.

Content from this work may be used under the terms of the Creative Commons Attribution 3.0 licence. Any further distribution of this work must maintain attribution to the author(s) and the title of the work, journal citation and DOI.
Published under licence by IOP Publishing Ltd

2. Analysis of literary data and raising of problem

The deterioration of underground railway networks and facilities serviced by rail requires the gradual replacement, modernization or construction of new utility networks, which imposes certain restrictions on builders [1, 2] and requires the simultaneous laying of several lines of the same type. Objects (LPOs) in protective cases through one common pore. For objects not related to train movement but only servicing the railway, restrictions are now imposed on the possibility of an open method of performing works for the construction of underground communications. Therefore, trenchless methods of laying underground communications are becoming widely used [3, 4]. The main disadvantage of these technologies is their cost, which comes from the high cost of drilling equipment and the cost of performing the work. Of the variety of methods available in the market for obtaining technological cavities in soil, the simplest and cheapest is the static puncture [5, 6].

The general regularities of soil puncture processes and the formation of horizontally directed boreholes are described in [6, 7, 8]. The influence of the process of piercing the soil on the ground [9, 10], rolling stock and its elements, as well as neighboring communications are considered in [11, 12,]. The results of studies to determine the effort of piercing a conical tip are described in [13, 14, 18] the influence of the design parameters of wedge-shaped working bodies on the process of soil destruction laid out in [15, 16]. Changes in soil quality in the surrounding well space during static soil puncture and their influence on the processes of formation of the well and its expansion have been considered in [10, 17]. Requirements for simultaneous laying of two or more communications or their protective cases through a common cavity of a cylindrical shape, as well as the possibility of controlling the trajectory of a static puncture are set out in regulatory documents [1, 2, 10]. Studies on the definition of rational parameters of the tip of the soil piercing working body of the conical shape are considered qualitatively in [8, 13], and options for controlling the trajectory of the cavity [19].

For the first time, this cross-sectional shape of the hole was considered and an analytical dependence of the resistance of movement of the working body on the shape of the working body and the number of linearly extended objects being laid simultaneously were created. The obtained dependences allow to state that in this form of aperture, depending on the number of communications that are laid simultaneously, it is possible to reduce the effort up to 2 times, as well as to reduce the area of structural changes in the soil around the received cavity in comparison with the traditional conical tip.

Purpose. It is known from [9] that puncture of soil by a conical tip can be effective within the diameter of the well, which is obtained from the condition of optimal resistance. Thus, it was found that the maximum reduction in puncture resistance occurs: for a solid loam at diameters m, for a semi-solid loam at diameters m, for a rigid plastic clay at diameters m. attention to the fact that their diameters are at least 100 mm. According to [2], the number of jointly laid protective cases in the form of steel or plastic pipes is from 3 to 8 pieces. However, the problem remains of the presence of cavities when laying several individual cases in one hole, which leads to soil subsidence, along the course of a puncture or collapses.

3. Influence of the wedge working equipment on the earthen cloth

We propose the construction of a soil perforation working equipment (Figure 1) for the simultaneous laying of two or more cases 3 for communications or drainage under embankments, consisting of a perforation wedge part 1 with half cones and a calibration part 2 in the shape of a rectangular cavity with a half-cylinder.

Determine the law of change in soil density in thickness d tip on the basis of equality of mass [13], which according to the scheme of action of normal soil pressure on the working equipment (Figure 1) has the form:

$$(n-1)d^2\rho_{nat} = (n-1)d(d-2x)\rho_x, \tag{1}$$

$$\rho_x = \frac{\rho_{nat}}{1-\dfrac{2x}{d}}, \tag{2}$$

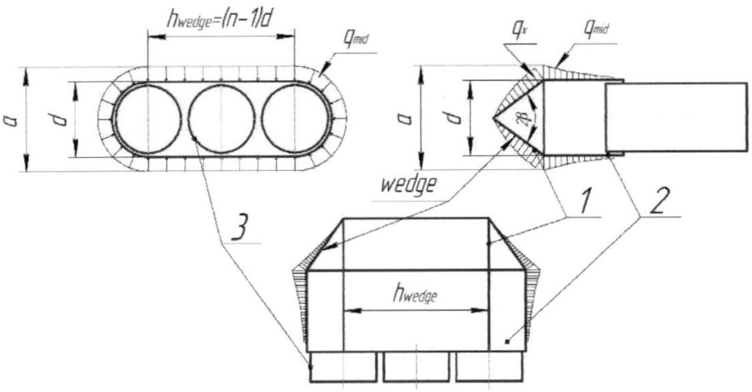

Figure 1. Cross-section replaceable pilot wedge-type and arrangement of loads on it: 1 – wedge-new piercing part with half cones; 2 – calibrating parts-on; 3 – Cases.

Change of normal soil pressure on the front surface of the tip [13] based on expressions (1) and (2) will be written in the form:

$$q_x = E_{soil}\frac{2x}{d}, \tag{3}$$

where $E_{soil} = \frac{1+\omega}{c_K}\frac{\rho_{hard}}{\rho_{nat}}$ – compression modulus of soil deformation; x – deformation of the soil by the front surface of the tip; d – the diameter of the linear-elongate element LEE; n – the number of LEE being laid simultaneously.

The resistance of the wedge tip consists of the resistance of the wedge width h_{wedge} and height d the resistances of the two extreme half cones equivalent to the resistance of one cone with a diameter d .

$$P_{\text{equiv. force}} = P_{\text{wedge}} + P_{cone}\,. \tag{4}$$

Determine the resistance of the wedge:

$$dP_{\text{wedge}} = q_x dF \sin\beta\left(1+f\cdot ctg\beta\right), \tag{5}$$

where dF – elemental area of the front surface of the wedge ($dF = dh\cdot\frac{dx}{\sin\beta}$); 2β – angle of sharpening of the wedge; f – wedge friction coefficient.

Then substituting the value (3) in (5) we obtain:

$$dP_{\text{wedge}} = E_{soil}\frac{2x}{d}dh\frac{dx}{\sin\beta}\sin\beta\left(1+fctg\beta\right), \tag{6}$$

$$P_{\text{wedge}} = E_{soil}\frac{4h_{\text{wedge}}}{d}\left(1+f\cdot ctg\beta\right)\int_0^{d/2} xdx = \frac{d\cdot h_{\text{wedge}}}{2}E_{soil}\left(1+f\cdot ctg\beta\right), \tag{7}$$

$$P_{\text{wedge}} = \frac{(n-1)d^2}{2}E_{soil}\left(1+fctg\beta\right). \tag{8}$$

Obviously, the analytical dependence of the total resistance on the total puncture resistance (4) consists of the resistance of the conical and wedge (6), (7) and (8) parts of the tip.

According to the previous work [13, 14, 15] of the authors the theoretically determined resistance to the indentation of the conical tip into the soil is equal to:

$$P_{\text{equiv. force wedge}} = P_{\text{wedge}} + P_{\text{equiv. force cone}} \text{ then } P_{\text{equiv. force cone}} = \frac{\pi}{8}D^2E_{soil}\left(1+f\,ctg\beta\right), \tag{9}$$

where $D = d$ – for wedge-shaped working body of rectangular shape.

4. Results

A mathematical model for calculating the total drag of the working fluid in the soil has been created. The total drag is equal (Figure 2):

$$P_{\text{equiv. force wedge}} = \frac{\pi + 4(n-1)}{8} d^2 E_{soil} \left(1 + fctg\beta\right). \tag{10}$$

To obtain a rational form and parameters of working equipment determine the ratio of the frontal perforation resistance of conical (9) and wedge (10) tips:

$$\frac{P_{\text{equiv. force } cone}}{P_{\text{equiv. force wedge}}} = \frac{\pi D^2}{\left[\pi + 4(n-1)\right] d^2}. \tag{11}$$

Determine the dependence of the size of the circle, which is described around several cases laid side by side. Consider several options for stretching pipes by number: 2, 3, and n pieces.

If $n = 2$, then $D = 2d$, then by the formula (11):

$$\frac{P_{\text{equiv. force } cone}}{P_{\text{equiv. force wedge}}} = \frac{\pi D^2}{\left[\pi + 4(2-1)\right] d^2} = \frac{\pi 4d^2}{\left[\pi + 4\right] d^2} = 1.76. \tag{11, a}$$

If $n = 3$, then $D = 2.155d$, then by the formula (11):

$$\frac{P_{\text{equiv. force } cone}}{P_{\text{equiv. force wedge}}} = \frac{\pi 4.644d^2}{\left[\pi + 8\right] d^2} = 1.31. \tag{11, b}$$

If $n = 4$, then $D = 2.41d$, then by the formula (11):

$$\frac{P_{\text{equiv. force } cone}}{P_{\text{equiv. force wedge}}} = \frac{\pi 5.808d^2}{\left[\pi + 12\right] d^2} = 1.20. \tag{11, c}$$

In general, for any number of cases the Figure 3:

Consider the triangles ΔAOB, ΔCOD and ΔKOD.

Angle $\angle KOD = \frac{\pi}{n}$, from where $sin\frac{\pi}{n} = \frac{KD}{OD}$, $OD = \frac{KD}{sin\left(\pi/_n\right)}$, $OB = OD + DB$.

$$D = OB = \frac{d}{2\sin\left(\pi/_n\right)} + \frac{d}{2} = \left(1 + \frac{1}{\sin \pi/_n}\right) d. \tag{12}$$

Then by formulas (11) and (12) we obtain:

$$\frac{P_{\text{equiv. force } cone}}{P_{\text{equiv. force wedge}}} = \frac{\pi \left(1 + \frac{1}{\sin \pi/n}\right)^2 d^2}{\left[\pi + 4(n-1)\right] d^2} = \frac{\pi \left(1 + \frac{1}{\sin \pi/n}\right)^2}{\pi + 4(n-1)}. \tag{13}$$

By formulas (11), (12) and (13) we obtain at $n = 5$.

$$\frac{P_{\text{equiv. force } cone}}{P_{\text{equiv. force wedge}}} = \frac{\pi \left(1 + \frac{1}{\sin \pi/5}\right)^2}{\pi + 4(5-1)} = \frac{\pi \left(1 + \frac{1}{\sin 36^\circ}\right)^2}{\pi + 16} = 1.2. \tag{13, a}$$

By formulas (11), (12) and (13) we obtain at $n = 10$.

$$\frac{P_{\text{equiv. force } cone}}{P_{\text{equiv. force wedge}}} = \frac{\pi \left(1 + \frac{1}{\sin \pi/10}\right)^2}{\pi + 4(10-1)} = \frac{\pi \left(1 + \frac{1}{\sin 18^\circ}\right)^2}{\pi + 36} = 1.44. \tag{13, b}$$

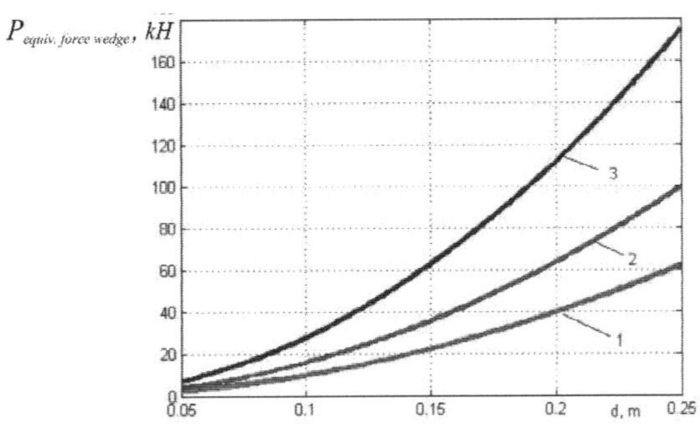

Figure 2. Dependence of the frontal resistance of the wedge drive on the diameter of the communication, offered: 1 – rigid plastic clay; 2 – semi-solid loam; 3 – solid loam.

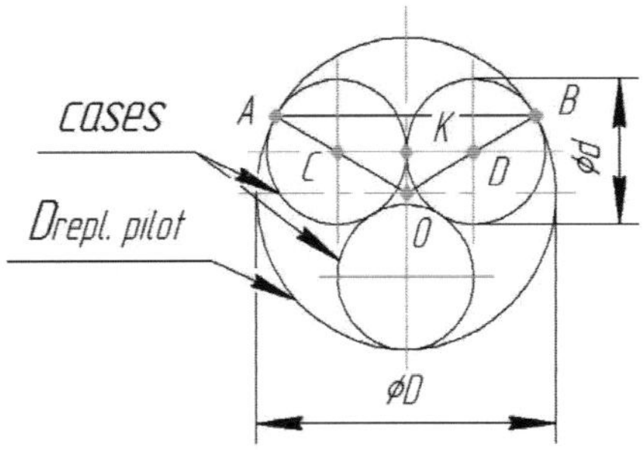

Figure 3. Determination of opening diameter is at a group gasket n- casing.

A lateral soil pressure is applied around the perimeter of the tip, which creates additional resistance from the forces of soil friction along the side walls and rounded edges of the tip.

The average soil pressure on the side walls and edges of the tip is equal:

$$q_{\text{mid}} = 2c \cdot tg\left(\frac{\pi}{4} + \frac{\varphi_0}{2}\right) + \frac{a_q}{2}h. \tag{14}$$

where c – soil adhesion coefficient; φ_0 – angle of internal friction of the soil; a_q – proportionality factor ($a_q = 0.026$ MPa/m); h – depth of replaceable pilot placement (depth of laying of communications).

The length of the perimeter of the tip housing is equal:

$$L_{perimeter} = 2(n-1)d + \pi d. \tag{15}$$

The resistance from friction forces is determined by the area of the lateral surface of the tip and the average pressure is equal to:

$$F_{\text{friction.}} = \left[\pi d + 2(n-1)d\right]h_{\text{wedge}}f\left[2c \cdot tg\left(\frac{\pi}{4} + \frac{\varphi_0}{2}\right) + \frac{a_q}{2}h\right], \tag{16}$$

where h_{wedge} – width of lateral surface replaceable pilot; f – coefficient of soil friction on the tip material.

Then the total drive resistance will have an overview:

$$\begin{aligned} P_\Sigma = {} & \frac{\pi + 2(n-1)}{8}d^2 E_{soil}\left(1 + fctg\beta\right) + \\ & + \left[\pi d + 2(n-1)d\right]h_{\text{wedge}}f\left[2c \cdot tg\left(\frac{\pi}{4} + \frac{\varphi_0}{2}\right) + \frac{a_q}{2}h\right]. \end{aligned} \tag{17}$$

Dependence P_Σ as a function of d for $n = 2$ shown in the figure 4 formulas (12, 15, 16, 17).

Dependence was obtained to determine the total soil resistance at its puncture by the proposed soil perforation working equipment taking into account the friction forces on its lateral surface.

The prospect of further research is to theoretically determine the equivalent dimensions of a wedge tip with rounded edges, to obtain rectangular openings, to remove the effect of friction on the cavity wall, and to determine the optimal diameter of the cases, which are laid on the soil type and its humidity.

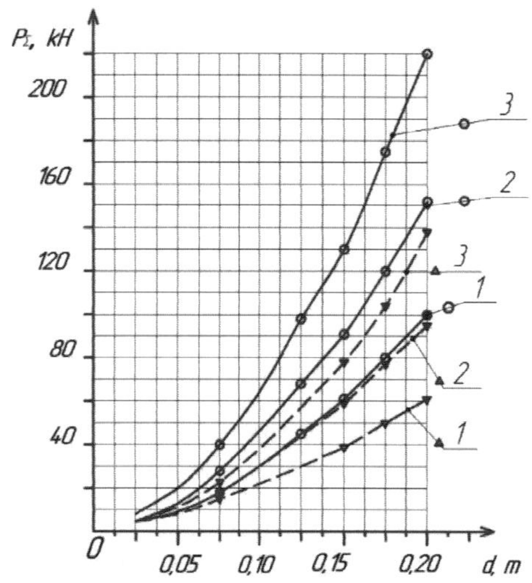

Figure 4. Dependence of the total resistance of the wedge tip on the diameter of the communications that are laid: 1 – rigid-plastic clay; 2 – semi-solid loam; 3 – solid loam; the point shows the results for the conical tip, and the triangle shows the results for the tip in the form of a wedge.

According to the graph in the figure 4 the total perforation resistance of the soil tip depends on the diameter of the tubes that are inserted into the caliber portion, the width of the rectangle of which lies within the optimum diameter of the conical puncture. Significant effort up to 2 times also differs from the type of soil.

5. Conclusions

It is established that with a static puncture of the soil with a conical tip, its diameter is limited by the calculated optimal value, which is not sufficient for the simultaneous laying of two or more linear-long underground communications.

The proposed design of the working equipment for static perforation of the soil and trenchless pro-laying simultaneously two or more underground linear-extended objects in the form of a tip consisting of a perforation e wedge-shaped part with semicons and a calibrating part in the form of a half-cylinder rectangle.

It is established that the ratio of the force of the frontal resistance of the soil to perforation conical tips can be more than 1.76 times compared to the wedge tip, depending on the number of cases simultaneously laid.

Acknowledgements

We would like to express our sincere gratitude to the scientific-industrial enterprise of LLC SPC "GAZTEHNIKA LTD" and personally to the director Vladimir Suponev for facilitating practical checks of the received theoretical researches and Director of LLC "MSK SYNERGYA" represented by Director Azimov Batyr for sponsorship, Chief Engineer Olexander Samokhvalov for technical support of research.

References

[1] State Construction Rules of Ukraine: 2008 SCR B.2.3-19-2008 (Kyiv, Ukraine: Rules of Ministry of Energy Construction of Ukraine) p 126

[2] Sectoral Building Standards of Ukraine: 2013 SBS B.2.5-00013741-72:2013. (Kyiv, Ukraine: Rules Ministry of Energy and Coal Industry of Ukraine) p 28

[3] Zhao J and Ling B 2014 Trenchless technology underground pipes *Machinery Industry Press* (Shanghai, China) p 134

[4] Cohen A and Ariaratnam S 2017 Developing a Successful Specification for Horizontal Directional Drilling Pipelines 2017: Planning and Design - Proceedings of Sessions of the Pipelines 2017 Conference 2017 (Phoenix, Arizona, USA: Pipelines) pp 553–563

[5] Esushutkin D, Smirnov Yu, Tsoi V and Isayev V 1990 High-performance hydropneumatic shock machines for laying engineering communications *Tutorial* (Moscow Russia: publishing

house Stroyizdat) p 171
[6] Kovanko V, Kravets S and Lukianchuk O 2015 Scientific bases of creation of soil-moving-tier machines and underground movable devices *Tutorial* (Rivne, Ukraine: National University of Water Management and Environmental Management) p 323
[7] Balezny S 2017 Features of processes of statistical puncture of the soil *Bulletin Kharkiv National Highway University* (Kharkov Ukraine: Kharkiv National Highway University) **76** p 141
[8] Romakin N and Malkova N 2007 Parameters of Working Tools for Soil Statistical Puncture *Tutorial Construction and Road Machines* (Moscow, Russia) pp 31–33
[9] Khachaturian S, Suponyev V and Oleksin V 2016 Investigation of the Process of Changing the State of the Soil Around a Horizontal Well After its Formation by the Method of Static Puncture of the Soil *Bulletin Kharkiv National Highway University* (Kharkov Ukraine: Kharkiv National Highway University) **73** pp 196–202
[10] Nilo Tsung, Mingming Zheng, Mohammad Najafi and Saleh Mehraban 2016 A *Comparative Study of Soil Pressure and Deformation of Pipes Installed by the Open-Cut Method and Trenchless Technology Pipelines 2016 Out of Sight, Out of Mind, Not Out of Risk* (Shanghai, China) p 135
[11] Lunys O, Neduzha L and Tatarinova V 2019 Stability research of the main-line locomotive movement *Proc. of the 23rd Int. Sci. Conf. Transport Means 2019 pt III* (Palanga, Lithuania: Kaunas Univ. of Technology) pp 1341–1345
[12] Raksha S, Anofriev P and Kuropiatnyk O 2019 Simulation modeling of the rolling stock axle test-bench *E3S Web of Conferences* 01032
[13] Kravets S, Posmituha O and Suponyev V 2017 Analytical Method for Determining Resistance Immersing Tapered Tip in Soil *Collection of scientific publications PSACEA* (Dnepro Ukraine: Dnieper Academy of Construction and Architecture) **97** 8 pp 91–98
[14] Posmituha O, Kravets S and Suponyev V 2017 Determining of the Equivalent and Optimum diameters of Cone-cylinder Tip With Soil Broaching Spurs *Science and transport progress* (Dnipro Ukraine: Dnieper National University of railway transport named of V Lazaryan) pp 89–98
[15] Posmituha O, Kravets S, Suponyev V and Glavatsky K 2018 Determination of equivalent and optimal sizes of wedge tip from flange for the static perforation of soil *MATEC Web of Conferences* 230 01011
[16] Kosyak O 2012 Cutting and cutting of soil with passive knives is analyzed *Collection of scientific publications PSACEA* (Dnipro Ukraine: Dnieper Academy of Construction and Architecture) **66** p 111
[17] Holovan V and Wolters O 2016 New annular working bodies of drilling machines *Bulletin Kharkiv National Highway University* (Kharkov Ukraine: Kharkiv National Highway University) **73** p 73
[18] Ziemskov V and Sudakov A 2005 *Analysis of the study of frontal resistance in trenchless pipeline laying by the puncture method. Tidings of the TSU. 6* (Tula Russia: Tula state University) p 38
[19] Gusev I and Chubarov F 2014 The use of controlled puncture of soil in trenchless pipe laying *The potential of modern science* (Kaluga: Russia) **2** p 33

IOP Conf. Series: Materials Science and Engineering **985** (2020) 012034 doi:10.1088/1757-899X/985/1/012034

Improvement of the open wagon for cargoes which imply loading with a "hat"

A Reidemeister[1,2], L Muradian[1], V Shaposhnyk[1], O Shykunov[1], O Kyryl'chuk[1] and V Kalashnyk[1]

[1]Dnipro National University of Railway Transport named after Academician V. Lazaryan, Department of Wagons, Lazaryana str, 2, 49010, Dnipro, Ukraine

[2]reidemeister.a@gmail.com

Abstract. The goal of the article is to evaluate the possibility of improving the design of an open wagon body in order to improve traffic safety when transporting cargoes whose height is beyond the upper belt of the side wall (cargo loading with a "hat"). To achieve it, the authors have proposed a variant of a pull-out bar in a wall rack. In lazy state such bars located inside racks, but during transportation of timber, pipes etc they are pulled out above the wall upper belt to prevent cargo from rolling out. These bars are quite light (mass of one pull-out bar is equal to 3 kg, that corresponds to 36 kg of an additional equipment per a wagon), but they considerably facilitate fixing cargo in an open wagon replacing additional wooden fixing racks. It also makes the transportation more reliable due to elimitation the possibility of skew and jamming of fixing equipment. Strength of the proposed element and the entire wagon side wall is estimated with the finite element method. The proposed bars withstand normative loads, and holes in the upper belt for them do not lead to desrease of the wall carrying capacity.

1. Summary

A sufficient amount of rolling stock influences the economic performance of the railway companies, and the technical condition of wagons influences the traffic safety. It is ensured not only by keeping the infrastructure and rolling stock in good working condition, but also by securing transported cargoes. Modern rolling stock should include new solutions that aim to increase the strength, dynamic performance, as well as to improve the design of the wagon in order to increase its versatility [1-4]. There is the highest level of demand for universal open wagons intended for transportation of coal, ore, timber, rolled metal, and other bulk and piece cargoes that do not need protection from precipitation. In works [5-10] it is pointed out the need to improve the rolling stock structures in order to increase its reliability and versatility.

2. Literature review

For transportation of a certain nomenclature of cargoes, such as timber, the additional fastening elements (wooden struts, wire ties) are used to fix cargo in a universal open car. When transporting boxes in several tiers, the height of the protruding above the upper belt part of the box should be no more than half its height, and the center of gravity of the protruding box should not overtop the upper belt, Figure

Content from this work may be used under the terms of the Creative Commons Attribution 3.0 licence. Any further distribution of this work must maintain attribution to the author(s) and the title of the work, journal citation and DOI.
Published under licence by IOP Publishing Ltd

1a [11, 12]. Most wagon models have a body height above rail heads from 3295 mm to 3800 mm. So, the situations can happen, when it is necessary to select the model of wagons according to the height of the body or to use additional means of fastening cargo for transportation. For timber transportation, round wooden posts with diameter of 120–140 mm in the lower cut, at least 90 mm in the upper one, and at least 100 mm at the level of the upper belt are additionally used. Wooden struts are passed through the timber claps and secured to the lower linking device by several threads of wire with diameter of at least 5 mm. Opposite racks should be tied with wire of diameter at least 6 mm laid in several threads, Figure 1b [11, 12].

(a) (b)

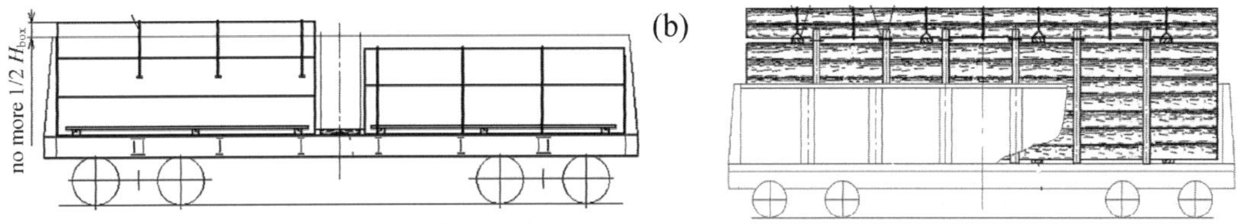

Figure 1. Placement and fixing cargo in an open wagon.

Figure 2. Transportation of timber in open wagons.

The main disadvantages of transportation of timber in the universal rolling stock (flat and open wagons, Figure 2) consist in the high consumption required for the transportation of equipment, its selection, complexity and risk of loading and unloading [13]. The data [14] show that at sites with intensive transportation of timber the number of detachments of wagons with timber cargo is up to 80 % of the total number. The main reasons for detachments of these wagons are misalignment of the wooden posts and the displacement of the cargo.

3. The pull-out bar in the side wall rack

To solve the above-mentioned problem of transportation in open cars cargoes loaded with a "hat", the authors proposed a telescopic pull-out bar shown in Figure 3. The cross-section of the bar consists of two welded angles No. 40 or square pipe. The weight of one bar is about 3 kg, or 36 kg per car with 12 pull-out bars in side walls. More bars can be installed in open wagons with racks in end walls.

The hole for pull-out bar is made in the upper belt at the point of its connection with the side wall rack. In idle position the pull-out bar is lowered into the wall rack and locked there. When necessary the pull-out bar is mounted at the desired height and securely locked in this position. The height of the pull-out bar in the working state is limited by the zonal loading gauge. At this stage of design the mechanism for fixing the bar is considered as a threaded rod with a screw-nut. When necessary to increase the height of the body the pull-out bar provide holes for bolt connection to fix the boards.

The theoretical assessment of the strength of the structure was carried out by the finite element method. The main loads on the side walls and pull-out bars are caused by lateral forces such as the inertial force in curved sections of the track and the force of bulk of rolling cargo bursting. These forces relate to III design mode. It is also required to check the strength of the side wall when unloading the wagon with the car dumper (I design mode). The mechanical properties of the material according to [15, Table. 12.1] are given in Table 1. The close analog of 09Г2С grade is the steel grade 13Mn6 (DIN 17145).

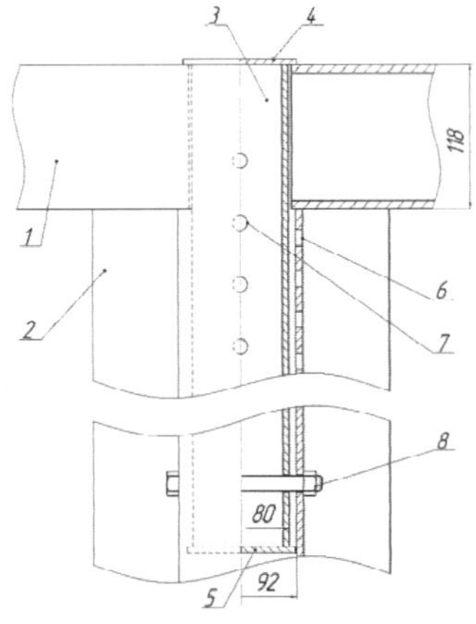

Figure 3. The pull-out bar in the side wall rack: 1 – upper belt; 2 – rack of the side wall; 3 – pull-out bar; 4 – patch; 5 – fastening; 6 – openings in the rack (for fixing the pull-out bar); 7 – holes in the pull-out bar (for mounting planks); 8 – fixing rod with a nut.

Table 1. Mechanical properties of material.

Part of a wagon	Steel grade	Stresses, MPa	
		permissible for I design mode $[\sigma_{\text{I}}]$	permissible for III design mode $[\sigma_{\text{III}}]$
Side wall of a body	09Г2Д, 09Г2С[a]	289	195

[a]The close analog of 09Г2С grade is the steel grade 13Mn6 (DIN 17145).

4. The strength of the side wall and the pull-out bar

The strength of the side wall and the pull-out bar is estimated separately. According to [15] the uniform load P_1 (the lateral inertial force of an upper part of the cargo in curved sections of a track) and the triangular load P_2 (the cargo bursting force) are applied to the pull-out bar (Figure 4). The height of the protruding part of the pull-out bar is 294 mm. We fix the model displacement at lines D_1 and D_2, it corresponds to resting on the threaded rod and on the hole in the upper belt.

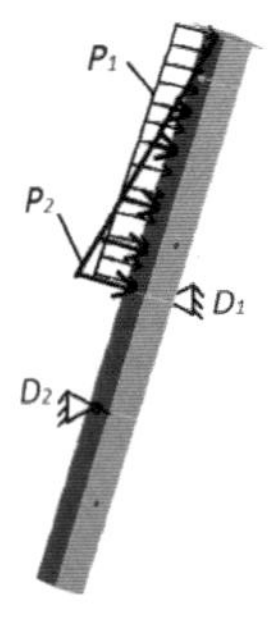

Figure 4. Loads acting on the pull-out bar.

We faply a similar load, namely a lateral inertial force and a force of cargo bursting, to the side wall. To the openings in the upper belt (Figure 5) we apply the forces corresponding to the reaction components R_{xi}, R_{yi} in the supports D_1 and D_2 of the pull-out bar (i is the number of the support D_i).

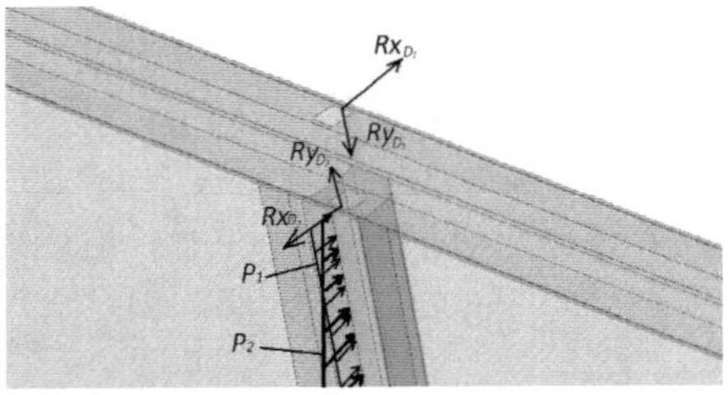

Figure 5. Loads of III design mode acting on the hole in the upper belt.

According to [15], for the conditions of unloading on wagons, the side wall of the open wagon was checked by I design mode for simultaneous action of vertical and horizontal loads, evenly distributed along the width of the upper belt at a length of 0.8 m (Figure 6). There a n such spots, each corresponds to a cur dumper clip.

The calculated vertical force, in Newtons, per plot was determined by the formula:

$$q = \frac{P}{n}, \quad (1)$$

where P is the gross weight of the wagon;

$n = 8$, i. e. there are 4 dumper clips per wall.

The lateral force is equal to $0.25q$.

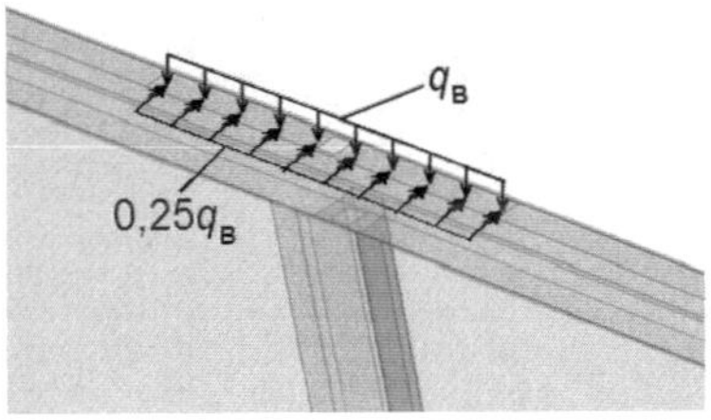

Figure 6. Loads on the hole in the upper belt acting during unloading on a car dumper.

Wall gravity is also taken into account.

Calculation carried out with the software package MSC.Patran/Nastran. The geometric model was meshed into finite elements with edge size of 10–50 mm for the side wall and 10 mm for the pull-out bar (Figure 7).

Figure 7. The part of a finite element mesh.

IOP Conf. Series: Materials Science and Engineering **985** (2020) 012034 doi:10.1088/1757-899X/985/1/012034

The equivalent stresses in the pull-out bar and in elements of the side wall under the loads of III design mode and in the side wall during unloading with the car dumper are shown in figures 8–10 and in Table 1.

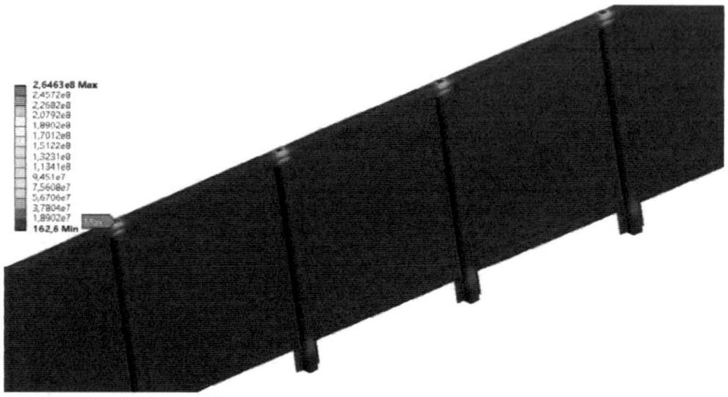

(a)

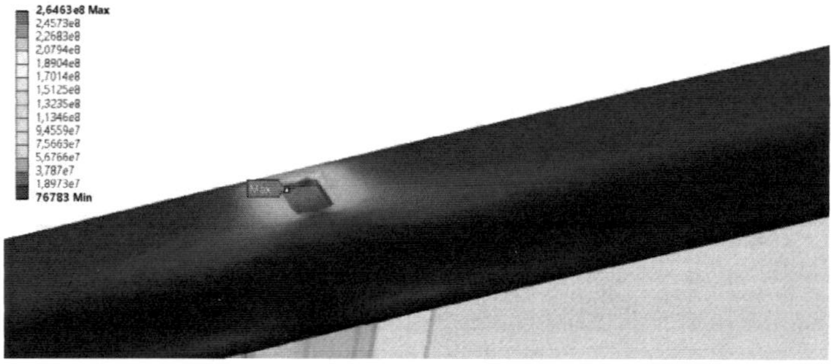

(b)

Figure 8. Stresses (von Mises) in the side wall when pull-out bars are used, Pa: (a) III design mode (general view); (b) the upper belt.

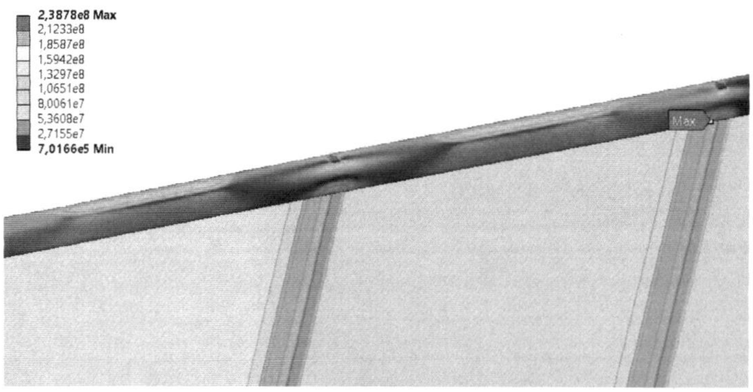

Figure 9. Stresses (von Mises) in the upper belt during unloading on a car dumper, Pa.

IOP Conf. Series: Materials Science and Engineering **985** (2020) 012034 doi:10.1088/1757-899X/985/1/012034

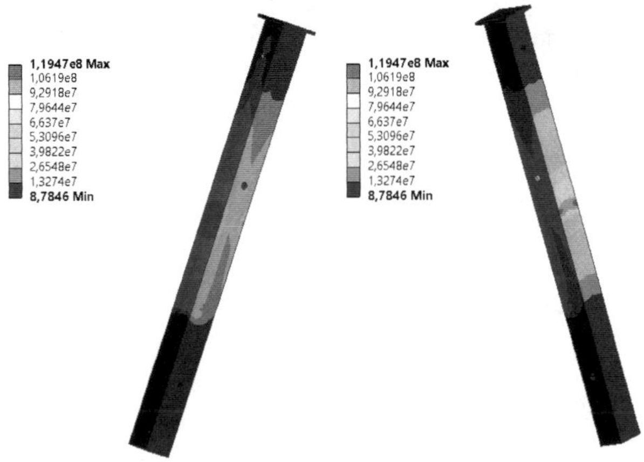

Figure 10. Stresses (von Mises) in the pull-out par, Pa.

Table 2. Maximum stresses in the side wall and pull-out bars.

Part of a wagon	Stresses, MPa				
	III design mode ($[\sigma_{III}]=195$ MPa)			unloading on a car dumper ($[\sigma_{I}]=289$ MPa)	
	below the upper belt	above the upper belt		upper belt	
		not reinforced	reinforced	not reinforced	reinforced
Rack of a side wall	110	87	87	245	245
Upper belt	63	265	158	238	238
Pull-out bar	—	120	120	—	—

As we can see, the stress $\sigma = 265$ MPa in the area of the hole in the upper belt, when bar is pull out, is higher than the permissible $[\sigma_{III}]=195$ MPa. So, this part of the upper bels needs reinforcement. As a reinforcement we offer a patch 54×54 mm and 4 mm thick. Stresses is the uppers belt with such reinforcement are shown in Table 2 and Figure 11.

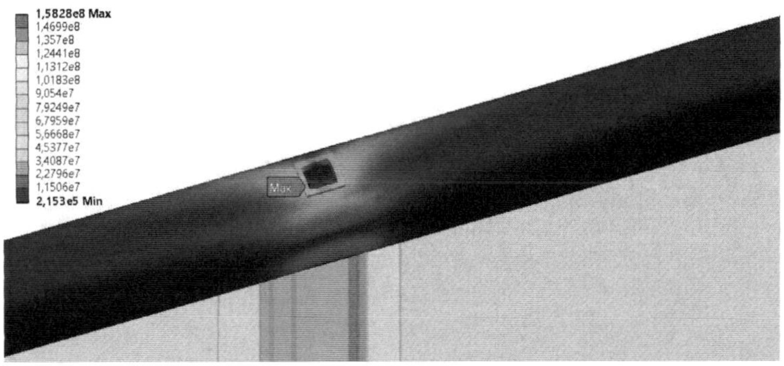

Figure 11. Stresses (von Mises) in the reinforced upper belt (III design mode), Pa.

5. Conclusions

The proposed pull-out bar of the semi-wagon increases not only its versatility, but also the safety of transportation when cargo is loaded with a "hat". It can be easily installed from a technological point of view, requires only small alterations of an upper belt and a rack, and does not lead to a noticeable increase in the wagon weight. The performed analysis has shown that the strength both the pull-out bar with selected dimensions and the side wall withstand loads the standard prescribes. The hole in the upper belt should not have sharp corners and requires reinforcement with a 4 mm thick patch. The mechanism of fixing of the pull-out bar needs more detailed consideration. It is advisable to consider the possibility of optimizing the proposed design by selecting the configuration of the cross section of the pull-out bar.

References

[1] Reidemeister A G and Shykunov A A 2015 Strength increase methods of the side frame of the bogie in three–piece trucks *Science and Transport Progress Bulletin of Dnipropetrovsk National University of Railway Transport* **59** pp 141–149

[2] Muradian L A, Shaposhnyk V Yu and Podosenov D O 2016 Improving the reliability of freight cars with new technologies of manufacturing and recovering of work surfaces *Electromagnetic compatibility and safety on railway transport* **11** pp 49–54

[3] Khilov I A 2012 *Improving the design of the body of a specialized open wagon* (St. Petersburg: PGUPS) p 166

[4] Fomin O 2015 Improvement of upper bundling of side wall of gondola cars of 12–9745 model *Metallurgical and Mining Industry* **1** pp 45–48

[5] Myamlin S, Lingaitis L P, Dailydka S, Bogdevičius M and Bureika G 2015 Determination of the dynamic characteristics of freight wagons with various bogie *Transport* **30** pp 88–92

[6] Kebal I 2016 Reducing metal consumption of gondola cars by increasing of strength characteristics of the supporting members of body *Railway Transport of Ukraine* **116/117** pp 42–46

[7] Kara S V 2018 Improvement of indicators of dynamics and strength of freight cars bogies by constructive changes *Visnik of the Volodymyr Dahl East Ukrainian National University* **235** pp 55–60

[8] Potapenko O 2017 Design features of frictional damper wedges of wagon running gears in Ukraine, CIS countries and North America *Wagon Park* **120–121** pp 13–16

[9] Myamlin S, Lunys O, Neduzha L and Kyryl'chuk O 2017 Mathematical Modeling of Dynamic Loading of Cassette Bearings for Freight Cars *Proc. of 21st Int. Sci. Conf. Transport Means 2017* pt. III (Juodkrante, Lithuania: Kaunas Univ. of Technology) pp 973–976

[10] Haraka S S, Sharma S C and Harsha S P 2014 Structural dynamic analysis of freight railway wagon using finite element method *Procedia Materials Science* **6** pp 1891–1898

[11] *Technical specifications for the placement and fastening of cargo: Appendix 3 to the Agreement on international railway freight traffic (SMGS)* 2019 p 820

[12] Turanov H T, Bondarenko A N and Vlasova N V 2006 *Fastenings of goods in wagons: a manual for universities of railway transport* (Yekaterinburg: Publishing House of Ural State Transport University) p 321

[13] Fedorov I V 2018 Improvement of cars designed to transport timbers *Bulletin of Scientific Research Results* 2 pp 105–126

[14] Korneev M V and Zorina Yu V 2017 Safety management of timber cargo transportation, *Transport Infrastructure of the Siberian Region* **1** pp 77–80

[15] DSTU 7598-2014. *The wagons. General requirements to calculation and design of new and modernized wagons of 1520 mm gauge track (non-self-propelled)* (Kiev: UkrNDNTs) p 157

IOP Conf. Series: Materials Science and Engineering **985** (2020) 012035 doi:10.1088/1757-899X/985/1/012035

Modernization of the buffer beam of PE2U traction unit electric locomotive

D Bannikov[1,2], **A Radkevich**[1] **and A Muntian**[1]

[1]Dnipro National University of Railway Transport named after Academician V. Lazaryan, Lazaryan str., 2, 49010, Dnipro, Ukraine

[2]bdo2020@yahoo.com

Abstract. The article presents the main results of a theoretical study on the analysis of the stress-strain state of the construction of the buffer bar of the electric locomotive for the control of the traction unit PE2U. The estimation of the possibility and effectiveness of attachment on the front wall of the buffer beam of a special securing bracket was considered as a separate task. In both cases, the finite element method based on the SCAD for Windows computing system was used. As a result, it was found that in general, the design of the buffer beam does not meet the requirements of the current standards. Up to 5 zones of high stress concentration are present in the structure, which, given the long service life of the machines (up to 40-50 years), are potential places for the development of fatigue damage. In order to balance the stress level, we recommended to strengthen the contact zone of the supports under the automatic coupling with the system of horizontal truss plates. The additional recommendation is to reinforce the lower sheet of the buffer beam structure up to and including its replacement by a sheet of greater thickness. Attachment to the front wall of the buffer beam of a special securing bracket for holding in emergency mode the coupling of the locomotive-car is possible provided that this zone is strengthened by the system of additional truss plates. The total increase in the mass of the buffer beam reaches 350 kg (25 % of the original weight of the buffer beam structure).

1. Introduction

Nowadays, a number of mining enterprises in Ukraine and the CIS countries continue to operate the PE2 series traction units. They have been developed and put into operation since the 1960s at the Dnepropetrovsk Electric Locomotive Building Plant (currently – SE Dniprovskiy Electric Locomotive Building Plant, Ukraine). Further, the traction unit was upgraded with a partial improvement of traction characteristics and was designated PE2U (Figure 1). It consisted of a DC (direct current) electric locomotive and two self-propelled dumping cars. In practice, the electric locomotive is often used as an independent traction unit of rolling stock, such as a locomotive.

As the life span of many of these machines is 40-50 years, more and more often their operation is accompanied by various failures and accidents due to the high level of abrasion of the structural elements. One such important structural element is the buffer beam, which directly receives the force of the automatic coupling. Therefore, special attention is paid to monitoring of its reliable operation.

Due to the widespread use of structural analysis software, mainly on the basis of the finite element method [1, 2], it has become possible to estimate the stress-strain state (SSS), including the buffer beam. Its calculations and design were done in its time using rather simplified models, and of course the spatial work of this construct was not taken into account.

Content from this work may be used under the terms of the Creative Commons Attribution 3.0 licence. Any further distribution of this work must maintain attribution to the author(s) and the title of the work, journal citation and DOI.
Published under licence by IOP Publishing Ltd

IOP Conf. Series: Materials Science and Engineering **985** (2020) 012035 doi:10.1088/1757-899X/985/1/012035

Figure 1. Traction unit PE2U.

Although the analysis of the SSS of various units of rolling stock is already quite commonplace both in domestic [3, 4] and in foreign practice [5, 6], however, there is no study of the construct of the PE2 series traction unit. The [7] presents the results of solution to the problem of balancing of this machine, yet the problem has been solved analytically without the use of the finite element method.

2. The purpose and objectives of the study

The main purpose of the conducted research was to estimate the SSS of the buffer beam of the electric locomotive of the PE2U traction unit with the provision of appropriate recommendations for improvement of its operation.

An additional practical task that was also considered during the research was to evaluate the possibility and effectiveness of attaching a special securing bracket to the front wall of the buffer beam. The function of this bracket is to keep the locomotive-car coupled in case of uncoupling when riding on steep inclines or in connection with abrasion and damage to the automatic coupling. The retention implied the presence of an additional chain, which was attached to the securing bracket. In fact, it is an additional emergency safety measure that has begun to be put into practice when operating electric locomotives, which is the variant of modernization of the buffer beam structure.

3. Research methodology

Structurally, the buffer beam is a complex spatial lamellar structure with overall dimensions of 2.1×1.5 m. On the sides, the beam joins the supporting side members of the locomotive underframe, which have a box section. The height of the buffer beam is 0.4 m. The top sheet has a thickness of 1 cm, the bottom sheet – 1.4 cm. In the structure of the buffer beam there are four vertical walls with a thickness of 0.8 cm, which connect the top and bottom sheets. The lower sheet and the vertical walls are provided with openings which somewhat lighten the structure. The central part of the buffer beam, which houses the automatic coupling system, has an additional system of special supports and fasteners. The total weight of the buffer beam structure reaches 1 400 kg.

The assessment of the SSS of the buffer beam structure was performed using the design-computing complex SCAD for Windows [8]. Although it is building-oriented software, it can be fairly correctly applied to the analysis of the work of machine-building structures, which is presented in [9].

A separate issue of the buffer beam structure modelling was that of taking into account a certain length of the supporting side members of the locomotive underframe, as it affects significantly the distribution of efforts in the structure. It was experimentally determined that modeling of beams at a length of 1.5 m from the back wall of the buffer beam (approximately 4 beam heights) is acceptable. This is sufficient to avoid the effect of the beam stiffness on the buffer beam structure SSS. The fastening of the model was made around the perimeter of the beam, as shown on Figures 2-4 in red lines. It was as rigid fastening according to results of researches presented in work [9].

The finite element model of the buffer beam structure was a plate system. The main defining type of loads, in accordance with the current norms [10, 11], was taken longitudinal compression force of 250 tons. When analysing the work of the securing bracket there was given a tensile load of 70 tons. These loads are shown on the Figure 2 and Figure 4.

Since the load was assumed to be symmetrical, half of the structure of the buffer beam was modeled geometrically with the task of the necessary conditions of symmetry of deformations (shown on Figures 2-4 in blue lines). As finite elements, we assumed a universal isoparametric shell finite element, available in the finite element base of used software complexes SCAD for Windows. Its use made it possible to reduce the total number of finite elements in the calculation model without losing the required accuracy of the obtained results [12]. Therefore, the overall dimension of the calculation model turned out to be relatively small – nodes within the order of 60,000, elements within the order of 65,000 with a total number of degrees of freedom within the order of 300.000.

4. Research results and their discussion

4.1. Stress-strain state of the buffer beam structure

Having assessed the stress-strain state (SSS) of the electric locomotive buffer beam structure of the PE2U traction unit, we should state the following. The most loaded element of the buffer beam is the internal vertical wall, which has supports that receive the load from the coupling. The stress level in it reached 330 MPa (zone G in Figure 2). You can also distinguish 3 more zones in the structure of the buffer beam, in which the stress level reaches 170-190 MPa. They are all located on the lower sheet. In other parts of the structure, the stress-strain state is more uniform and the stresses do not exceed the level of 100 MPa in total (Table 1). Due to the out-of-center application of the load from the coupling, we observed a vertical deformation of the buffer beam structure, which reached 18 mm.

It is worth noting that the allowable stress level for the buffer beam structure made from steel 295-09Г2С according to norm [10, table 5.1] is 265 MPa (for mode 1) and 195 MPa (for mode 2). Therefore, given the long life of PE2U machines, it is recommended to carry out special fatigue tests on samples taken from specified stress concentration points especially for zone G, where the level of stresses is higher than allowable stress level.

Such SSS required some measures to strengthen the structure of the buffer beam. For this purpose, we analyzed 6 structural variants of reinforcement of the support zone with the truss plates system. In this case, both their vertical and horizontal positions were considered. The most effective option was the setting of four horizontal truss plates, 16 mm thick each, which reduced the level of tension in this zone almost twice (Figure 3, Table 1). Additional local structural changes were also proposed to balance SSS of the buffer beam structure. The stress state of it is shown on Figure 4 and Figure 5 for variants № 1 and № 4 according to the Table 1.

4.2. Stress-strain state of the buffer beam structure with a securing bracket

Attaching of a special securing bracket, as shown by the simulation results, significantly worsens the SSS of the front wall of the buffer beam. The stress level in individual areas increases to 370 MPa (Figure 6, Table 2).

For SSS balancing, we analysed the system of special reinforcing truss in vertical, horizontal, fan-shaped and combined positions. As a result, the most rational variant was determined, which, however, led to increased weight of the buffer beam structure (together with the securing bracket) by almost 350 kg. Also, according to the results of the assessment of the SSS of the securing bracket itself, a number of recommendations were given to improve and modernize its structural concept.

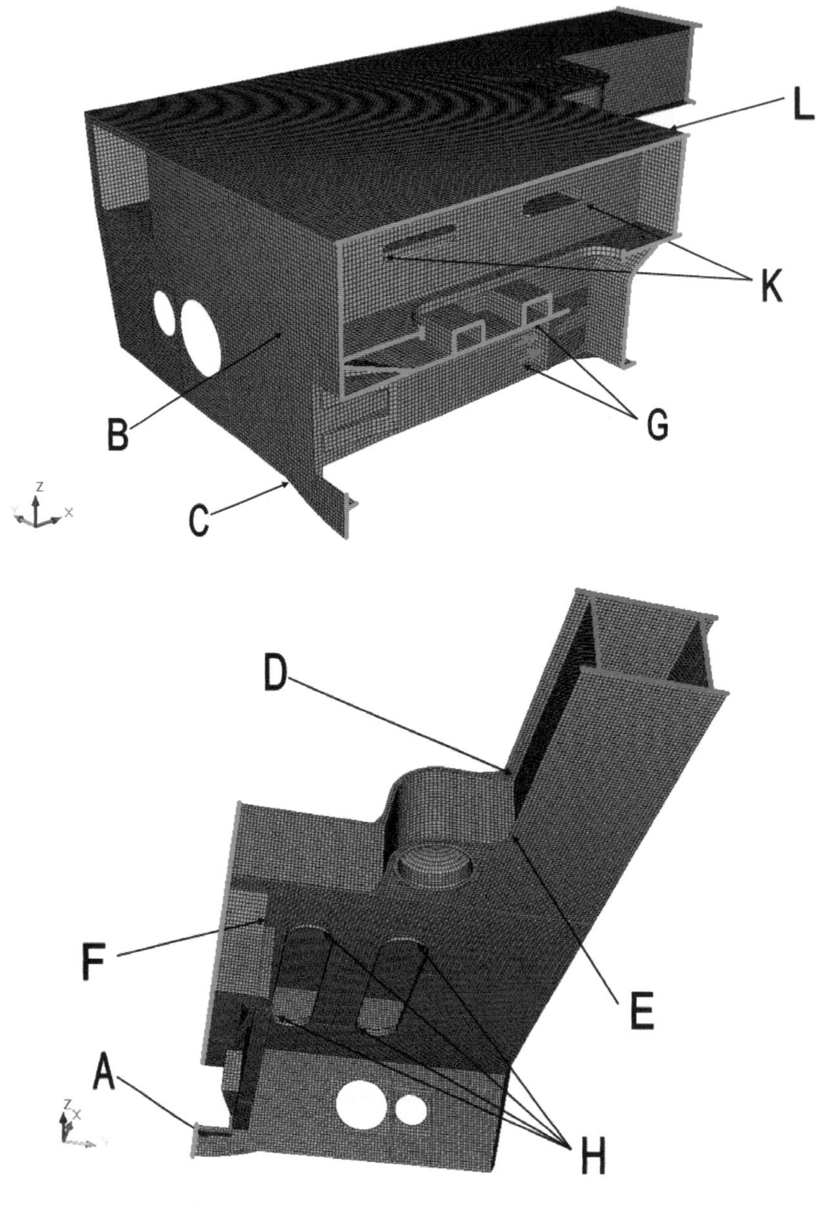

Figure 2. Stress concentration zones in buffer beam structure of PE2U traction unit.

Table 1. Level of stresses in the buffer beam structure of the PE2U traction unit (MPa).

Stress concentration zone									
A	B	C	D	E	F	G	H	K	L
1.Initial structure variant									
51	78	97	86	184	172	327	170	53	85
2.Vertical truss plate structure variant									
49	77	97	86	184	169	297	169	51	85
3. Horizontal truss plate structure variant (shortened)									
46	78	98	86	184	168	248	170	50	85
4. Horizontal truss plate structure variant (full length)									
45	78	98	86	184	162	170	170	51	85

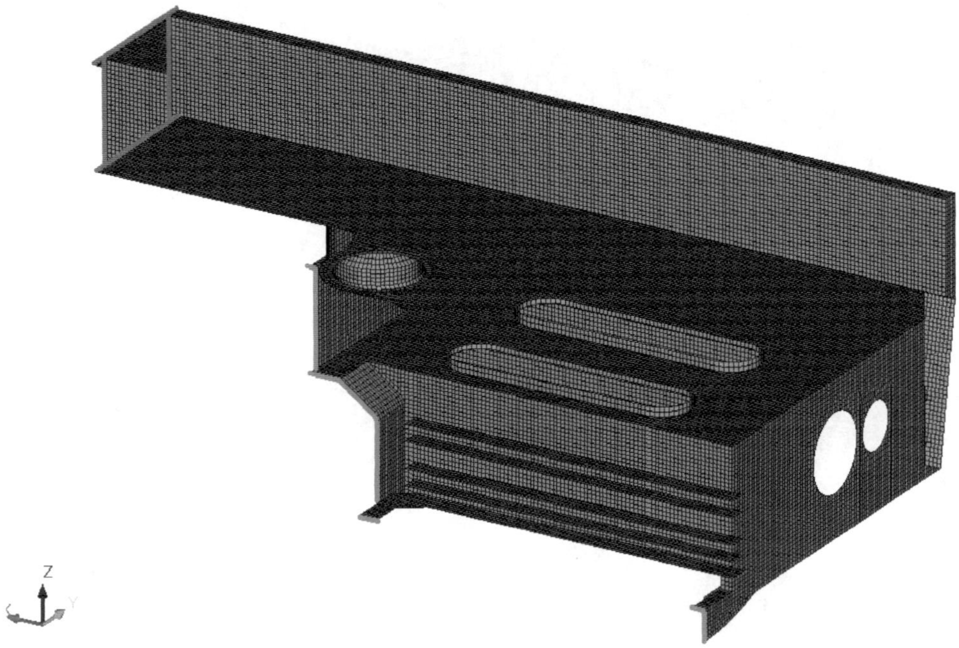

Figure 3. Strengthening of support zone with truss plates system in the buffer beam structure of PE2U traction unit.

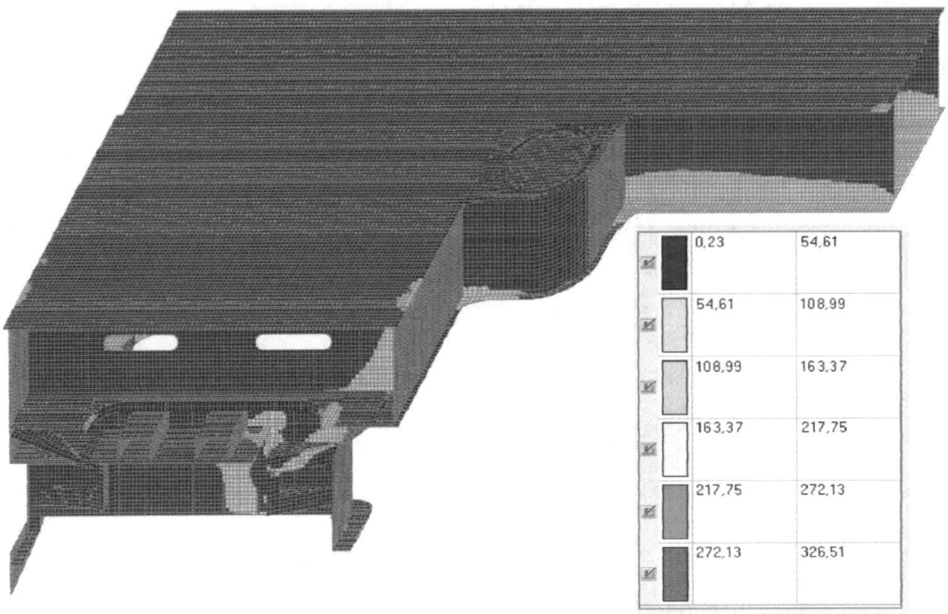

Figure 4. Stress state (MPa) of initial structure variant for the buffer beam structure of PE2U traction unit.

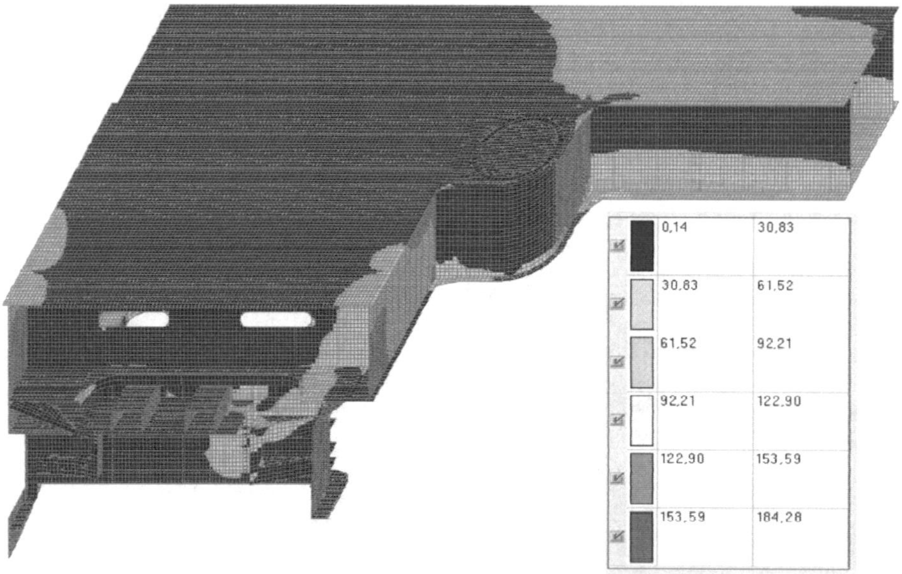

Figure 5. Stress state (MPa) of horizontal truss plate structure variant (full length) for the buffer beam structure of PE2U traction unit.

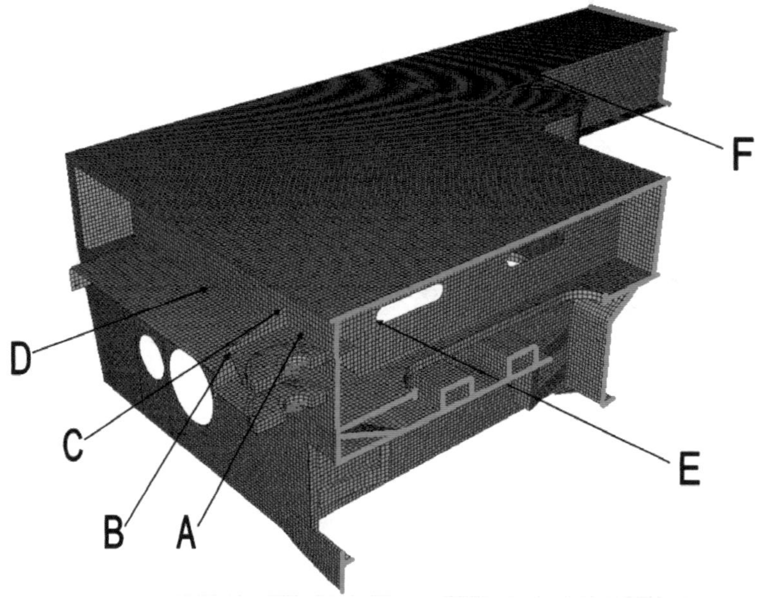

Figure 6. Stress concentration zones in buffer beam structure of PE2U traction unit with securing bracket.

Table 2. Level of stresses in the buffer beam structure of the PE2U traction unit with the securing bracket (MPa).

Stress concentration zone					
A	B	C	D	E	F
1. Initial variant of the structure with bracket					
368	330	230	184	121	112
2. Final variant of the structure with bracket					
160	177	123	102	85	74

5. Conclusions

The performed studies of the assessment of the stress-strain state of the locomotive buffer beam structure of the PE2U traction unit allowed us to make the following conclusions:

- In general, the design of the buffer beam does not meet the requirements of the current standards. Up to 5 zones of high stress concentration are present in the structure, which, given the long service life of the machines (up to 40-50 years), are potential places for the development of fatigue damage.
- In order to balance the stress level, we recommended to strengthen the contact zone of the supports under the automatic coupling with the system of horizontal truss plates. The additional recommendation is to reinforce the lower sheet of the buffer beam structure up to and including its replacement by a sheet of greater thickness.
- Attachment to the front wall of the buffer beam of a special securing bracket for holding in emergency mode the coupling of the locomotive-car is possible provided that this zone is strengthened by the system of additional truss plates. The total increase in the mass of the buffer beam reaches 350 kg (25 % of the original weight of the buffer beam structure).

References

[1] Kovtunenko V A and Kunisch K 2018 Revisiting generalized FEM: a Petrov–Galerkin enrichment based FEM interpolation for Helmholtz problem *Springer Science and Business Media LLC in Calcolo* **55** 38

[2] Perig A and Matveyev I 2019 FEM-based deformation regression analysis of ECAE strains *FME Transactions* **47** 4 pp 851–855

[3] Myamlin S, Lunys O, Neduzha L and Kyryl'chuk O 2017 Mathematical Modeling of Dynamic Loading of Cassette Bearings for Freight Cars *Proc. of 21st Int. Sci. Conf. Transport Means 2017* pt. III (Juodkrante, Lithuania: Kaunas Univ. of Technology) pp 973–976

[4] Kostritsa S, Pshinko A, Ursulyak L, Kuzyshyn A, Kramarenko M and Grechkin A 2019 Stress-strain state analysis of the leading car body of DPKr-2 diesel train under action of design and operational loads *MATEC Web of Conferences* **294** 03018

[5] Slavchev S, Stoilov V and Purgic S 2015 Static strength analysis of the body of a wagon *Journal of the Balkan Tribological Association* **21** 1 pp 38–57

[6] Shukri A, Willem D and Anas Y 2016 Modeling aspects of the design of railway vehicle structures and their crashworthiness *Proceedings of the Institution of Mechanical Engineers, Part F: Journal of Rail and Rapid Transit* **230** 6 pp 1575–1589

[7] Bannikov D and Radkevich A 2019 Analytical method for compiling and applying a ballast map for the traction unit PE2U *Eastern-European Journal of Enterprise Technologies: Engineering technological systems* **2-1** 98 pp 6–14

[8] Karpilovskij V S, Kriksunov Je Z, Maljarenko A A, Fialko S Ju, Perel'muter A V and Perel'muter M A 2015 SCAD Office. Version 21. Computing complex SCAD++ Moscow: SCAD SOFT p 848

[9] Bannikov D 2018 Usage of construction-oriented software SCAD for analysis of work of machine-building structures *Science and Transport Progress. Bulletin of Dnipropetrovsk National University of Railway Transport* **1** 73 pp 98–111

[10] Norms for calculating and evaluating the strength of load-bearing elements, dynamic qualities and impacts on the track of the crew of locomotives of railways of the Ministry of Railways of the Russian Federation of 1520 mm gauge 1998 Moscow: VNIIZHT p 145

[11] TU U 35.2-32495626-010-20 Traction units. Technical specifications 2003 Dnepropetrovsk: DEVZ p 48

[12] Nastran J, Sircelj B, Gams M and Bokal D 2019 Sensitivity analysis of computational models for solving paradox *Journal of Networking technology* **1** 10 pp 18–24

Analysis of physical and chemical transformations during thermal spraying of coatings based on carbides of tungsten and chromium

Yu O Kharlamov[1,2], V I Sokolov[1], O S Krol[1] and O V Romanchenko[1]

[1]Machinery Engineering and Applied Mechanics Department of Volodymyr Dahl East Ukrainian National University, Pr. Central 59-a, Severodonetsk, Ukraine, 93400

[2]krolos.snu.edu@gmail.com

Abstract. The formation of structure and properties of thermal coatings is largely determined by the transformation intensity of sprayed materials at all stages of obtaining of coating. It is shown that during D – gun spraying, in order to ensure optimal conditions for leading transformations and phenomena that have a decisive influence on structure and properties of sprayed coatings, composition of combustible mixture, geometry and dimensions of barrel, conditions for introducing powder into the barrel and a unit dose of powder, spraying distance, timing of spraying cycle and other technological methods can be used. The possibility of transformation in powders of tungsten carbide and chromium under different methods of thermal spraying is analyzed and the structure and properties of coatings are obtained. The influence of the conditions of D – gun spraying on spraying distances, gas flow rates, type of powder, etc.; phase composition and properties of coatings is studied. The basic physicochemical transformations that accompany D – gun spraying of coatings of alloy WC – C, such as oxidation and reduction of tungsten carbide in high – temperature oxygen – containing media, interaction of composite coating components are considered problems of modelling of bonding of particles with a part surface during thermal spraying of coatings are considered.

1. Introduction

To improve the reliability and durability of elements and units of machines and mechanisms [1-3], it is reasonable to apply thermal spraying of coatings [4-6]. Due to the use of protective coatings it is possible to solve a number of problems that allow to reduce material and energy consumption of production and operation of machines, to increase their durability, develop and introduce new types of modern machinery [7-9]. Therefore, it is relevant to use protective coatings to increase the strength and reliability of parts and structures of railway rolling stock [10-12].

The thermal spraying of coating encompasses a group of technologies [13-15] and equipment [16-18] that provide heating and acceleration of dispersed particles of material by high – temperature streams of gas and subsequent collision of particles with surface of sprayed part. A significant advantage of thermal spraying method is the ability to control the composition, structure and, accordingly, properties of obtained materials and coatings, including through the use of various powder compositions [19-21]. However, technological systems of thermal spraying are quite complex [22-24], the process of obtaining of coating is accompanied by various effects on particles of initial material of coating as a result of interaction of particles with environment gas, the processes of

Content from this work may be used under the terms of the Creative Commons Attribution 3.0 licence. Any further distribution of this work must maintain attribution to the author(s) and the title of the work, journal citation and DOI.
Published under licence by IOP Publishing Ltd

interaction of components of composite powders when heated, as well as the interaction of particles with part surface of and between themselves and the subsequent cooling [25-27].

Therefore, obtaining of coatings with high performance properties requires an understanding of the transformation processes with the material that sprayed by thermal spraying technology. Carbides of tungsten and chromium are used as a major component of many thermal sprayed coatings. Regularities and tendencies of change of structure of coatings $WC-Co$ and others. depending on complex of technological parameters of the thermal spraying technology have not been sufficiently studied. Under the action of high temperatures, oxidizing environment and shock interaction of particles of WC with part surface, the transition of $WC \rightarrow W_2C \rightarrow W$ is carried out. Intermetallides and double tungsten carbides are also found in coatings [28–30].

The objective of this paper is to make a comparative analysis of possible transformations in coatings based on carbides of tungsten and chromium by various methods of thermal spraying, as well as to study transformations occurring during thermal spraying of such materials.

2. Physico–chemical transformations during thermal spraying of coatings based on carbides of tungsten and chromium

The most widespread thermal spraying methods include gas flame, plasma and high speed methods.

At gas – flame spraying, acetylene – oxygen flame (or gases – acetylene substitutes) is used as a high – temperature heat source. The advantages of method are high productivity, locality of processing, a small influence on substrate, ability to apply coatings on products of large sizes, etc. The use of these technologies for spraying of carbides of tungsten and chromium is limited due to poor quality of obtained coatings.

Plasma spraying allows spraying of coatings of almost any type of material, although to optimize the process of spraying of powders with different melting points, some modification of plasmatron design may be required. During plasma spraying, tungsten monocarbide (WC) is decarbonised to tungsten sub-carbide (W_2C), as well as metallic tungsten. The presence of oxygen in plasma stream contributes to generation of oxycarbides in a significant amount, which is undesirable for wear resistance of coatings. In addition, high porosity and lower hardness are major drawbacks of plasma coatings for some applications. High – speed processes of thermal spraying, that based on the use of mechanical activation, minimize the decomposition of carbide phase due to lower heat enthalpy and shorter duration of coating formation process. Higher particle velocities provide a number of advantages to coating, such as lower porosity, higher adhesion strength and hardness.

Among carbides the WC has the highest wettability of metals which is used as bonding. This, along with its relatively high viscosity, makes WC the most widely used carbide for sintered alloys. Cobalt is the most commonly used binder because it has excellent wetting and adhesive properties. WC does not have a stable molten phase, but is easily converted to other phases when the carbon content is non – stoichiometric. Therefore, WC is a non – technological material for treatment at high temperatures, which lead to oxidation / decarburization during thermal spraying, especially in plasma.

The high – velocity spraying methods make it possible to significantly expand the capabilities of traditional thermal spraying methods. Here, the residence time of particles in the high – temperature stream is reduced and their overheating is reduced and, accordingly, the oxidation of particles and, accordingly, the content of oxides in coatings are reduced. Increasing the kinetic energy of particles leads to an increase of coatings density and adhesive strength of coatings with the substrate. The characteristic features of coatings obtained by high – velocity spraying are low porosity (less than $1{,}5\ \%$) and preservation of chemical composition of original powder, as well as high adhesion strength (more than $50...70\ MPa$). High – speed coating methods include D – gun spraying, high velocity plasma gas air spraying (HVPGA) and high velocity oxygen fuel spraying (HVOF). The particle velocity at spraying is: up to 1000 m/s at D – gun spraying, $500-1000$ m/s at HVPGA spraying and up to 800 m/s at HVOF spraying.

High durability of thermal coatings of $WC-Co$ is ensured by combination of WC as a solid component and cobalt as a plastic binder. In the case of spraying of $WC-Co$ coatings, high – speed methods are better than other methods of thermal spraying, as the high velocities and lower temperatures of powder particles reduce the degree of decomposition of WC during spraying process, thus reducing the hardness and wear resistance. Powders of $WC-Co$ with content of $6-7$ % of cobalt obtained by various methods are used for spraying. To increase the resistance of coatings of $WC-Co$ to corrosion, powders of $WC-Co$ are alloyed by chromium, since $Co-Cr$ matrix provides higher corrosion resistance compared to $WC-Co$ materials. Chromium content in powders is $4-8$ % ($WC-10\ \%\ Co-4\ \%\ Cr$; $WC-6\ \%\ Co-8\ \%\ Cr$). Depending on powder composition, type of equipment and the modes of spraying, microhardness of obtained $WC-Co$ coatings is $6-12\ MPa$ at D – gun spraying, $10-15\ MPa$ at HVPGA and HVOF spraying. The porosity of such coatings obtained by high – speed spraying techniques does not typically exceed 1.5 %. In the structure of coating obtained by D –gun method, there are oxide layers (about 5 %).

The microhardness of coatings of $WC\ -\ 9\ \%\ Co-9\ \%\ Cr$ obtained by the HVOF and HVPGA method is $11{,}0-11{,}7\ MPa$; the microhardness of the D – gun coating is $8{,}5\ GPa$. Reduced (by about $3\ GPa$) microhardness of D – gun coating and presence of oxide layers in microstructure are associated with application of oxidizing environment of detonation products of oxygen – propane – butane mixture during spraying process, which leads to development of oxidation processes of sprayed material and as a result of partial carbon loss and oxidation of matrix material.

In the study of structure and properties of composite coatings based on carbides of tungsten and chromium as a source used following powders: 1. Mechanical standard mixtures of WC – $8\%Co$ powders (VK8); 2. Conglomerated powders of alloy $WC-15$ % of alloy $NiCrSiB$ (VSNGN – 85); 3. Conglomerated powders of alloy $WC-25\ \%\ Co$ (VK25); 4. Conglomerated powders of Cr_xC_y-Ni alloy (KHN15C); 5. Mechanical mixtures of powders of chromium carbides and nickel of KHN type.

The chemical composition of D – gun sprayed coatings of standard VK8 powder, depending on spraying distance, was studied by method of local X – ray spectroscopic microanalysis (Table 1). The change of carbon content of coatings may be related to both the dissociation and interaction of WC with oxygen and nitrogen that present in high temperature stream. The monocarbide WC melts with decomposition at $2600\ °C$. Sub–carbide W_2C is thermally resistant to $2750\ °C$, i.e. practically to melting point ($2730\pm15\ °C$).

Table 1. The chemical composition of D – gun sprayed coatings of VK8 powder (in mass, %).

	Spraying distance, mm		
	50	100	150
Co	9.3; 9.06; 9.22	9.05; 11.82	9.55
C	3.05; 5.72; 5.98	2.5; 2.59	3.42; 3.37
W	87.65; 85.22; 84.8	88.44; 85.59	87.03; 86.71

The influence of spraying distance, cost of working gas flow, powder and transport gas on phase composition of coatings of VK8 was studied (Tables 2, 3). At the consumption of acetylene $179\ см^3/cycle$ ($\gamma = O_2/C_2H_2 = 1{,}07$) the coatings are containing W, W_2C, WC, Co_7W_6. At the consumption of acetylene $183\ cm^3/cycle$ ($\gamma = 1{,}04$) coatings consist of W_2C, $Co_3W_9C_4$, Co_3W_3C, WC.

With excess in the gas phase of oxygen, the following reactions of its interaction with WC are thermodynamically possible [29]:

$$WC + 2O_2 \Leftrightarrow WO_2 + CO \quad (1)$$

$$WC + 2O_2 \Leftrightarrow WO_2 + CO; \tag{2}$$

$$WC + 2O_2 \Leftrightarrow WO_2 + CO_2; \tag{3}$$

$$WC + 15/8O_2 \Leftrightarrow 1/4W_4O_{11} + CO; \tag{4}$$

$$WC + 19/8O_2 \Leftrightarrow 1/4W_4O_{11} + CO; \tag{5}$$

$$WC + 2O_2 \Leftrightarrow WO_3 + CO; \tag{6}$$

$$WC + 5/2O_2 \Leftrightarrow WO_3 + CO_2. \tag{7}$$

Table 2. Spraying modes of coatings of VK8 powder.

No	$Q_{C_2H_2}$ $cm^3/cycle$	l, mm	n	t_c, mm	t_s, μm
1	200	120	30	0.5	16.7
2	183	120	80	0.56	7
3	179	120	50	0.42	8.4
4	163	150	80	1.04	13
5	179	150	50	0.71	14.2
6	183	150	30	0.41	13.7
7	183	180	30	0.33	11
8	200	180	50	0.31	6.2
9	163	180	50	0.34	6.8

Note: $Q_{C_2H_2}$ – consumption of acetylene per cycle; l – is the spraying distance; n – is the number of single spraying cycles; t_c – is the total thickness of spraying spot; t_s – is the thickness of a single sprayed spot

Table 3. Phase composition and modulus of elasticity of D – gun sprayed coatings of VK8 powder at different spraying modes.

N	Phases	E, MN/m^2
1	W, W_2C, WC, $C_{o3}W_3C$	$1.99 \cdot 10^5$
2	W, W_2C, $C_{o3}W_3C$ WC	$3.0 \cdot 10^5$
3	W, W_2C, WC, $C_{o7}W_6$	$3.2 \cdot 10^5$
4	W, W_2C, WC, $C_{o7}W_6$	$2.3 \cdot 10^5$
5	W, WC, $C_{o7}W_6$	$3.4 \cdot 10^5$
6	W, W_2C, WC, $C_{o3}W_3C$	$3.04 \cdot 10^5$
7	W_2C, $C_{o3}W_9C_4$, $C_{o3}W_3C$	$2.33 \cdot 10^5$
8	W_2C, WC, $C_{o7}W_6$	$2.6 \cdot 10^5$

The interaction reactions occurring with the formation of CO_2 are thermodynamically more likely than similar reactions with the formation of CO. The most likely is the last reaction. The thin powder WC starts to oxidize rapidly at 500 – 520 °C, the sub–carbide W_2C at the same temperature is oxidized to WO_3 [30].

Tungsten carbide is decarburized at temperatures above 750^oC:

$$WC + H_2O \Leftrightarrow W + H_2 + CO(CO_2) \quad (8)$$

As water vapor content in gas environment increases the carbon losses increase sharply.

Some of cobalt grains of output powder are in an oxidized state. Cobalt oxides can be completely recovered by hydrogen or carbon. At temperatures above 190 °C a reaction is possible:

$$Co_3O_4 + H_2 \Leftrightarrow Co + H_2O. \quad (9)$$

In temperature range of $200 - 400\ ^oC$:

$$CoO + H_2 \Leftrightarrow Co + H_2O \quad (10)$$

If soot (free carbon) is formed during combustion of combustible mixture, then if it is contacted with particles of oxidized cobalt at temperatures above $500\ ^oC$, a reaction is possible:

$$2Co + 2C = 2Co + CO_2 \quad (11)$$

The process of carbonation of tungsten is possible mainly due to content of carbon at gas environment. Carbidization reactions can be represented by following equations:

$$2C + H_2 = C_2H_2; \quad (12)$$

$$2W + C_2H_2 = 2WC + H_2; \quad (13)$$

$$C + CO_2 = 2CO; \quad (14)$$

$$2CO + W = WC + CO_2; \quad (15)$$

$$2W + CH_4 = W_2C + 2H_2; \quad (16)$$

$$W_2C + CH_4 = 2WC + 2H_2 \quad (17)$$

$$2W + 2CO = W_2C + CO_2 \quad (18)$$

$$W_2C + 2CO = WC + CO_2 \quad (19)$$

The diffuse layer formed on particle surface of W consists of two phases – WC and W_2C, and initially a WC phase is formed on the surface, from which carbon diffuses inside with formation of W_2C phase layer.

A similar mechanism is also manifested during decarburization of tungsten carbide particles. Initially, a W_2C phase is formed on surface into which carbon from the particle nucleus diffuses.

If the particles of WC and Co are in contact during movement, reaction at a temperature above $530\ ^oC$ is possible:

$$2CoO + WC = CO_2 + 2Co + W \quad (20)$$

At temperatures above 850 ^oC a reaction is possible:

$$3CoO + 3WC = W_3Co_3C + CO + CO_2 \quad (21)$$

It is also possible to dissolve of carbon in cobalt at direct contact of particles and through the gas phase at a temperature above $600\ ^oC$, eutectic melting of $Co + WC$ at a temperature of $1300 - 1370\ ^oC$.

3. Discussion

Physico – chemical transformations in powder particles are due not only to composition of detonating mixture of gases, but also to a number of other technological factors (Table 2). First of all, it is necessary to take into account features of output placement in the barrel of a fresh charge – detonation mixture and portions of powder of sprayed material, which determine the nature of dynamic, thermal and chemical interaction of particles with combustion products and then with the surface of workpiece. This is evidenced by significant dependence of coatings hardness from consumption of acetylene (Figure 1).

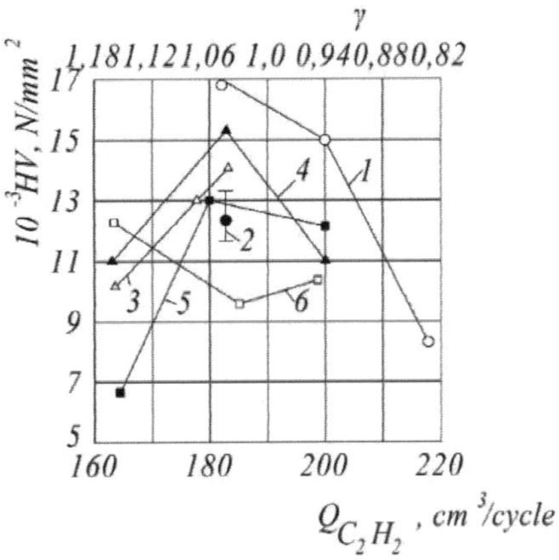

Figure 1. The dependence of coatings hardness of powder VK8 from consumption of acetylene. Spraying distance, mm: 120 $(1,2)$, 150 $(3,4)$, 180 $(5,6)$. Consumption of transporting gas, cm³/cycle: 35 $(1,3,5)$, 45 $(2,4,6)$.

For sintered hard alloys of $WC\ -8\ \%\ Co$ the HV hardness at size of main mass of carbide grains of $1 - 2\ \mu m$ is $12160\ MPa$, and at size of main mass of carbide grains of $2 - 5\ \mu m$ is $-10300\ MPa$. The W_2C sub-carbide has a slightly higher hardness compared to the WC monocarbide, but has a lower strength. As the consumption of acetylene decreases, the total consumption of fuel mixture for spraying cycle is also reduced, as well as powder cloud is moved to closed end of the barrel. This leads to a longer residence of powder in hot detonation products and the intensification of the $WC \rightarrow W_2C \rightarrow W$ conversion. The decrease of coatings hardness with increasing consumption of acetylene to $200{-}218\ cm^3/cycle$, associated with increase of content of free carbon coatings.

With increase of mass of a single dose of powder introduced into the barrel, and thus the thickness of single coating layer, the hardness of coatings decreases, their composition increases content of tungsten (Figure 2). The transformation phase of output of VK8 powder is significantly influenced by the type of transporting gas. The use of air intensifies the process of transformation $WC \rightarrow W_2C \rightarrow W$.

Similar phase transformations are also observed in coating from conglomerated powder – VK25 ($WC\ -\ 25\ \%\ Co$). In the output powder the main phases are monocarbide WC and Co, there are a small amount of W_2C carbides. On the radiograph of coating, the major phases are W_2C sub-carbide, Co, and a small amount of WC monocarbide is contained.

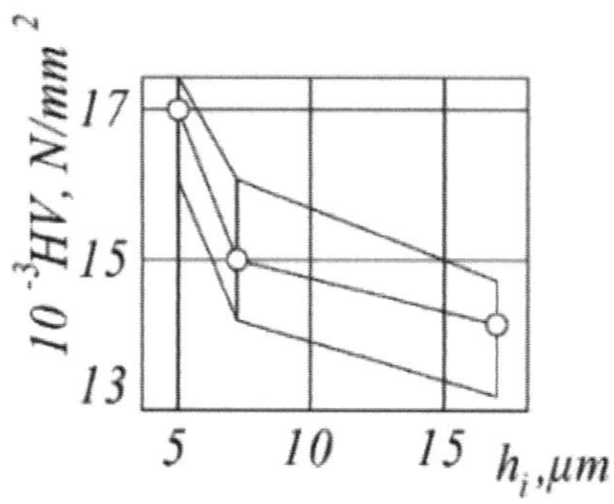

Figure 2. Dependence of coatings hardness of VK8 powder from thickness of a single layer.

Similar patterns are evident during spraying of coating from output conglomerated powder – VSNGN – 85 (WC – 15 % of alloy SNGN). The radiograph of output powder contains only lines of WC monocarbide. Two new phases appear in the coatings – W_2C sub-carbide and W tungsten. As the spraying distance increases, the relative intensity of W_2C and W lines increases, it means that the content of these phases increases.

The change of hardness and other characteristics of chromium carbide coatings is also associated with a change of intensity of transition of higher carbides to lower ones ($Cr_3C_2 \rightarrow Cr_7C_3$, $Cr_{23}C_6$), whose hardness is much lower than Cr_3C_2 carbide. The X–ray structural studies of phase composition of coatings from conglomerated powder – KHN15 (chromium carbides – 15% $NiCr$) were carried out. The output powder contains the Cr_3C_2, Cr_7C_3, $Cr_{23}C_6$, Ni, Cr phases.

At a spraying distance of 150 mm the coatings containing Cr_7C_3 phases (hexagonal), Cr_3C_2, Cr_3Ni_2, $NiCr$, small amount of Cr_7C_3 (triclinic) and Cr_2C, Cr_7C_3 (cubic), $Cr_{23}C_6$. When the spraying distance is increased to 200 – 250 mm that means the increase of residence time of powder particles in high – temperature stream, the lines of Cr_2C, Cr_3Ni_2, Cr_7C_3 (cubic), $Cr_{23}C_6$, are disappear and the traces of Cr_2O_3 oxide appear. From position of thermodynamic the most stable is higher chromium carbide – Cr_3C_2. By reducing of oxygen environment stability, carbides of VI group form a variety of Cr_3C_2, WC, Mo_2C, Cr_7C_3, $Cr_{23}C_6$.

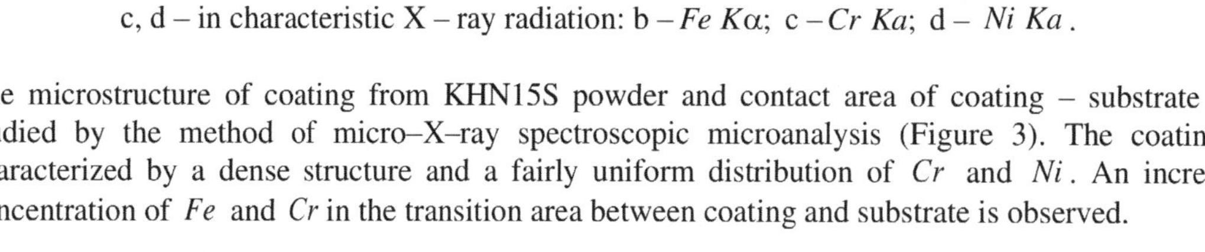

Figure 3. The microstructure of coatings of KHN15S powder: a – in secondary electrons, ×500; b, c, d – in characteristic X – ray radiation: b – $Fe\ K\alpha$; c – $Cr\ Ka$; d – $Ni\ Ka$.

The microstructure of coating from KHN15S powder and contact area of coating – substrate was studied by the method of micro–X–ray spectroscopic microanalysis (Figure 3). The coating is characterized by a dense structure and a fairly uniform distribution of Cr and Ni. An increased concentration of Fe and Cr in the transition area between coating and substrate is observed.

4. Summary

- The structure and properties of thermal coatings are determined by nature and intensity of processes of physico–chemical transformations of powder particles and their interaction with working environment at all stages of coating obtaining, including manufacture and preparation of powder. At D – gun spraying to provide optimal conditions for the flow of conductive transformations and phenomena, that have a decisive influence on structure and properties of sprayed coatings, the regulations of such parameters as composition of combustible mixture, geometry and dimensions of barrel, conditions of introduction of powder into the barrel, a single powder dose, spraying distance, timing of spraying cycle and other technological techniques can be applied.

- The basic physico–chemical transformations accompanying D – gun spraying of coatings from powders of $WC - Co$ alloy are studied: oxidation and reduction of tungsten carbide in high – temperature oxygen environment, interaction of components of composite coating.
- The general principle of design of optimal operating technology of D – gun spraying of coatings is to provide a homogeneous heating conditions, composition of surrounding gas environment and same residence time of particles in spraying stream.
- To obtain coatings based on transition carbides with high adhesion and cohesion strength and preservation of initial phase composition, it is recommended to ensure their formation from solid particles heated to pre – melting temperatures, at collision velocities with substrate of $300 - 600\ m/s$ and cooling velocities of single layers of molded coatings exceeding $5 \times 10^3\ K/s$, that excluding carbide decomposition.
- The use of protective coatings should be recommended to increase the strength and reliability of parts and structures of railway rolling stock.

References

[1] Okorokov A, Fomin O, Lovska A, Vernigora R, Zhuravel I and Fomin V 2018 Research into a possibility to prolong the time of operation of universal semi-wagon bodies that have exhausted their standard resource *Eastern-European journal of enterprise technologies* **3/7** 93 pp 20–26

[2] Krol O and Sokolov V 2019 Parametric Modeling of Transverse Layout for Machine Tool Gearboxes *Advances in Manufacturing II. Lecture Notes in Mechanical Engineering* **4** pp 122–130

[3] Krol O and Sokolov V 2019 Parametric Modeling of Gear Cutting Tools *Advances in Manufacturing II. Lecture Notes in Mechanical Engineering* **4** pp 3–11

[4] Davis J 2004 *Handbook of Thermal Spray Technology* (Materials Park, OH: ASM International) p 337

[5] Pierre L, Joachim V R, Maher I 2014 *Thermal Spray Fundamentals: From Powder to Part* (New York: Springer Science + Business Media) p 1587

[6] Budag'janc N A 1998 *D–Gun processes in industry* (Lugansk: Publishing house of VUGU) p 223

[7] Fesenko A, Basova Y, Ivanov V, Ivanova M, Yevsiukova F and Gasanov M 2019 Increasing of Equipment Efficiency by Intensification of Technological Processes *Periodica Polytechnica Mechanical Engineering* **63** 1 pp 67–73

[8] Sokolov V, Krol O and Stepanova O 2018 Automatic Control System for Electrohydraulic Drive of Production Equipment *International Russian Automation Conference IEEE*

[9] Sokolov V, Krol O and Stepanova O 2020 Choice of Correcting Link for Electrohydraulic Servo Drive of Technological Equipment *Advances in Design, Simulation and Manufacturing II. DSMIE 2019. Lecture Notes in Mechanical Engineering* pp 702– 710

[10] Fomin O, Gerlici J, Lovska A, Kravchenko K, Fomina Yu and Lack T 2019 Determination of the strength of the containers fittings of a flat wagon loaded with containers during shunting *IOP Conf Ser: Mater Sci Eng* **659** 012056

[11] Bondarenko I, Lunys O, Neduzha L and Keršys R Dynamic track irregularities modeling when studying rolling stock dynamics *Proc. of the 23rd Intern Sci Conf Transport Means 2019* pt **II** (Palanga: Kaunas Univ. of Technology) pp 1014–1019

[12] Zelenko Yu, Lunys O, Neduzha L and Steišūnas S 2019 The assessment of negative impact of oil products on railroad track and rolling stock constructions *Proc of the 23rd Intern Sci Conf Transport Means 2019* pt **III** (Palanga: Kaunas Univ. of Technology) pp 1300–1306
[13] Pawlowski L 2008 *The Science and Engineering of Thermal Spray Coatings* (Chichester: John Wiley & Sons, Ltd) p 647
[14] Karpus V, Ivanov V, Dehtiarov I, Zajac J, Kurochkina V 2019 Technological assurance of complex parts manufacturing *Advances in Design, Simulation and Manufacturing. DSMIE 2018. Lecture Notes in Mechanical Engineering* pp 51–61
[15] Rogovyi A 2018 Energy performances of the vortex chamber supercharger *Energy* **163** pp 52–60
[16] Pavlenko I, Simonovskiy V und Demianenko M 2017 Dynamic analysis of centrifugal machines rotors supported on ball bearings by combined application of 3D and beam finite element models *IOP Conf Ser: Mater Sci Eng* **233** 012053
[17] Sokolov V, Krol O and Stepanova O 2019 Nonlinear simulation of electrohydraulic drive for technological equipment *Journal of Physics: Conference Series* **1278** 012003
[18] Fedorovich V and Mitsyk A 2014 Mathematical simulation of kinematics of vibrating boiling granular medium at treatment in the oscillating reservoir *Key Engineering Materials* **581** pp 456–461
[19] Korobov Yu S 2016 *Analysis of properties of thermal coatings. Part 2. Evaluation of coating parameters* (Yekaterinburg: Publishing House of Ural University) p 92
[20] Rjahovskij A, Kosenko V and Vlasenko V 2012 Features of estimation of adhesion strength of D–Gun spraying coatings *Weapons systems and military equipment* **3** 31 pp 215–217
[21] Faradzhallah M, Panarin V and Bys' S 2011 The problem of technological residual stresses in improving the quality of D–Gun spraying coatings *Visnik of Khmelnytskyi National University* **2** pp 14–18
[22] Ilyushchenko A, Okovityj V, Kundas S and Formanek B 2002 *Formation of thermal sprauing coatings: theory and practice* (Minsk: Bestprint) p 480
[23] Sokolov V 2019 Diffusion of Circular Source in the Channels of Ventilation Systems *Advances in Engineering Research and Application. ICERA 2018. Lecture Notes in Networks and Systems* **63** pp 278–283
[24] Rogovyi A, Khovanskyy S, Grechka I and Pitel J 2020 The Wall Erosion in a Vortex Chamber Supercharger Due to Pumping Abrasive Mediums *Advances in Design, Simulation and Manufacturing II. DSMIE 2019. Lecture Notes in Mechanical Engineering* pp 682–691
[25] Movshovich A, Chernaja Yu and Ishchenko G 2011 Investigation of influence of physico – mechanical characteristics of D–Gun spraying coatings on wear resistance of cutting elements of reworkable dies *Vestnik NTU «HPI»: collection of scientific papers, thematic release: New solutions in modern technology* **45** pp 22–28
[26] Maksimovich G G, Shatinskij V F and Kopylov V I 1983 *Physicochemical processes during plasma spraying and destruction of coated materials* (Kyiv: Nauk. Dumka) p 264
[27] Lin C K and Berndt C C 1994 Measurement and analysis of adhesion strength for thermally sprayed coatings *Journal of Thermal Spray Technology* **3** 1 pp 75–104
[28] Heintze G N and McPherson R 1988 Fracture toughness of plasma sprayed zirconia coatings *Journal of Surface Coating Technology* **134** pp 15–23
[29] Samsonov G V, Vitrjanjuk V K and Chaplygin F I 1974 *Tungsten carbides* (Kyiv: Nauk. Dumka) p 175
[30] Babich M M 1975 *Hard alloys heterogeneity by carbon content and its elimination* (Kyiv: Nauk. Dumka) p 175

IOP Conf. Series: Materials Science and Engineering **985** (2020) 012037 doi:10.1088/1757-899X/985/1/012037

Research of the mechanism of particles bonding with substrate during thermal spraying of coating

Yu O Khrlamov[1,2], V I Sokolov[1], O S Krol[1] and O V Romanchenko[1]

[1]Machinery Engineering and Applied Mechanics Department of Volodymyr Dahl East Ukrainian National University, Pr. Central 59-a, Severodonetsk, Ukraine, 93400

[2]krolos.snu.edu@gmail.com

Abstract. The problems of modelling of bonding of particles with a part surface during thermal spraying of coatings are considered. The formation of adhesive and cohesive strength of thermal coatings is considered from the standpoint of the theory of welding of materials in solid phase. The formation of interatomic connections between particle materials and parts is accompanied by deformation processes in contact zone, which facilitate the unloading and rupture of connections in the metal – oxygen system, loading and connections between atoms of interacting materials. The kinetics of formation of interatomic connections between materials of particle and substrate at high collision velocities is satisfactorily described by a modified semi – empirical equation that uses experimentally measured quantities: the magnitude of tensile deformation of interatomic connection, the Young's modulus, the coefficient of thermal linear expansion. Possible influence of nanotopography of substrate surface on the interaction with sprayed particles and formation of coatings with high adhesive and cohesive strength is shown.

1. Introduction

An important reserve for increasing of strength and reliability of the elements and units of machines and mechanisms [1-3] is the application in the technology of their manufacture of thermal spraying of coatings [4-6]. The application of protective coatings allows to solve various problems related with the reduction of material and energy consumption of production and operation of machines, increase their durability, creation and development of new products of modern technology [7-9]. In this regard, the use of protective coatings is becoming relevant to increase the strength and reliability of parts and structures of railway rolling stock [10-12].

Systematic analysis of physicochemical processes and phenomena during thermal spraying showed the important role of bonding and forming of strong connections in process of coatings formation. Adhesive and cohesive strength of thermal coatings are one of their most important properties that determine their practical application [13-15].

The problem of modeling of particle bonding with a part surface is one of the basic in theory and practice of thermal spraying technologies of coatings (TST). Quite a detailed analysis of it is contained in many papers [16-18]. The chemical interaction between the contacting materials is possible when the metal – oxygen ($Me - O$) connections are broken from each of contacting surfaces and deleting from zone of oxygen connection. Possible channels of activation leading to rupture of interatomic connections in system ($Me - O$) are: 1. Thermal; 2. Mechanical; 3. The channel associated with release of energy upon exit to surface as a result of plastic deformation of structural defects such as

Content from this work may be used under the terms of the Creative Commons Attribution 3.0 licence. Any further distribution of this work must maintain attribution to the author(s) and the title of the work, journal citation and DOI.
Published under licence by IOP Publishing Ltd

IOP Conf. Series: Materials Science and Engineering **985** (2020) 012037 doi:10.1088/1757-899X/985/1/012037

dislocations, vacancies, interstitial atoms; 4. Chemical activation (the interaction of oxide layer or chemically adsorbed oxygen with a substance by mechanism of reduction reaction).

There are also other ideas about the formation of strong thermal coatings. Many researchers prefer the mechanical nature of bonding, but there are no theories that allow to calculate the strength of bonding that formed. More justified is application of the theory of bonding of materials during welding of materials in solid phase [4, 5, 13]. According to these ideas, the interaction of substrate materials and particles during thermal spraying on local areas of surface is represented as three successive stages: 1) formation of physical contact between substrate and particles, at a distance close to value of lattice parameter; 2) activation of contacting surfaces and chemical interaction of materials at the interface of phase separation; 3) the volumetric development of interaction, which leads to mutual penetration of particle and substrate materials due to diffusion mass transfer [19-21] across the boundary with already established chemical connections.

The formation of physical contact precedes chemical interaction, which in formed coating determines the actual contact area on which strong connections are formed. The interaction of created physical contact of single particles with a smooth surface is well studied [22-24]. The need to take into account the influence of real microrelief of substrate surface on the formation of contact strong bonding with sprayed particles has been considered in a number of papers [5, 6, 13], where the model of deformation of a spherical micro – projection upon collision with a particle and its subsequent spreading is considered. In some works, the roughness of substrate is also modeled by prismatic projections of rectangular and triangular shape. However, not all of these works take into account the existence of nanoscale subsurface on real surfaces.

The objective of this work is to analyze and find new approaches to mechanisms of bonding and to determine the possible influence of nanotopography of substrate surface on interaction with sprayed particles and the creation of coatings with high adhesion and cohesion strength.

2. Investigation of mechanism of bonding of particles with substrate during thermal spraying of coatings

During thermal spraying, the coating is formed from a large number of particles of starting material. Thus, more than $10^{10} - 10^{11}$ particles per 1 m^2 can be used to form a coating of 0,5 mm thick. Moreover, the corrosion protection properties of coating depend mainly on composition of coating material, and its thermomechanical properties not only on material composition, but also largely on the final size and shape of crystals, interparticle contacts and nature of bonding (adhesion) on contact surfaces. Therefore, to obtain coatings of high quality with necessary chemical, physical and mechanical properties, the design engineers of coatings and technologies for their manufacturing must provide for formation of coatings from particles of optimal size and shape, and strong bonding between the particles in coating and particles with basic material of product.

When the particles collide with surface, sprayed particles of various shapes are formed, which strongly depends on properties of particles materials and surface, the state of surface or pre–sprayed layers, as well as on critical parameters of the process: particle temperature, particle velocity (mainly its normal component), and substrate surface temperature. During the particles impact with substrate in their contact with substrate cells of bonding with dimensions in the transverse direction of 200 – 700 μm are forming. As the chemical reaction develops, the number of cells increases, and with a bonding strength close to maximum, they practically merge, filling the entire central region of deformed particle. The diameter of the chemical interaction zone of splat is usually slightly larger than initial diameter of sprayed particle. The mechanism of particle bonding is considered from standpoint of theory of topochemical reactions during solid phase welding. At high velocities of impact of particles with surface it is possible to use the equation for the velocity of chemical reaction in the following form [4]:

$$\frac{dx}{dt} = (N_0 - x)\upsilon \exp\left(-\frac{U_0 - \gamma\sigma}{kT_k}\right) \quad (1)$$

where x – is number of atoms that have reacted during time t ; N_0 – is number of atoms on surface of a solid or particle in area of their mutual contact; υ – is frequency of natural oscillations of atoms; N_0 – is activation energy of formation of chemical connections; T_k – absolute contact temperature; k – is the Boltzmann constant; σ – stresses acting in contact area during chemical reaction; γ – is a structurally sensitive coefficient proposed in kinetic theory of strength of solids and estimates the activation volume and overstresses on interatomic connections compared to average stress in sample.

The stresses reduce energy barrier U_0 required to break connections in solid surface – oxygen ($Me - O$) system by value of γs , and as a result of barrier reduction the probability of connections breaking is sharply increases. However, the absence of theory of coefficient γ, dependence of activation energy U_0 on temperature, stresses and other factors complicate the practical use of equation (1) in engineering calculations.

The values of U_0 and γ are expressed through constants of solid, experimentally measured quantities: atomic heat capacity C , coefficient of thermal linear expansion α, and Young's modulus E [25, 26]

$$\gamma = \frac{C}{\alpha E},\ U_0 = \varepsilon * \frac{C}{\alpha} \tag{2}$$

where $C = \Sigma / \sigma$ – is the coefficient of local overload, Σ and σ – are respectively local and average stress, $\varepsilon*$ – is the relative elongation at which interatomic connections lose their stability and are broken ($\varepsilon* = 0.2$). Taking for monoatomic crystals $C = 3k$, the Zhurkov S.N. equation [25, 26] is as follows:

$$\tau = \tau_0 \exp\left[\frac{3(\varepsilon * - \chi\sigma / E)}{q\alpha T}\right] \tag{3}$$

Here $q = a_s / a$, a_s – is coefficient of thermal expansion on surface of solid; τ – durability of a solid under mechanical loading; $\chi = \upsilon^{-1} \approx 10^{-13}$ sec. Using modern physical methods: slow electron diffraction and surface spectroscopy show that α_s is several times larger than α. At room temperature for Al: $q = 2.5$, for Mo: $q = 10$ [27, 28].

According to the ideas developed in [25, 26], the parameter $\varepsilon*$ characterizes theoretical strength of interatomic connections, qaT – is fluctuating thermal pressure, and the interatomic connections are stretched. The $\gamma\sigma / E$ parameter displays the external loads. With proper application of load, the external load deforms the interatomic connections and breaks them. The consideration of nonlinearity of interatomic forces radically changes the situation in the problem of bonding of materials. The strength of formed bonds depends on thermo–fluctuation factor (4) and mechanical factor (5):

$$\exp\frac{3\varepsilon^*}{q\alpha T_k} = \beta \tag{4}$$

$$\exp\frac{3\chi\sigma / E}{q\alpha T_k} \tag{5}$$

An increase of adhesion strength of particles with part surface can be achieved by a decrease in of thermo–fluctuation factor β or an increase of mechanical factor θ. The first is achieved by increasing temperature in contact T_k or by coefficient α. The second is achieved by increase of stresses applied to interatomic bonds.

Using equations (2) and (3), equation (1) can be reduced to the form:

$$\frac{dx}{dt} = (N_0 - x)\nu \exp\left[-\frac{3(\varepsilon^* - \chi\sigma / E)}{q\alpha T_k}\right]. \tag{6}$$

After integration (6) when $T_k = const$ and substitution ($t = 0,\ x = 0,\ t = t_0,\ x = N_0$) we obtain the duration of reaction during which N atoms will react:

$$t_\alpha = -\frac{1}{\nu}\ln\left[1 - \frac{N}{N_0}\right]\exp\left[\frac{3(\varepsilon^* - \chi\sigma / E)}{q\alpha T_k}\right]. \tag{7}$$

Theoretical equation of change of relative strength of bonding of particles with surface as a result of chemical reaction in contact can be obtained from equation:

$$\frac{N(t)}{N} = \frac{\sigma_a(t)}{\sigma_{a\max}} = 1 - \exp\left[-\frac{\nu t_a}{\exp\left[3(\varepsilon^* - \chi\sigma / E) / q\alpha T_k\right]}\right], \tag{8}$$

where $\sigma_a(t)$ – is bonding strength achieved during time t_a; σ_{amax} – maximum bonding strength; $N(t)$ – is number of atoms that have reacted during time t.

The elastic–deformed state of contact area of particle with part is deeply specific and can be characterized by analogy with external friction process [27, 28] with following features: 1) High value of ratio of surface to deformed volume. Here, a direct force effect on substructure of surface is realized in area of actual contact with particle, so the plastic deformation is localized in the thinnest surface layers. This effect is greatly enhanced by presence of soft films and layers on contact surfaces; 2) High uniformity (homogeneity) of plastic deformation and abnormal plasticity of surface layers.

If under volumetric load distribution of stresses on submicroscopic level is associated with property of defective structure to concentrate stresses and relax them due to implementation of easy movement elements, then during friction distribution of stresses is determined by physical contact, which makes it almost equally possible to move all elements of structure. This allows us to accept the value of local overload coefficient $\chi = 1$.

Real surfaces are characterized by presence of a large number of crystallographic defects, the main ones being dislocation exit points, traces of grain boundary exit, etc. Therefore, when particles are spreading by surface, the exit of fresh defects on contact surface, the addition of tensile loads on individual interatomic connections of adsorbed atoms is possible, and, therefore, the considered thermo–fluctuation model of bonding takes place subject to the simultaneous occurrence of acts of atomic rearrangements. The most favorable conditions for this are created in peripheral annular areas of contact of particle with part.

In Figure 1 presented calculated dependences (according to (8)) of relative bonding strength of particles of Cu with surface of sample with Mo at different values of collision velocities (100; 500; 700; 1000 m/s). The experimental dependences obtained by converting the temperature of substrate T_s on contact temperature T_k are also shown there. In calculations, the following parameter values were taken: $q = 10$; $\alpha^0 = 0.56 \cdot 10^{-5}$; E in MN/m^2 at a temperature in K^0: 300 – $3.23 \cdot 10^5$; 400 – $3.17 \cdot 10^5$; 500 – $3.11 \cdot 10^5$; 600 – $3.05 \cdot 10^5$; 700 – $2.99 \cdot 10^5$; 800 – $2.93 \cdot 10^5$; 900 – $2.87 \cdot 10^5$; 1000 – $2.81 \cdot 10^5$; $\sigma = 0$ at $V_0 = 100\ m/s$; $\sigma = 2400$ MN/m^2 at $V_0 = 500\ m/s$; $\sigma = 4000$ MN/m^2 at $V_0 = 700\ m/s$; and $\sigma = 9000$ MN/m^2 at $V_0 = 1000\ m/s$.

3. Discussion

From the analysis of calculated dependences, we have very important conclusions:

1. The nature of the calculated dependences for relative strength of bonding of particles with surface at low collision velocities from equation (8) coincides with similar calculated dependencies of known first–order topochemical reaction equation;

2. The relative bonding strength of particles with surface depends essentially on diameter of particles. At the same impact velocity, reducing the particle diameter reduces collision time and requires higher heating of part surface to reach uniform collision.

For the considered combination of materials, reducing of particle diameter (and collision time) by an order to obtain a uniform connection the contact temperature should be increased by more than 100 C and the substrate temperature by more than 150°C.

3. The increase of particles velocity on relative bonding strength that formed by the thermo–fluctuation mechanism, is ambiguous. At collision velocities that do not provide a sharp increase of pressure and, accordingly, the mechanical factor θ, with increasing of particles velocity at the same contact temperature, the relative strength of bonding decreases. This is explained by decrease of duration of reaction (collision time). When collision velocities that provide a sharp increase of pressure and a corresponding increase of mechanical (athermic) factor θ, further increase of particles velocity at the same contact temperature leads to an increase of bonding strength compared to low collision velocities. Moreover, the larger the particles diameter, the less the critical value of particles velocity, which increases relative bonding strength. In this case, for particles with a diameter of 100 μm, $V_{cr} \approx 700\ m/s$, and at $d_p = 10\mu m$, $V_{cr} > 1000\ m/s$.

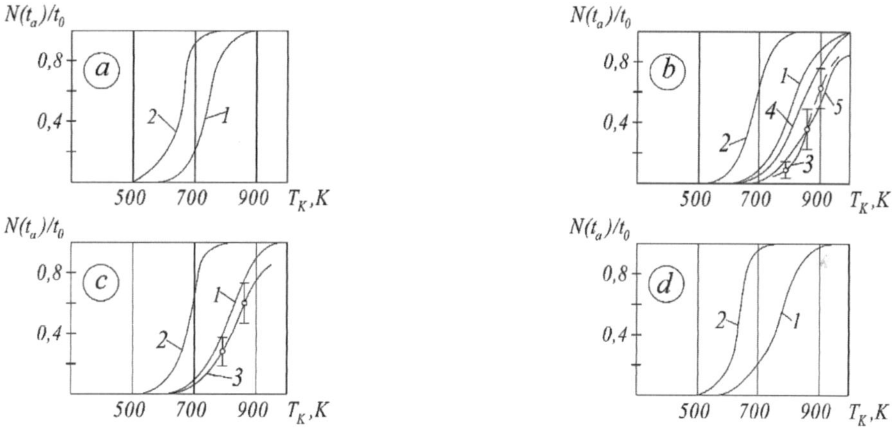

Figure 1. The calculated dependences of the relative bonding strength of copper particles with the surface M_o from contact temperature. Particle diameter, μm: 10 (curves 1), 100 (curves 2); velocity of particles V_0, m/s: 100 (*a*), 500 (*b*), 700 (*c*), 1000 (*d*), for curves 4 and 5 $V_0 = 0$ m/s. Activation time t_a, s, $a - 1\times10^{-7}(1)$; $1\times10^{-6}(2)$; $b - 1\times10^{-8}(5)$, $2\times10^{-8}(1{,}4)$, $2\times10^{-7}(2)$; $c - 1.43\times10^{-8}(1)$, $1.43\times10^{-7}(2)$; $d - 1\times10^{-8}(1)$, $1\times10^{-7}(2)$. Curve 3 was obtained experimentally.

4. At high collision velocities, the slope of middle sections of curves of relative strength for different particle diameters varies. The larger diameter of particles, the steeper the relative strength curves, the smaller the difference in values between $N(t_a)/N_o = 0$ and $N(t_a)/N_0 = 1$. As the particle velocity increases, the steepness of these curves also increases.

5. The experimental curves for relative bonding strength close to the calculated ones, also were considered that σ equal to border of tensile strength (for copper about 150 MN/m^2).

This model allows to take into account such basic parameters of technological operation process of spraying that affect the bond of particle with substrate surface, such as surface temperature, velocity and temperature of particles, but does not take into account a number of important parameters, and above all the topography of substrate surface, the state of its surface layer, etc. It is problematic to estimate a number of physical parameters included in calculation equations. The presence in reality of shape variations of different scale level leads to significant errors in calculation of contact characteristics, if they do not take into account their effect on deformation of sprayed particles and the local area of substrate. First, the presence of nanoscale leads to discreteness of contact of particle with substrate at the initial moment of collision. Instead of circular contact area the collision of particle with substrate occurs on the vertices of contacting irregularities, that forming an ensemble of small contact areas located within the larger so – called contour area of contact. Such discreteness of contact of particle with substrate allows to consider differently the process of formation of bonding focuses. The process of substrate deformation, generation and exit of dislocations, which are active centers, begins at the vertices of nanoscales.

Reducing of characteristic dimensions of an object or its structure to $R^* \leq 1\ \mu m$ (in at least one of three dimensions) entails a significant change in its mechanical properties. Particularly strong dimensional effects occur at $R^* \leq 100\ \mu m$, and at $R^* \leq 10\ \mu m$ their character can change radically again. It follows that the mechanical behavior of solids in nanoscale cannot be deduced by simple extrapolation from the macro – or micro – area, which significantly affects the mechanism and kinetics of bonding of particle – to – base with substrate.

4. Summary

1. The bonding of sprayed particles with product surface is accompanied by a complicated complex of physicochemical processes and phenomena. The adhesive and cohesive strength of coatings depends on a large number of factors, but their impact is manifested by changes of character and intensity of deformation and thermal processes in contact area of interacting materials. It can be assumed that, depending on multifactor conditions of spraying the bonding of particles with substrate may occur by different mechanisms, some of which may be driving and others concomitant.

2. The formation of interatomic connections between materials of particles and parts is accompanied by deformation processes in contact area, which contribute to the unloading and breaking of connections in metal – oxygen system, loading and formation of connections between atoms of interacting materials.

3. The kinetics of formation of interatomic connections between materials of particle and substrate at high collision velocities are satisfactorily described by the modified equation of S.N. Zhurkov, using experimentally measured values: the value of tensile strain of interatomic connection ($\varepsilon^* \geq 0.2$), the Young's modulus, the coefficient of thermal linear expansion.

4. It is promising to study the effect of surface nanotopography on interaction with sprayed particles, the mechanism of their bonding and on the basis of their results, development of recommendations for preparation of various surfaces for spraying.

5. The use of protective coatings will increase the strength and reliability of parts and structures of railway rolling stock.

References

[1] Fomin O, Gerlici J, Lovska A, Kravchenko K, Fomina Yu and Lack T 2019 Determination of the strength of the containers fittings of a flat wagon loaded with containers during shunting *IOP Conf Ser: Mater Sci Eng* **659** 012056

[2] Krol O and Sokolov V 2019 Parametric Modeling of Transverse Layout for Machine Tool Gearboxes *Advances in Manufacturing II. Lecture Notes in Mechanical Engineering* **4** pp 122–130

[3] Krol O and Sokolov V 2019 Parametric Modeling of Gear Cutting Tools *Advances in Manufacturing II. Lecture Notes in Mechanical Engineering* **4** pp 3–11

[4] Budag'janc N A 1998 *D–Gun processes in industry* (Lugansk: VUGU) p 223

[5] Davis J 2004 *Handbook of Thermal Spray Technology* (Materials Park: ASM International) p 337

[6] Pierre L, Joachim V R, Maher I 2014 *Thermal Spray Fundamentals: From Powder to Part* (New York: Springer Science + Business Media) p 1587

[7] Fesenko A, Basova Y, Ivanov V, Ivanova M, Yevsiukova F and Gasanov M 2019 Increasing of Equipment Efficiency by Intensification of Technological Processes *Periodica Polytechnica Mechanical Engineering* **63** 1 pp 67–73

[8] Sokolov V, Krol O and Stepanova O 2020 Choice of Correcting Link for Electrohydraulic Servo Drive of Technological Equipment *Advances in Design Simulation and Manufacturing II DSMIE 2019. Lecture Notes in Mechanical Engineering* pp 702–710

[9] Sokolov V, Krol O and Stepanova O 2018 Automatic Control System for Electrohydraulic Drive of Production Equipment *International Russian Automation Conference IEEE*

[10] Okorokov A, Fomin O, Lovska A, Vernigora R, Zhuravel I and Fomin V 2018 Research into a possibility to prolong the time of operation of universal semi-wagon bodies that have exhausted their standard resource *Eastern-European journal of enterprise technologies* **3/7** 93 pp 20–26

[11] Zelenko Yu, Lunys O, Neduzha L and Steišūnas S 2019 The assessment of negative impact of oil products on railroad track and rolling stock constructions *Proc. of the 23rd Intern. Sci. Conf. Transport Means 2019* pt **III** (Palanga: Kaunas Univ. of Technology) pp 1300–1306

[12] Bondarenko I, Lunys O, Neduzha L and Keršys R Dynamic track irregularities modelling when studying rolling stock dynamics *Proc. of the 23rd Intern. Sci. Conf. Transport Means 2019* pt. **II** (Palanga: Kaunas Univ. of Technology) pp 1014–1019

[13] Pawlowski L 2008 *The Science and Engineering of Thermal Spray Coatings* (Chichester: John Wiley & Sons, Ltd) p 647

[14] Rjahovskij A, Kosenko V and Vlasenko V 2012 Features of estimation of adhesion strength of D–Gun spraying coatings *Weapons systems and military equipment* **3** 31 pp 215–217

[15] Faradzhallah M, Panarin V and Bys' S 2011 The problem of technological residual stresses in improving the quality of D–Gun spraying coatings *Visnik of Khmelnytskyi National University* **2** pp 14–18

[16] Lin C and Berndt C 1994 Measurement and analysis of adhesion strength for thermally sprayed coatings *Journal of Thermal Spray Technology* **3** 1 pp 75–104

[17] Samsonov G, Vitrjanjuk V and Chaplygin F 1974 *Tungsten carbides* (Kyiv: Nauk. Dumka) p 175

[18] Babich M 1975 *Hard alloys heterogeneity by carbon content and its elimination* (Kyiv: Nauk. Dumka) p 175

[19] Korobov Yu S 2016 *Analysis of properties of thermal coatings. Part 2. Evaluation of coating parameters* (Yekaterinburg: Publishing House of Ural University) p 92

[20] Sokolov V 2019 Diffusion of Circular Source in the Channels of Ventilation Systems *Advances in Engineering Research and Application. ICERA 2018. Lecture Notes in Networks and Systems* (Cham: Springer) **63** pp 278–283

[21] Sokolov V, Krol O and Stepanova O 2019 Nonlinear simulation of electrohydraulic drive for technological equipment *Journal of Physics: Conference Series* **1278** 012003

[22] Ilyushchenko A, Okovityj V, Kundas S and Formanek B 2002 *Formation of thermal spraying coatings: theory and practice* (Minsk: Bestprint) p 480

[23] Movshovich A, Chernaja Yu and Ishchenko G 2011 Investigation of influence of physico – mechanical characteristics of D–Gun spraying coatings on wear resistance of cutting elements of reworkable dies *Vestnik NTU «HPI»* **45** pp 22–28

[24] Maksimovich G, Shatinskij V and Kopylov V 1983 *Physicochemical processes during plasma spraying and destruction of coated materials* (Kyiv: Nauk. Dumka) p 264

[25] Zhurkov S 1980 To the question of the physical basis of strength *Solid body physics* **22** 11 pp 3344–3349

[26] Zhurkov S and Petrov V 1978 The physical basis of the temperature – time dependence of strength of solids *Reports of the Academy of Sciences of the USSR* **239** 6 pp 1316–1319

[27] Chichinadze A, Braun Je and Bushe N 2001 *Fundamentals of tribology (friction, wear, lubrication)* (Moscow: Mechanical Engineering) p 664

[28] Heintze G N and McPherson R 1988 Fracture toughness of plasma sprayed zirconia coatings *Journal of Surface Coating Technology* **134** pp 15–23

IOP Conf. Series: Materials Science and Engineering **985** (2020) 012038 doi:10.1088/1757-899X/985/1/012038

Energy-efficient excavation of the soil of the lower track structure by bulldozers with a combined knife system

K Hlavatskyi[1,2], **S Raksha**[1,3] **and Y Gorbenko**[1,4]

[1]Dnipro National University of Railway Transport named after Academician V. Lazaryan, Lazaryan St., 2, Dnipro, Ukraine, 49010

[2]kazimir.glavatskij@gmail.com, [3]raksha@ukr.net, [4]yuriygorbenko1984@gmail.com

Abstract. The motivation of the research and further development of the new type combined knife system on the non-return bulldozer blade and the results of the first series of laboratory studies of the process of the soil digging with the non-return blade of the bulldozer with the first set of its physical models, which indicate the feasibility of continuing research in this direction is given. The proposed combined knife system makes it possible to realize oblique digging of the soil on the non-return blade of the bulldozer and maximally complete its orientation in the prism of the soil before the blade in order to increase the productivity of the bulldozer with non-return blade by decreasing soil losses in the side rollers, decreasing the time of filling and digging the soil by decreasing the coefficient of specific soil resistance to digging. The paper substantiates the technology of more efficient use of bulldozers with non-return blade compared to their traditional counterparts, which is possible due to the use of the new combined knife system, and argues in favor of expanding the technological capabilities of the bulldozer with non-return blade, equipped with the proposed engineering solutions of the new type knife system with different digging angles and spatial inclination of cutting edges.

1. Problem statement

Ground transportation vehicles, to which bulldozers belong, are widely used in the building of earthen structures, in particular, the lower structure of roads and railways. Bulldozers are also widely used in the construction of building sites at the initial and final stages. Therefore, they must be highly efficient and productive when performing technological operations.

Important parameters when evaluating the energy efficiency of a bulldozer are the specific coefficient of soil resistance to digging, which minimization is one of the directions of scientific research during all the time of development and creation of new equipment in our country, as well as the productivity, energy intensity and material intensity of the technological processes performed by it. Taking into account the general tendency of achieving the best performance of bulldozers due to the development of new technical solutions of their working equipment, the main attention is paid to improving the design of the main working body of the bulldozer - the blade and its knife system.

2. Analysis of recent research

The study of bulldozer equipment and its knife systems has been reflected in a number of papers, including classic publications [1, 2, 3], patents [4, 5] and dissertations [6, 7]. However, the study of issues related to changing the profile of the knife system for soil digging intensification is relevant, as it is directly related to improving the productivity of bulldozers, their efficient use and energy saving.

Content from this work may be used under the terms of the Creative Commons Attribution 3.0 licence. Any further distribution of this work must maintain attribution to the author(s) and the title of the work, journal citation and DOI.
Published under licence by IOP Publishing Ltd

IOP Conf. Series: Materials Science and Engineering **985** (2020) 012038 doi:10.1088/1757-899X/985/1/012038

Previously investigated flat and space knife systems based on bulldozer equipment with non-return blade (Figure 1) provide the reduction of specific energy intensity of soil digging due to replacement of blocked digging by semi-free or free one and, for this purpose, have constructive execution of knife systems, combined from separate sections and allow one to divide the straight cutting edge of a traditional flat knife into separate proportional segments and to move them relatively vertically, horizontally and at an angle. At the same time loosening teeth are set on separate plates of the knife system for the intensification of solid soil digging.

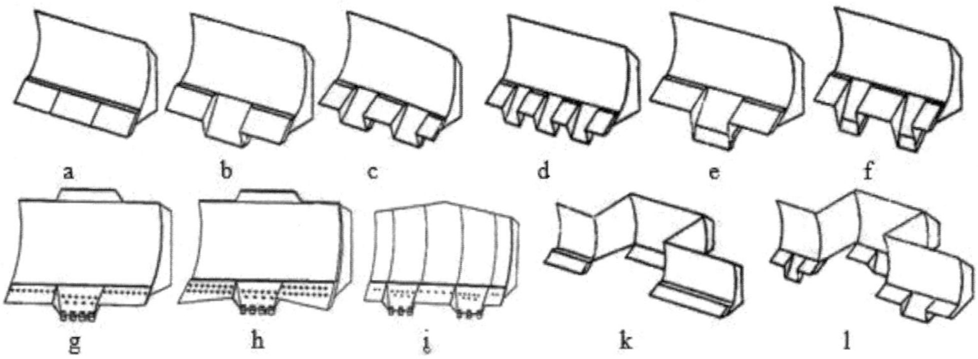

Figure 1. Schemes of known knife systems for the bulldozer blade: a –traditional one with a straight blade; b –with a medium projecting knife; c –with two projecting knives; d –with three projecting knives; e –with a medium knife and a variable cutting angle; f –with two projecting knives and a variable cutting angle; g –with a medium trapezoidal knife; h –with a medium trapezoidal knife and sloping side blades; i–with two medium trapezoidal knives; k – multisection blade; l –multi-section blade with projecting knives.

When the soil digging with a bulldozer with a non-rotary blade and the traditional knife system (Figure 2, a, b), the significant symmetrical loss of cut soil in the side rollers is observed, which reduces the performance of the bulldozer by increasing the path of full prism recruitment and the corresponding increase in cycle time. Symmetrical soil losses can be avoided by using a rotary blade (Figure 2, c) and directing the soil to the desired direction.

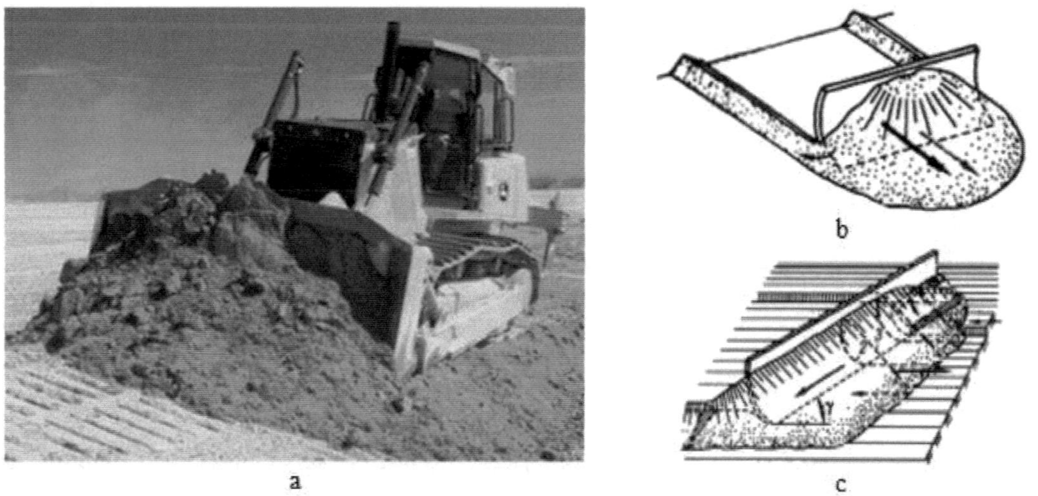

Figure 2. Loss of soil to the sides when digging by a bulldozer: a, b – with a traditional blade; c– with a rotary blade.

To significantly improve productivity, the simultaneous work of bulldozers with non-return blade (Figure 3, a) and with rotary blade (Figure 3, b) should be used, which has significantly lower soil losses in the side rollers, or these losses are absent until the full prism of ground.

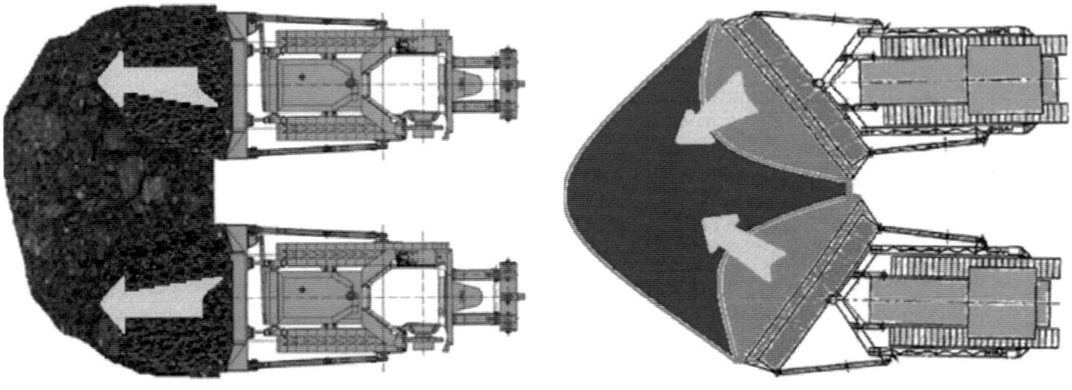

a b

Figure 3. Technology flow charts of the bulldozers simultaneous work: a – with a non-return blade; b – with a rotary blade.

3. The purpose of the work

Is to justify the conditions of energy-efficient soil digging by a non-return blade of the bulldozer with a combined new type knife system, through which it is possible to increase the efficiency and productivity of the bulldozer.

4. Basic material

Achieving this goal is ensured by the installation on the non-return blade of the bulldozer with a combined knife system, which uses exclusively oblique soil digging, creates conditions for directing soil chips separated from the array inside the prism of the soil before the blade, replaces the blocked digging of the soil with a semi-free one [8, 9].

In the combined knife system of the new type, the cutting knives are made of separate plates and are connected in pairs and with blade so that the cutting edges are symmetrical relative to the longitudinal vertical plane of the blade symmetry at a given angle between themselves in the front and vertical projections, and the knife plates are installed at an acute angle to the horizontal in such a way as to ensure exclusively oblique digging of the soil and its orientation in the prism before the blade with minimal losses (Figure 4). When using different number of cutting blades pairs it is possible to get an analogy of technological scheme for bulldozers with rotary blades simultaneous work (Figure 3, b).

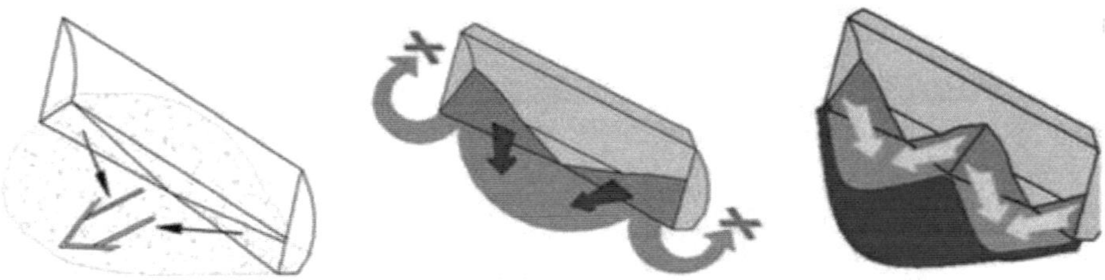

a b c

Figure 4. Schematic diagrams of the proposed spatial knife systems: a - with one pair of knives located within the height of the traditional knife system with a continuous straight cutting edge installation; b - with one pair of knives mounted at an angle equal to the angle of the traditional knife system installation; c - with two pairs of knives mounted at an angle, similar to the scheme (b).

A new-type spatial knife system with a non-return blade can solve a number of technological problems which are usual for basic bulldozer analogues (Figure 5), used in the development of the technical solution, namely: to effectively destroy a solid soil similar to a mulch (Figure 5, a) with the tips of knives tilting the blade forward; to carry out exclusively oblique soil digging similar to a bulldozer with a rotary blade (Figure 5, b); to dig the soil more effectively than the traditional non-return bulldozer blade (Figure 5, c).

a b c

Figure 5. Similar bulldozers for the development of a new type spatial knife system: a – a snubber; b – with a rotary blade; c – with a non-return blade.

Schematic diagrams of the proposed technical solution of the non-return blade of the bulldozer with the new type combined blade system (Figure 6) can be performed in two main variants: with the placement of the proposed combined blade system within the height of the traditional flat blade system with a straight cutting edge (Figure 6, a) and with its knives installed at an angle equal to the knife installation angle of the traditional knife system. In this case, the combined knife system will extend beyond the height of the traditional knife and will rest not only on the support surface of the traditional knife system, but also on the front blade leaf (Figure 6, b). This approach to the development of a new type combined knife system on a non-return blade is conditioned by the possibility of comparing the efficiency of its technological application with the traditional blade.

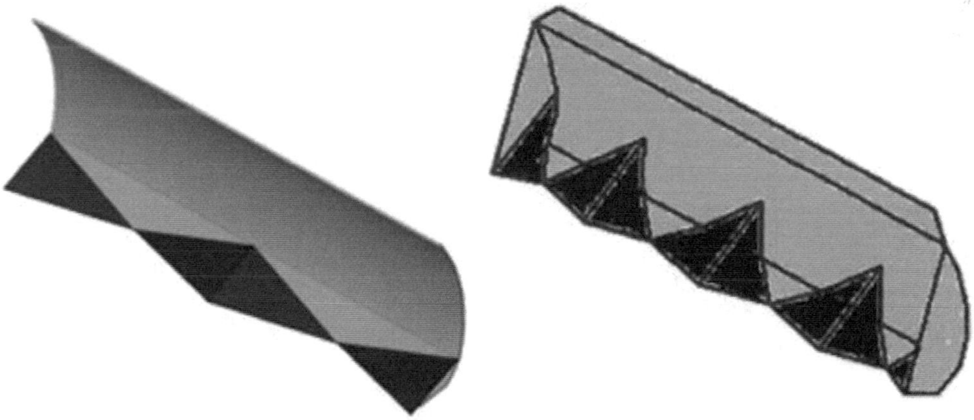

a b

Figure 6. Schematic diagrams of the non-return blade of the bulldozer with the combined knife system: a – located within the height of the traditional knife system; b – with knives at the same cutting angle as for the traditional knife system.

Investigations of technological schemes of work of the non-return blade of a bulldozer with the new type combined knife system involves establishing the efficiency of its work depending on the location of the vertices of the cutting edges of the knives that make up its construction (Figure 7, points 1, 2, 3).

From the experience of similar studies conducting [1] it is obvious that the displacement of points 2 and 3 in height will have an impact on the energy intensity of the digging process.

From the point of view of the proposed knife system technology application, such displacement of points 2 and 3 will lead to two essential types of the soil surface topography: a flat surface (Figure 8, a) and a ridge surface (Figure 8, b).

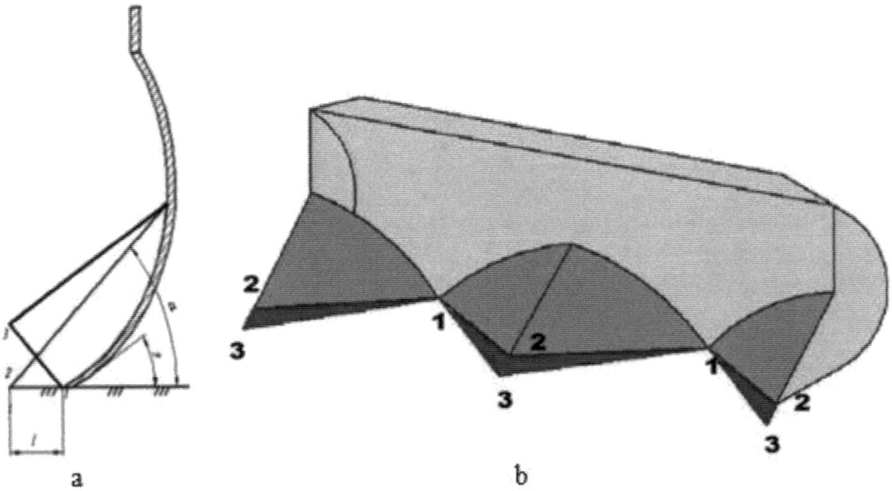

Figure 7. Scheme of mutual arrangement of vertices of cutting edges of the combined knife system: a – a side view; b – an axonometry.

In this case, the flat surface can be formed by the proposed knife system when placing points 1 and 2 in one horizontal plane, and the comb surface when displacing points 2 above or below point 1.

The first series of scientific researches performed on physical models (Figure 9) proves their effectiveness in comparison with the traditional non-return blade as a whole [10].

To compare the results of theoretical and experimental studies of the new type combined knife system on the non-return blade of the bulldozer from the correctness point of view, the non-return blade of the bulldozer with the most common traditional knife system, which performance indicators are well known and undeniable, was adopted.

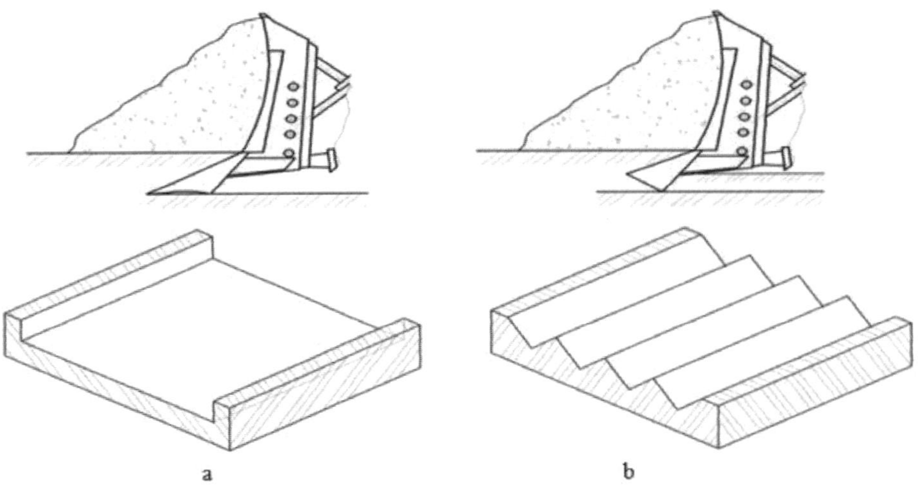

Figure 8. Schemes of soil digging and types of a soil surface relief: a - flat at the placement of knives edges points 1-2-3 in one plane; b - crest at vertical displacement of points 2 (3) relative to point 1.

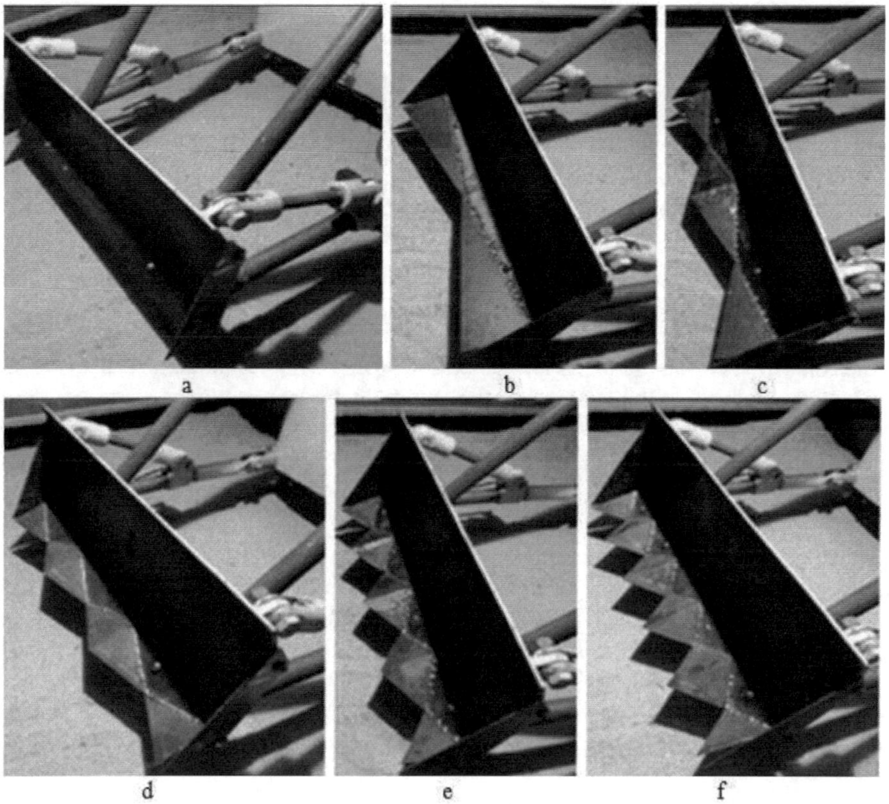

Figure 9. Physical models of traditional and combined knife system on a non-return blade: a – a bulldozer blade with a traditional knife system; b, c, d, e, f - with a combined knife systems with one, two, three, four and five pairs of knives.

5. Conclusions

The investigated physical model of a non-return blade of a bulldozer with the combined knife system has significant advantages over the non-return blade of a bulldozer with the traditional knife system in all basic experimental parameters, namely: the average productivity has increased by 5%, the average digging power has decreased by 30%, the average specific energy consumption decreased by 41% and the average specific drag coefficient has decreased by 40% [10].

The highest percentage of the indicated coefficients positive change corresponds to the physical model of the non-return blade of the bulldozer with the combined knife system with the longitudinal displacement of the knives edges up to 50 mm and one pair of knives.

The smallest percentage of positive change of the specified parameters corresponds to the physical model of the non-return blade of the bulldozer with the combined knife system with the 55 mm longitudinal displacement of the blades edges and two pairs of knives, and in this case the average digging force of the physical model of the non-return blade system is 3.4% higher than for the corresponding bulldozer blade with the traditional knife system, and productivity is 35% down for the bulldozer with the traditional blade system.

An alternative to the physical models of a bulldozer with a non-return blade and a combined knife system in the second place is a model with a 50 mm longitudinal offset of knife edges and 5 knife pairs, in the third place a 50 mm long knife edge offset and 4 pairs of knifes and in the fourth place is a model with a longitudinal offset of the knives edges equal to 50 mm and the number of knives pairs equal to 3 [10].

In [10] research on the methods of combined knife systems has been carried out to determine their rational parameters.

It should be noted that the increase in the prism of soil drag before the blade is also due to the installation of side scroll blades of the enlarged size.

The overall reduction of the specific energy consumption of the soil excavation process is also due to the replacement of the blocked soil excavation by a semi-free one.

In further research, the authors intend to compare the results of their research with the latest scientific advances in knife systems on a non-return blade, in particular with a flat knife system with a protruding middle knife, a flat knife system with protruding knives and a side plate, as well as with trapezoid protruding knifes and side plates. Such a comparison will allow us to draw conclusions about the relative energy efficiency and other major technological indicators that affect the performance of the bulldozer and establish the area of its rational use.

From the point of view of ensuring the manufacturability and application of the proposed technical solution of the combined knife system, its implementation on the in-situ machine is provided in the form of separate modules, which in the first stage of implementation in production and operation will not violate the integrity of the basic design of the blade, and in the future, in the case of demand for this knife system, it is planned to develop a new design of the blade in order to change only the part of the knife system.

In further studies of the combined blade system on the non-return blade, on the basis of the obtained research results, it is planned to construct regression models to determine the ranges of rational and optimal parameters of the soil digging process with the non-return blade of the bulldozer with the combined knife system, as well as to compare the results of theoretical and experimental research.

References

[1] Khmara L, Kravets S, Nichke L, Nazarov V, Sabluk M and Nikitin V 2010 The soil-moving machine *Tutorial* (Kharkov Ukraine:Kharkiv National Automobile & Highway University) p 557

[2] Khmara L, Kravets S, Skoblyuk M, Nikitin V, Derevyanchuk M and Suponev V 2014 Soil-moving machine *Tutorial* (Kharkov Ukraine:Kharkiv National Automobile & Highway University) p 548

[3] Blokhin S and Malich N 2006 The main parameters of technological machines *Tutorialmachines for work with soil* (Kyiv, Ukraine: publishing house Vyshcha Shkola) p 497

[4] Khmara L, Talalay V and Sokolov I 2006 Bulldozer blade with rotary knife system: *Patent № 112205* (Dnipro Ukraine:Dnipro Academy of Building and Architecture)

[5] Khmara L, Talalay V and Sokolov I 2006 Bulldozer blade *Patent № 116566* (Dnipro Ukraine:Dnipro Academy of Building and Architecture)

[6] Korotkikh V 1995 Research and development of bulldozer with protruding knives and side gussets *Author's abstract dissertation Candidate of Technical science* (Dnipro Ukraine: Dnipro Academy of Building and Architecture) p 21

[7] Talalay V 2008 Intensification of soil development by bulldozer blade improvement of the parameters of the knife system *Dissertation Candidate of Technical since* (Dnipro Ukraine: Dnipro Academy of Building and Architecture) p 186

[8] Bohomaz V, Hlavatskyi K, Dorohokuplya M, Krasnoshchok S, Proskurnya V and Sereda O 2016 Bulldozer blade with combined knife system *Patent № 112205* (Dnipro Ukraine: Dnipro National University of Railway Transport named after Academician V. Lazaryan)

[9] Glavatskyi K, GorbenkoY and Anofriyev P 2019 Investigation of the process of digging the soil with a bulldozer non-return blade with a bulk knife system *Bulletinof DNUZT* (Dnipro Ukraine:Dnipro National University of Railway Transport named after Academician V. Lazaryan) **79** p 230

[10] Raksha S, Glavatskyi K and GorbenkoY 2020 Investigation of the process of digging the soil by a physical model of bulldozer equipped with a non-rotating blade with a variable combined three-dimensional knife system *Bulletin of Kharkiv National Automobile & Highway University* (Kharkiv, Ukraine: Kharkiv National Automobile & Highway University) **88** 2 pp 86–92

IOP Conf. Series: Materials Science and Engineering **985** (2020) 012039 doi:10.1088/1757-899X/985/1/012039

Technology of soil compaction of the base of the track with vibrating static machines of blocking action

K Hlavatskyi[1,2], V Cherkudinov[1,3] and O Posmitiukha[1,4]

[1]Dnipro National University of Railway Transport named after Academician V. Lazaryan, 2, Lazaryan St., Dnipro, 49010, Ukraine

[2]kazimir.glavatskij@gmail.com, [3]volodymyrcherkudinov@gmail.com, [4]AleksandrP@3g.ua

Abstract. The directions of combination of working bodies in the existing soil compactors and machines of the new generation and the technologies of their application for productive and effective soil compaction of the lower track structure, soil structures of roads and other construction objects are given. The basis of the research is the principle of accelerated soil compaction due to the creation on the surface of the contact with the working bodies of the sealing machines pressure, which exceeds the limit of plasticity of the soil and at the same time limiting the slipping of the sides from under the working sealing surface. It is proposed that the working sealing surface modular type that can change its profile depending on the technological requirements. Further research is aimed at creating the required profile of the working body based on the desired compaction core and the structural formula of the working body of the soil compactor. In this case, replacing the elements in the working bodies with the corresponding mathematical models, we obtain a common mathematical model of the working body of the soil compactor, as well as the profile of its working body, based on the specified energy consumption and technological requirements.

1. Problem statement

Soil compacting machines used at the modern level of construction or reconstruction of the track structure mostly have working elements of the traditional type with a simple or embossed profile of the working surface. As a rule, such working elements have a vibrating contour with a constant vector of vibration and adjustable disturbing force, which intensifies the compaction of the soil.

Technologically compacting machines are used in the traditional sequence (light, medium and heavy) as the soil is compacted and depending on its physical and mechanical properties (for loose or weakly connected soils – vibration action machines; for connected machines – static machines, shock and shock).

In the operation of such machines, their working equipment is limited in its forceful action on the soil by its physical and mechanical properties and the ability to slip out from under the working equipment in case of failure to balance the conditions of working stresses at the boundary of contact of working equipment-soil and soil strength.

2. Analysis of recent research

The research of the issues mentioned above has been reflected in a number of works, including dissertations [1, 2, 3], educational publications [4, 5, 6], materials on technological parameters of machines and technology of performance of soil compaction works [7, 8], and building codes [9]. But

Content from this work may be used under the terms of the Creative Commons Attribution 3.0 licence. Any further distribution of this work must maintain attribution to the author(s) and the title of the work, journal citation and DOI.
Published under licence by IOP Publishing Ltd

the study of issues related to changing the profile of the working surface of the GUM RO, the force vector and its magnitude, a rational combination in one GUM RO of different types, and creating the conditions for the intensification of soil compaction is relevant, since it is directly related to improving productivity GUM, their efficient use and economy of energy resources.

3. The purpose of the work

Is to justify the conditions of accelerated compaction of the soil by blocking its dislocation from under the working equipment during accelerated compaction, and reducing the number of sealing machines in the technological process.

4. The main section

Achieving this goal is ensured by the following sequence of material presentation: setting criteria for the selection of parameters of work equipment and evaluation of research results; proposals for basic solutions of work equipment; schemes of working equipment; sketches of technical solutions of working equipment and sealing machines; conclusions and directions for further research.

When finding rational and optimal solutions, the criteria are taken into account: maximum productivity with the maximum volume of dense soil zone, called the compaction core; optimum energy consumption; minimum cost of work within the specified time with restrictions: by engine power; the speed of movement on a given section of the seal; by overall dimensions of the work equipment.

For the effective execution of the soil compaction process, it is necessary that the work equipment (rollers, fiberboards) adapt to the necessary technological conditions of the work, namely, quickly change their weight, dimensions, surface profile in order to maximize the efficiency of soil compaction in one pass of the work equipment. for the soil in its present condition. Thus, the construction of work equipment with a smooth surface can be taken as a basis for comparison, and on this basis it can be divided into components in the form of, for example, type work equipment from disks, sections, etc., which have different profiles, and also lay in design of work equipment with an elastic frame the possibility of change of internal pressure in each element or its rigidity. Depending on the size of the work equipment elements, compared to the width of the soil sealing strip, they can be used as standalone work equipment or by combining several elements into one work equipment. Schemes of possible forms of work equipment are shown in Figure 1, and schemes of options for work equipment are shown in table 1.

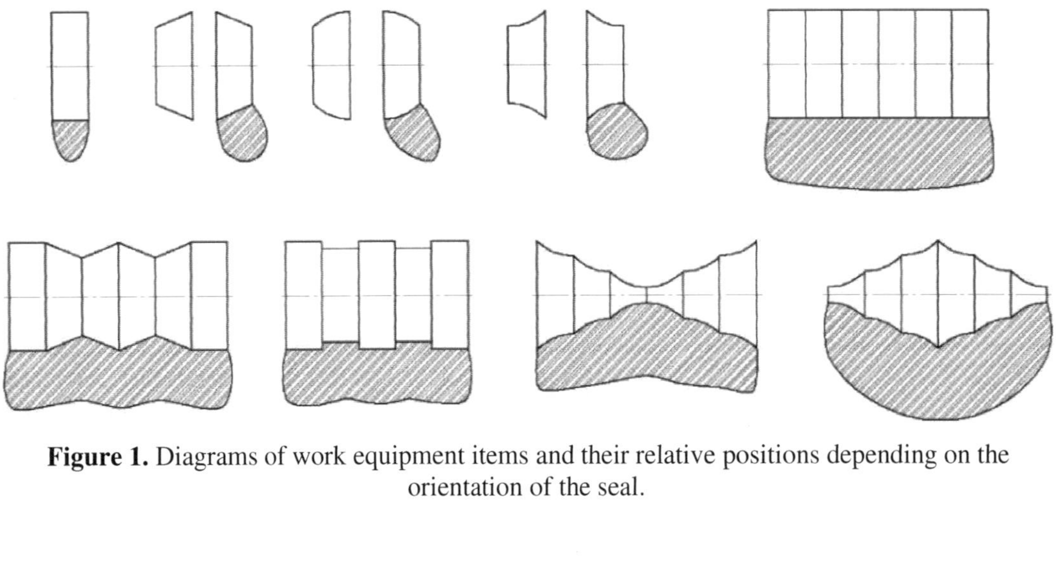

Figure 1. Diagrams of work equipment items and their relative positions depending on the orientation of the seal.

When rolling a freshly ground soil with a heavy two-roller, it is advisable to pre-compact the soil with a vibrating plate that has an adapter working surface of the appropriate shape (Figure 2, a). Given that cohesive soils are compacted by vibration at shallow depth and are prone to sideways sliding under the influence of a heavy cat when used in the first stage of compacting loose soil, the vibrating plate will play the role of a light cat, but will be more effective on the soil and will have the ability to act on the ground change the magnitude and vector of the force of excitement as needed. As a result, the soil sealing layer will gain sufficient strength and will not collapse under the action of a roller of 2 or 3 heavy rollers. The vibrating plate 1 is mounted on a hinged frame 4 driven by hydraulic cylinders 5.

Table 1. Options for combining work equipment to accelerate effective soil compaction.

Scheme of work equipment	The result of technological application
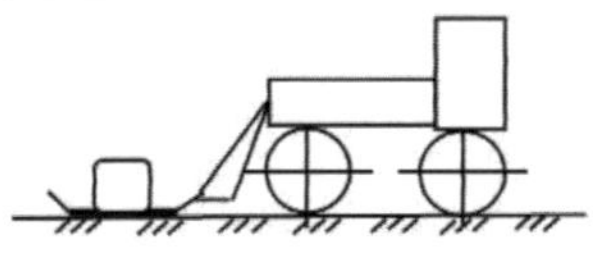	1. Pre-compacting the soil, loading the cat with a vibrating plate. Three stages of sealing in one pass due to the implementation of the disturbing forces with separate working equipment.
	2. Compaction of the topsoil, loading of the smooth roller with a vibrating plate, cam roller or vibrating roller.
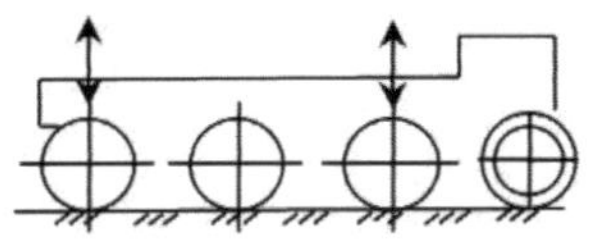	3. Changing the area of contact of work equipment with soil by gradually selectively withdrawing from contact with the soil of its elements and (or) using vibrations in them.
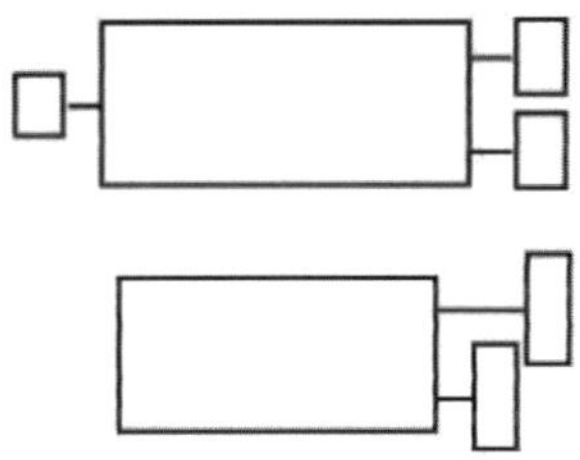	4. Absence of non-sealed areas after the passage of the ramming machine due to the overlapping of the ramming zones by: - installation of the third plate; - placement of plates with overlapping stripes; - regulation and joint selection of the impact force of the plates, their overall dimensions, the surface area of contact of work equipment with the soil (telescopic plate, plate with sliding sections); increasing the impact force by providing initial acceleration; providing each plate with amplitude and frequency.
	5. Smoothing the surface due to the vibrating plate; vibratory roller, static action cat attached to the ramming machine, their use in the last passage.
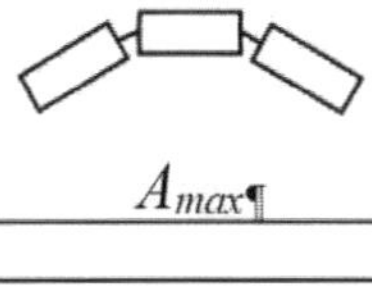	6. Blocking of soil from its displacement to the side due to: - the shape of the surface of the work equipment; - location of work equipment; - the operation of the equipment.
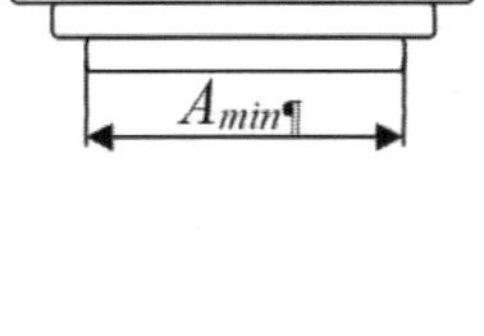	7. The gradual approximation of the working equipment pressure to the soil to the soil strength limit by changing the working equipment area. Automatic change of contact area of work equipment with soil depending on soil density at intermediate stages of compaction; A_{min}, A_{max} – sealing area.

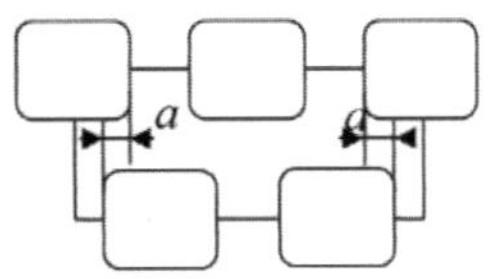

8. Increasing the width of the sealing strip and reducing the number of passes due to the clutch of several vibration plates similar in characteristics. Automatic change of the speed of movement of the plate depending on the density of the soil; *a* – overlapping area.

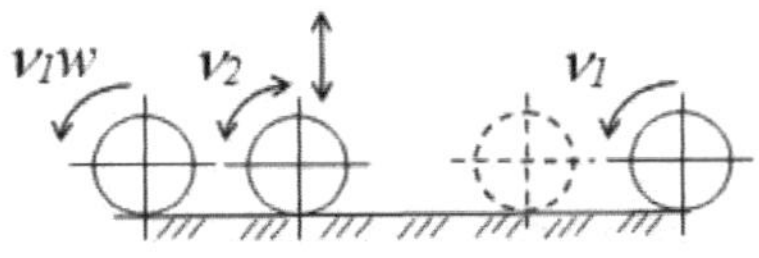

9. Reducing the number of passages of sealing machines due to the multiple reciprocating movement of the middle roller between the extreme rollers at speed $v_2 \gg v_1$; w - angular velocity.

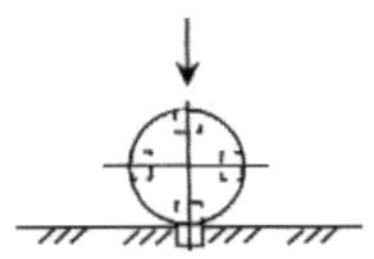

10. Absence of soil destruction at the exit of its cams due to the control of the moment of entry - the exit of the elements of the working equipment from the cat into the soil array (with the use of momentum).

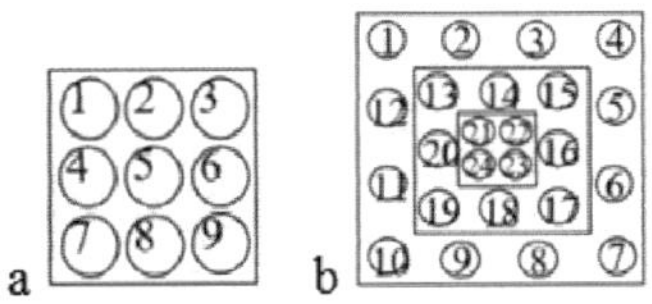

11. Different form of working equipment in contact with the soil, their number and the order of immersion (first 1… 12, then 13… 20, then 21… 24, or 13… 20, then 1… 12, then 21… 21, etc.).

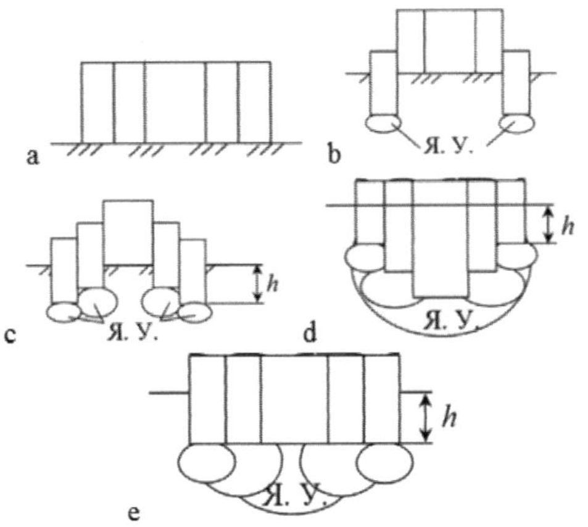

12. Flowchart of the soil compaction process with a roller working equipment or vibration and impact plate with a variable adapter profile of the working surface to illustrate the enlargement of the compacted zone by blocking the soil by sliding sections of working equipment of a certain shape and size (as an option, these profiles can be made in the form of rotating bodies) involves the use of soil properties for better compaction: a – working equipment of the traditional profile; b – blocking of the soil parting by the outer sections; c – sealing of blocked soil with internal sections of work equipment; d, e – variants of mutual arrangement of sections by working equipment depending on the necessary technological conditions and physical and mechanical properties of the soil; *h* – sealing depth.

The possibility of realizing the sealing in one pass due to the gradual increase of the disturbing forces of individual work equipment is achieved by the use of adjustable vibrating circuits. Also, the vibrating plate is a delivery to the cat, and can be used when working in compressed conditions.

It is known that when the soil is compacted with a cam roll, the surface of the sealing upper layer remains slightly loosened.

To eliminate this drawback, compacting and smoothing the soil layer can be installed on the vibrating plate 1 sealing machines (Figure 2, b).

When performing work where roller seals are not possible in the area, for example, insufficient width of the embankment, the vibrating plate 1 can be used as a support around which, by weighing, the sealing machines will be able to return to the desired angle by staying in place (for example, due to a pivot-mounted circle on sealing machines). Also, if the soil is lumpy or frozen, it is advisable to use a vibrating circuit inside the cam roller, and if the soil is loose, then vibration can be used to create additional static force.

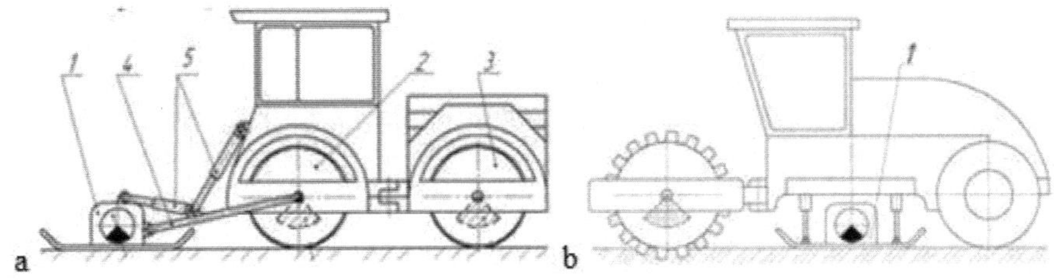

Figure 2. Double roller with vibratory plate (a) and cam roller with support plate (b).

In order to create a universal roller with rollers of different types, in terms of adjusting the values of the contact pressure of the working equipment on the soil within wide limits, a scheme of sealing machines with four rollers can be proposed (Figure 3, a). As a result, we will be able to effectively compact the soil at different stages due to the selective output from contact with the soil rollers 1, 2, 3, or their pairs 1–3, 2–3, and to use vibration circuits installed in them at certain stages and modes. The acquired strength of the soil at a certain stage of its compaction.

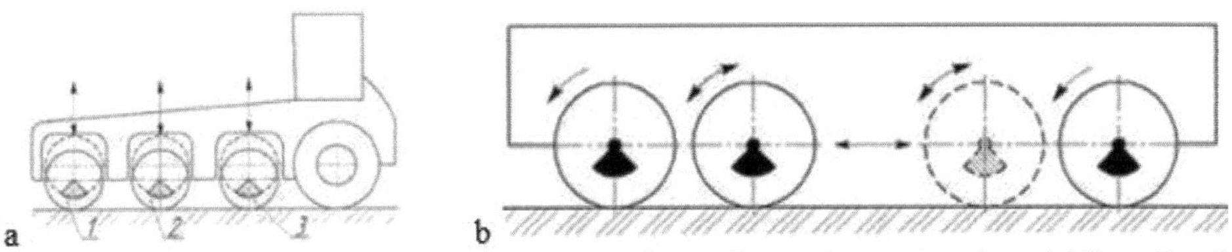

Figure 3. Sealing machine with three reciprocating rollers (a), moving the middle roller (b).

In order to reduce the number of passes of sealing machines along the sealing section, sealing machines with three working equipment of roller type of one or more of them, relative to the base of sealing machines in the longitudinal or longitudinal direction, may be used to achieve the required soil density (Figure 3, b). This is achieved by repeatedly originating one of the working equipment (for example, the middle) of the sealing section, which is located among the other working equipment sealing machines. That is, its angular velocity must be much greater than the angular velocity of the latter and is multiple of it. As the working equipment of sealing machines can act as vibrating smooth and cam rollers, and pneumatic wheels.

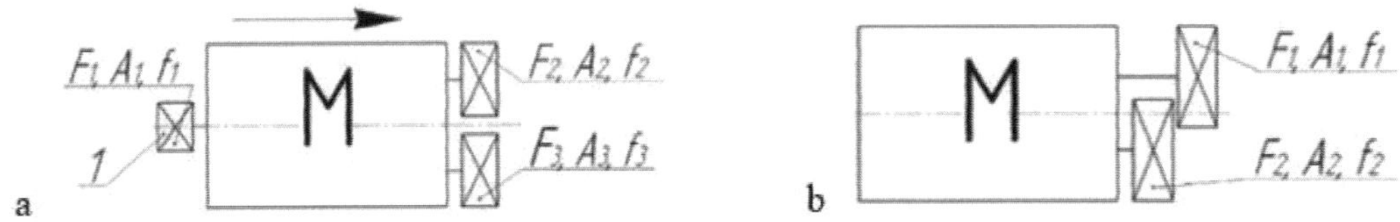

Figure 4. Placement of tractor ram plates (plates are conventionally shown by diagonally crossed rectangles, and the tractor unit is a rectangle with the letter "M"): F – disturbing force.

When the soil is compacted by tractor rammers with falling plates, the non-compacted zone of the soil layer between the lateral inner planes of the plates is inevitably formed. This problem can be solved by installing a third plate 1 (Figure 4, a), or by placing the plates so that one plate overlaps the sealing strip of another (Figure 4, b).

Moreover, the impact force of the first plate is less than the forces of the second and third equal to each other (Figure 4, a) and the overall dimensions of the plates must be reduced accordingly. It is also possible to adjust the surface area of contact of the surface of the slab with the soil, for example, by installing a telescopic plate or a plate with sliding sections. In this case, it is possible to obtain an increase in the impact force by providing an initial acceleration for each of the plates with its individual amplitude A and the frequency of impacts f (Figure 4). On the other hand, the problem is solved by attaching to the tractor-tractor of the vibrating plate (Figure 5, a), or the vibrating roller (Figure 5, b). Not only do they additionally seal the soil layer, they also smooth it out. By changing the direction of travel, you can effectively compact both loose and loose soils of varying degrees of loosening.

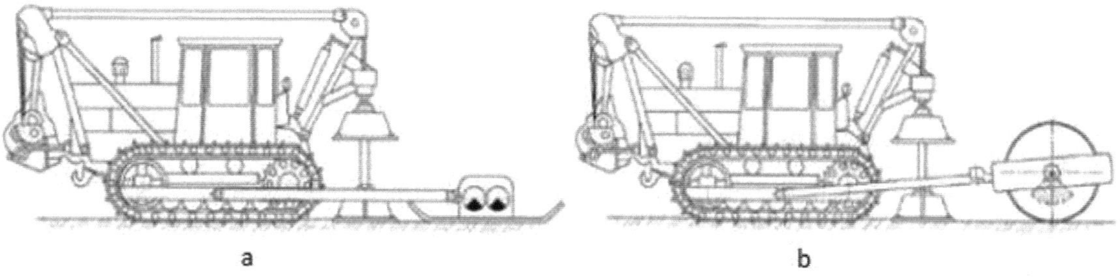

Figure 5. Combined action rubber.

For compacting the entire width of the embankment, along with slopes, relatively large and road-specific sealing machines (Figure 6, a) and ramming, vibrating or combination plates (Figure 6, b) can be created. Thus, conditions are created for blocking the soil from its displacement from under the working equipment to the side due to the shape of the surface of the working equipment and their location.

The possibility of regulating the contact pressure of the working equipment of the vibrating plate on the soil is achieved by telescopic execution of its surface (Figure 6, c), namely: at a constant mass of the vibrating plate, when the contact area of its working equipment is maximum, the contact pressure on the soil will be minimal. Conversely, with a minimum contact area of the work equipment, the contact pressure is maximum. Thus, effective soil compaction is ensured by automatically changing the contact area of the work equipment with the soil, depending on the density of the soil at intermediate stages of its compaction.

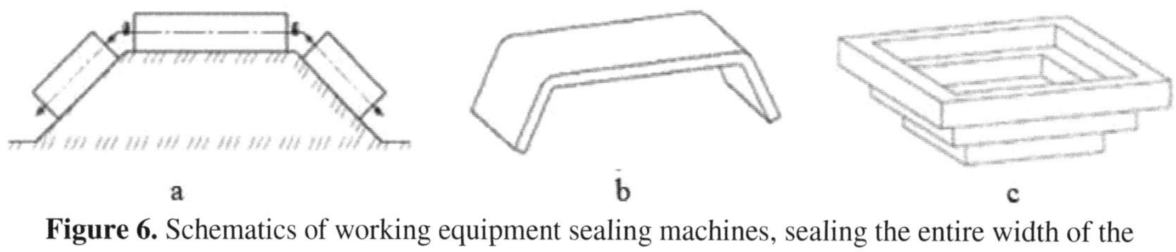

Figure 6. Schematics of working equipment sealing machines, sealing the entire width of the embankment in one pass.

To increase the core of the seal can be used plate with sliding sections-plates (Figure 7). First, the extreme sections are immersed in the soil, followed by alternate pairwise lowering of the other sections, ending with the middle one. Thus, moving towards the middle, the sealing soil is between opposite plates-sections that prevent it from diverging from the side of the working body, and as a consequence, the soil moves the maximum volume to the center of the sealing core in the sealed zone by the shortest possible path. by compacting and forming the core of the compaction, as well as being the transfer link between the working body and the non-compressed soil.

This effect is also achieved by a plate with sliding sections of different shape and their mutual arrangement. Sections can move relative to each other in different combinations, depending on the desired shape and volume of the sealing core. An embodiment of such a plate is presented in Figure 7, b. Initially, the slab compacts the soil only with its support plane, without extending sections. Then the rows of sections located along the outer perimeter are immersed in the soil, and with the subsequent immersion of the rows of sections of the next relatively smaller perimeter, and to the smallest. Such operations result in a concentrated sealing core. In the general case, a certain group of working equipment of sealing machines of roller and flat type that would keep the soil from sliding out from under the working surfaces, can be formed on the basis of working equipment of sealing machines of the traditional form by installing on their working surfaces variable elements (Figure 8).

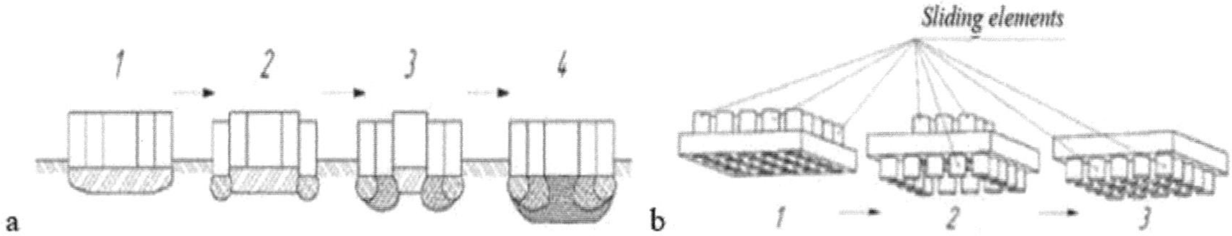

Figure 7. Slab with sliding sections (a) and sliding elements (b).

In this case, the replaceable elements can be designed so that the working surface of the working equipment of sealing machines on which they will be installed remains unchanged, or of another form suitable for mounting and dismantling when designing working equipment of sealing machines of the new generation.

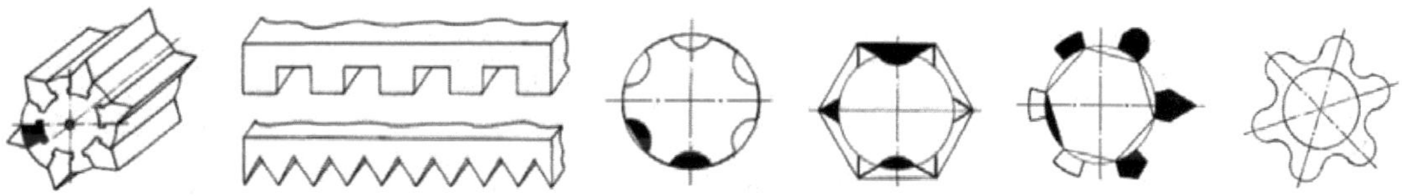

Figure 8. Working equipment of sealing machines with interchangeable elements.

Components of working equipment of sealing machines in the general case can be: the shape of the main surface (cylindrical, profile, multifaceted, flat, combined); shape of interchangeable elements (cams, plates, combined); a way of installing interchangeable elements (permanent, solid and detachable bandages, sliding, insert sections, combined).

Further research is aimed at creating the required profile of working equipment based on the desired NAV and the structural formula of working equipment of sealing machines. In this case, by replacing the elements in the working equipment with the appropriate mathematical models, you can get a common mathematical model of working equipment of sealing machines, as well as a profile of working equipment of sealing machines, based on the specified energy consumption and technological requirements.

5. Conclusions

Thus, justified conditions of accelerated compaction of the soil by blocking its sliding from under the GUM during accelerated compaction and reducing the amount of GUM in the technological process will allow to accelerate the process of compaction of the bulk soil and to increase the productivity of machines for compaction.

The results of the research will be useful for scientists and engineers involved in rolling stock brakes and calculations in terms of the effect of soil base compaction on the dynamics of rolling stock braking [10, 11].

References

[1] Savelyev S 2004 Substantiation of regime parameters of vibrating hydraulic roller for soil Compaction: *Author's abstract. dissertation Candidate of Technical since* (Omsk, Russia:

Siberian Highway University) p 19

[2] Burhovich M 2007 Improvement of operational efficiency of pneumatic tires of machines for soil and road works: *Author's abstract. dissertation Candidate of Technical since* (Poltava, Ukraine: Poltava National technical university Y. Kondratyuk) p 21

[3] Manakin E 2007 Improving the efficiency of compaction of burned rocks by choosing rational parameters of a vibrating cat: *Author's abstract. dissertation Candidate of Technical since* (Dnipro, Ukraine: Dnipro Academy of Construction and Architecture) p 20

[4] Khmara L, Kravets S, Skoblyuk M, Nikitin V, Derevyanchuk M and Suponev V 2014 Soil-moving machine: *Tutorial* (Kharkov, Ukraine: Kharkiv National Highway University) p 548

[5] Khmara L, Kravets S, Nichke L, Nazarov V, Sabluk M and Nikitin V 2010 The soil-moving machine: *Tutorial* (Kharkov, Ukraine: Kharkiv National Highway University) p 557

[6] Kharhuta N and Vasiliev S 1975 Strength, stability and compaction of soil subgrade of roads: *Tutorial* (Moscow, Russia: publishing house Transport) p 288

[7] Blokhin S and Malich N 2006 The main parameters of technological machines: *Tutorial machines for work with soil* (Kyiv, Ukraine: publishing house Vyshcha Shkola) p 497

[8] Process of soil compaction back-fill in the cramped conditions of industrial and civil engineering: *Issues I and II. Bureau of the Introduction of the USSR* 1978 (Moskow, Russia: Central Research and Design and Experimental Institute of Organization, Mechanization and Technical Assistance for Construction) p 203

[9] Guidelines for the design of subgrade of roads 1973 SN 449-72 (Moskow, Russia: publishing house Stroyizdat) p 113

[10] Klimenko I, Černiauskaite L, Neduzha L and Ochkasov O 2018. Mathematical Simulation of Spatial Oscillations of the "Underframe-Track" System Interaction *Proc. of 12th Int. Conf. ITELMS'2018* (Panevėžysm, Lithuania: Kaunas Univ. of Techn.) pp 105–114

[11] Bondarenko I, Lunys O, Neduzha L and Keršys R 2019 Dynamic track irregularities modeling when studying rolling stock dynamics *Proc. of the 23rd Intern. Sci. Conf. Transport Means 2019* pt II (Palanga, Lithuania: Kaunas Univ. of Techn.) pp 1014–1019

A new method to create of the plane section standards of grain metal structures

T Danilenko

State University of Infrastructure and Technologies, Ivan Ohienko str, 19, 03049 Kyiv, Ukraine

danilenko.stereo.7@gmail.com

Abstract. A new method to create of the plane section standards of grain metal structures with a previously known size distribution of 3D grains is proposed. A quantitative analysis of G3, G4, G5 indices of the standard ISO 643:2012 was performed. The stereological modeling method taking into account the polyhedral shape of 3D grains was used to create new versions of plane section standards. Versions of 3D grains structures as the diameter size distributions of 3D grains, plane sections of which correspond to the main quantitative characteristics of existing G3, G4, G5 indices have been created. It was shown that these versions should contain 3D grain more than 1.5 times the mean diameter of the corresponding plane sections. Choosing the diameter size distributions of 3D grains, necessary for the standard, its possible to create a visual index as a picture, as it done on the existing standard, and also to give the parameters of 3D structure, plane section of which corresponds to this standard. Using a new approach to creating standards will increase the accuracy of the grain structure parameters determining.

1. Introduction

Grain metal structure is most often examined on a metallographic section, where it is represented by a system of polygons that are plane sections (let's call these sections 2D grains) of 3D grain-polyhedrons. The methods for determining the grain size presented in the existing standards of DSTU ISO 643: 2009 [1], ISO 643: 2012 [2], allow us to estimate the average size and some other parameters of a set of 2D grains, but do not provide information about the 3D structure, which really affects the properties of the material. Estimation of a 3D structure by its plane section only in studies establishing the influence of structure parameters on the properties of the metal material may give incorrect results.

For such studies it is important not the average grain size only, but also other characteristics: the maximum and modal diameter of 3D grains, the number of grains per unit volume (in 1 mm^3), the characteristics of the size heterogeneity – the standard deviation from the average size, the coefficient of variation K_V of the size distribution grains. These characteristics of the 3D structure can be determined by establishing the grain diameter size distribution $P(D_k)$. Information on the distribution of grain sizes may be more important than information on their average size, for example, when predicting the development of a crack in a material and its destruction [3].

In the works [4-6] the distribution $P(D_k)$ was established by the method of successive serial sections, which is considerable laboriousness and leads to the destruction of the sample analyzed, that limits its application.

Content from this work may be used under the terms of the Creative Commons Attribution 3.0 licence. Any further distribution of this work must maintain attribution to the author(s) and the title of the work, journal citation and DOI.
Published under licence by IOP Publishing Ltd

There are also methods for the stereological reconstruction the distribution $P(D_k)$ [7-11]. The initial information in these methods is the diameter size distribution of 2D grains $P_k(d)$ or chords $P_k(l)$ obtained on a metallographic section. However, these stereological methods and their relations between distributions $P(D_k)$ and $P_k(d)$, $P_k(l)$ are not used to create plane section standards of grain structures. An important obstacle to their use is that most of them are created for spherical structural components [12]. Some studies [13, 14] apply such methods to grain polyhedral structures, however it should be assumed that differences in the shapes of the sphere and polyhedron (circle and polygon, such as plane sections of 3D grains) introduce significant errors in the results of such stereological calculations [15].

Based on the foregoing, a new method to create of the plane section standards of grain metal structures proposed in this paper is the method that would provide information on the parameters of a 3D structure, the plane section of which is presented on the standard, and in which the polyhedral shape of the grains would be taken into account.

2. Methodology

On the standards [1, 2] the quantitative parameters of 2-D structures are: the number of 2D grains m per 1 mm^2; average grain area $\bar{a}$, mm^2; average grain diameter $\bar{d}$, mm; average number of grain intersections $\overline{N}_L$ per 1 mm of the measuring line; average grain intersection length l, mm.

Based on the parameters required and in order to obtain distributions $P_k(d)$, indices $G5$, $G4$, $G3$ [1, 2] were analyzed in detail. For that their schemes were created in the Compass program and the area of each 2D grain is measured. Based on the area the diameter of each grain was calculated and distributed on a scale of 10 uniform intervals. The scale was chosen based on the maximum 2D grain visible on the index. The 2D grain diameter was calculated as isometric circle diameter from the equation:

$$d_S = \sqrt{4S/\pi}. \tag{1}$$

The obtained distributions $P_k(d)$ of the indices $G5$, $G4$, $G3$ are in the Table 1.

Table 1. Distributions $P_k(d)$ of the indices $G5$, $G4$, $G3$ of ISO 643:2012 on the scale of 10 uniform intervals.

Size range №	$G5$		$G4$		$G3$	
	Scale, μm	$P_k(d)$, %	Scale, μm	$P_k(d)$, %	Scale, μm	$P_k(d)$, %
1	0…11.7	0	0…14.9	0	0…22	0
2	11.7…23.4	8.33	14.9…29.8	1.27	22…44	2.63
3	23.4…35.1	13.10	29.8…44.7	7.59	44…66	15.79
4	35.1…46.8	18.45	44.7…59.6	15.19	66…88	13.16
5	46.8…58.5	21.43	59.6…74.5	11.39	88…110	5.26
6	58.5…70.2	19.64	74.5…89.4	16.46	110…132	15.79
7	70.2…81.9	10.71	89.4…104.3	10.13	132…154	15.79
8	81.9…93.6	5.95	104.3…119.2	22.78	154…176	13.16
9	93.6…105.3	1.79	119.2…134.1	6.33	176…198	5.26
10	105.3…117	0.60	134.1…149.0	8.86	198…220	13.16

To choose the distribution $P_k(d)$ for the 2D structure standards, the method described in [16] was used. It is based on the reconstruction of $P_k(d)$ for a structure with a previously known distribution

$P(D_k)$ by solving linear equations in which the polyhedral shape of 3D grains is taken into account by using shape factors α_i:

$$P_k'(d) = \alpha_a P(D_k)\sqrt{D_k^2 - d_{k-1}^2} + \alpha_b P(D_{k+1})\left(\sqrt{D_{k+1}^2 - d_{k-1}^2} - \sqrt{D_{k+1}^2 - d_k^2}\right) + \dots$$
$$+ \alpha_n P(D_n)\left(\sqrt{D_n^2 - d_{k-1}^2} - \sqrt{D_n^2 - d_k^2}\right). \quad (2)$$

Researches have shown that the obtained $P_k'(d)$ should be normalized in accordance with the equation:

$$P_k(d) = P_k'(d) \Big/ \sum_{k=1}^{n} P_k'(d). \quad (3)$$

Since the standards [1, 2] include the method of drawing cut lines on the image of 2D structure and determining the average number of grain intersections $\overline{N}_L$ per 1 mm and the average grain intersection length l (chords), distribution $P_k(l)$ of chord lengths were reconstructed according to [16] by solving equations in which polygonal shape of the 2D grains is taken into account using shape factors β_i:

$$P_k'(l) = \beta_a P_k(d)\sqrt{d_k^2 - l_{k-1}^2} + \beta_b P_{k+1}(d)\left(\sqrt{d_{k+1}^2 - l_{k-1}^2} - \sqrt{d_{k+1}^2 - l_k^2}\right) + \dots$$
$$+ \beta_n P_n(d)\left(\sqrt{d_n^2 - l_{k-1}^2} - \sqrt{d_n^2 - l_k^2}\right). \quad (4)$$

Researches have shown that the obtained $P_k'(l)$ should be normalized in accordance with the equation:

$$P_k(l) = P_k'(l) \Big/ \sum_{k=1}^{n} P_k'(l). \quad (5)$$

The number of equations (2) and (4) is equal to the number of size intervals n in the size scale. To obtain a certain distribution $P(D_k)$ one can calculate the corresponding distributions $P_k(d)$ and $P_k(l)$, thus choosing $P(D_k)$ that gives a plane section with quantitative characteristics corresponding to the required 2D standard structure.

The method to obtain the shape factors is described in detail in the work [17]. It is based on the fact that shape factors α_i and β_i are intended to account the differences in the distributions $P_k(d)$ from the sphere and grain of the average shape, and the differences in the distributions $P_k(l)$ from the circle and plane section of the grain. So, the shape factors were found from the equations:

$$\alpha_i = P_k(d)_{grain} / P_k(d)_{sphere}, \quad (6)$$

$$\beta_i = P_k(l)_{grain} / P_k(l)_{circle}. \quad (7)$$

Note that in all calculations by this method the diameter of 2D grain was calculated as isometric circle diameter from the equation (1) and the diameter of 3D grains was taken to be the diameter of an isometric sphere. But on the standards [1, 2] the mean diameter of 2D grain is calculated from the equation:

$$\bar{d} = \sqrt{\bar{a}}. \quad (8)$$

So, the ratio between d_S and $\bar{d}$ is: $d_S = 1.1287\bar{d}$ and this is taken into account further.

3. Results and discussion

To create the standard index versions, it was suggested that the diameter of the largest 2D grain of the standard is equal to the diameter of the largest 3D grain. Distributions $P_k(d)$ of $G5$, $G4$, $G3$ indices, modeling distributions $P(D_k)$ for their versions, corresponding calculated distributions $P_k(d)$ are presented in Figure 1, 2, 3 and the distribution parameters $P(D_k)$, $P_k(d)$, $P_k(l)$ are presented in Tables 2, 3, 4. It was assumed that the grains of any size interval have the size of the interval middle.

On the $G5$ index the maximum 2D grain has $d = 117$ μm. If the 3D structure is represented by grains with a diameter $D = 117$ μm only, then on a plane section structure has $l = 56.4$ μm, $\bar{N}_L = 17.7$ pcs, that is close to the $G5$ index parameters, but it has $\bar{d} = 65.7$ μm, that is slightly more than the standard requires ($\bar{d} = 62.5$ μm). Because $\bar{d}$ is the main parameter on the standard, $P(D_k)$ version are proposed, for which $\bar{d} = 62.5$ μm. For them maximum diameter of 3D grains has sizes of 120, 125, 135, 145 μm (see Figure 1, Table 2). For these $P(D_k)$ options was obtained that 31...48 % of 3D grains have $D > 117$ μm (maximum grain diameter on a plane section); $\bar{D} = 96.7...104.3$ μm, i.e. 3D structure has $\bar{D}$, that is at least 1.5 times larger than $\bar{d}$ on the plane section ($\bar{d} = 62.5$ μm), and 3D structure is more homogeneous ($K_V = 0.11...0.29$, $K_P = 0.40...0.46$). And version *d* ($D_{max} = 145$ μm) is most suitable for the $G5$ index parameters.

On $G4$ index the maximum 2D grain has $d = 149$ μm. If the 3D structure is represented by grains with $D = 149$ μm only, then on a plane section 2D grains have $\bar{d} = 83.6$ μm, $l = 71.8$ μm, $\bar{N}_L = 13.9$ pcs, i.e. this 2D structure is finer-grained than on the $G4$ index (see Table 3), so 3D structure should contain grains with $D > 149$ μm. The approximately necessary 2D structure are given by the 3D structure, all grains of which have diameter no less 157.5 μm. Options that also meet the required parameters of the 2D structure are shown in Figure 2 and Table 3. For these options it was suggested that the maximum diameter 3D grains have size of 170, 180, 185, 205 μm. For them $P_k(d)$ was obtained for which 47...68 % of 3D grains have $D > 149$ μm; $\bar{D} = 139.4...147.6$ μm, i.e. 3D structure has $\bar{D}$, that is at least 1.6 times larger than $\bar{d}$ on the plane section, and 3D structure is more homogeneous ($K_V = 0.11...0.25$, $K_P = 0.40...0.45$). And version *d* ($D_{max} = 205$ μm) is most suitable for the $G4$ index parameters.

On $G3$ index the maximum 2D grain has $d = 220$ μm. If the 3D structure is represented by grains with $D = 220$ μm only, then on a plane section 2D grains have $\bar{d} = 124$ μm, that is close to the $G3$ index, but $l = 106$ μm and $\bar{N}_L = 9.4$ pcs, that does not corresponds to $G3$ index (see Table 4). So, $P(D_k)$ versions are proposed, in which maximum diameter of 3D grains have sizes of 240, 250, 280, 300 μm. In these cases received $P(D_k)$ with 42...63 % of 3D grains, that have $D > 220$ μm; $\bar{D} = 199...208$ μm, i.e. 3D structure has $\bar{D}$, that is at least 1.6 times larger than $\bar{d}$ of $G3$ index and the 3D structure is more homogeneous, than its plane section ($K_V = 0.12...0.23$, $K_P = 0.40...0.44$). And version *d* ($D_{max} = 300$ μm) is most suitable for the $G3$ index parameters.

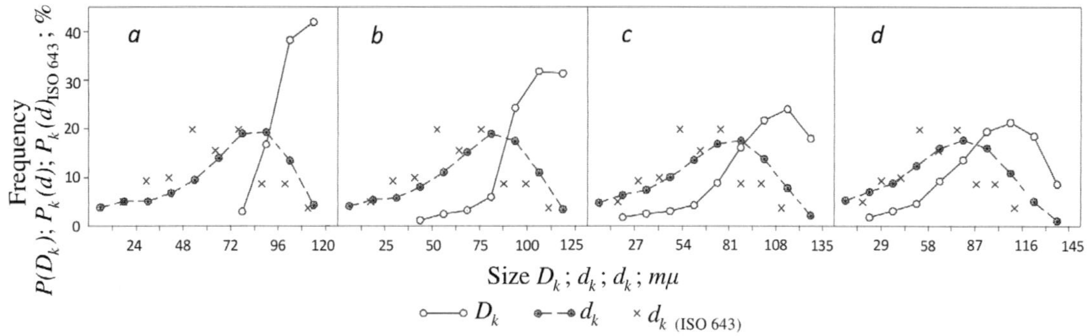

Figure 1. Distribution $P_k(d)_{ISO\ 643}$ of the $G5$ index of ISO 643 and models of the distribution $P(D_k)$ (*a*, *b*, *c*, *d*), giving the distributions of plane sections $P_k(d)$, corresponding to the quantitative characteristics of the $G5$ index of ISO 643.

Table 2. Structure parameters of the $G5$ index of ISO 643 and version parameters $P(D_k)$, giving the distribution sections $P_k(d)$ and $P_k(l)$, corresponding to the $G5$ index parameters.

Analysis object	Structure parameters	Values of parameters $G5$ index for versions			
		a	*b*	*c*	*d*
Distribution of grain diameters, $P(D_k)$	Mean diameter of grains $\overline{D}$, µm	104.3	102.5	98.3	96.7
	Mean square deviation from mean diameter, µm	11.5	17.3	26.3	28.3
	Coefficient of variation K_V of the diameters D	0.11	0.17	0.27	0.29
	Modal diameter of grains, µm	114	106.3	114.8	108.8
	Maximum diameter of grains D_{max}, µm	120	125	135	145
	Number of grains N_V per 1 mm^3, pc.	1684	1774	2012	2113
Distribution of plane section diameters, $P_k(d)$	Mean diameter of plane grains $\overline{d}$, µm	62.5	62.5	62.5	62.5
	Mean diameter of plane grains d_s, µm	70.5	70.5	70.5	70.5
	Mean square deviation from mean diameter, µm	28.3	29.2	31.2	32.3
	Coefficient of variation K_P of diameters d	0.40	0.41	0.44	0.46
	Modal diameter, µm	90.0	81.3	87.8	79.8
	Mean area of plane grain $\overline{a}$, µm^2	3902	3902	3902	3902
	Maximum diameter of plane grains, µm	120	125	135	145
	Number of plain grains m per 1 mm^2, pc.	256	256	256	256
Chords length distribution, $P_k(l)$	Mean grain intersection (chord) length l, µm	54.2	54.7	55.9	56.6
	Mean square deviation from mean l, µm	26.0	26.7	28.1	29.1
	Coefficient of variation K_L of chord lengths l	0.48	0.49	0.50	0.51
	Modal size intersection length l, µm	66	56.3	60.8	65.3
	Mean number of chords $\overline{N_L}$ per 1 mm of line, pc.	18.5	18.3	17.9	17.7
Structure parameters of the $G5$ index of ISO 643	Mean diameter of plane grains $\overline{d}$, µm	62.5			
	Number of plain grains m per 1 µm^2, pc.	256			
	Mean area of plane grain $\overline{a}$, µm^2	3900			
	Mean number of chords $\overline{N_L}$ per 1 mm of line, pc.	17.7			
	Mean grain intersection (chord) length l, µm	56.6			

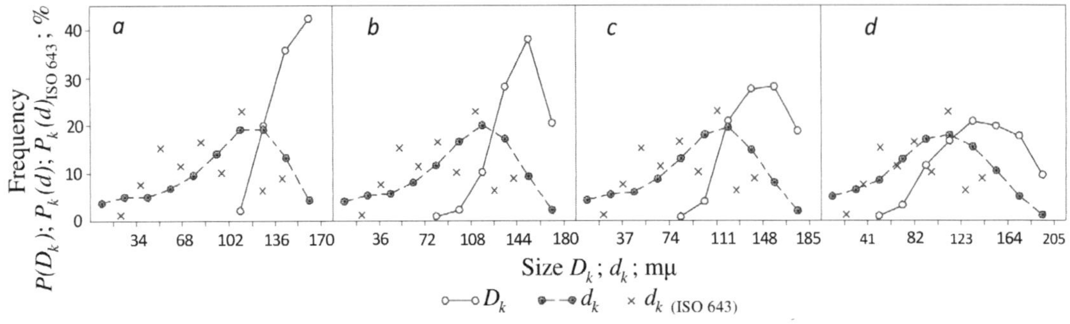

Figure 2. Distribution $P_k(d)_{ISO\ 643}$ of the $G4$ index of ISO 643 and models of the distribution $P(D_k)$ (*a, b, c, d*), giving the distributions of plane sections $P_k(d)$, corresponding to the quantitative characteristics of the $G4$ index of ISO 643.

Table 3. Structure parameters of the $G4$ index of ISO 643 and version parameters $P(D_k)$, giving the distribution sections $P_k(d)$ and $P_k(l)$, corresponding to the $G4$ index parameters.

Analysis object	Structure parameters	Values of parameters $G4$ index for versions			
		a	*b*	*c*	*d*
Distribution of grain diameters, $P(D_k)$	Mean diameter of grains $\overline{D}$, μm	147.6	146.0	145.1	139.4
	Mean square deviation from mean diameter, μm	16.4	21.0	23.5	35.5
	Coefficient of variation K_V of the diameters D	0.11	0.14	0.16	0.25
	Modal diameter of grains, μm	161.5	153.0	157.3	133.3
	Maximum diameter of grains D_{max}, μm	170.0	180.0	185.5	205.0
	Number of grains N_V per 1 mm^3, pc.	594	614	625	705
Distribution of plane section diameters, $P_k(d)$	Mean diameter of plane grains $\overline{d}$, μm	88.4	88.4	88.4	88.4
	Mean diameter of plane grains d_S, μm	99.8	99.8	99.8	99.8
	Mean square deviation from mean diameter, μm	40.0	40.9	41.7	45.1
	Coefficient of variation K_P of diameters d	0.40	0.41	0.42	0.45
	Modal diameter, μm	128.5	117.0	120.3	112.8
	Mean area of plane grain $\overline{a}$, μm^2	7819	7819	7819	7819
	Maximum diameter of plane grains, μm	170.0	180.0	185.0	205.0
	Number of plain grains m per 1 mm^2, pc.	128	128	128	128
Chords length distribution, $P_k(l)$	Mean grain intersection (chord) length l, μm	76.7	77.3	77.8	79.9
	Mean square deviation from mean l, μm	36.8	37.6	38.2	41.0
	Coefficient of variation K_L of chord lengths l	0.48	0.49	0.49	0.51
	Modal size intersection length l, μm	93.5	81.0	81.0	92.3
	Mean number of chords $\overline{N_L}$ per 1 mm of line, pc.	13.0	12.9	12.9	12.5
Structure parameters of the $G4$ index of ISO 643	Mean diameter of plane grains $\overline{d}$, μm	88.4			
	Number of plain grains m per 1 μm^2, pc.	128			
	Mean area of plane grain $\overline{a}$, μm^2	7810			
	Mean number of chords $\overline{N_L}$ per 1 mm of line, pc.	12.5			
	Mean grain intersection (chord) length l, μm	80			

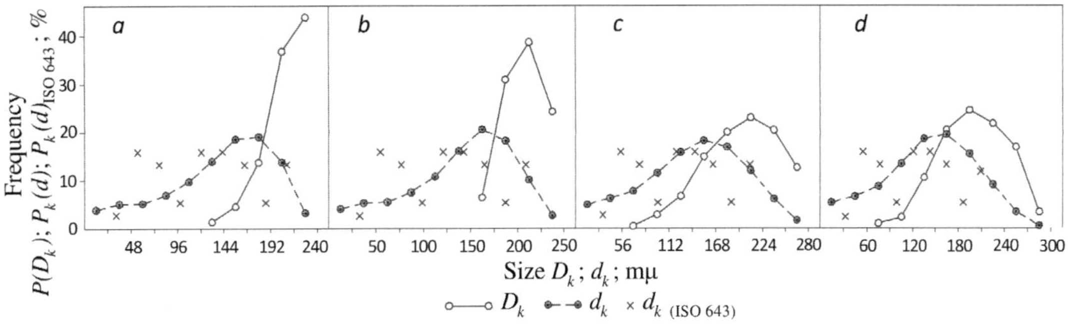

Figure 3. Distribution $P_k(d)_{ISO\ 643}$ of the $G3$ index of ISO 643 and models of the distribution $P(D_k)$ (*a*, *b*, *c*, *d*), giving the distributions of plane sections $P_k(d)$, corresponding to the quantitative characteristics of the $G3$ index of ISO 643.

Table 4. Structure parameters of the $G3$ index of ISO 643 and version parameters $P(D_k)$, giving the distribution sections $P_k(d)$ and $P_k(l)$, corresponding to the $G3$ index parameters.

Analysis object	Structure parameters	Values of parameters $G3$ index for versions			
		a	*b*	*c*	*d*
Distribution of grain diameters, $P(D_k)$	Mean diameter of grains $\overline{D}$, μm	208	208	200	199
	Mean square deviation from mean diameter, μm	25	25	46	46
	Coefficient of variation K_V of diameters D	0.12	0.12	0.23	0.23
	Modal diameter of grains, μm	228	213	210	195
	Maximum diameter of grains D_{max}, μm	240	250	280	300
	Number of grains N_V per 1 mm^3, pc.	212	212	239	242
Distribution of plane section diameters, $P_k(d)$	Mean diameter of plane grains $\overline{d}$, μm	125	125	125	125
	Mean diameter of plane grains d_S, μm	141	141	141	141
	Mean square deviation from mean diameter, μm	57	57	62	63
	Coefficient of variation K_P of diameters d	0.40	0.40	0.44	0.44
	Modal diameter, μm	180	163	154	165
	Mean area of plane grain $\overline{a}$, μm^2	15607	15607	15607	15607
	Maximum diameter of plane grains, μm	240	250	280	300
	Number of plain grains m per 1 mm^2, pc.	64	64	64	64
Chords length distribution, $P_k(l)$	Mean grain intersection (chord) length l, μm	108	109	112	113
	Mean square deviation from mean l, μm	52	52	57	58
	Coefficient of variation K_L of chord lengths l	0.48	0.48	0.51	0.51
	Modal size intersection length l, μm	132	113	126	135
	Mean number of chords $\overline{N}_L$ per 1 mm of line, pc.	9.20	9.19	8.92	8.84
Structure parameters of the $G3$ index of ISO 643	Mean diameter of plane grains $\overline{d}$, μm	125			
	Number of plain grains m per 1 μm^2, pc.	64			
	Mean area of plane grain $\overline{a}$, μm^2	15600			
	Mean number of chords $\overline{N}_L$ per 1 mm of line, pc.	8.84			
	Mean grain intersection (chord) length l, μm	113			

IOP Conf. Series: Materials Science and Engineering **985** (2020) 012040 doi:10.1088/1757-899X/985/1/012040

4. Conclusions

A detailed analysis of the indices $G5$, $G4$, $G3$ [1, 2] using the stereological method showed that 3D grain structure, the plane section of which has the parameters of the standard, should contain 3D grain more than 1.5 times the diameter $\bar{d}$ of the corresponding index. It can be argued approximately that the 3D structure consists more than 30% of grains with a diameter larger than the maximum 2D grain diameter on the plane section. It is important that the 3D structure is much more homogeneous than its plane section. The new stereological modeling method, taking into account the polyhedral shape of 3D grains, can be effectively used to create of plane section standards with a previously known size distribution of 3D grains. Choosing the distribution $P_k(d)$ necessary for the standard, it's possible to create a visual index as a picture, as it done on the standards [1, 2], and also to give the parameters of 3D structure. Using a new approach to creating standards will increase the accuracy of determining of the grain structure parameters and their effect on properties of metal, including metal structures of railway transport, which experience significant stress during operation.

References

[1] Steels. Micrographic determination of the apparent grain size 2009 DSTU ISO 643:2009 p 30
[2] Steels. Micrographic determination of the apparent grain size 2012 BS EN ISO 643:2012 p 35
[3] Wasen J and Waren R 1989 True grain size distribution of WC–Co alloy *Materials Science and Technology* **5** pp 222–226
[4] Liu G and Yu H 2000 Experimental evaluation of stereological methods for determining 3D grain size and topological distributions *Image Anal Stereol* **19** pp 91–97
[5] Liu G and Yu H 2000 On the sampling of serial sectioning technique for three-dimensional space-filling grain structures *Image Anal Stereol* **19** pp 81–84
[6] Tewari A and Gokhale A 2001 Estimation of three-dimensional grain size distribution from microstructural serial sections *Materials Characterization* **46** pp 329–335
[7] Saltykov S A 1976 *Stereometric metallography* (Moscow: Metallurgy) p 270
[8] Underwood E 1970 *Quantitative stereology* (Reading MA: Addison-Wesley, Massachusetts) p 274
[9] Russ J and Dehoff R 2012 *Practical Stereology* (Springer) p 382
[10] Exner H 2004 Stereology and 3D microscopy: useful alternatives or competitors in the quantitative analysis of microstructures *Image Anal Stereol* **23** pp 73–82
[11] Cruz-Orive L 2017 Stereology: a historical survey *Image Anal Stereol* **36** pp 153–177
[12] Exner H 1987 Shape – a key problem in quantifying microstructures *Acta Stereol* **6/III** pp 1023–1028
[13] Tucker J, Chan L, Rohrer G, Groeberc M and Rollett A 2012 Comparison of grain size distributions in a Ni-based superalloy in three and two dimensions using the Saltykov method *Scripta Materialia* **66** pp 554–557
[14] Gulbin Y 2008 On estimation and hypothesis testing of the grain size distribution by the Saltykov method *Image Anal Stereol* **27** pp 163–174
[15] Danilenko T 2010 Zastosuvannya stereolohyyi do analyzu metalevikh zerennikh struktur *Metaloznavstvo ta obrobka metaliv* **3** pp 35–42
[16] Danilenko T and Danilenko E 2016 Patent for invention No 113112 Ukraine *Promyslova vlasnist* **23**:**3** pp 64–65
[17] Danilenko T 2018 Method for the simulation and reconstruction of 3D particle size distributions in metal structures *MATEC Web of Conf. Transbud-2018* **230** pp 01002-1–01002-8

Higher efficiency of control over functional status of locomotive crew members

V Brusentsov[1], V Puzyr[2], M Vorozhbiian[3], M Ivashchenko[3,4] and Yu Datsun[2]

[1]Department Occupational Safety and Environment Protection, Ukrainian State University of Railway Transport, Kharkiv, Feierbakh sq., 7, Ukraine
[2]Department Maintenance and Repair of Rolling Stock, Ukrainian State University of Railway Transport, Kharkiv, Feierbakh sq., 7, Ukraine
[3]Department Occupational and Life Safety, O M Beketov National University of Urban Economy in Kharkiv, Kharkiv, Marshala Bazhanova str., 17, Ukraine

[4]Maryna.Ivashchenko@kname.edu.ua

Abstract. The objective of the study is research into psycho-physiological methods of control over the functional status of a person for higher efficiency of a pre-trip control of locomotive crew members. It is related to the fact that today's medical methods do not reveal some states which decrease a functional reliability level, particularly, a fatigue level and borderline states. The authors selected parameters which, according to the functional system theory, give information on conditions of two regulation levels in the human body, physiological and higher nervous activity, and correspond to the formulated requirements in terms of the procedure. The study experimentally proved the informative value of 27 psycho-physiological parameters for estimation of exhaustion and borderline states. For a state of exhaustion the hemodynamic parameters, including those of arrhythmia, both sinus and slow waves, were of the most informative value. The study revealed the parameter of arterial pressure pulse, the value of which in the post-trip group decreased by 25%. While identifying existence of borderline states by sinus arrhythmia parameters, it turned out that much more people with borderline states were in the exhaustion zone, rather than in the control group (42% and 16% respectively).

1. Introduction

Higher professional reliance of locomotive crew members is a pressing challenge, as its lower level leads to emergencies and accidents. The most important component of the professional reliance of an employee is functional reliability, specifically, the functional status, as the most labile element identifying the general level. Nowadays the functional status of a locomotive crew member is controlled through a pre-trip examination, which is often ineffective, as it does not reveal certain statuses, which decrease a functional reliance level while remaining within the physical standards. It refers to such statuses as fatigue, stresses, borderline states, etc. The article deals with some extra parameters which may reveal the above-mentioned statuses during a pre-trip control.

2. The analysis of publications and problem statement

The priority task for railway transportation is safety, and nowadays 80% of safe transportation depends on the human factor [1-3]. One of the most important components of the human factor is the professional reliability level of an employee. The 'professional reliance' concept is a system with a number of sub-systems, namely, biomedical, psycho-physiological, education and qualification,

Content from this work may be used under the terms of the Creative Commons Attribution 3.0 licence. Any further distribution of this work must maintain attribution to the author(s) and the title of the work, journal citation and DOI.
Published under licence by IOP Publishing Ltd

functional, etc.[2].But the functional reliability is underestimated, though it is the reason for 90% of erroneous actions of an operator. And the most important component of functional reliability is the functional status. Taking into account its great significance, the functional status should be carefully controlled. Today the functional status of locomotive crewmembers in railway transport is controlled through a pre-trip medical examination [4]. When implemented such an examination yielded positive results in terms of greater transportation safety, but it turned out insufficient, as it had a number of serious shortcomings, namely:

1) strong influence of the human factor (the examiner's);

2) use of merely traditional diagnostics methods which could not reveal the statuses decreasing a performance level though remaining within the physical standards.

Therefore, there is a need to equip a pre-trip medical examination with some methods which may solve the problem. Control of the functional status of an operator includes a series of approaches, particularly, one-parameter and poly-parameter ones, each with positive and negative aspects.In the former case only one parameter is under control; it considerably facilitates the procedure and technical support [5]. However, such an ease leads to poor reliability in terms of results, it does not consider the vegetal dynamic stereotype, i.e. individual response reactions.

The poly-parametrical approach means the use of some parameters; it overcomplicates the procedure and technical support, but produces more reliable results. The functional status assessment integrates certain factors of functional systems. Thus, the assessment is a measured analysis of the parameters from functional systems. A great amount of studies stress the importance of the poly-parametrical approach for evaluating the functional status [5, 6]. It was mentioned that such an approach is extremely efficient for diagnostic procedures. However, a choice of parameters characterizing different functional systems is not substantiated [6].

It is evident that the objective evaluation of the functional status is dependent on the informative value of parameters under control. And the integral estimate should be reasonably made with a certain massive of data [7]. Among the existing methodological approaches to the functional status diagnostics of a person the most adequate is the functional system theory according to which the body is considered as a complex hierarchical self-regulation system, evolutionarily formed of several levels. The experts distinguish from two to six such hierarchical levels [8].

A choice of diagnostic methods, providing information on the levels selected, is of primary importance. One of the methods for simultaneous estimation of several regulation levels is the psychomotor response measurement. These measurements characterize the overall condition of the body, as the motion analyzer is an integrator; all neurophysiologic mechanisms which relate to the effective performance of an operator affect this integrator. It also refers to the functional status of the central nervous system and the psychomotor apparatus.

Deeper levels are characterized by hemodynamic parameters, particularly, ones of heart rhythm variability. They are more sensible indicators of the health status. Moreover, these parameters can indicate a high risk of sudden ischemic death which is typical for transport industry [9].

Skin electrical activity parameters give great diagnostic opportunities for identifying statuses which are not included in the pathological zone. These parameters are essential in diagnostics of emotional states, especially stresses [10].

Therefore, there is a need to integrate in a pre-trip examination psycho-physiological parameters for identification of states directly influencing a professional reliability level and ability to solve problems in the specific conditions in which locomotive crews work.

3. The objective and research tasks

The objective of the study is efficiency of control techniques used for identification of the functional status of an operator, based on the functional system theory, in order to improve efficiency of a pre-trip control.

The following tasks were set to achieve the objective:

1. Formation of a complex of methods which allow the health worker to obtain information on performance of two regulation levels of the body, physiological and higher neurological, which make it possible to perform diagnostics in real conditions of rail industry.

2. Determination of the informative value of suggested methods for revealing states which decrease a functional reliability level of an operator, i.e. fatigue and borderline states.

4. Materials and methods

The methodological basis of the study is the concept that diagnostics of the actual functional status of an operator is limited to a level of certain state development from a given number of situations [11, 12]. These states, first of all, are fatigue and borderline ones.

For obtaining needed information the authors selected examination methods according to the functional system theory. And decision was taken to consider two regulation levels, physiological and higher nervous function, which united the psychomotor apparatus and higher psychic functions (perception and brainwork).

The study considered a specific character of diagnostics conditions which imposed some restrictions. Thus, the study formulated requirements for the methods with consideration of existing restrictions. Besides, a number of mutually conflicting conditions were taken into consideration: on the one hand, strict time frame and on the other – information awareness. It required some information which could be obtained without any extra time and examination.

The authors analyzed the methods used for the functional status diagnostics and selected those which met the formulated requirements more sufficiently. The selected methods were divided into two groups: basic and extra. The parameters from the first group characterized the basic performance level. They included heart activity parameters which characterized the life support system, and were considered as a sensitive indicator of the overall condition of the body.

The second group of parameters was complementary. These parameters could substantially complement the basic information, though they were obtained without any additional time and procedures.

During the research in the field the computer appliance Gamma was developed. It is intended for functional status diagnostics within a pre-trip examination [7]. The database was formed for each test person, which made it possible to analyze results and forecast statuses impacting a human operator's reliability.

5. The results of research into the informative value of methods for a pre-trip control over locomotive crew members

The efficiency of the suggested diagnostics methods for identifying an actual state of locomotive crew members was tested in several locomotive depots. The original assumption was to obtain some needed information through a pre-trip examination by a health worker with the system under development. This information was used for an argumentative conclusion on a person's potential workability at a specific moment, and also for forecasting on a period of a shift. Besides, the health worker evaluated possible statuses in a test person, which were especially risky for an engineman. The conditions under identification were, first of all, fatigue and borderline states. It was supposed that, as far as probability characteristics were concerned, with time and statistical data, artificial intelligence elements, particularly neuronal networks, could be applied.

The first study defined the ability of the developed system to reveal a fatigue level accumulated during a working shift. It was supposed that a fatigue level, achieved in the course of a working shift, was unacceptable for most workers. Besides, individual characteristics were taken into account. The same scope of work could differently affect people; some of them became unreliable, while the others could maintain a sufficient level of reliability. Therefore, it referred to probable parameters. Thus, as far as the parameters might distinguish the pre-trip and post-trip groups, they also could be used to identify an appropriate fatigue level.

Generally 74 enginemen were examined before and after a trip for 27 psychological parameters. The results were later compared and analyzed. Some parameters differentiated, though the average values differentiated slightly; there were areas where probability of one state was considerably higher than probability of the other. A patterns of the distribution obtained is presented in Figure 1 and Figure 2.

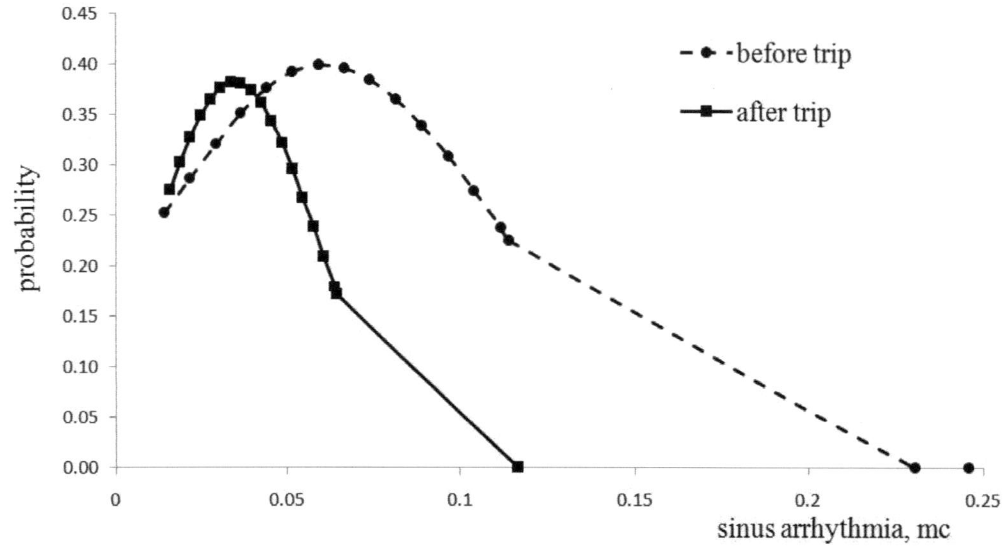

Figure 1. Distribution of the sinus arrhythmia parameter.

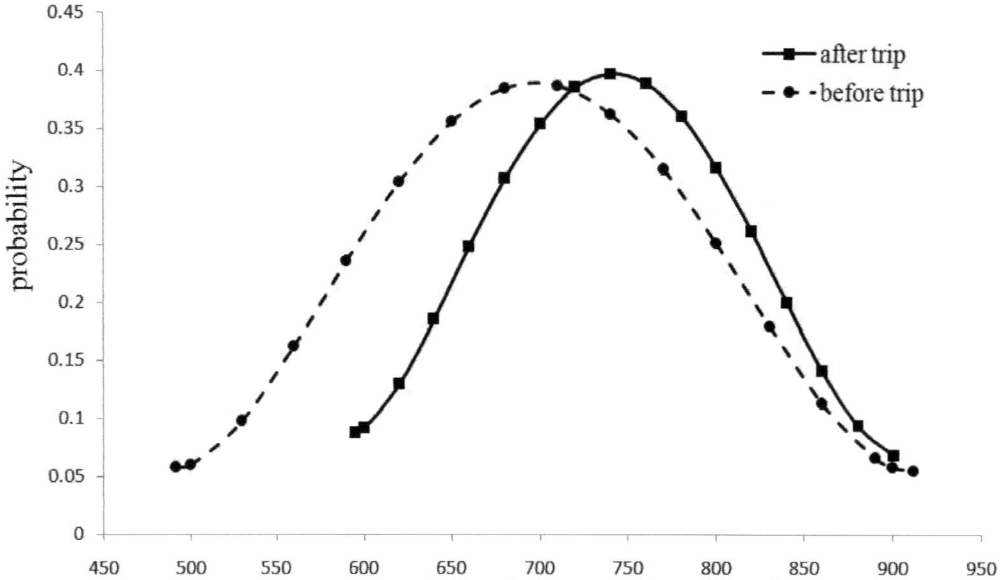

Figure 2. Distribution of cardio intervals duration.

For example, an average heart rate interval in two groups differentiated by 6%, but members of one group could get into a zone of 600 msect ice as often as members of the other group. The same pattern could be traced with parameters of the heart rhythm variability; and different arrhythmias had different dynamics.It was interestingly that the distribution of breath and sinus arrhythmias had zones which differentiated from ones of the normal distribution. It can be explained by the fact that these arrhythmias were physiological and pathological, and the later might give such a zone.

Considerable variations were in the pulse pressure parameter, which decreased by 25% in the post-trip group. It can be explained by the fact that the parameter characterizes a stroke rush of the cardiac muscle, and naturally, in the state of fatigue the heart worked less efficiently.

All these finest details may influence the result and reveal a certain state of a person.

The pattern of dependencies and the ratio of data before and after a trip present a possibility, with a comprehensive approach, to reveal a certain state and, thus, to forecast the functional reliability in performing responsible professional operations.

The second study dealt with an ability of the suggested methods to identify borderline states (lower standard of health).During the study some patients with a light form of tuberculosis from a TB dispensary (33 men of the borderline state group) were examined. The results were compared with those of healthy persons (a control group included 28 men).The most valuable information was in heart rhythm variations, first of all, in sinus arrhythmia. It is known that sinus arrhythmia has a series of zones: optimum zone, stress zone and exhaustion zone. Much more people with the borderline state were in the exhaustion zone ($V_{RR} \leq 0.04$) than in the control group (42 % and 16 %, respectively).An opposite situation was in the stress zone ($V_{RR} \geq 0.08$).This group included much more healthy and rested people (26 % and 12 %, respectively).These data can use probable parameters from various groups for objective functional state diagnostics.

The distinctions manifested themselves more obvious with a slight physical activity. Even a change in position from sitting to standing moved 58 % persons from the borderline state group and 32 %from the healthy group to the exhaustion group. The situation was opposite with the stress group ($V_{RR} \geq 0.08$) – 12 % and 36 % respectively.

6. The discussion of results of the research into methods of control over the functional state of locomotive crew members

The obtained results testify that the chosen methodological approach can control locomotive crew members in real working conditions. It gives information for estimation of existence and intensity of real states which ensure the required functional reliability level. Particularly, it refers to fatigue and borderline states. Fatigue states cannot be persuasively confirmed by parameters of the cardiovascular system. Thus, in the pre-trip group the pulse was higher than that in the post-trip group (86 and 80.8 BPM respectively), though it would be reasonable to expect the opposite. It can be explained by the examination effect. Many enginemen took the examination very emotionally, although such behavior is typical, with time it can be taken as a routine. In that particular case it was intensified by the fact that many took the examination as a hidden agenda, which might affect their occupational status. Therefore, the pulse of some enginemen reached up to 120 BPM.

A high excitation level bordering a stress was confirmed by the skin electric impedance parameters (the coefficient of variation by the skin electric impedance 15.28 in the pre-trip group versus 8.9 in the post-trip group).After a trip enginemen did not have that emotional state, besides, they were not afraid to be suspended from a trip; therefore the examination effect was less evident.

The pulse pressure parameter, which characterizes the heart performance efficiency and equals the difference between systolic and diastolic pressure, had a high informative value. Most engineman had 40 mmHg before a trip and about 30 mmHg after a trip. A level of fatigue can be efficiently evaluated at an individual level with appropriate mathematical methods, for example, a prior probability (Bayer's theorem).

Thus, the study suggests a complex of parameters for a pre-trip control which have certain advantages, and by applying these psycho-physiological parameters the health worker can reveal states decreasing a functional reliability level though remaining out of the pathological zone, and, therefore, undetectable by an existing control.

7. Conclusions

1. The authors substantiated methods of control over an operator's state (including psycho-physiological ones), which can be realized in real conditions of rail industry and related to serious industrial obstacles. Methods for assessment of the functional state are divided into two levels: basic and extra. The basic methods include sensomotor reactions and hemodynamic parameters, and extra methods include parameters of heart rhythm variation, skin electric activity and bilateral asymmetry.

2. The study experimentally proved the informative value of hemodynamic parameters including those of arrhythmia, sinus and slow waves for estimation of a fatigue level. The authors singled out the pulse pressure parameter, which was changed by 25% in the post-trip group.

3. The authors proved the informational value of the suggested parameters for estimation of a depressed health level. The most valuable information was observed in parameters of heart rhythm variation, first of all, in intensity of sinus arrhythmia. The exhaustion zone ($V_{RR} \leq 0.04$) included 42 % of borderline states and 16 % of the control group; the stress zone ($V_{RR} \geq 0.08$) had an opposite situation – 12 % and 36 % respectively. Such differences were more obvious with physical activity.About 58% of the borderline state group and 32 % of the healthy group joined the exhaustion group while changing a pose from sitting to standing. The situation was opposite in the stress group – 12 % and 36 % respectively.

References

[1] Report on Railway Safety and Interoperability in the EU 2018 *European Union Agency for Railways*(Luxembourg: Publications Office of the European Union)
[2] Oboznov A A, Bessonova Yu V, Petrovych D L, Enyna E S and Serykov V V 2014 Safety Culture in Rail Transport *J. The human factor: problems of psychology and ergonomics* (Tver: Interregional public organization "Ergonomic Association") **4** 7 pp 45–52
[3] Smyrnova I L and Pryz O O 2016 The concept of professional reliability in the context of training future air traffic controllers *Bulletin of the National Aviation University* **9** pp 156–160
[4] Brusentsov V H, Vorozhbiian M I, Puzyr V H and Brusentsov O V 2016 Ergonomic support of the activities of drivers in high-speed traffic *Compilation of Scientific PapersUkrainian State University of Railway Transport* **160** pp 120–124
[5] Nowakowski T, Mlynczak M, Jodejko-Pietruczuk A and Werbinska-Wojciechowska S 2014 Safety and Reliability: Methodology and Applications (London: CRC Press Published) p 408
[6] Felice F D and Petrillo A 2011 Methodological Approach for Performing Human Reliability and Error Analysis in Railway Transportation System *International Journal of Engineering and Technology* **3** 5 pp 341–353
[7] Brusentsov V H and Puzyr V H 2012 Implementation of a system for monitoring the functional reliability of railway operators *Visnik of the Volodymyr Dahl East Ukrainian National University* **5** 176 chapter 2 pp 167–169
[8] Red'ko V G, Prokhorov D V and Burtsev M S 2004 Theory of Functional Systems, Adaptive Critics and Neural Networks *In Proceedings of IJCNN* (Budapest: IJCNN) pp 1787–1792
[9] Hillebrand S, Gast K B, de Mutsert R, Swenne C A, Jukema J W, Middeldorp S, Rosendaal F R and Dekkers O M 2013 Heart rate variability and first cardiovascular event in populations without known cardiovascular disease: meta-analysis and dose-response meta-regression *Europace* **15** 5 pp 742–749
[10] Sterlyhova O P 2015 *Fundamentals of the psychophysiology of professional activity: a manual* (Moscow: Moscow State University of Railway Transport) p 86
[11] Serykov V V, Zakrevskaia A A and Koliahyn V Ia 2014 Dynamics of professional and functional reliability of locomotive crew workers in shift work*The human factor: problems of psychology and ergonomics* (Tver: Interregional public organization "Ergonomic Association") **4** 71 pp 56–62
[12] Walls L, Revie M and Bedford T 2016 Risk, Reliability and Safety: Innovating Theory and Practice *Proceedings of ESREL 2016* (Glasgow: CRC Press Published) p 486

AUTHOR INDEX